Metal Ions in Life Sciences

Volume 23

Molecular Bio-Sensors and the Role of Metal Ions

Guest Editor
Thomas J. Meade
Departments of Chemistry, Molecular Biosciences, Neurobiology, and Radiology
Northwestern University
2145 Sheridan Road
Evanston, IL 60208, USA
<tmeade@northwestern.edu>

Series Editors
Astrid Sigel, Helmut Sigel
Department of Chemistry
Inorganic Chemistry
University of Basel
St. Johannsring 19
CH-4056 Basel
Switzerland
<astrid.sigel@unibas.ch>
<helmut.sigel@unibas.ch>

Eva Freisinger, Roland K.O. Sigel
Department of Chemistry
University of Zürich
Winterthurerstrasse 190
CH-8057 Zürich
Switzerland
<freisinger@chem.uzh.ch>
<roland.sigel@chem.uzh.ch>

CRC Press
Taylor & Francis Group
Boca Raton London New York

CRC Press is an imprint of the
Taylor & Francis Group, an **informa** business

Cover illustration: The figure is part of Figure 1 in Chapter 6 and was prepared by Jong Seung Kim.

First edition published 2022
by CRC Press
6000 Broken Sound Parkway NW, Suite 300, Boca Raton, FL 33487-2742

and by CRC Press
4 Park Square, Milton Park, Abingdon, Oxon, OX14 4RN

CRC Press is an imprint of Taylor & Francis Group, LLC

Library of Congress Cataloging-in-Publication Data
Title: Molecular bio-sensors and the role of metal ions / guess editors, Thomas J. Meade.
Description: First edition. | Boca Raton : CRC Press, 2022. |
Series: Metal ions in life sciences, 1559-0836; volume 23 |
Includes bibliographical references and index.
Identifiers: LCCN 2022012374 (print) | LCCN 2022012375 (ebook) |
ISBN 9781032135786 (hbk) | ISBN 9781032135861 (pbk) | ISBN 9781003229971 (ebk)
Subjects: LCSH: Biosensors. | Chemical detectors. | Metal ions. |
Molecular recognition. | Biochemical engineering.
Classification: LCC R857.B54 M65 2022 (print) | LCC R857.B54 M65 2022 (print) |
LCC R857.B54 (ebook) | DDC 610.28/4—dc23/eng/20220706
LC record available at https://lccn.loc.gov/2022012374
LC ebook record available at https://lccn.loc.gov/2022012375

ISBN: 9781032135786 (hbk)
ISBN: 9781032135861 (pbk)
ISBN: 9781003229971 (ebk)

ISSN 1559-0836 e-ISSN 1868-0402

DOI: 10.1201/9781003229971

Typeset in Times New Roman
by codeMantra

About the Editors

Thomas J. Meade completed his Ph.D. at The Ohio State University (1985), was a NIH Fellow in Harvard Medical School (1985–1987), was a Postdoctoral Fellow in CalTech (1988–1990). Currently, he is a Professor in the Department of Chemistry, Northwestern University, Evanston, Illinois; Eileen M. Foell Professor of Cancer Research, Director, Center for Advanced Molecular Imaging; Professor of Weinberg College of Arts and Sciences and Radiology. Focus of work: The Meade Lab research focuses on inorganic coordination chemistry for the study of molecular imaging of *in vivo* gene expression and intracellular messengers, transition metal enzyme inhibitors, and electronic bio-sensors. The design, synthesis, and physical properties of transition metal and lanthanide coordination complexes are the foundation of our research. Academic Focus Program Area(s): Bioinorganic Chemistry; Nanochemistry; Organometallic & Coordination ComplexesInterest(s): Organic, Synthetic, Chemical Biology, Inorganic, Bioinorganic, Physical, Biophysical, Nanoscience.

Astrid Sigel has studied languages; she was an Editor of the *Metal Ions in Biological Systems* (*MIBS*) series (until Volume 44) and also of the *Handbook on Toxicity of Inorganic Compounds* (1988), the *Handbook on Metals in Clinical and Analytical Chemistry* (1994; both with H.G. Seiler and H.S.), and *Handbook on Metalloproteins* (2001; with Ivano Bertini and H.S.). She is also an Editor of the *MILS* series from Volume 1 on, and she has co-authored more than 50 papers on topics in Bioinorganic Chemistry.

Helmut Sigel is Emeritus Professor (2003) of Inorganic Chemistry at the University of Basel, Switzerland. He is a Co-editor of the series *Metal Ions in Biological Systems* (1973–2005; 44 volumes) as well as of the Sigels' new series *Metal Ions in Life Sciences* (since 2006). He has also co-edited three handbooks and published over 350 articles on metal ion complexes of nucleotides, amino acids, coenzymes, and other bio-ligands. Together with Ivano Bertini, Harry B. Gray, and Bo G. Malmström, he founded (1983) the International Conferences on Biological Inorganic Chemistry (ICBICs). He has lectured worldwide and was named *Protagonist in Chemistry* (2002) by *Inorganica Chimica Acta* (issue 339). Among Endowed Lectureships, appointments as Visiting Professor (e.g., Austria, China, Japan, Kuwait, UK), and further honors, he received the P. Ray Award (Indian Chemical Society, of which he is also a Honorary Fellow), the Alfred Werner Award (Swiss Chemical Society), and a Doctor of Science honoris causa degree (Kalyani University, India). He is also a Honorary Member of SBIC (Society of Biological Inorganic Chemistry).

Eva Freisinger is Associate Professor for Bioinorganic Chemistry and Chemical Biology (2018) at the Department of Chemistry at the University of Zürich, Switzerland. She obtained her doctoral degree (2000) from the University of Dortmund, Germany, working with Bernhard Lippert and spent 3 years as a postdoc at SUNY Stony

Brook, New York, with Caroline Kisker. Since 2003, she performs independent research at the University of Zürich, where she held a Förderungsprofessur of the Swiss National Science Foundation from 2008 to 2014. In 2014, she received her *Habilitation* in Bioinorganic Chemistry. Her research is focused on the study of plant metallothioneins with an additional interest in the sequence-specific modification of nucleic acids. Together with Roland Sigel, she chaired the 12th European Biological Inorganic Chemistry Conference (2014 in Zürich, Switzerland) as well as the 19th International Conference on Biological Inorganic Chemistry (2019 in Interlaken, Switzerland). She also serves on a number of Advisory Boards for international conference series; since 2014, she is the Secretary of the European Biological Inorganic Chemistry Conferences (EuroBICs), and is currently co-Director of the Department of Chemistry. She joined the group of Editors of the *MILS* series from Volume 18 on.

Roland K. O. Sigel is Full Professor (2016) of Chemistry at the University of Zürich, Switzerland. In the same year, he became Vice Dean of Studies (B.Sc./M.Sc.), and in 2017, he was elected Dean of the Faculty of Science. From 2003 to 2008, he was endowed with a Förderungsprofessur of the Swiss National Science Foundation, and he is the recipient of an ERC Starting Grant 2010. He received his doctoral degree summa cum laude (1999) from the University of Dortmund, Germany, working with Bernhard Lippert. Thereafter, he spent nearly 3 years at Columbia University, New York, with Anna Marie Pyle (now Yale University). During the 6 years abroad, he received several prestigious fellowships from various sources, and he was awarded the EuroBIC Medal in 2008 and the Alfred Werner Prize (SCS) in 2009. From 2015 to 2019, he was the Secretary of the Society of Biological Inorganic Chemistry (SBIC), and since 2018, he is the Secretary of the International Conferences on Biological Inorganic Chemistry (ICBICs). His research focuses on the structural and functional role of metal ions in ribozymes, especially group II introns, regulatory RNAs, and on related topics. He is also an Editor of Volumes 43 and 44 of the *MIBS* series and of the *MILS* series from Volume 1 on.

Historical Development and Perspectives of the Series *Metal Ions in Life Sciences*[*]

It is an old wisdom that metals are indispensable for life. Indeed, several of them, like sodium, potassium, and calcium, are easily discovered in living matter. However, the role of metals and their impact on life remained largely hidden until inorganic chemistry and coordination chemistry experienced a pronounced revival in the 1950s. The experimental and theoretical tools created in this period and their application to biochemical problems led to the development of the field or discipline now known as *Bioinorganic Chemistry*, *Inorganic Biochemistry*, or more recently also often addressed as *Biological Inorganic Chemistry*.

By 1970, *Bioinorganic Chemistry* was established and further promoted by the book series *Metal Ions in Biological Systems* founded in 1973 (edited by H.S., who was soon joined by A.S.) and published by Marcel Dekker, Inc., New York, for more than 30 years. After this company ceased to be a family endeavor and its acquisition by another company, we decided, after having edited 44 volumes of the *MIBS* series (the last two together with R.K.O.S.), to launch a new and broader minded series to cover today's needs in the *Life Sciences*. Therefore, the Sigels' new series is entitled

"Metal Ions in Life Sciences".

After the publication of 22 volumes (since 2006), we are happy to join forces from Volume 23 on in this still growing endeavor with Taylor & Francis, London, UK, a most experienced publisher in the *Sciences*.

The development of *Biological Inorganic Chemistry* during the past 40 years was and still is driven by several factors; among these are (i) attempts to reveal the interplay between metal ions and hormones or vitamins, etc.; (ii) efforts regarding the understanding of accumulation, transport, metabolism, and toxicity of metal ions; (iii) the development and application of metal-based drugs; (iv) biomimetic syntheses with the aim to understand biological processes as well as to create efficient catalysts; (v) the determination of high-resolution structures of proteins, nucleic acids, and other biomolecules; (vi) the utilization of powerful spectroscopic tools allowing studies of structures and dynamics; and (vii) more recently, the widespread use of macromolecular engineering to create new biologically relevant structures at will. All this and more are reflected in the volumes of the series *Metal Ions in Life Sciences*.

The importance of metal ions to the vital functions of living organisms, hence, to their health and well-being, is nowadays well accepted. However, in spite of all the progress made, we are still only at the brink of understanding these processes.

---

[*] Reproduced with some alterations by permission of John Wiley & Sons, Ltd., Chichester, UK (copyright 2006), from pages v and vi of Volume 1 of the series *Metal Ions in Life Sciences* (MILS-1)

Therefore, the series *Metal Ions in Life Sciences* links coordination chemistry and biochemistry in their widest sense. Despite the evident expectation that a great deal of future outstanding discoveries will be made in the interdisciplinary areas of science, there are still "language" barriers between the historically separate spheres of chemistry, biology, medicine, and physics. Thus, it is one of the aims of this series to catalyze mutual "understanding."

It is our hope that *Metal Ions in Life Sciences* continues to prove a stimulus for new activities in the fascinating "field" of *Biological Inorganic Chemistry.* If so, it will well serve its purpose and be a rewarding result for the efforts spent by the authors.

Astrid Sigel and Helmut Sigel
Department of Chemistry, Inorganic Chemistry
University of Basel, CH-4056 Basel, Switzerland

Eva Freisinger and Roland K. O. Sigel
Department of Chemistry
University of Zürich, CH-8057 Zürich, Switzerland

October 2005 and March 2022

Preface to Volume 23
Molecular Bio-Sensors and the Role of Metal Ions

Over the years, substantial evidence has accumulated for the diverse roles that metal ions play in biology. Aside from *a priori* toxic metal ions, essential ones can also be harmful and therefore their level needs to be carefully controlled. For example, calcium(II) ions occur in bones and cartilages, but they are also involved in many intra- and extracellular processes; therefore, generally speaking, transport and homeostasis are relevant. This reflects to some extent the role of *Metal Ions in Bio-Imaging Techniques*, an analytical approach and the topic of the preceding Volume 22 in the *Metal Ions in Life Sciences* (MILS) series. The present book, entitled *Molecular Bio-Sensors and the Role of Metal Ions*, complements and widens the topic.

The discussion of molecular bio-sensors involving metal ions begins with metalloid-sensing of transcriptional regulatory proteins. As noted in *Chapter 1*, environmental changes, such as exposure to arsenic, have had a devastating impact on living organisms. As a result, microbes have evolved resistance determinants for coping with toxicity, including transport systems and pathways for biotransformation. Advancements in arsenic detention and bioremediation are encouraging, as well as our increased understanding of substrate binding sites and the molecular mechanisms of transcriptional control.

Recent progress made to develop MRI bio-sensors for calcium(II) is highlighted in *Chapter 2*. The ability of MR imaging to detect calcium fluctuations at the molecular and cellular level and on a large scale makes this technique useful for the visualization of critical biological processes. These sensors have facilitated the first *in vivo* monitoring of intra- and extracellular Ca(II) concentration changes. While calcium (Ca^{2+}) regulates numerous biological processes, its mishandling correlates with numerous diseases. *Chapter 3* focuses on the critical advances made in sensing calcium dynamics and signaling, and its use in the facilitation of drug discovery.

The detection of Mn(II) and Fe(II) is challenging because both bind weakly to biological and synthetic molecules. There are significant challenges regarding the design of molecules that can selectively bind and enable detection and quantification. *Chapter 4* explores fluorescent bio-sensors for these metals and promising ways to address the challenges they pose. Zinc, the subject of *Chapter 5*, is simultaneously helpful and harmful. However, it is also an environmental pollutant implicated in physiological processes associated with many diseases, such as diabetes and Alzheimer's. Studies are now developing methods for rapid and efficient detection of zinc in water, soil, and biological samples.

Copper is well known to play a critical role in chemical, environmental, and biological systems. The detection of copper species is essential for plants, animals, and humans, as controlled copper levels are necessary for optimal health. *Chapter 6* underscores the importance of monitoring Cu(I) and Cu(II) ions in living organisms using chemo- and bio-sensors. Another toxic metal for living systems is cadmium, which needs to be detected, quantified, and imaged. As *Chapter 7* discusses, small changes in probe structure reverse fluorescence selectivity between zinc and cadmium, two group 12 elements that are difficult to distinguish. There has been noteworthy progress in developing cadmium-specific fluorescent sensors concerning signal-switching machinery and metal ion selectivity.

Chapter 8 takes us back in time to survey vanadium through a historical lens. Vanadium was first isolated in 1801, but it was not correctly described or analyzed until later. The benefits and toxic effects of vanadium in biological systems are now understood. Today, vanadium is an essential metal ion in the construction of bio-sensors.

As we look toward the future, stem cell therapy offers a unique approach to regenerative medicine. While there is clinical interest in such therapies to treat cancer, cardiovascular diseases, neurological disorders, and more, clinically available non-invasive imaging techniques will be imperative to monitor advanced therapeutic products. *Chapter 9* reviews imaging modalities for stem cell tracking and the metal ion-containing contrast agents that promote specific detection of stem cell populations *in vivo.*

Bio-sensors are used for metal ion detection and analysis in many areas, including the study of biological processes, environmental monitoring, and quality control in the food industry. A significant number of sensors rely heavily on metals or metal ions as the source or amplifier of the sensing signal. *Chapter 10* looks at bio-sensors that take advantage of the properties of metal compounds to quantitate biologically relevant species or metal ions *in situ* or *in vivo* and how sensing metal ions and luminescence sensing is utilized in neurobiology.

To conclude, this volume, devoted to *Molecular Bio-Sensors and the Role of Metal Ions*, is rich on specific information and thus a valuable source for the specialist and the generalist as well. MILS-23 updates our knowledge and provides deep insights into the new research frontiers in the fast-growing field of bio-sensors. It is a *must* for all researchers working in medicinal chemistry and related fields and beyond. It is also an ideal source for teachers giving courses on bio-sensing or in analytical chemistry in a wider sense.

Thomas J. Meade

Contents

Contributors to Volume 23

Goran Angelovski
Laboratory of Molecular and Cellular Neuroimaging, International Center for Primate Brain Research (ICPBR), Center for Excellence in Brain Science and Intelligence Technology (CEBSIT)
Chinese Academy of Sciences (CAS)
Shanghai 200031, People's Republic of China

Patrick S. Barber
Department of Chemistry
University of West Florida
Pensacola, FL 32514, USA

Jian Chen
Department of Cellular Biology and Pharmacology
Herbert Wertheim College of Medicine
Florida International University
Miami, FL 33199, USA

Xiaoqiang Chen
State Key Laboratory of Materials-Oriented Chemical Engineering, College of Chemical Engineering, Jiangsu National Synergetic Innovation Center for Advanced Materials (SICAM)
Nanjing Tech University
Nanjing 211816, People's Republic of China

Joseph A. Cotruvo, Jr.
Department of Chemistry
The Pennsylvania State University
University Park, PA 16802, USA

Debbie C. Crans
Department of Chemistry and Cell and Molecular Biology Program
Colorado State University
Fort Collins, CO 80523, USA

Xiaonan Deng
Department of Chemistry, Center for Diagnostics and Therapeutics and Advanced Translational Imaging Facility
Georgia State University
Atlanta, GA 30303, USA

Ana de Bettencourt-Dias
Department of Chemistry
University of Nevada Reno
Reno, NV 89557, USA

Meghan W. Dukes
Departments of Chemistry, Molecular Biosciences, Neurobiology, and Radiology
Northwestern University
2145 Sheridan Rd
Evanston, IL 60208, USA

Paramesh Jangili
Department of Chemistry
Korea University
Seoul 02841, South Korea

Ilwha Kim
Department of Chemistry
Korea University
Seoul 02841, South Korea

Jong Seung Kim
Department of Chemistry
Korea University
Seoul 02841, South Korea

Michael Kirberger
Department of Chemistry, Center for Diagnostics and Therapeutics and Advanced Translational Imaging Facility
Georgia State University
Atlanta, GA 30303, USA
and
School of Science and Technology
Georgia Gwinnett College
Lawrenceville, GA 30043, USA

Susan Kleinfelter
Department of Chemistry, Center for Diagnostics and Therapeutics and Advanced Translational Imaging Facility
Georgia State University
Atlanta, GA 30303, USA

Nahyun Kwon
Department of Chemistry and Nano Science
Ewha Womans University
Seoul 03760, South Korea

Thomas J. Meade
Departments of Chemistry, Molecular Biosciences, Neurobiology, and Radiology
Northwestern University
2145 Sheridan Rd
Evanston, IL 60208, USA

Yuji Mikata
Department of Chemistry, Biology, and Environmental Sciences, Faculty of Science
Nara Women's University
Nara 630–8506, Japan

Michel Modo
Departments of Radiology and Biochemistry, McGowan Institute for Regenerative Medicine
Centre for Neural Basis of Cognition
University of Pittsburgh
3025 East Carson Street
Pittsburgh, PA 15203, USA

Jorge H.S.K. Monteiro
Department of Chemistry
Humboldt State University
Arcata, CA 95521, USA

Jennifer Park
Department of Chemistry
The Pennsylvania State University
University Park, PA 16802, USA

Deborah A. Roess
Department of Biomedical Science
Colorado State University
Fort Collins, CO 80523, USA

Barry P. Rosen
Department of Cellular Biology and Pharmacology
Herbert Wertheim College of Medicine
Florida International University
Miami, FL 33199, USA

Nuttaporn Samart
Department of Chemistry
Colorado State University
Fort Collins, CO 80523, USA
Department of Chemistry
and
Rajabhat Rajanagarindra University
Chachoengsao, Thailand

Nem Singh
Department of Chemistry
Korea University
Seoul 02841, South Korea

Li Tian
Department of Chemistry, Center for Diagnostics and Therapeutics and Advanced Translational Imaging Facility
Georgia State University
Atlanta, GA 30303, USA

Tingwen Wei
State Key Laboratory of Materials-Oriented Chemical Engineering, College of Chemical Engineering, Jiangsu National Synergetic Innovation Center for Advanced Materials (SICAM)
Nanjing Tech University
Nanjing 211816, People's Republic of China

Jiansong Xu
Department of Chemistry
The Pennsylvania State University
University Park, PA 16802, USA

Jenny Yang
Department of Chemistry, Center for Diagnostics and Therapeutics and Advanced Translational Imaging Facility
Georgia State University
Atlanta, GA 30303, USA

Juyoung Yoon
Department of Chemistry and Nano Science
Ewha Womans University
Seoul 03760, South Korea

Zehra Zunbul
Department of Chemistry
Korea University
Seoul 02841, South Korea

Handbooks and Book Series Published and (Co-)edited by the SIGELs

"Handbook on Toxicity of Inorganic Compounds" (ISBN: 0-8247-7727-1) Eds H. G. Seiler, H. Sigel, A. Sigel; Dekker, Inc.; New York; 1988; 1069 pp.

"Handbook on Metals in Clinical and Analytical Chemistry" (ISBN: 0-8247-9094-4) Eds Hans G. Seiler, Astrid Sigel, Helmut Sigel; Dekker, Inc.; New York, Basel, Hong Kong; 1994; 753 pp.

"Handbook on Metalloproteins" (ISBN: 0-8247-0520-3) Eds I. Bertini, A. Sigel, H. Sigel; Marcel Dekker, Inc.; New York, Basel; 2001; 1182 pp.

Metal Ions in Biological Systems

Volumes 1–44

<https://www.routledge.com/Metal-Ions-in-Biological-Systems/book-series/IHCMEIOBISY>

(see also the website given below)

Metal Ions in Life Sciences

Volumes 1–22

Details about all books (series) edited by the SIGELs, including the Guest Editors, can be found at

<http://www.bioinorganic-chemistry.org/mils>

Foreword

Peter J. Sadler
Department of Chemistry, University of Warwick, UK
P.J.Sadler@warwick.ac.uk

The sensing of metals and use of metals as sensors are very important topics in the Life Sciences.

At least ten metals are essential in the human body. Their distribution, oxidation states, coordination numbers, coordination geometries, and ligand types are carefully controlled to suit their functions, be they charge carriers, structural control centers or triggers, redox centers, or catalysts.

Probably the biochemistry of most of them is controlled by genes and feedback loops, but we are not sure. Their special and temporal behavior needs to be mapped on timescales of nanoseconds to hours, resolutions of nanometers to meters, and concentrations of molar to picomolar and below.

Moreover, non-essential metals can also enter the body, from the environment, from medicines, and from diagnostic agents. Micro-organisms may have different requirements for metals compared to humans. The symbiotic relationship of human life with over 30 trillion microbial cells in the body cannot be overlooked.

Also a range of metals have favorable and useful sensing properties related to, for example, their electronic or nuclear composition, their optical and X-ray absorptions and emissions, or their radioactive emissions. In theory, individual isotopes of metals need to be tracked. Heavier isotopes slow down reactions.

This volume illustrates some of the exciting advances being made in the metal-sensing field.

Regulation of the uptake, transport, and activity of metals in biochemical pathways is not only important for essential metals but also for metallodrugs. The metalloid arsenic is an effective drug for treatment of promyelocytic leukemia. About 250 As-regulated proteins have been identified in leukemia cells, and there is a huge family of >500 arsenic resistance regulator proteins in bacteria.

Tracking Ca^{2+} at micro- to milli-molar concentrations, and the structural changes it induces in proteins on a sub-millisecond timescale, is critical to understanding a variety of cellular events. The optical and magnetic probe properties of lanthanide 3+ ions, of similar size ligand (oxygen) preferences as Ca^{2+}, find application in a wide range of bio-sensors. Gd^{3+}, for example, can modulate the relaxation of protons creating contrast in magnetic resonance imaging (MRI), even on a submillimeter scale for studies of cellular organization in tissues.

Remarkable is a Tb^{3+}-specific bio-sensor based on the recently discovered bacterial protein lanmodulin, which binds lanthanide ions with picomolar affinity. Interestingly, Er^{3+} complexes can be designed, which upconvert infrared radiation to red and green emissions.

Riboswitches, which control gene expression by transcriptional or translational attenuation, can be adapted as bio-sensors for Fe^{2+} and Mn^{2+} within the range of concentrations maintained by metalloregulators. Turn-on emissive probes for Zn^{2+} can sense free zinc ions released during apoptosis, and Cu^{2+} sensors for understanding the subcellular compartmentalization and roles of copper. Importantly, fluorescence bio-sensors can distinguish between Zn^{2+} and Cd^{2+}.

Bio-sensors for ubiquitous metals, such as vanadium, are likely to provide insight into their potentially essential physiological roles, not only in the body, but also in micro-organisms.

There is perhaps no more important area than sensing and understanding the roles of metals in the brain. Neuroscience has traditionally been the territory of only organic chemistry. Metal bio-sensing is now beginning to reshape that landscape.

1 Metalloid-Sensing Transcriptional Regulatory Proteins

*Jian Chen and Barry P. Rosen**
Department of Cellular Biology and Pharmacology, Herbert Wertheim College of Medicine, Florida International University, Miami, FL 33199, USA

CONTENTS

* Correspondence: Barry P. Rosen

DOI: 10.1201/9781003229971-1

ABSTRACT

A changing environment poses continuing challenges to living organisms. Arsenic is a toxic element that widely occurs in the Earth's crust. From the origin of life, the first organisms were exposed to arsenic, which exerted a strong selective pressure on primordial microbial communities. As a result, microbes evolved resistance determinants for coping with arsenic toxicity, including transport systems and pathways for biotransformation such as methylation. In bacteria, archaea, and fungi, a variety of transcriptional regulatory proteins have been identified that control expression of genes that confer resistance and biotransformation of arsenic and the related metalloid antimony. These are specialized allosteric proteins in which direct binding of trivalent arsenicals or antimonials results in a conformational change in the regulator and distorts the promoter DNA to control transcription. Considerable progress has been made over the past few decades, elucidating the structure and function of As(III)/Sb(III)-metalloregulatory proteins. Here, we summarize recent advances in understanding substrate-binding sites and the molecular mechanisms of transcriptional control. Applications of these metalloregulatory proteins for arsenic detection and bioremediation will be briefly reviewed.

KEYWORDS

Arsenic Biosensor; Arsenite; Antimonite; Methylarsenite; Roxarsone; ArsR Repressor; MSMA Herbicide

1 INTRODUCTION

Many metals and metalloids are essential for life and possess important physiological roles. They are indispensable for nearly all aspects of metabolism [1]. In cells, key bioenergetic and biogeochemical processes, including respiration, photosynthesis, and nitrogen assimilation, are dependent on metal ion co-factors [2]. In excess, many of these metals and metalloids can be toxic [3]. Metalloregulatory proteins regulate expression of gene-encoding proteins that contribute to intracellular homeostasis, the delicate balance of intracellular concentration, and the oxidation states of various metal species, which ensures the proper functioning of essential enzymes while preventing metal intoxication [4]. On the other hand, some metals and metalloids are not essential and possess risks to human health and the environment, including arsenic, antimony, cadmium and lead [5]. For this reason, many organisms utilize regulatory proteins for reduction of intracellular concentrations of these metals.

Arsenic is a ubiquitous naturally occurring metalloid widely distributed in the environment through natural and anthropogenic sources [6]. Depending on the physicochemical conditions and microbial activity, arsenic can be found in several different oxidation and methylation states, with various levels of toxicity and bioavailability [7]. In general, inorganic arsenic (e.g., arsenate (As(V)) and arsenite (As(III)) are the most common forms found in the environment. Humans are routinely exposed to inorganic arsenic in food, water, air, and soil. Long-term exposure to low levels of arsenic present in drinking water and food increases the risk of various forms of cancer and other serious health effects, including diabetes, and cardiovascular and neurological disorders [8]. To cope with arsenic stress, microbes have evolved numerous detoxification systems that lower the intracellular levels of arsenic or reduce its toxicity by enzymatic biotransformations [9]. In most cases, expression of these systems is controlled at the level of transcription by regulatory proteins that serve as biosensors of environmental arsenic and sense arsenic by direct coordination. In bacteria and archaea, the genes for arsenic resistance cluster in *ars* operons [10]. A minimal *ars* operon, such as that found in the *Escherichia coli* chromosome [11] or *Staphylococcus aureus* plasmid pI258 [12], has just three genes, *arsRBC*, encoding the ArsR transcriptional repressor, the ArsB arsenite efflux permease, and the ArsC arsenate reductase. ArsR is an As(III)-responsive repressor metalloregulatory protein that binds to the *ars* promoter and represses transcription. When it binds As(III) or Sb(III), there is a conformational change that results in ArsR dissociation from the promoter, derepressing expression of the *arsRBC* genes. ArsB catalyzes efflux of As(III) or Sb(III), the simplest way to confer resistance. ArsC expands the resistance to include As(V) by reducing it to As(III), the ArsB substrate. Since the discovery of *ars* operons, many other *ars* genes have been identified; indeed, every letter in the English alphabet has been assigned to an *ars* gene, and additional *ars* genes await assignments. For example, the first well-characterized *ars* operon, that from *E. coli* plasmid R773 [13], has two additional *ars* genes, forming a relatively common five-gene *arsRDABC* operon [14]. ArsA is an As(III)-stimulated ATPase that associates with the ArsB permease, converting it into the ATP-coupled ArsAB extrusion pump, which confers much resistance to high levels of As(III) or Sb(III) than ArsB alone [15]. ArsD is a metallochaperone that delivers As(III) or Sb(III) to the ArsA ATPase [16]. Groups of *ars* genes and operons are often clustered together to form "*arsenic gene islands*" [17]. *ars* operons can be found in bacterial chromosomes and also on plasmids or transposons [7]. It is interesting to note that many microorganisms possess multiple, and even redundant, arsenic resistance systems on chromosomes and plasmids that synergistically increase the level of resistance or expand the substrate specificity to include organoarsenicals. The distribution of arsenic resistance genes is a reflection of the ubiquitous environmental presence of arsenic.

ArsR was the first identified member of the ArsR/SmtB family of metalloregulatory proteins [18]. Subsequently, other members of this family were identified that have specificity for a wide variety of metals, including Zn, Cd, Pb, Zn, Co [19, 20]. To date more than 3,000 individual members of the ubiquitous ArsR/SmtB family have been identified in the NCBI database. ArsR/SmtB family members are homodimers with a highly conserved DNA recognition helix-turn-helix (HTH) motif. They bind to their cognate operator/promoter (O/P) region (Figure 1), repressing expression in absence of metal ions, and derepressing in the presence of high concentrations of toxic metal ions, allowing microbes to survive in challenging environments [21]. This review will not only focus on ArsR repressors but will also describe additional As(III)-responsive regulatory proteins from both bacteria and fungi. It will summarize recent biochemical and structural studies of arsenic transcriptional repressors and their potential application in arsenic biosensing and bioremediation.

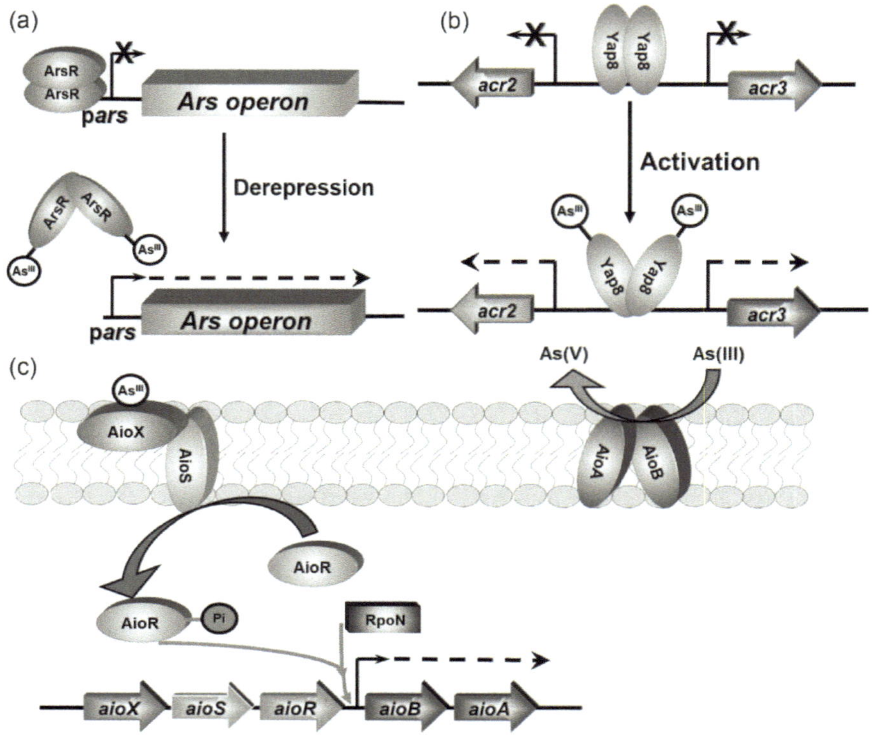

FIGURE 1 Principles of regulation of arsenic resistance genes. (a) The ArsR/SmtB family protein, represented by *E. coli* ArsR, regulates expression of *ars* genes. In the absence of As(III), a basal level of ArsR is synthesized, which binds to the operator of the ArsR promoter (P*ars*), preventing transcription. When As(III) is present, it binds to ArsR, producing a conformation change that results in dissociation from P*ars* and transcription of *ars* genes. (b) Yap8 from *S. cerevisiae* is a member of the AP-1 family of activator proteins and is the only known eukaryotic As(III)-responsive transcription factor. In yeast, Yap8 regulates the expression of the *acr2* and *acr3* genes by binding to the promoter. In the absence of As(III), homodimeric Yap8 prevents transcriptional activity. When As(III) binds to Yap8 cysteine residues Cys132, Cys137, and Cys274, a conformational change drives the expression of *acr2* and *acr3*. (c) The three-component AioXSR system regulates expression of the arsenite oxidase *aioAB* genes. Periplasmic AioX binds As(III) and transmits the signal to the AioS sensor kinase, which phosphorylates the AioR response regulator. Phosphorylated AioR associates with RpoN to bind to a palindrome-like DNA sequence, activating transcription of *aioBA*.

2 FAMILIES OF METALLOREGULATORY PROTEINS

2.1 The ArsR Family

The ArsR/SmtB family is a widespread group of transcriptional factors characterized by the ability to repress the expression of operons linked to stress-inducing concentrations of di- and multivalent heavy metal ions [19]. An evolutionary analysis, coupled with comparative structural and spectroscopic studies of ArsR/SmtB family members, reveals they have similar ligand-responsive transcriptional mechanisms and structure models, but individual members have evolved distinct metal selectivity profiles by alteration of one or both of two structurally distinct metal coordination sites. Well-characterized ArsR/SmtB proteins include the As(III)/Sb(III)-responsive ArsR from plasmid R773, the first identified member of this family [18], the Cd(II)/Pb(II)/Zn(II)-responsive CadC from *S. aureus* plasmid pI258 [22], and the Zn(II)-responsive SmtB from *Synechococcus* sp. [23].

The 1.9 Å X-ray crystal structure CadC was the first to be determined with a two-fold axis of symmetry consisting of six α-helices and two β-strands arranged into an α1-α2-α3-α4-α5-β1-β2-α6 fold [24] (Figure 2). CadC has two different types

(a)

(b)

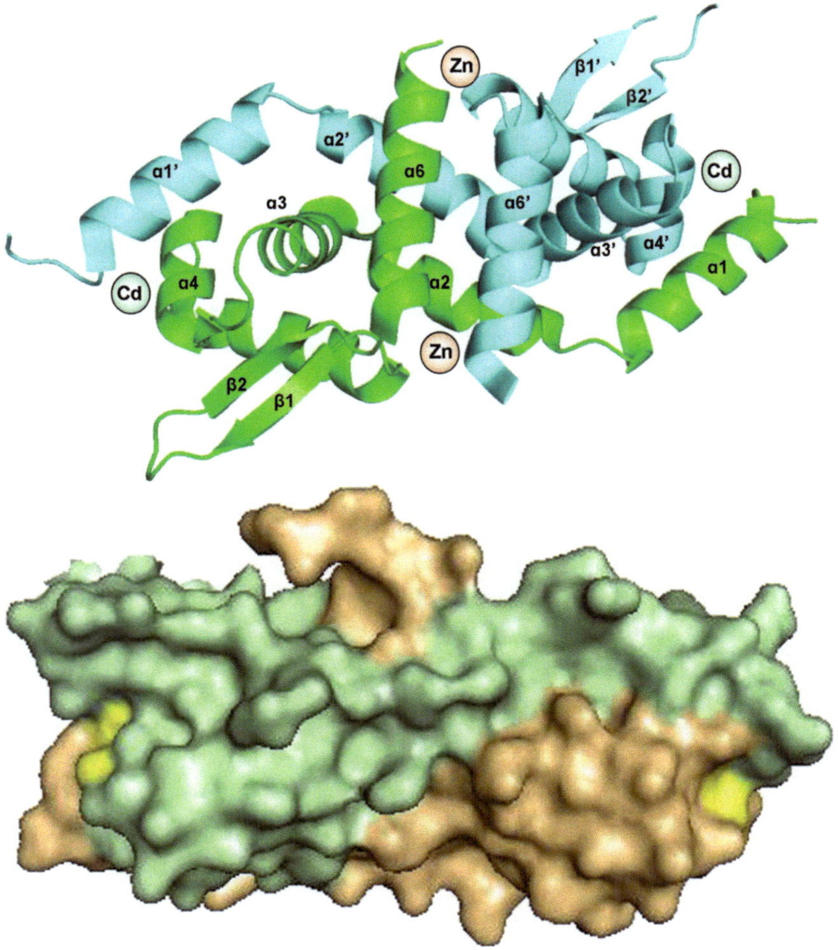

FIGURE 2 Structure of the CadC Cd(II)/Pb(II)/Zn(II)-responsive repressor. (a) Ribbon diagram of the *S. aureus* pI258 CadC apo-repressor with secondary structural units N-α1-α2-α3-β1-α4-α5-β2-β3-α6-C. The two monomers are colored in green and cyan, respectively. CadC exhibits both a regulatory Cd(II)-binding site at the DNA-binding site and a structural Zn(II)-binding site at the dimer interface. Cd (light green spheres) and Zn (orange spheres) atoms were manually added near the binding sites predicted from biochemical and molecular biological analysis. (b) Surface diagram of the CadC dimer. Metal(loid)-binding sites required for metalloregulation are highlighted. The two dimer subunits are colored in green and orange, respectively, with the Cd(II)-binding sites colored in yellow.

of metal-binding sites. It has a four-coordinate Cd(II)/Pb(II)/Zn(II)-binding site composed of $Cys7_a$ and $Cys11_a$ from the N-terminus of one monomer and $Cys58_b$ and Cys60b in or near the DNA-binding site of the other monomer. These residues form the metal-sensing module of the repressor that results in transcription of the *cad* operon genes when metal is bound [25]. There is a second binding site for Zn(II) located at the dimer interface. In the dimer, there are two Zn(II)-binding sites composed of $Asp101_a$ and $His103_a$ from one monomer and $His114_b$ and $Glu117_b$ from the other monomer. This site does not appear to be a regulatory site but may have a structural role [25]. It is congruent with the regulatory Zn(II)-binding site of *Synechococcus* PCC 6301 SmtB. The crystal structures of CadC with bound Zn(II) and of a mutant lacking the Zn(II)-binding site were compared with the structures of the related SmtB Zn(II)-responsive repressor with and without bound Zn(II). In CadC, a glycine residue near the binding site corresponds to an arginine residue in SmtB that is involved in Zn(II) regulation. We proposed that a glycine residue was present in an ancestral Zn(II)-binding CadC-like repressor, and that acquisition of regulatory ability in SmtB was a more recent evolutionary event resulting from a Gly-to-Arg substitution.

Structures of a number of ArsR/SmtB family members have been subsequently reported [19, 26, 27]. They all have the common backbone of winged-helix DNA proteins with consensus DNA-binding motifs, with a similar arrangement of structural elements, although the number of α-helices differs. From structure-function analysis of various ArsR/SmtB repressors, it is apparent that the metal(loid)-binding sites arose independently by convergent evolution on the common backbone of the DNA-binding protein, most likely as a result of selective pressure from exposure to different metal(loid)s. This is clearly illustrated in the multiple alignments of representative repressors, which shows that identified sulfur, nitrogen, and oxygen ligands to metal(loids) vary, even though their overall relatedness is clear (Figure 3). This is visually obvious when the surface of the CadC homodimer is used to show the approximate location of the binding sites for different metal(loid)s in selected ArsR/SmtB family members, including CadC (Cd(II)/Pb(II)/Zn(II) and second Zn(II)) site) [24], R773 ArsR (As(III)/Sb(III)) [10], *Acidobacillus ferrooxidans* AfArsR (As(III)/Sb(III)) [28], *Corynebacterium glutamicum* CgArsR (As(III)/Sb(III)) [29], *Comamonas testosteroni* JL40 AntR (Sb(III)) [30], *Shewanella putrefaciens* (MAs(III) [31], *Synechococcus* PCC 6301 SmtB (Zn(II)) [23], and *Mycobacterium tuberculosis* CmtR (Cd(III)/Pb(II)) [32] (Figure 4). Metal-binding sites are located near the DNA-binding site or at the dimer interface. They are two-, three-, or four-coordinate composed of cysteine thiolates, especially in the case of metalloids and softer metals, or nitrogen or oxygen ligands for harder metals. They are formed from residues from just one subunit of the homodimer or residues contributed by both monomers. Based on their evolutionary relatedness, the ArsR members of the family fall into five clades that correspond with the location of the metalloid-binding sites (Figure 5).

2.1.1 Clade1: ArsR from *E. coli* Plasmid R773, the Patriarch of the ArsR/SmtB Family

ArsR encoded by *E. coli* plasmid R773 was the first member of the family to be identified [10] and is essentially identical to the *E. coli* chromosomal protein, with

```
R773      1  ---------------------------------------------MLQLTPQLFKNLSDE
EcArsR    1  ---------------------------------------------MSFLLPIQLFKILADE
CgArsR    1  --MT---------TLHTIQLANPTECC-----TLATG---PLSSDESEHYADLFKVLGDP
AfArsR    1  ------------------------------------M---EP-LQDPAQIVARLEALASP
SpArsR    1  -------------------------------MNIADM---NVADMNVENAAKVLKELGHP
AntR      1  MALEKRNELPACSLKPSLQDRD------------------LITSAEAGEVVVLFKVLAND
SmtB      1  ------------MTKPVLQDGETVVCQGTHA--AIASELQAIAPEVAQSLAEFFAVLADP
CadC      1  --MK-------------KKDTCEIFCYDEEKVNRIQGD---LQTVDISGVSQILKAIADE

R773     17  TRLGIVLLLREM--GELCVCDLCMALDQSQPKISRHLAMLRESGILLDRKQGKWVHYRLS
EcArsR   17  TRLGIVLLLSEL--GELCVCDLCTALDQSQPKISRHLALLRESGLLLDRKQGKWVHYRLS
CgArsR   42  VRLRILSQLAAGGCGPVSVNELTDLMGLSQPTISHHLKKMTEAGFLDRVPEGRVVLHRVR
AfArsR   21  VRLEIFRLLVEQEPTGLVSGDIAEHLGQPHNGISFHLKNLQHAGLVTVQREGRYQRYRAA
SpArsR   27  TRLALFRLLVKGGYTGVAVGQLQEALQIPGSTLSHHISALMSAGIISQRREGRVLYCVPD
AntR     43  TRLRLLHALARS--GGLCVTDLAAAVGMKPQAVSNQLQRLADRRILRAARCGNNIHYRIV
SmtB     47  NRLRLLSLLAR---SELCVGDLAQAIGVSESAVSHQLRSLRNLRLVSYRKQGRHVYYQLQ
CadC     43  NRAKITYALCQD--EELCVCDIANILGVTIANASHHLRTLYKQGVVNFRKEGKLALYSLG

R773     75  PHIPSWAAQIIEQAWLSQQDDVQVIARKLASVNCSGSSKAVCI
EcArsR   75  PHIPAWAAKIIDEAWRCEQEKVQAIVRNLARQNCSGDSKNICS
CgArsR  102  PELFAELRTVLQIGS-MEL
AfArsR   81  MPVVRALVAYLTENCCHGTRDCALSGETRSPSVQEGNQ
SpArsR   87  YELLQGLVHFLQDQCCSGQ
AntR    101  DPCVLR---MLELGLCLIEEAEQQAGG
SmtB    104  DHHIVA---LYQNAL-DHLQECR
CadC    101  DEHIRQ---IMMIAL-AHKKEVKVNV
```

FIGURE 3 Multiple sequence alignments of ArsR/SmtB repressors. Multiple sequence alignments of selected members of the ArsR//SmtB family. Representative ArsR orthologs (accession numbers in parentheses) are from: *E. coli* plasmid R773 ArsR (CAA34168); *E. coli* chromosomal EcArsR (WP_089623100.1); *C. glutamicum* CgArsR1 (CAF21518); *A. ferrooxidans* AfArsR (ACK80311); *S. putrefaciens* SpArsR (ADV53698); *C. testosterone* JL40 AntR (WP034375793.1); *Synechococcus elongatus*, PCC 7942 SmtB (CAA45872); and *S. aureus* plasmid pI258 CadC (P20047).

a highly conserved three-coordinate As(III)/Sb(III)-binding motif composed of residues $C_{32}VC_{34}DLC_{37}$ [33]. The difference between the R773 *arsRDABC* and chromosomal *arsRBC* operons is the presence of genes for the ArsA As(III)-stimulated ATPase and ArsD As(III) chaperone in the R773 operon that confers much higher arsenic resistance than the chromosomal operon. A homology model of *E. coli* ArsR built on the CadC crystal structure indicates that each ArsR monomer has a fold of five α-helices and a pair of antiparallel β-strands with an α1-α2-α3-α4-β1-β2-α5 topology. In each monomer, helices 3 and 4 comprise the HTH DNA-binding motif, and the β-sheets form the "wings" of these winged-helix DNA-binding proteins [34]. The sulfur atoms of Cys32, Cys34, and Cys37 in each subunit form the "arsenic sensor" module, a very specific three-coordinate binding site for the trivalent metalloids As(III) and Sb(III) (and perhaps Bi(III)). ArsR lacks the first helix of CadC, which contains two cysteine residues that form part of the four-coordinate Cd(II)/Pb(II)/Zn(II)-binding site. The conserved $ELC_{32}VC_{34}DLC_{37}$ As(III)-binding motif of ArsR is in helix 3, as part of the projected α3-turn-α4 DNA-binding motif. Helix 4 is proposed to be the recognition helix that binds to the DNA major groove. The dimer interface is formed by the two antiparallel helix 5s of the two monomers. Binding of metalloid to ArsR is presumed to induce a conformational change, leading to

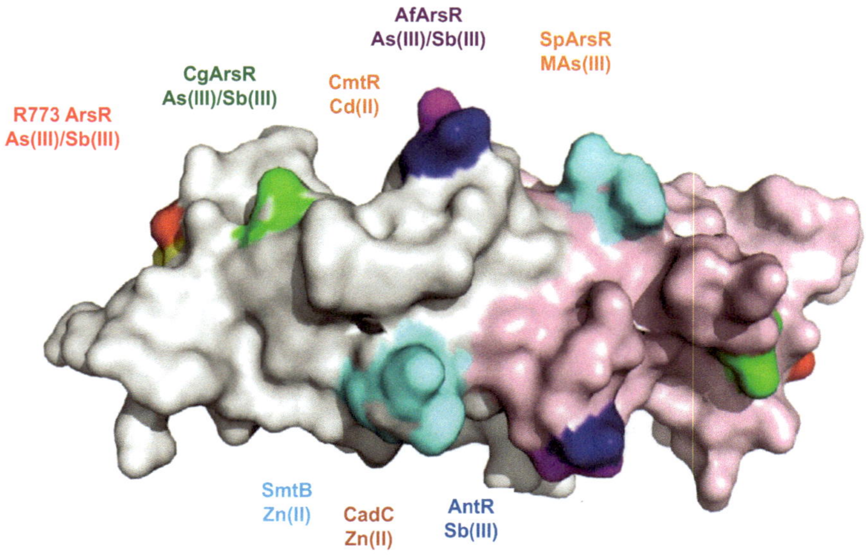

FIGURE 4 Location of metal(loid)-binding sites in ArsR/SmtB repressors. Metal(loid)-binding sites in members of ArsR/SmtB family of repressor proteins are shown on a surface model of the CadC aporepressor structure by coloring CadC residues. The S_3 As(III)-binding site of the R773 ArsR (red) formed within each monomer overlaps with the corresponding S_4 Cd(II) binding site of CadC (yellow) formed between the N-terminus of one subunit and the DNA-binding domain of the other subunit. The S_3-binding sites of CgArsR1 (green) include residues in the DNA-binding site. The Zn(II)-binding sites of CadC and SmtB (cyan) are formed between the antiparallel C-terminal α6 helices. The S_3 As(III)-binding site of AfArsR (purple) and the S_2 MAs(III)-binding site of SpArsR (blue) differ by a single cysteine residue. The variety of the location of metal(loid)-binding sites distributed over the surface of the respective repressors demonstrates the plasticity of evolutionary solutions to similar environmental stresses. (Adapted from [31].)

dissociation from the DNA, allowing transcription of the *ars* genes. However, these predictions were based on homology modeling and not actual structures. It took another 25 years to solve the crystal structures of two genuine ArsR repressors, AfArsR and CgArsR [27].

2.1.2 Clade 2: *Acidithiobacillus ferrooxidans* ArsR, a Quite Different As(III)-Binding Site

A second ArsR that has a completely different As(III)-binding site from R733 ArsR is AfArsR from *Acidithiobacillus ferrooxidans*. This organism is used for bioleaching of metals, such as gold, zinc, and copper, in the mining industry and for bioremediation of mine drainage [35]. It produces sulfuric acid from iron sulfides and dissolves metals for recovery. Arsenic is a common contaminant in ores and is liberated by *A. ferrooxidans* along with other metals. In the chromosome of *A. ferrooxidans* are two divergent *ars* operons, *arsCR* and *arsBH*, that are both transcriptionally

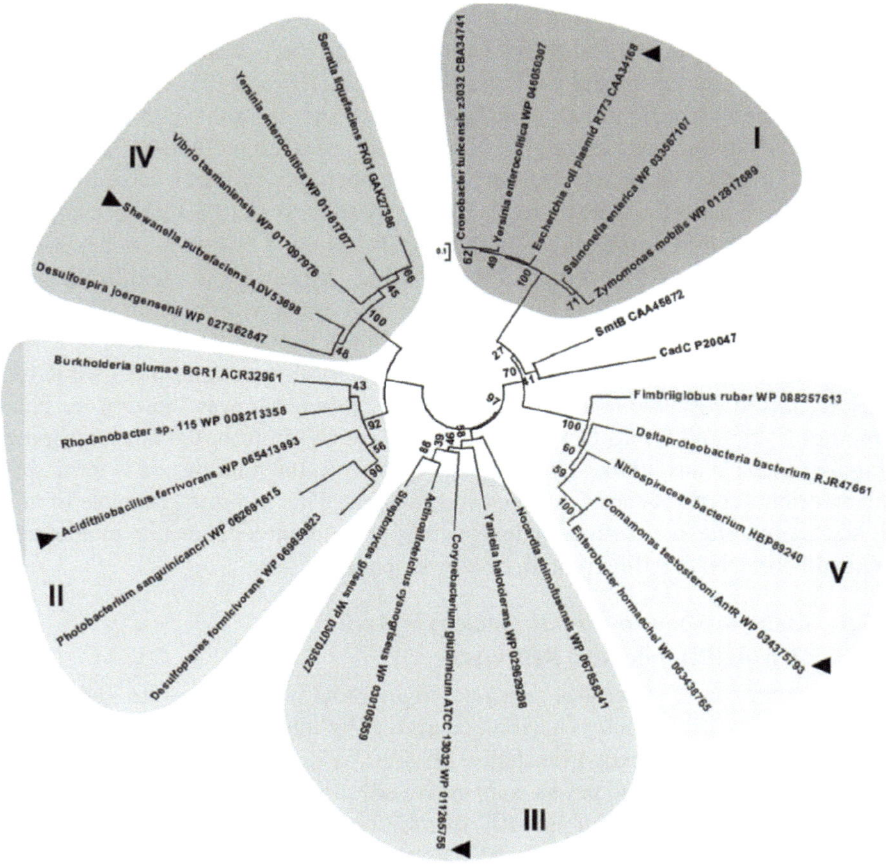

FIGURE 5 Evolutionary relatedness of ArsR repressors. A neighbor-joining phylogenetic tree shows that there are five clades of ArsR repressors with different placement of As(III)-, Sb(III)-, or MAs(III)-binding cysteine residues. ArsR repressors with typical CVCXXC As(III)-binding motif, such as *E. coli* R773 ArsR, are clustered in clade I. ArsR repressors with three C-terminal conserved cysteine residues, such as *A. ferrooxidans* AfArsR, are clustered in clade II. ArsR repressors with three N-terminal conserved cysteine residues, such as *C. glutamicum* CgArsR, are clustered in clade III. MAs(III)-responsive ArsR repressors with two C-terminal conserved cysteine residues, such as *S. putrefacience SpArsR*, are close to AfArsR group and clustered in clade IV. ArsR repressors with selectivity for Sb(III) and distinct cysteine residues, such as *C. testosterone* AntR, have a binding site that is more closely related to that of the Zn(II)-responsive repressor SmtB than to As(III)/Sb(III)-responsive ArsR repressors and are clustered in clade V. The well-characterized ArsR repressors in each clade are indicated by black triangles.

controlled by AfArsR. These two *ars* operons protect the acidophilic bacterium from the toxic arsenic it produces [36]. The 1.86 Å X-ray crystal structure of AfArsR was recently solved with an As atom bound in one monomer to the sulfur atoms of Cys95, Cys96, and Cys101 [27]. Like R733 ArsR, all three As(III) ligands are contributed by a single subunit.

2.1.3 Clade 3: *Corynebacterium glutamicum* CgArsR, Variation on a Theme

C. glutamicum is one of the most arsenic-resistant microorganisms described to date and is able to grow in media with up to 12 mM As(III) and 0.5 M As(V) [37]. In the *C. glutamicum* genome are two operons (*ars1* and *ars2*) that confer resistance to arsenic. Both operons have a gene encoding a CgArsR repressor, an ArsB efflux permease, and an ArsC arsenate reductase. One operon contains an additional arsenate reductase gene. CgArsR responds to As(III) and less so to Sb(III) because it is taken up poorly by *C. glutamicum* [29]. CgArsR has three cysteine residues, Cys15 and Cys16 near the N-terminus, and Cys55, which is near but not within the putative DNA-binding domain that does not align with cysteine residues in other ArsR repressors (Figure 3). From mutagenic analysis, those three residues were proposed to form the As(III)-binding site. The 1.6 Å X-ray crystal structure of CgArsR was recently solved with two bound As atoms. Each binding site was shown to be composed of Cys15 and Cys16 from one monomer and Cys55 from the other monomer (Figure 6) [27]. Thus, unlike R773 ArsR and AfArsR, the binding site is formed by the contribution of residues from opposite subunits. This is a third example of how S_3 arsenic binding sites formed independently in different locations in an ancestral winged-helix DNA-binding protein by convergent evolution.

2.1.4 Clade 4: *Shewanella putrefaciens* SpArsR, an MAs(III)-Selective Repressor

In the environment, microbes generate highly toxic methylarsenite (MAs(III)) by methylation of arsenite (As(III)) catalyzed by the ArsM arsenite (As(III)) S-adenosylmethionine methyltransferase [38, 39]. We proposed that bacteria evolved the ability to use arsenic to gain a competitive advantage over other bacteria by producing MAs(III) as an antibiotic [40]. In addition to biogenic MAs(III), MAs(V), which has been used as the herbicide MSMA (monosodium methylarsenate) as an herbicide for many decades, can be reduced to MAs(III) by soil bacteria [41]. Although not currently allowed for use on food crops, MSMA is still used in the United States and other countries on cotton fields or for post-emergence control of turf weeds and crabgrass [42]. Unlike As(III), which forms three-coordinate complexes with sulfur ligands, MAs(III) forms two-coordinate complexes. Although they can bind to ArsR repressors that have S_3 sites, those repressors lack specificity for trivalent organoarsenicals. Perhaps in response to environmental exposure to organoarsenicals, bacteria evolved ArsR repressors with S_2-binding sites that are selective for MAs(III) over As(III).

One such MAs(III)-selective ArsR transcriptional repressor is encoded in the genome of *Shewanella putrefaciens* 200 [31]. SpArsR responds to MAs(III) but not As(III). From a phylogenetic analysis of ArsR orthologs with different arsenic binding sites, SpArsR appears to be closely related to *A. ferrooxidans* AfArsR (Figure 5). From comparative sequence analysis, SpArsR has two conserved cysteine residues, Cys101 and Cys102, located near the C-terminus. These correspond to Cys95 and Cys96 in AfArsR, which are two As(III) ligands (Figure 7), but SpArsR lacks a cysteine residue corresponding to AfArsR Cys102, which contributes to the third sulfur

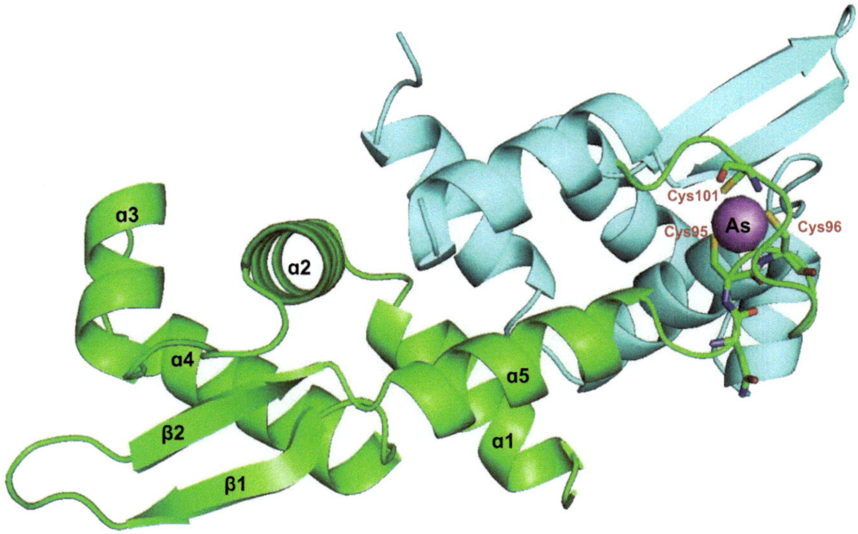

FIGURE 6 Structure of the CgArsR As(III)-responsive repressor. Schematic representation of CgArsR in its As(III)-bound form. The two monomers are colored in cyan and green, respectively. CgArsR shows arsenic atoms (purple spheres) bound to $Cys15_a$ and $Cys16_a$ in one chain and $Cys55_b$ from the other chain.

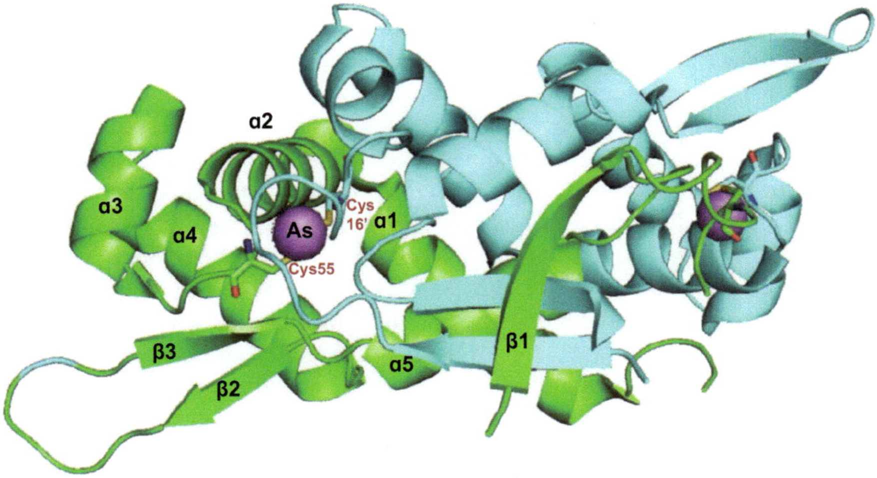

FIGURE 7 Structure of the AfArsR As(III)-responsive repressor. Schematic representation of AfArsR with bound As(III). The two monomers are colored in cyan and green, respectively. AfArsR shows an arsenic atom (purple spheres) bound to N-terminal cysteine residues Cys95, Cys96, and Cys102.

ligand in the S_3 As(III)-binding site. The results of the mutagenic analysis support the argument that the Cys95–Cys96 vicinal pair is a MAs(III)-selective binding site.

2.1.5 Clade 5: *Comamonas testosteroni* AntR, an Sb(III)-Selective Repressor

Recently a novel member of ArsR/SmtB family, AntR, was identified that responds to metalloids in the order Sb(III) = MAs(III) >> As(III) [30]. Although it responds similarly to Sb(III) and MAs(III), there is reason to consider that it may have an Sb(III) selective function. The gene for AntR is encoded in the *antRCA* antimony resistance operon in the chromosome of the Gram-negative soil bacterium *C. testosteroni* JL40 [43]. This microbe was isolated from an antimony mine in Lengshuijiang, Hunan Province, China and is resistant to high levels of environmental antimony. The *antA* gene encodes the only identified Sb(III)-translocating P1B-type ATPase that confers resistance by extrusion of Sb(III) from the cells. AntC is a small protein that is proposed to be an Sb(III) chaperone for the AntA efflux pump. AntR is an ArsR-type negative repressor that regulates the *ant* operon. Cells expressing *antR* respond to Sb(III) and MAs(III) but only weakly to As(III) and not at all to Sb(V) or As(V). The results of isothermal titration calorimetry and transcriptional induction experiments demonstrated that AntR is an Sb(III)/MAs(III)-selective repressor, with little response to As(III). Since most metalloregulatory proteins can sense both Sb(III) and As(III), these were surprising results—the first report of a transcriptional repressor AntR with high selectivity for environmental antimony.

What confers the ability of AntR to distinguish between Sb(III) and As(III)? As discussed above, SpArsR has a reduced affinity for As(III) compared with MAs(III) by reducing the number of sulfur ligands from the three in AfArsR to two, so it was reasonable to predict that AntR might have an S_2 site Sb(III)-binding site. AntR has five cysteine residues in locations that differ from other ArsRs (Figure 3). Members of the AntR clade appear to be more closely related to CadC and SmtB than to ArsR repressors in the other four clades. Therefore, the location of the ligand-binding residues could not be predicted from comparison with other ArsR repressors. From the results of mutagenesis, only two cysteine residues, Cys103 and Cys113, appear to be required for Sb(III) sensing [30]. The crystal structure of AntR was recently solved at 2.1 Å (Figure 8). It adopts a classical ArsR/SmtB topology architecture, where the overall structure has the same winged-helix backbone as other ArsR repressors. All five cysteine residues are visible in both chains. The two conserved cysteines, Cys103 and Cys113, are visible in both α6 helices at the dimer interface, with Cys103a juxtaposed to Cys113b, and Cys113a to Cys103b. Although the structure does not have bound Sb(III), those residues are the only two close enough to each other to form an S_2 site with the antimony atom. The other three non-conserved cysteines are too far away to participate in Sb(III) binding. Typical Sb–S bond lengths fall in the range of 2.4–2.5 Å, so the two sulfur atoms could not be further than 5 Å from each other. However, in the apo-AntR structure, the distance between the Cys103a sulfur atom in one monomer and the Cys113b sulfur atom in the other monomer of the apo-AntR structure is approximately 10–11 Å, which implies that the two helices would have to move closer to each other to form an S_2-binding site for Sb(III). We propose that a conformational change occurs stochastically that shortens the distance between the two helices at the dimer interface, bringing the two cysteine residues close enough to

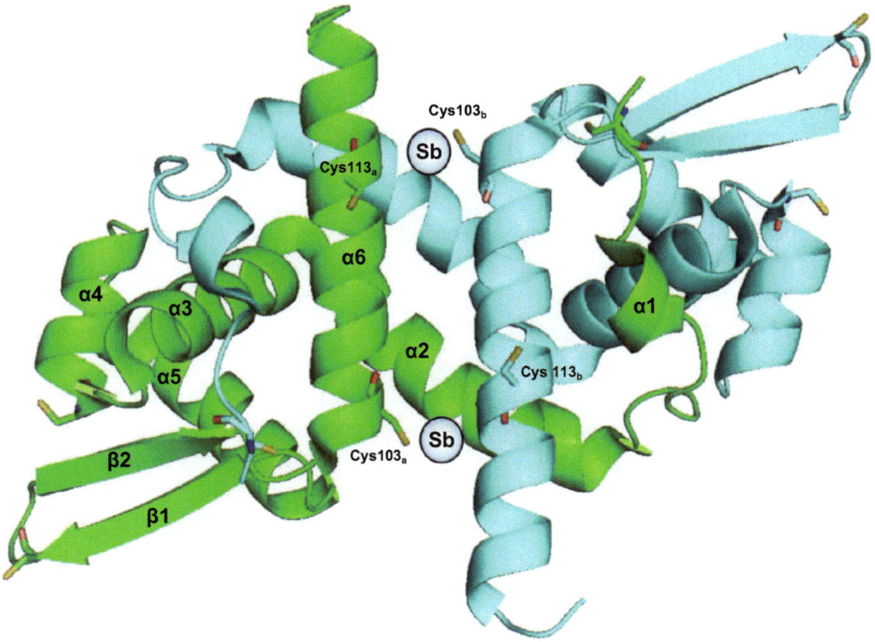

FIGURE 8 Structure of the AntR Sb(III)-responsive repressor. Schematic representation of apo-AntR. The two monomers are colored in cyan and green, respectively. Sb atoms (light blue spheres) were manually added near the predicted binding site of $Cys103_a$ from one chain and $Cys113_b$ from the other chain.

each other to form the binding site. Clearly, additional structural data are required to test this hypothesis, but from a combination of the results of mutagenesis and structure, this is not an unreasonable proposal.

2.2 *Saccharomyces cerevisiae* Yap8: A Eukaryotic As(III)-Responsive Transcription Factor

ArsR metalloregulatory proteins are widespread in bacteria and archaea but are absent in higher eukaryotes. To cope with exposure to toxic arsenic, the yeast *S. cerevisiae* uses Yap8 (also called Acrl or Arrl), a homolog of the stress-responsive transcriptional regulator AP-1 [44], as a different type of arsenic sensor [45]. *S. cerevisiae* has genes for eight different AP-1-like proteins: Yapl–Yap8. These proteins contain a basic leucine zipper (bZIP) DNA-binding domain as well as a conserved cysteine-rich domain [46]. Two of these transcriptional regulators, Yapl and Yap8, respond to arsenic stress. Yapl is crucial for general oxidative stress tolerance in yeast. It responds to peroxides and other oxidants, chemicals with electrophilic properties, and metals by controlling the expression of about 70 genes, with functions in oxidative stress defense and sulfur metabolism in response to As(III) exposure [47]. Yapl stimulates gene expression in response to As(III) or Sb(III) exposure, and the stress signals activate Yapl through distinct mechanisms that involve disulfide bond formation and/or covalent modifications of conserved cysteines [48]. Whether As(III) sensing by Yapl involves a direct interaction between protein and ligand or through oxidative modifications remains unknown. In contrast, Yap8 is specific for arsenic resistance [49] through transcriptional regulation of two genes: *acr2*, encoding an As(V) reductase, and *acr3*, encoding an As(III) exporter [45]. Yap8 has eight cysteine residues, but only Cys132, Cys137, and Cys274 are conserved. Yap8 is nuclear localized, where it binds to the Acr2/Acr3 promoter as a homodimer, repressing transcription. Yap8 is constitutively expressed and is regulated at the posttranslational level. In the absence of As(III), Yap8 levels in the nucleus are low as a result of degradation in the ubiquitin-proteasome system. As(III) binding produces a conformational change in Yap8 that prevents degradation and increases levels in the nucleus. Yap8 levels are not affected by oxidative stress with peroxide or paraquat. From mutational analyses, it appears that Cys132, Cys137, and Cys274 are involved in Yap8 stabilization and transcriptional activation [50]. Using purified Yap8, direct binding of As(III) as a three-coordinate complex with sulfur atoms was shown using X-ray absorption spectroscopy, while a Yap8-C132A/C274A mutant protein did not bind As(III) [49]. These results are consistent with the binding of As(III) to Cys132, Cys137, and Cys274 serving as a molecular switch that converts inactive Yap8 into an active transcriptional regulator (Figure 1b).

2.3 *Agrobacterium tumefaciens* AioR

Microbial response to arsenic has resulted in the evolution of diverse resistance mechanisms [7]. Compared to As(V), As(III) shows higher toxicity and is more mobile. A very common environmental transformation process oxidation of As(III) to less toxic As(V), so microbial arsenite oxidation can be considered a detoxification process. Additionally, As(III)-oxidizing microorganisms can also utilize As(III) as a sole electron donor, deriving energy from arsenite oxidation [51]. Aerobic As(III) oxidation is catalyzed by the Aio As(III) oxidase (Aio), which couples oxidation of As(III) to the reduction of oxygen to water, generating ATP for energy and NADH for carbon dioxide fixation [52]. Aio consists of two subunits, AioA and AioB. AioA is a catalytic molybdopterin guanine dinucleotide-containing subunit that contains a 3Fe-4S cluster, and AioB is a small Rieske 2Fe-2S subunit [53]. Regulation of expression of the *aio* genes is controlled by a three-component As(III)-binding protein/sensor/regulator signal transduction system, AioXSR. This complex has been proposed to induce the *aioBA* operon in the presence of As(III) in *Agrobacterium tumefaciens*, *Herminiimonas arsenitoxydans*, and *Rhizobium* sp. strain NT-26 [54–56] (Figure 1c). In *A. tumefaciens* 5A, the *aioXSR* genes are upstream of the *aioBA* operon and regulate its expression. In this As(III)-sensing system, regulator AioR does not directly interact with As(III). AioX is a periplasmic As(III)-binding protein that senses As(III) and transduces the signal to AioS. AioS is a sensor histidine kinase and phosphorylates the response regulator AioR [57]. Phosphorylated AioR then activates transcription of *aioBA* in association with the RpoN promoter. The regulator AioR belongs to the NtrC $\sigma 54$

RNA transcriptional activator family. It is likely to be a specific co-activator with σ54 in the initiation of *aioBA* operon transcription [55]. AioX from *A. tumefaciens* has only one cysteine residue, Cys108, that is conserved in all AioX-like proteins. Mutation of Cys108 to alanine or serine resulted in the loss of As(III) binding and As(III) oxidation [58]. Unlike many other arsenic sensor proteins, AioX does not respond to Sb(III) [59]. Thus, the AioXSR system represents a third evolutionary alternative to ArsR and Yap8 with a quite different mechanism of transcriptional activation and metalloid selectivity.

3 APPLICATIONS FOR ARSENIC SENSING AND BIOREMEDIATION

Arsenic is released into the environment from both natural and anthropogenic sources. The contamination of groundwater and crops is of major concern worldwide. Millions of people in countries such as Bangladesh and Chile are exposed to high levels of arsenic in groundwater contaminated by geological sources [60]. In addition, humans introduce organoarsenicals as herbicides, pesticides, and animal growth enhancers. As mentioned above, MSMA was used as an herbicide for half a century, although the U.S. Environmental Protection Agency prohibited its use on food crops in 2014. It is currently allowed in the United States for use on cotton fields, golf courses, sod farms and highway medians. In addition to MSMA, the synthetic pentavalent aromatic arsenicals roxarsone (4-hydroxy-3-nitrophenylarsonate), nitarsone (4-nitrophenylarsenate), atoxyl (arsanilic acid or aminophenylarsenate), and carbarsone (4-(carbamoylamino)phenylarsenate) were extensively used since the early and mid-1900s as antimicrobial and antiprotozoal drugs for controlling infections in poultry and swine [61]. They were banned in the United States in 1991, but some, like nitarsone, remained available for controlling histomoniasis (blackhead disease) in turkeys until 2015 [62]. Their use is prohibited in many countries but is still available in others [63], and compliance with the ban is not always uniformly enforced [64]. Even though they are no longer used, they remain persistent soil contaminants. Microbes gradually degrade them into toxic and carcinogen inorganic arsenic that enters our food and water supplies [65]. These issues have created a demand for new technologies for arsenic detection and bioremediation. Biological arsenic bio-sensors and bioremediation methodologies have advantages in comparison with conventional physicochemical methods. Biological treatment can be cost-effective and does not require expensive instrumentation run by highly trained staff. Rather, microbes provide sensors and can be used for removal [66, 67]. ArsR metalloregulatory proteins can be used as sensors for arsenic detection, which can lead to new methods for bioremediation.

3.1 Application of Metalloregulatory Protein in Arsenic Detection

ArsR has high selectivity as well as high affinity for arsenic. It senses and directly binds As(III), allosterically regulating binding to the *ars* O/P DNA. This induces dissociation of ArsR from the promoter, activating gene expression. This concept of a negatively autoregulated ArsR circuit was widely applied in bio-sensor development for sensing arsenic in soil and water [68]. These transcriptional switches of reporter genes are used in a number of bio-sensors to monitor arsenic contamination. Coupling the *arsR* gene to a reporter gene has made it possible for cells to emit a measurable signal in response to low concentrations of environmental arsenic. When As(III) enters cells through an aquaglyceroporin, such as GlpF [69], and binds to ArsR, it can control the expression of a reporter, such as GFP (green fluorescent protein) [70], luciferase [71], or β-galactosidase [72]. Most commonly the R773 or *E. coli arsR* genes are used for the construction of whole-cell biosensors [73–75]. ArsR-based biosensors are commercially available as field test kits such as the ARSOlux Bio-sensor [76]. This kit can detect arsenic amounts lower than 10 μg/L, the current limit recommended by the World Health Organization for arsenic in drinking water [66]. In addition to biosensors that use *E. coli* R773 ArsR, AfArsR was used to construct a whole-cell biosensor using GFP fluorescence [77] (Figure 9). The advantage of AfArsR is that it binds not only As(III) but also trivalent organoarsenicals, including the reduced forms of the herbicide MSMA and synthetic aromatic arsenicals, such as roxarsone. In their pentavalent oxidation states, those organoarsenicals are relatively nontoxic. They are activated to the active (toxic) species by reduction, so there is value to having biosensors for trivalent organoarsenicals. The AfArsR-GFP biosensor senses the reduced forms of MSMA and synthetic aromatic arsenicals, including roxarsone and PhAs(III). The selectivity for organoarsenicals over inorganic As(III) to phenylarsenite and MAs(III)

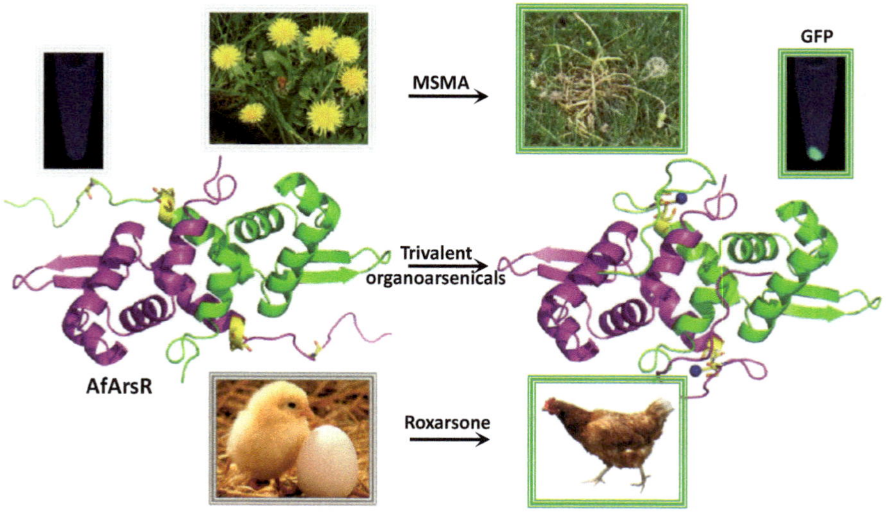

FIGURE 9 A biosensor for trivalent methylated and aromatic arsenicals. An *E. coli* cell-based biosensor was constructed that reports sensing of environmental organoarsenicals by generation of GFP fluorescence. The components of the biosensor are the gene for the AfArsR repressor, shown as the structural model, and the gene for GFP, shown by evolving fluorescence in the Eppendorf tubes. The biosensor responds to trivalent aromatic The biosensor responds to trivalent aromatic arsenicals. It responds with highest affinity to the reduced form of roxarsone. It responds with intermediate affinity to the reduced form of the herbicide MSMA. Finally, it responds with very low affinity to inorganic As(III). (Adapted from [66].)

can be modulated by tuning the intracellular expression levels of ArsR [77]. In addition, the biosensor could be genetically engineered to change the specificity from As(III) to MAs(III) by mutagenesis of Cys102 [66]. Thus, new biosensors can be constructed by engineering old ones.

3.2 Applications of Metalloregulatory Proteins in Arsenic Bioremediation

Bioremediation technology has become popular because of its ecological and environmental sustainability [67]. Thus, *in situ* bioremediation of arsenic by microorganisms has been widely hailed because of their potential advantages in cost-effectiveness and environmental friendliness. Arsenic can be methylated to trivalent species which are further oxidized in air into less toxic pentavalent methylated arsenicals such as methyl-, dimethyl-, and trimethylarsenate, or can be converted into volatile arsines [38]. Biovolatilization could potentially remove arsenic from polluted sites and provide a potential scientific basis for bioremediation of soil arsenic. ArsM methyltransferases have been exploited for arsenic removal in bacteria and plants [78–80].

In addition, arsenic sequestration inside cells provides a potential method for bioremediation. Overexpressing *E. coli* ArsR in cells of *E. coli* produces a 60-fold increase in As(III) and five-fold increase in As(V) accumulation compared to control cells [81]. On the other hand, engineered *E. coli* with ArsR showed high selectivity toward arsenic. No significant increase in accumulation was observed with Cd(II) or Zn(II), demonstrating that ArsR overexpression confers selective sequestration of arsenic. Besides inorganic arsenic, ArsR has also been exploited for its potential in the uptake of organic arsenic species and their removal or remediation [82]. Overexpression of *E. coli* ArsR in *E. coli* cells showed significant improvements in the accumulation of both MAs(V) and DMAs(V). The accumulation of methylated arsenic is virtually not affected by the presence of competing heavy metal species, such as Cd(II) and Pb(II). Commercial methods for arsenic remediation cannot easily distinguish between trivalent and pentavalent arsenicals. The trivalent species are more mobile but less easily removed by ion exchange methods since they are neutral. Although ArsR does not respond to pentavalent arsenicals such as As(V), MAs(V), or DMAs(V) [83], those species can be reduced *in vivo* by many types of bacteria, including *E. coli* [84]. Trivalent MAs(III) and DMAs(III) exhibit higher affinity toward intracellular thiols than As(III) [85], so that overexpression of ArsR in *E. coli* cells can remove not only trivalent but also pentavalent arsenicals from aqueous environments. Sequestration of either inorganic or organic arsenicals in *E. coli* genetically engineered to overproduce ArsR demonstrates the potential to be a useful strategy for arsenic bioremediation.

4 CONCLUSIONS

Metalloregulatory proteins exhibit remarkably precise control of cytoplasmic metal(loid) homeostasis. Arsenic metalloregulatory proteins allow for arsenic sensing and resistance in both eukaryotic and prokaryotic microbes. In response to environmental pressure by arsenic, a diverse number of arsenic metalloregulatory proteins evolved, including ArsR, Yap8, and AioX. There are at least five

ArsR clades. Although they all evolved from a common ancestral winged-helix DNA-binding homodimer, the various ArsRs have convergently evolved recognition sites for As(III), MAs(III), or Sb(III) in different locations in the protein. Some are two coordinates, and some are three coordinates. Did one evolve from the other? SpArsR and AfArsR have closely related binding sites that differ by a single cysteine residue. Yet SpArsR binds As(III) to Cys95, Cy96 and Cys102, while SpArsR binds only MAs(III) to Cys101 and Cys102, which correspond to Cys95 and Cys96 in AfArsR. However, SpArsR is a slightly smaller protein that lacks the last 19 C-terminal residues of AfArsR. Thus, SpArsR lacks a residue corresponding to AfArsR Cys102, the third arsenic ligand. Which type of binding site did their common ancestor have? Interesting evolutionary questions like this are difficult to answer but can be approached by genetically engineering binding sites. As mentioned above, by mutagenesis of Cys102 to a serine residue, AfArsR was converted from an As(III)-responsive repressor into a MAs(III)-selective sensor [66] similar to SpArsR. The C102S mutant had greatly reduced affinity for As(III) without affecting MAs(III) binding. Thus, changing a single residue can change the selectivity of an ArsR sensor. The high affinity and specificity of ArsR have been exploited for the development of whole-cell biosensors. ArsR repressors discriminate effectively against sulfate, cobalt, and cadmium [86]. Could ArsR biosensors be engineered to increase sensitivity or change selectivity by changing the number and spatial location of cysteine residues? Could the three-coordinate As(III)-binding site of R773 ArsR be converted into a four-coordinate Cd(II)-binding site like the CadC repressor by addition of an N-terminal α-helix containing two additional cysteine residues? These hypotheses are testable. Structures of more ArsR proteins with different types of binding sites will expand our knowledge about the evolution of regulatory metal-binding sites and allow engineering of new repressors that will be useful for sensing and bioremediation of toxic metals.

ACKNOWLEDGMENT

This work was supported by NIH grants R01 GM55425 and R35 GM136211.

ABBREVIATIONS AND DEFINITIONS

ars	arsenic resistance genes
As(III)	arsenite
As(V)	arsenate
DMAs(V)	dimethylarsenate
GFP	green fluorescent protein
HTH	helix-turn-helix
ITC	isothermal titration calorimetry
MAs(III)	methylarsenite
MAs(V)	methylarsenate
Rox(III)	roxarsone with trivalent arsenic
Roxarsone	3-nitro-4-hydroxybenzenearsenate or Rox(V)
Sb(III)	antimonite

REFERENCES

1. P. R. Chen, C. He, *Curr. Opin. Chem. Biol.* **2008**, *12*, 214–221.
2. P. Chandrangsu, C. Rensing, J. D. Helmann, *Nat. Rev. Microbiol.* **2017**, *15*, 338–350.
3. K. S. Egorova, V. P. Ananikov, *Organometallics* **2017**, *36*, 4071–4090.
4. K. J. Waldron, J. C. Rutherford, D. Ford, N. J. Robinson, *Nature* **2009**, *460*, 823–830.
5. M. Jaishankar, T. Tseten, N. Anbalagan, B. B. Mathew, K. N. Beeregowda, *Interdiscip. Toxicol.* **2014**, *7*, 60–72.
6. Y. G. Zhu, M. Yoshinaga, F. J. Zhao, B. P. Rosen, *Annu. Rev. Earth Planet Sci.* **2014**, *42*, 443–467.
7. J. Andres, P. N. Bertin, *FEMS Microbiol. Rev.* **2016**, *40*, 299–322.
8. B. K. Mandal, K. T. Suzuki, *Talanta* **2002**, *58*, 201–235.
9. I. Ben Fekih, C. K. Zhang, Y. P. Li, Y. Zhao, H. A. Alwathnani, Q. Saquib, C. Rensing, C. Cervantes, *Front. Microbiol.* **2018**, *9*, 2473.
10. J. H. Wu, B. P. Rosen, *J. Biol. Chem.* **1993**, *268*, 52–58.
11. A. Carlin, W. Shi, S. Dey, B. P. Rosen,*J. Bacteriol.* **1995**, *177*, 981–986.
12. S. Silver, K. Budd, K. M. Leahy, W. V. Shaw, D. Hammond, R. P. Novick, G. R. Willsky, M. H. Malamy, H. Rosenberg, *J. Bacteriol.* **1981**, *146*, 983–996.
13. H. L. T. Mobley, B. P. Rosen, *Proc. Nat. Acad. Sci. U. S. A. Biol. Sci.* **1982**, *79*, 6119–6122.
14. Y. F. Lin, J. B. Yang, B. P. Rosen, *J. Bioenerg. Biomembr.* **2007**, *39*, 453–458.
15. B. P. Rosen, *FEBS Lett.* **2002**, *529*, 86–92.
16. J. B. Yang, S. Rawat, T. L. Stemmler, B. P. Rosen, *Biochemistry* **2010**, *49*, 3658–3666.
17. S. Silver, L. T. Phung, *Appl. Environ. Microbiol.* **2005**, *71*, 599–608.
18. M. J. D. San Francisco, C. L. Hope, J. B. Owolabi, L. S. Tisa, B. P. Rosen, *Nucl. Acids Res.* **1990**, *18*, 619–624.
19. L. S. Busenlehner, M. A. Pennella, D. P. Giedroc, *FEMS Microbiol. Rev.* **2003**, *27*, 131–143.
20. D. Osman, J. S. Cavet, *Nat. Prod. Rep.* **2010**, *27*, 668–680.
21. R. P. Saha, S. Samanta, S. Patra, D. Sarkar, A. Saha, M. K. Singh, *Biometals* **2017**, *30*, 459–503.
22. M. D. Wong, Y. F. Lin, B. P. Rosen, *J. Biol. Chem.* **2002**, *277*, 40930–40936.
23. M. L. VanZile, N. J. Cosper, R. A. Scott, D. P. Giedroc, *Biochemistry* **2000**, *39*, 11818–11829.
24. J. Ye, A. Kandegedara, P. Martin, B. P. Rosen, *J. Bacteriol.* **2005**, *187*, 4214–4221.
25. Y. Sun, M. D. Wong, B. P. Rosen, *J. Biol. Chem.* **2001**, *276*, 14955–14960.
26. I. Arunkumar, G. C. Campanello, D. P. Giedroc, *Proc. Natl. Acad. Sci. U. S. A.* **2009**, *106*, 18177–18182.
27. C. Prabaharan, P. Kandavelu, C. Packianathan, B. P. Rosen, S. Thiyagarajana, *J. Struct. Biol.* **2019**, *207*, 209–217.
28. J. Qin, H. L. Fu, J. Ye, K. Z. Bencze, T. L. Stemmler, D. E. Rawlings, B. P. Rosen, *J. Biol. Chem.* **2007**, *282*, 34346–34355.
29. E. Ordonez, S. Thiyagarajan, J. D. Cook, T. L. Stemmler, J. A. Gil, L. M. Mateos, B. P. Rosen, *J. Biol. Chem.* **2008**, *283*, 25706–25714.
30. T. Viswanathan, J. Chen, M. Wu, L. An, P. Kandavelu, B. Sankaran, M. Radhakrishnan, M. Li, B. P. Rosen, *Mol. Microbiol.* **2021**, *116*, 427–437.
31. J. Chen, V. S. Nadar, B. P. Rosen, *Mol. Microbiol.* **2017**, *106*, 469–478.
32. S. Chauhan, A. Kumar, A. Singhal, J. S. Tyagi, H. Krishna Prasad, *FEBS J.* **2009**, *276*, 3428–3439.
33. C. Xu, W. Shi, B. P. Rosen, *J. Biol. Chem.* **1996**, *271*, 2427–2432.
34. N. L. Brown, J. V. Stoyanov, S. P. Kidd, J. L. Hobman, *FEMS Microbiol. Rev.* **2003**, *27*, 145–163.

35. D. E. Rawlings, D. B. Johnson, *Microbiology* **2007**, *153*, 315–324.
36. B. G. Butcher, D. E. Rawlings, *Microbiology* **2002**, *148*, 3983–3992.
37. E. Ordonez, M. Letek, N. Valbuena, J. A. Gil, L. M. Mateos, *Appl. Environ. Microbiol.* **2005**, *71*, 6206–6215.
38. J. Qin, B. P. Rosen, Y. Zhang, G. Wang, S. Franke, C. Rensing, *Proc. Natl. Acad. Sci. U. S. A.* **2006**, *103*, 2075–2080.
39. J. Qin, C. R. Lehr, C. G. Yuan, X. C. Le, T. R. McDermott, B. P. Rosen, *Proc. Natl. Acad. Sci. U. S. A.* **2009**, *106*, 5213–5217.
40. J. Chen, M. Yoshinaga, B. P. Rosen, *Mol. Microbiol.* **2019a**, *111*, 487–494.
41. M. Yoshinaga, Y. Cai, B. P. Rosen, *Environ. Microbiol.* **2011**, *13*, 1205–1215.
42. J. Chen, L. D. Garbinski, B. P. Rosen, J. Zhang, P. Xiang, L. Q. Ma, *Crit. Rev. Environ. Sci. Technol.* **2020,** *50*, 217–243.
43. L. An, X. Luo, M. Wu, L. Feng, K. Shi, G. Wang, B. P. Rosen, M. Li, Sci. Total. Environ. **2021**, *754*, 142393.
44. L. M. Del Razo, B. Quintanilla-Vega, E. Brambila-Colombres, E. S. Calderon-Aranda, M. Manno, A. Albores, *Toxicol. Appl. Pharmacol.* **2001**, *177*, 132–148.
45. R. Wysocki, M. J. Tamas, *FEMS Microbiol. Rev.* **2010**, *34*, 925–951.
46. R. Wysocki, P. K. Fortier, E. Maciaszczyk, M. Thorsen, A. Leduc, A. Odhagen, G. Owsianik, S. Ulaszewski, D. Ramotar, M. J. Tamás, *Mol. Biol. Cell.* **2004**, *15*, 2049–2060.
47. M. Thorsen, G. Lagniel, E. Kristiansson, C. Junot, O. Nerman, J. Labarre, M. J. Tamas, *Physiol. Genom.* **2007**, *30*, 35–43.
48. K. A. Morano, C. M. Grant, W. S. Moye-Rowley, *Genetics* **2012**, *190*, 1157–1195.
49. N. V. Kumar, J. Yang, J. K. Pillai, S. Rawat, C. Solano, A. Kumar, M. Grøtli, T. L. Stemmler, B. P. Rosen, M. J. Tamás, *Mol. Cell. Biol.* **2015**, *36*, 913–922.
50. R. A. Menezes, C. Amaral, A. Delaunay, M. Toledano, C. Rodrigues-Pousada, *FEBS Lett.* **2004**, *566*, 141–146.
51. D. Paez-Espino, J. Tamames, V. de Lorenzo, D. Canovas, *Biometals* **2009**, *22*, 117–130.
52. C. Watson, D. Niks, R. Hille, M. Vieira, B. Schoepp-Cothenet, A. T. Marques, M. J. Romão, T. Santos-Silva, J. M. Santini, *Biochim. Biophys. Acta Bioenerg.* **2017**, *1858*, 865–872.
53. C. Badilla, T. H. Osborne, A. Cole, C. Watson, S. Djordjevic, J. M. Santini, *Sci. Rep.* **2018**, *8*, 1–12.
54. Y. S. Kang, B. Bothner, C. Rensing, T. R. McDermott, *Appl. Environ. Microbiol.* **2012**, *78*, 5638–5645.
55. S. Koechler, J. Cleiss-Arnold, C. Proux, O. Sismeiro, M. A. Dillies, F. Goulhen-Chollet, F. Hommais, D. Lièvremont, F. Arsène-Ploetze, J. Y. Coppée, P. N. Bertin, *BMC Microbiol.* **2010**, *10*, 53.
56. S. Sardiwal, J. M. Santini, T. H. Osborne, S. Djordjevic, *FEMS Microbiol. Lett.* **2010**, *313*, 20–28.
57. R. A. Rawle, Y. S. Kang, B. Bothner, G. Wang, T. R. McDermott, *Environ. Microbiol.* **2019**, *21*, 2659–2676.
58. G. Liu, M. Liu, E. H. Kim, W. S. Maaty, B. Bothner, B. Lei, C. Rensing, G. Wang, T. R. McDermott, *Environ. Microbiol.* **2012**, *14*, 1624–1634.
59. Q. Wang, T. P. Warelow, Y. S. Kang, C. Romano, T. H. Osborne, C. R. Lehr, B. Bothner, T. R. McDermott, J. M. Santini, G. Wang, *Appl. Environ. Microbiol.* **2015**, *81*, 1959–1965.
60. S. Shankar, U. Shanker, *Sci. World J.* **2014**, 304524.
61. K. E. Nachman, P. A. Baron, G. Raber, K. A. Francesconi, A. Navas-Acien, D. C. Love, *Environ. Health Perspect.* **2013**, *121*, 818–824.
62. S. Noack, H. D. Chapman, P. M. Selzer, *Parasitol. Res.* **2019**, *118*, 2009–2026.
63. L. Konkel, *Environ. Health Perspect.* **2016**, *124*, A150–A150.

64. D. Zhao, J. Y. Wang, D. X. Yin, M. Y. Li, X. Q. Chen, A. L. Juhasz, A. Navas-Acien, H. Li, L. Q. Ma, *J. Hazard Mater.* **2020**, *383*, 121178.
65. D. W. Rutherford, A. J. Bednar, J. R. Garbarino, R. Needham, K. W. Staver, R. L. Wershaw, *Environ. Sci. Technol.* **2003**, *37*, 1515–1520.
66. J. Chen, B. P. Rosen, *Biosensors* **2014**, *4*, 494–512.
67. A. I. Zouboulis, I. A. Katsoyiannis, *Environ. Int.* **2005**, *31*, 213–219.
68. D. Merulla, N. Buffi, S. Beggah, F. Truffer, M. Geiser, P. Renaud, J. R. van der Meer, *Curr. Opin. Biotechnol.* **2013**, *24*, 534–541.
69. O. I. Sanders, C. Rensing, M. Kuroda, B. Mitra, B. P. Rosen, *J. Bacteriol.* **1997**, *179*, 3365–3367.
70. Y. Kawakami, M. S. Siddiki, K. Inoue, H. Otabayashi, K. Yoshida, S. Ueda, H. Miyasaka, I. Maeda, *Biosens. Bioelectron.* **2010**, *26*, 1466–1473.
71. S. Tauriainen, M. Karp, W. Chang, M. Virta, *Appl. Environ. Microbiol.* **1997**, *63*, 4456–4461.
72. S. Ramanathan, W. P. Shi, B. P. Rosen, S. Daunert, *Anal. Chem.* **1997**, *69*, 3380–3384.
73. F. Cortes-Salazar, S. Beggah, J. R. van der Meer, H. H. Girault, *Biosens. Bioelectr.* **2013**, *47*, 237–242.
74. X. Jia, R. Bu, T. Zhao, K. Wu, *Appl. Environ. Microbiol.* **2019**, *85*, e00694-19.
75. J. Stocker, D. Balluch, M. Gsell, H. Harms, J. Feliciano, S. Daunert, K. A. Malik, J. R. Van der Meer, *Environ. Sci. Technol.* **2003**, *37*, 4743–4750.
76. K. Siegfried, S. Hahn-Tomer, A. Koelsch, E. Osterwalder, J. Mattusch, H. J. Staerk, J. M. Meichtry, G. E. De Seta, F. D. Reina, C. Panigatti, M. I. Litter, *Int. J. Environ. Res. Public. Health* **2015**, *12*, 5465–5482.
77. J. Chen, Y. G. Zhu, B. P. Rosen, *Appl. Environ. Microbiol.* **2012**, *78*, 7145–7147.
78. J. Chen, G. X. Sun, X. X. Wang, V. Lorenzo, B. P. Rosen, Y. G. Zhu, *Environ. Sci. Technol.* **2014b**, *48*, 10337–10344.
79. X. Y. Meng, J. Qin, L. H. Wang, G. L. Duan, G. X. Sun, H. L. Wu, *N. Phytol.* **2011**, *191*, 49–56.
80. J. Zhang, Y. Xu, T. Cao, J. Chen, B. P. Rosen, F. J. Zhao, *Plant Soil* **2017**, *416*, 259–269.
81. J. Kostal, R. Yang, C. H. Wu, A. Mulchandani, W. Chen, *Appl. Environ. Microbiol.* **2004**, *70*, 4582–4587.
82. T. Yang, J. W. Liu, C. Gu, M. L. Chen, J. H. Wang, *ACS Appl. Mater. Interf.* **2013**, *5*, 2767–2772.
83. J. Chen, S. Sun, C. Z. Li, Y. G. Zhu, B. P. Rosen, *Environ. Sci. Technol.* **2014a**, *48*, 1141–1147.
84. Z. W. Wang, H. Q. Zhang, X. F. Li, X. C. Le, *Rapid Commun. Mass Spectr.* **2007**, *21*, 3658–3666.
85. M. L. Lu, H. L. Wang, X. F. Li, X. F. Lu, W. R. Cullen, L. L. Arnold, S. M. Cohen, X. C. Le, *Chem. Res. Toxicol.* **2004**, *17*, 1733–1742.
86. D. L. Scott, S. Ramanathan, W. Shi, B. P. Rosen, S. Daunert, *Anal. Chem.* **1997**, *69*, 16–20.

2 Magnetic Resonance Imaging Bio-Sensors for Calcium(II)

Goran Angelovski
Laboratory of Molecular and Cellular Neuroimaging, International Center for Primate Brain Research (ICPBR), Center for Excellence in Brain Science and Intelligence Technology (CEBSIT), Chinese Academy of Sciences (CAS), Shanghai 200031, People's Republic of China

CONTENTS

ABSTRACT

Calcium is an essential ion for life due to its involvement in cellular signaling. The ability of magnetic resonance imaging (MRI) to follow its fluctuations at the molecular and cellular level and on a large scale makes this technique very helpful for the visualization and understanding of critical biological processes. This can successfully be

DOI: 10.1201/9781003229971-2

achieved through the development of effective MRI contrast agents that are sensitive to Ca(II). A great number of probes have been investigated in the past two decades that use different MRI methodologies based on relaxation processes, chemical exchange saturation transfer, and use of nuclei other than ^{1}H or hyperpolarized MRI. The progress has been spectacular, resulting in a few very successful sensors that have allowed the first *in vivo* monitoring of intra- and extracellular Ca(II) concentration changes. This chapter summarizes the current state-of-the-art applications in this field and provides some practical insights that may be relevant for the future advancement of this exciting field and the development of functional molecular imaging.

KEYWORDS

Calcium; Chelates; Lanthanides; MRI; Sensors

1 INTRODUCTION

Calcium is one of the most biologically relevant metal cations with a critical physiological role. It acts as a second messenger during the cellular signaling processes; hence, the normal functioning of any living organism heavily depends on fluctuations of calcium ions. In our body, Ca(II) is in charge of regulating muscle contraction and relaxation, whereas its involvement in neuronal signal transduction is essential for normal function of the brain and the nervous system.

Successful monitoring of Ca(II) fluctuations is a key challenge for understanding the physiology as well as the pathology of many biological processes. A number of classical analytical chemistry methods for detection of Ca(II) were developed over many decades, mostly for application *in vitro* [1]. Conversely, modern molecular imaging techniques strive to detect, visualize, monitor, and quantify Ca(II) in biological samples and tissues, ideally in the intact living organism.

Depending on the intrinsic properties of a particular method, the imaging of Ca(II) can be done on a microscopic or macroscopic scale. Optical methods were widely used for the former type of investigation: particularly, fluorescence imaging of Ca(II) experienced great expansion due to the substantial technological advances in the field of optics in the recent few decades. Here, two types of Ca(II)-sensitive indicators were developed depending on the chemical nature of the sensor [2]. The first group consists of fluorescent synthetic organic molecules mainly based on BAPTA, the well-known Ca-chelator [3–5]. The second group is represented by a series of genetically encoded calcium indicators. Their development rapidly grew upon the discovery of a green fluorescence protein and the follow-up advances in genetic engineering techniques that allowed an easy optimization and adjustment of their properties (see also Chapter 3 of this book) [6–8].

When discussing the imaging of Ca(II) on a macroscopic scale, molecular imaging techniques such as magnetic resonance imaging (MRI) are more appropriate. Unlike the optical methods, high penetration through the tissue and large field of view makes MRI suitable for investigations of large volumes. This opens new realms for understanding the biological role of Ca(II) in whole organs or even organisms, which is among the ultimate aims of molecular imaging.

To achieve this goal, the field of Ca(II) sensors for MRI has been successfully developing over the last two decades. Given the lower sensitivity, but also the complexity and diversity of MRI methods, it is expected that progress in the results is somewhat slower compared to other imaging methods. Nevertheless, the current achievements are already striking, and initial *in vivo* results have shown great prospects for further development of this field. The aim of this chapter is to summarize the current progress in the development of MRI bio-sensors for Ca(II). Initially, a brief overview of the different types of MRI contrast agents is provided, followed by an analysis of examples of MRI probes sensitive to Ca(II) that have been reported to date.

2 TYPES OF MRI CONTRASTS AND THEIR CHEMICAL PROBES

MRI can be performed at any frequency of any nuclear magnetic resonance (NMR)-active nuclei provided that sufficient signal can be obtained and that technical prerequisites are met, i.e., suitable hardware is installed. The most common MRI measurements are done using the most abundant NMR-active nucleus in the body, ^{1}H; hence, a great deal of MRI studies use the intrinsic signal coming from water in the living organisms. Nevertheless, changes in water signals among tissues and organs are often insufficient to provide accurate diagnosis or functional information. MRI contrast agents alleviate these problems by enhancing the contrast or improving the quality of the recorded MR image and subsequently the specificity, thereby aiding better conclusions of these measurements [9]. Additionally, heteronuclear MRI experiments require the use of specific probes, as the majority of nuclei other than ^{1}H are either not available in the body (e.g., ^{19}F) or are available in an insufficient amount to provide a suitable signal (e.g., ^{13}C, ^{15}N) [10]. No matter which kind of nucleus is used in the experiment, contrast in MRI signal is mainly generated by taking advantage of the longitudinal or transverse relaxation times (T_1 or T_2, respectively), or the chemical exchange saturation transfer (CEST) effect.

2.1 T_1-WEIGHTED MRI CONTRAST AGENTS

The so-called T_1-weighted MRI contrast agents are typically suitable for ^{1}H MRI. Although the abundance of ^{1}H is sufficiently high in living organisms to generate the MR signal without using the CAs, differences in water (proton) density among different tissues are often not enough to generate a good contrast. Changes in relaxation times, particularly in T_1, are more adequate to generate it. Nevertheless, the T_1 in tissue is often long, requiring long acquisition times. Using T_1-weighted CAs, the longitudinal relaxation rates of water protons that surround the CA are enhanced. This results in a shorter T_1 and a much faster generation of the MRI signal. These agents are typically based on complexes with the metal cations Gd(III) and Mn(II), which are highly paramagnetic [9, 11, 12]. The cations are chelated with acyclic or macrocyclic multidentate ligands in order to form stable complexes and reduce their amount in a free cationic form, which is toxic to biological tissue. The design of the complex should be considered carefully, taking into account the necessity of forming a highly stable complex, while still leaving a vacant position for direct coordination of a water molecule to the paramagnetic metal ion. The presence of coordinated water allows the paramagnetic effect to be efficiently transferred from the metal ion to the water of tissues in which the CA distributes, giving rise to shorter T_1 values and higher contrast in the MR image. The quantitative parameter that describes the efficacy of the CA to shorten the T_1 is called longitudinal relaxivity, r_1. This is the value normalized per unit concentration of CA (1 mM) and expresses the shortening of the relaxation rate (parameter inverse of the relaxation time) that is achieved by this defined amount of CA. The r_1 for the above-mentioned complexes depends on several parameters, among which are the number of water molecules bound to the paramagnetic ion (q number), the exchange rate of water molecules in the inner coordination sphere with the bulk, or the rotation of entire paramagnetic system (complex) either locally (rotation around the paramagnetic complex) or globally (in case the complex is conjugated with a macromolecule). The strategies that lead to r_1 changes in Ca(II)-sensitive MRI CAs rely on alterations of these parameters caused by the presence of Ca(II) (see below).

2.2 T_2-Weighted MRI Contrast Agents

Following the same analogy as for the above-mentioned class of CAs, the T_2-weighted MRI contrast agents shorten the transverse relaxation time of water. This leads to loss of the MRI signal, making the tissue regions affected with the T_2-weighted CA darker. Strong T_2 effect is achieved by using superparamagnetic CAs, which are typically formed from iron-oxide nanoparticles of different size: MIONs (>1 μm), SPIOs (>50 nm), or USPIOs (<50 nm) [13]. In this case, the efficacy of the superparamagnetic CA is expressed with transverse relaxivity, r_2, which depends on the amount of paramagnetic species incorporated into the nanoparticle, the size of the particle, and also the shape and interaction (diffusion) of the water molecules on the surface of the particle. Changes in r_2 were also implemented, albeit to a lesser extent, to develop Ca(II)-sensitive agents (see Section 3.2). Recording the dynamics of the physiological processes using the endogenous contrast agent in the blood has already been achieved and is currently an extremely valuable research and diagnostic imaging tool. Namely, the functional variant of MRI (fMRI) capitalizes on the local changes in blood oxygenation during the neuronal activity, which changes the ratio of paramagnetic and diamagnetic iron, resulting in the local alterations of the 'effective' T_2 (so-called T_2*) [14]. The discovery of the T_2* effect from iron in blood gave spectacular momentum to the further development of MRI, particularly in the field of neuroimaging [15].

2.3 CEST MRI Contrast Agents

The CEST MRI capitalizes on the signal generated by the magnetization transfer caused by the chemical exchange [16]. Here one pool of the NMR-active nuclei of a particular frequency exchanges with a second pool resonating at a different frequency. The presaturation RF pulse that automatically decreases the net magnetization (MR signal) of the former pool is also reducing the signal of the latter pool due to the chemical exchange (and hence the magnetization transfer) between the two pools.

Typically, CEST MRI is performed at the ^{1}H frequency. The saturated pool of protons is on the CEST-active molecule, which then exchanges with bulk water. In this case, the net magnetization of water reduces and the CEST MRI signal practically reports on the intensity of the CEST effect on water. The CEST-active probes can be endogenous biomolecules, such as the abundant proteins, whereas exogenous molecules should also contain the pool of protons that is in slow to intermediate exchange with water. These can be purely organic molecules. For example, slowly exchanging protons from the amide or alcohol groups or metal ion complexes (so-called paraCEST agents) cause a frequency shift of the exchanging pool, thereby creating the CEST signal sufficiently distant from the bulk water frequency [17]. Besides ^{1}H, nuclei such as ^{19}F and ^{129}Xe were also used in the respective ^{19}F and ^{129}Xe CEST MRI experiments (see below).

2.4 Fluorinated MRI Probes

Fluorinated probes for ^{19}F MRI are the second most abundantly used in MRI, following those suitable for the different types of ^{1}H MRI modalities [18, 19]. There are a few reasons for this. The abundance of naturally available and NMR-active isotope ^{19}F is 100%, while its gyromagnetic ratio is very close to that of ^{1}H ($\gamma(^{19}F)/\gamma(^{1}H)=0.94$), allowing the use of the same scanners with minimal alterations in the supporting imaging hardware (RF coils). Since fluorine is absent in the soft tissue, there is consequently no background signal, which allows much easier quantification of the ^{19}F MRI signal and the ^{19}F probe. For the same reason, the ^{19}F probes need to be applied in a higher amount than the typical ^{1}H MRI CAs, which led to the development of different types of perfluorinated probes [19, 20]. ^{19}F often exhibits long T_1 as well; combined with a lower signal generated by these probes, ^{19}F MRI acquisition times consequently tend to be longer than those of ^{1}H MRI. Analogous to ^{1}H MRI, similar attempts exist to enhance the ^{19}F T_1 by combining this NMR-active nucleus with the paramagnetic metal ion, often within the same conjugate, to take advantage of the PRE effect. The ^{19}F conjugates with paramagnetic cations commonly use this effect to alter both ^{19}F T_1 and T_2 in order to change the recorded ^{19}F MRI signal.

As mentioned previously, the ^{19}F probes can also be used for generating CEST effect. In this case, two pools of NMR-active nuclei in exchange are from the same fluorinated group; however, due to the high NMR sensitivity of ^{19}F and large spectral window (>200 ppm), its frequency can change substantially, leading to the formation of two distinct pools of nuclei that can be assessed by means of CEST.

2.5 Hyperpolarized MRI Probes

The most recent progress in the field of MRI contrast agents concerns hyperpolarized probes. These are based on different types of isotope-enriched heteronuclei that are present in biomolecules (^{13}C, ^{15}N, ^{31}P), although the noble gas ^{129}Xe can also be used in combination with the appropriate supramolecular host molecules. A specific feature of these probes is that they undergo an unusual spin polarization procedure that causes a change in the nuclear spin population in the hyperpolarized sample [21]. As a result, the sample exhibits an enhanced MRI signal of up to a few orders of magnitude for a short period of time before Boltzmann distribution is restored. Depending on the type of nucleus, different hyperpolarization methods have been developed, such as dissolution dynamic nuclear polarization, spin exchange optical pumping, and parahydrogen-induced polarization. Hyperpolarized MRI is an excellent choice for heteronuclei that have intrinsically low MRI signals, thus significantly increasing the specificity of MRI and expanding its scope of applications. However, fast relaxation times limit the wider practical use of hyperpolarized probes in molecular imaging.

3 CALCIUM-SENSITIVE MRI PROBES

Given the essential role of Ca(II) in many biological processes and the unique features of MRI, there is quite an interest in developing an imaging methodology that can monitor the fluctuations of this cation. The pioneering steps in developing a Ca-sensitive molecule suitable for NMR and consequently MRI investigations were made based on the success of BAPTA derivatives and their application as fluorescence indicators. The preliminary works date back to the early 1980s, with the development of fluorinated BAPTA chelators for Ca(II), which showed decent Ca-induced shifts in ^{19}F NMR [22]. Further progress in this direction was halted as it became clear that the design of suitable and active MRI sensors for Ca(II) must be performed considering many physicochemical specificities of MRI as a technique, which cannot be directly translated from the experiences obtained while developing the fluorescence indicators. The work on MRI sensors for Ca(II) required an approach on its own, taking into account many relevant aspects of NMR and MRI physics. In the following sections, an overview of different approaches and the progress to current date is provided, classifying the Ca-sensitive probes based on the type of signal that is generated, as well as the chemistry of these probes.

3.1 Ca-Responsive Probes Suitable for T_1-Weighted MRI

An MRI sensor must be made by carefully analyzing the most important aspects that influence the MRI signal generated by the probe. Considering the complexity and versatility of the MRI methods, there are many approaches suitable to induce the signal changes; however, they are almost always specific for a single MRI method, and rarely suitable for more than one of them. Designing a T_1-weighted sensor for MRI means that the parameters determining the r_1 (Section 2.1) must be affected; in the case of preparing a Ca(II) sensor, this would mean causing r_1 changes upon the interaction of this cation with the MRI probe.

The foundations for the development of Ca(II) sensors for MRI and the field of bioresponsive MRI probes were established in the late 1990s. Taking an already extensively explored and selective BAPTA chelator for Ca(II), and combining it with a couple of macrocyclic Gd(III) chelators based on DO3A, the Meade group designed a triggering mechanism that proved to be very efficient and served as the basic principle for causing the r_1 changes in a series of the follow-up studies (Figure 1). The major idea for inducing the r_1 alterations consisted of combining the Ca(II) and Gd(III) chelators with the linker that keeps the distance between these two coordination environments close. In absence of Ca(II), the carboxylate arms

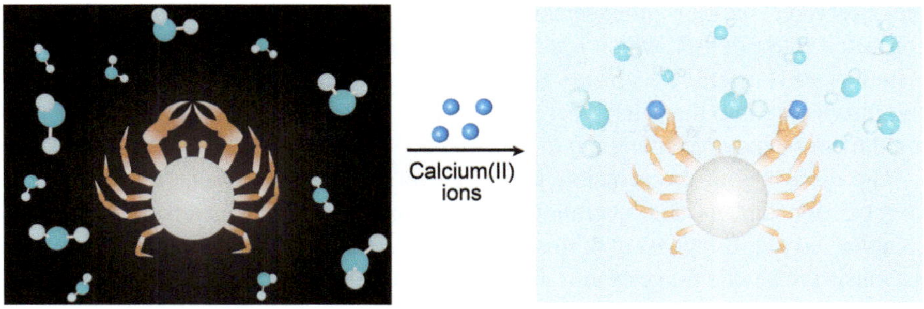

FIGURE 1 The mechanism responsible for changes in T_1: the access of water to the chelated paramagnetic ion is restricted in the absence of the target ion (left); in the presence of the target ion, the functional group(s) from the MR sensor interact with it, which grants the access of water to the paramagnetic center.

from the BAPTA chelator participate in coordination with Gd(III), thus precluding the presence of inner-sphere water molecules and consequently reducing r_1. Once the probe starts interacting with Ca(II), the above-mentioned carboxylate groups do not coordinate with Gd(III) anymore, but with Ca(II). This gives access to water molecules to coordinate with Gd(III), which results in an overall shortening of the relaxation time of the solution.

The first Ca(II) sensor **1** (Figure 2) exhibited a maximal r_1 change of ~80% when saturated with Ca(II), as the r_1 value increased from 3.26 to 5.76 $mM^{-1}s^{-1}$ upon Ca(II) addition (11.7 T, 25°C) [23]. Luminescence lifetime experiments performed on the Tb(III) analog indicated that the q number doubles from 0.5 to 1.0 upon the addition of Ca(II), while the extensive relaxometric studies by means of NMRD experiments showed that the contribution to r_1 of second-sphere molecules is the major cause for

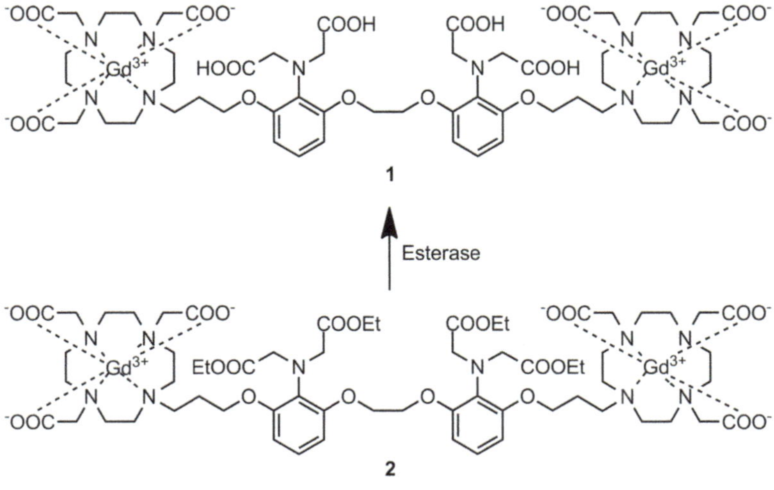

FIGURE 2 Structures of the complexes **1** and **2**. The esterase hydrolyzes the ethyl esters in **2**, which activate the Ca-chelating part of the sensing moiety to give the Ca-responsive probe **1**.

the observed relaxation enhancement [24]. In a follow-up study, this system could be converted into probe **2**, which has the carboxylates from the Ca-chelating part protected as ethyl esters. This probe is expected to easily penetrate the cell membrane and internalize into the cells, where the esterase will activate the probe by hydrolyzing the esters and converting **2** into **1** [25].

The discovery of the principle that can lead to relaxivity changes induced by a target cation and its demonstration on this dinuclear and bismacrocyclic Gd(III) complex led to the expansion of this research. Different types of MRI probes that are responsive to a wide range of ions and molecules were developed [26]. Following the principle established for **1**, Angelovski and colleagues prepared and studied a series of bismacrocyclic Gd(III) complexes that contained different chelators for Ca(II), namely BAPTA, EDTA, DTPA, or EGTA, which were modified and incorporated into the sensor molecule (Figure 3). The rationale for their modification was to reduce the binding affinity of Ca(II) with the sensor by converting one of the carboxylic groups available for coordination to an amide, which links the fragments of the probe designed to chelate Gd(III) and Ca(II). Additional modifications were implemented by varying the linker length, which allowed investigating whether changing the distance between the Gd(III) and Ca(II) cages has an effect on the responsiveness of the probe in terms of r_1 changes.

The probes **3** and **4**, which contained a BAPTA-modified chelator, did not offer significant changes in r_1 upon Ca(II) addition, and the q number did not change [27]. The probes based on the DTPA- (**5–6**) or EDTA-modified chelators (**7**) already showed more response, although the changes were either moderate (all complexes) or not selective for Ca(II) over Mg(II) (probe **7**) [28]. The highest changes by far were observed for systems **8** and **9**, which bear an EGTA-derived chelator. Similar to the case of **5** and **6**, the Ca(II) and MR-reporting moieties are connected with an ethyl or propyl linker. Both of the complexes display a Ca(II)-induced relaxivity change comparable to **1**, with the r_1 value increasing from 4.05 to 6.86 $mM^{-1}s^{-1}$ (69%)

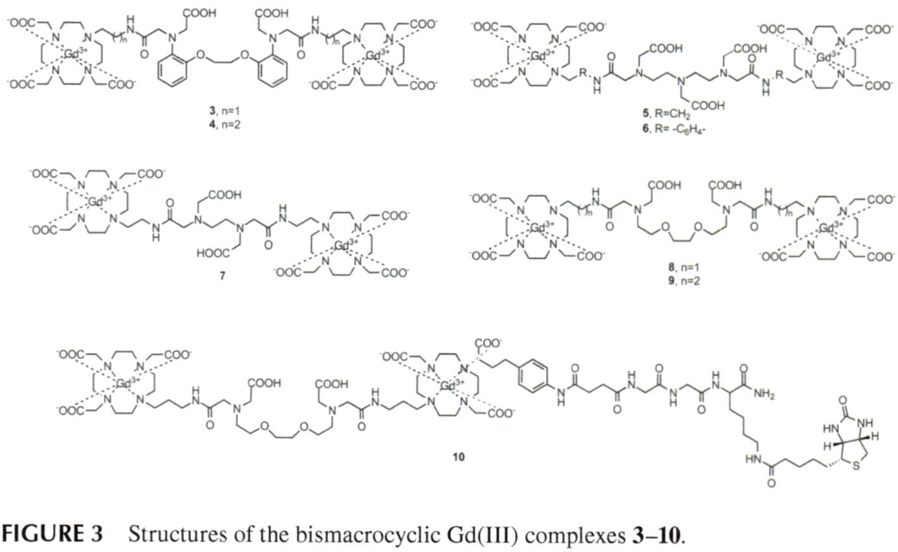

FIGURE 3 Structures of the bismacrocyclic Gd(III) complexes **3–10**.

and from 3.44 to 6.29 mM^{-1}s^{-1} (83%) for **8** and **9**, respectively (all values at 11.75 T, 25°C). The luminescence lifetime experiments performed on the Eu(III) analogues confirmed that an increase in the hydration number of $\Delta q = 0.4$ is the main cause of the observed r_1 changes, as previously established for the other probes of this class [29]. In follow-up studies, these systems were more thoroughly investigated to better assess their properties for potential application in functional imaging studies. The results from investigations on cell cultures and rat brains *in vivo* are summarized in Section 3.7. The outstanding performance of these EGTA-derived bismacrocycles was also a motivation to develop a synthetic strategy that could yield these complexes functionalized with other functional groups. To this end, the prototype of the target-specific and bioresponsive bismacrocyclic complex **10** was prepared using a solid-phase synthetic procedure [30]. The obtained biotinylated probe still exhibited strong activity toward Ca(II), opening new potential pathways for the development of probes that could expand the scope of functional MRI studies.

In parallel to developing bismacrocyclic Ca(II) complexes, several attempts were focused on monomacrocyclic probes as well. Due to their slightly simpler synthesis, it was easier to prepare a greater number of examples and subsequently study the relation between the structures and the r_1 response provided by these complexes in presence of Ca(II). A series of complexes that contain EDTA- (**11–12**), EGTA- (**13**), and pyro-EGTA-derived (**14–16**) complexes was prepared, using different linkers to couple the macrocycle and the Ca(II)-chelating part (Figure 4) [31]. The achieved r_1 changes were low to moderate, with the probe **16** having comparable enhancement per Gd(III) as the one observed for bismacrocycles [32]. It should be noted that bismacrocycles can produce double enhancement per molecule due to the higher amount of Gd(III) present in the latter.

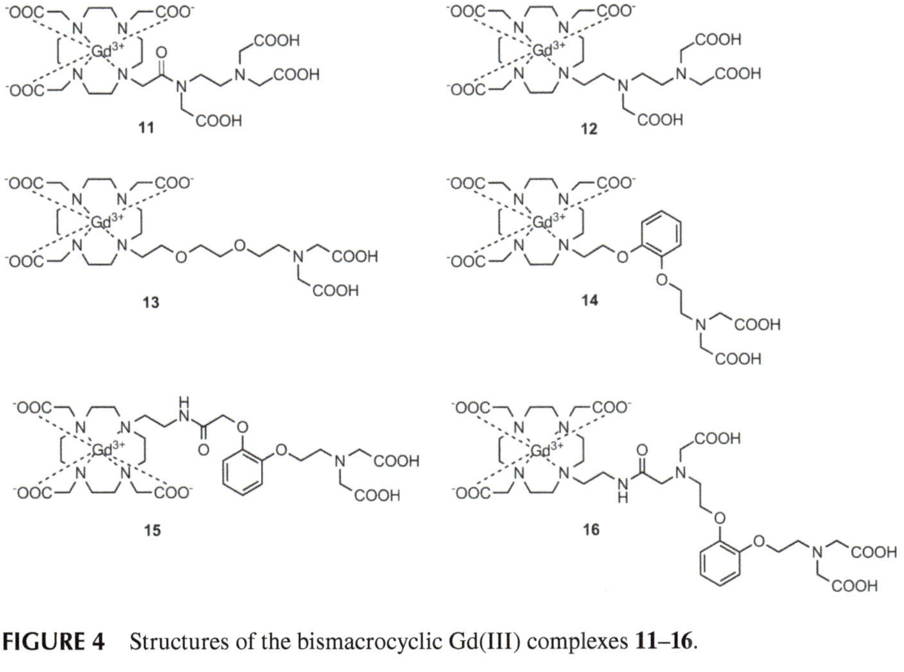

FIGURE 4 Structures of the bismacrocyclic Gd(III) complexes **11–16**.

In a similar manner, a series of APTRA-derived complexes **17–21** were prepared and studied (Figure 5). Here the linker length, the position of the binding oxygen to improve stability of the probe (phenyl or benzyl ether), as well as the nature of the group at the APTRA-type chelators was changed in order to adjust the affinity and sensitivity toward Ca(II) [33, 34]. While the short linker resulted in a probe without any change in r_1 upon Ca(II) addition (**18**), all remaining complexes exhibited high r_1 changes, with probes **19–21** exceeding 100% relaxivity enhancements.

Finally, another strategy relied on using the same building blocks as the bismacrocyclic complexes to make the monomacrocyclic analogues, providing the series of complexes **22–24** (Figure 5). These were expected to show activity toward Ca(II) [35, 36]. The triggering mechanism in these probes was supposed to be preserved from the one established for **8** and **9**. Instead of the second macrocycle with the Gd(III) chelator, a reactive moiety was introduced to allow further synthetic modification and selective coupling of these macrocyclic systems to a vector or nanosized carrier (for properties of the nanosized systems, see Section 3.6). This was achieved by introducing an aryl nitro group on the opposite side of the Ca-chelator relative to the position of the macrocycle, which can be converted to an aryl-isothiocyanate group suitable for selective coupling reactions with amino groups. Moreover, the effect of linker elongation was assessed by preparing the complexes **23** and **24**, which contain aliphatic butyl and pentyl groups as linkers between the macrocyclic Gd(III) and EGTA-derived Ca(II) chelators, respectively. The r_1 values for **22–24** at least doubled when saturating with Ca(II) and increased up to 131% for probe **24**, which was likely due to increased flexibility of the entire system (relative to the bismacrocycles) [36].

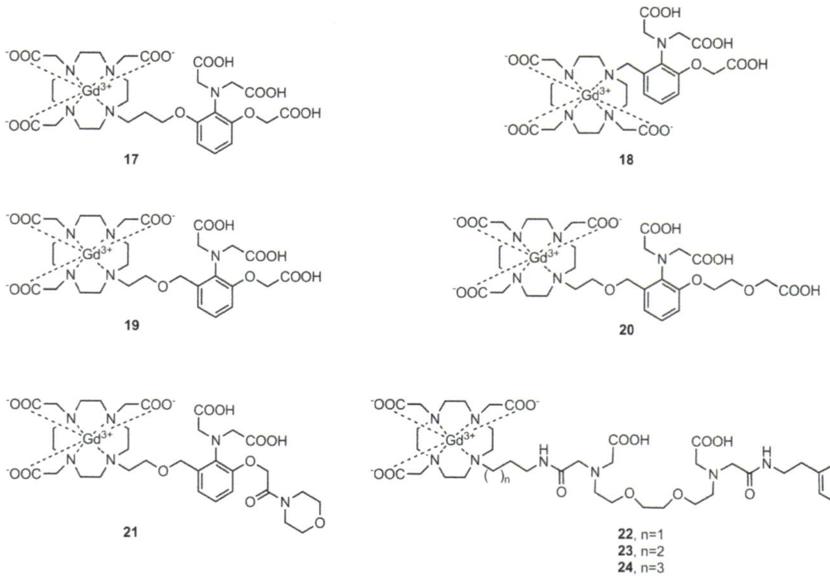

FIGURE 5 Structures of the monomacrocyclic Gd(III) complexes **17–24**.

Following another path, two different groups of phosphonate-containing Ca(II)-responsive complexes were prepared and analyzed (Figure 6). The four aminobisphosphonates containing DO3A units **25–28** were synthesized and the relaxometric behavior of their Gd(III) complexes was tested in presence of Ca(II). The r_1 relaxivity of the three complexes with the longer aliphatic linker chain (**26–28**) reduced upon the addition of Ca(II), which is caused by the formation of Ca-induced aggregates. Complex **25** did not behave this way since its phosphonate groups coordinate to Gd(III). In any case, these complexes displayed entirely different response mechanisms compared to those described above based on BAPTA-, EDTA-, DTPA-, or EGTA-modified chelators. Instead of the carboxylate and potentially amide groups swapping from Gd(III) to Ca(II), the aggregate formation due to the coordination of Ca(II) with the bisphosphonate groups reduces the access of water to the coordination sphere of the paramagnetic ion, thus reducing r_1 [37]. Similarly, the Gd(III) complex **29**, which bears a bisphosphonate group, exhibited a large r_1 increase at low magnetic fields (0.47 T) and basic conditions (pH > 7) in presence of Ca(II), likely due to a similar aggregate formation [38]. Indeed, r_1 tends to be very sensitive to the rotational dynamics of the complex at intermediate magnetic fields (~0.25–3 T) [39], and thus any significant increase in the size of complexes (e.g., through the formation of aggregates like in this case) can lead to an increase in r_1.

Besides Gd(III) complexes, an example of a Ca-sensitive probe based on a paramagnetic Mn(III) probe has also been reported (Figure 7). Here the design relied on a series of 1,2-phenylenediamine-derived ligands that form cell-permeable paramagnetic Mn(III) complexes [40]. In a subsequent optimization, a BAPTA chelator was coupled with one of these ligands to induce the Ca(II) sensitivity.

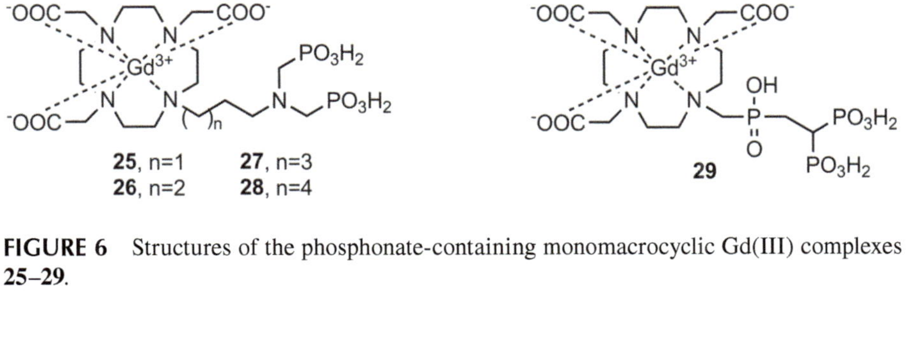

FIGURE 6 Structures of the phosphonate-containing monomacrocyclic Gd(III) complexes **25–29**.

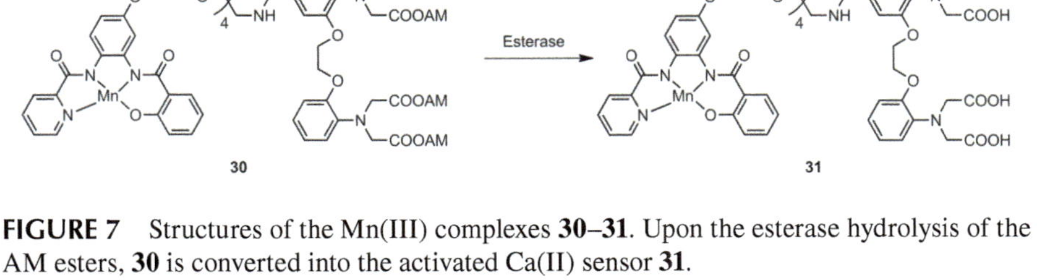

FIGURE 7 Structures of the Mn(III) complexes **30–31**. Upon the esterase hydrolysis of the AM esters, **30** is converted into the activated Ca(II) sensor **31**.

Initially, the carboxylate groups on prepared probe **30** were appended with the AM esters to give a neutral probe that can enhance internalization into the cells. Following the esterase-induced hydrolysis of the AM esters into the cell cytosol, the formed probe **31** is activated to sense Ca(II), increasing the r_1 from 3.6 to 5.1 $mM^{-1}s^{-1}$, while remaining insensitive toward Mg(II) [41]. Subsequently, this probe was used to perform a functional MRI of neural activation *in vivo* (more details in Section 3.7).

3.2 Ca-Responsive Probes Suitable for T_2-Weighted MRI

Since the chemistry of T_2-weighted MRI probes differs substantially from those suitable for T_1-weighted MRI, the development of Ca-sensitive probes of this class must follow a different approach for their design. Although iron nanoparticles typically have a larger size than their T_1-weighted analogues, their influence on T_2 and thus the r_2 relaxivity is affected if this size is further altered. This leads to a significant change in the particle surface size. It affects the diffusion of water molecules in the vicinity of the surface of the superparamagnetic probe, and consequently the transverse relaxation.

Ca-selective r_2 changes with T_2-weighted probes have been achieved by using appropriate coatings that interact with this ion, for instance, Ca-binding polymers or proteins. In the presence of Ca(II), complexation occurs at the surface of the particle leading to agglomeration of the particles and change in r_2 (Figure 8).

There are several examples of T_2-weighted Ca-sensitive probes that were developed and reported. A typical polymer-based representative is a sensor coated with alignates, biocompatible polysaccharides that interact with Ca(II). When the alignate-coated NPs are prepared, their initial hydrodynamic size of about 70 nm increases up to 200 nm upon Ca(II)-triggered aggregation. This also leads to a significant change in the T_2-weighted signal in the subsequent MRI experiments [42]. A few additional examples of such probes with protein-based and Ca-binding coatings have also been developed, and some of them have been successfully used in molecular fMRI studies *in vivo* (see Section 3.7) [43–46]. Following the valuable knowledge gained from the studies on Ca-binding proteins and genetically encoded fluorescent Ca(II) indicators (see Chapter 3 of this book), the first reported protein-based MRI sensor for Ca(II) was based on calmodulin [43], while the improved version of this probe consisting

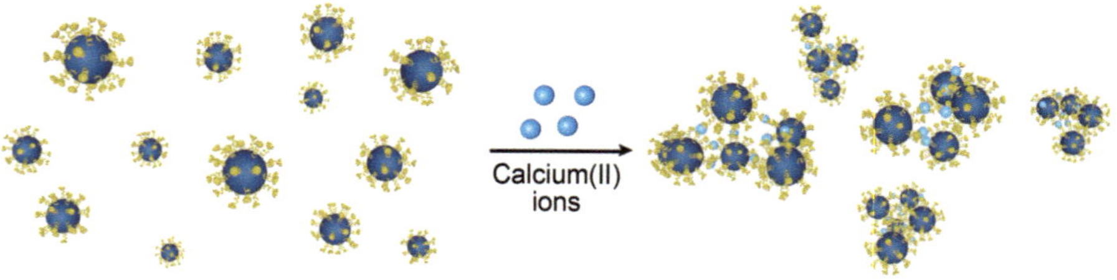

FIGURE 8 The mechanism that leads to changes in r_2 relaxivity in case of T_2-weighted Ca-sensitive probes: iron nanoparticles coated with Ca-binding proteins cause nanoparticle aggregation upon interaction with Ca(II). Formation of agglomerates, i.e., larger particles further reduces the T_2, allowing performance of the T_2-weighted MRI.

of smaller and lipid-coated NPs showed faster kinetics [44]. The best performances in terms of the r_2 response that can be generated within seconds and is suitable for monitoring the extracellular Ca(II) signaling processes in the brain (0.1–01.0 mM), were achieved with the MRI probe engineered using synaptotagmin proteins [45]. The most recently reported example in this class is the genetically encoded MRI sensor based on calprotectin, which coordinates paramagnetic Mn(II). Initial investigations of this probe showed that both T_1 and T_2 alter upon mixing with biologically relevant Ca(II) concentrations [46].

3.3 CEST-Based Ca-Responsive Probes

Although the CEST effect has a different origin than the MR effect arising from T_1 and T_2 relaxation processes, it also heavily depends on a few parameters that determine and influence the amplitude of the response of the probe. The exchange rate between the two pools of nuclei, k_{ex}, is certainly the most relevant parameter that can be chemically controlled. Proton exchange is usually changed to a lesser or greater extent with pH, resulting in remarkable CEST effect changes.

Realization of a Ca-sensitive CEST probe should follow the same strategy used for T_1 agents, i.e., it should involve designing an agent that can undergo k_{ex} changes induced by Ca(II). This can be done by combining a Ca-binding group with the CEST-active molecule, usually a metal complex that has paramagnetically shifted proton signals involved in a slow chemical exchange with bulk water (Figure 9).

Alternatively, the binding of Ca(II) with its host can generate the second exchanging pool, which provides signal at a different frequency. This mechanism is typical for other NMR-sensitive nuclei, such as ^{19}F, and was also exploited to make the Ca-sensitive probes based on CEST (see below).

The best examples of the probes that follow the typical mechanism for CEST agents are complexes **32** and **33** (Figure 10). These were prepared as Eu(III) and Yb(III) complexes with the same macrocyclic chelator that bears four imino(diacetate) moieties to bind Ca(II), coupled via amide groups to the macrocyclic rim. The interaction of Ca(II) with either of the complexes leads to the k_{ex} reduction at the Eu(III)-bound water molecule with the bulk water in **32**, or the amide protons in **33** that are paramagnetically shifted by Yb(III) [47].

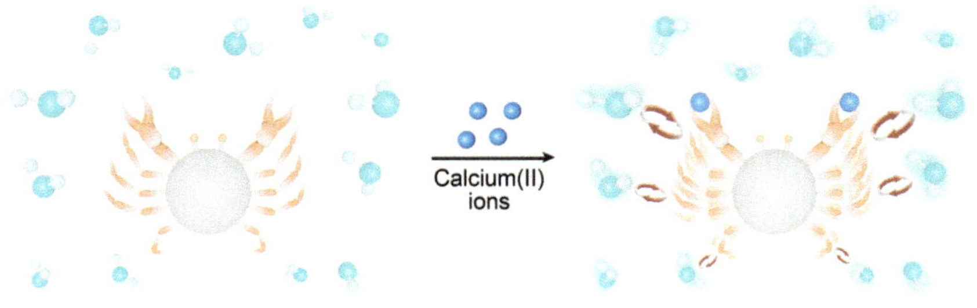

FIGURE 9 Ca-responsive CEST-based probe: interaction of Ca(II) with the probe is affecting the exchange rate of the protons (in the case of ^{1}H CEST probes). Subsequently, the alteration of the exchange rate also causes the change in amplitude of the CEST effect, which results in different CEST MRI signal.

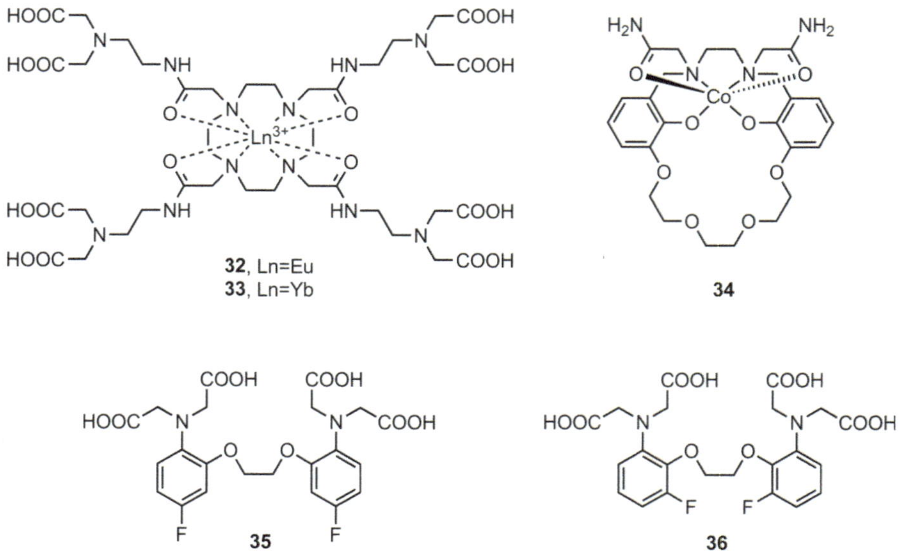

FIGURE 10 Structures of the CEST sensor MRI probes **32–36**.

Using another chemical approach, the oxa-aza macrocycle **34** contains the transition metal cation Co(II) coordinated in one end of the macrocyclic cavity. Its amide protons experience a large paramagnetic shift and its CEST effect can be observed at 77 ppm. Following the complexation of Ca(II) with the etheric oxygens of **34**, the CEST effect shifts further to 80 ppm and a reduction in the exchange rate is observed. The binding of Ca(II) causes a slight increase in the CEST effect (decrease in bulk water signal). Combined with the CEST signal at 67 ppm of the complex bound to Na(I), this effect can be used to set the basis for the development of the analytical method to quantify Ca(II) concentration under physiological conditions [48].

Besides the CEST studies that are typically done at ^{1}H frequency, this mechanism for generation of the viable MR signal was used at the ^{19}F frequency, too. Here the 5,5′-difluoro derivative of BAPTA, 5F-BAPTA (**35**), which was already known to display a remarkable Ca-induced shift in ^{19}F NMR [22], was used to generate the ^{19}F CEST signal. Namely, the binding of **35** with Ca(II) results in the appearance of an additional signal in the ^{19}F NMR spectrum, corresponding to the Ca–**35** complex. Since the free and bound forms of the ligand **35** are in slow exchange, the application of a presaturation pulse at 6.2 ppm (peak of Ca–**35**) relative to the peak of the free **35** results in the signal decrease of the latter [49]. This very intelligent approach was used to sense other metal cations in addition to Ca(II), such as Zn(II), Mg(II) or Fe(II), since both **35** and its 6,6′-difuoro analog **36** display different ^{19}F NMR shifts for each of these cations [50].

3.4 Fluorinated Ca-Responsive Probes

Given the high NMR sensitivity of the ^{19}F nucleus, there are a few efficient ways to affect its signal in the presence of a metal cation, such as Ca(II). As mentioned in

Section 3.3, the first exploitation of the fluorinated agents for sensing Ca(II) was done on the ^{19}F-labeled BAPTA-derivative **35**. This probe exhibited an NMR shift of a few ppm between the free and Ca-bound forms, which could be used for the development of a quantitative method to determine the intracellular free Ca(II) concentration in the cell culture loaded with this molecule. The difference in the NMR shift and slow exchange was also used to generate a Ca-sensitive ^{19}F CEST signal (see Section 3.3).

The second way of altering the ^{19}F signal is based on the use of paramagnetic complexes. This is done by exploiting the PRE, an effect commonly used to enhance the relaxation of protons in water for ^{1}H MRI: an NMR-active nucleus is brought in the vicinity of the paramagnetic agent to enhance its relaxation and thus increase the NMR or MRI signal. A change in ^{19}F MRI signal under the PRE effect can efficiently be done by changing the distance between the ^{19}F nucleus and the paramagnetic moiety: at a longer distance the PRE effect will be weaker and vice versa, it will become more pronounced when the distance gets shorter (Figure 11).

This principle has successfully been applied to develop a couple of fluorinated probes sensitive to Ca(II) and suitable for ^{19}F MRI applications. A dual-frequency responsive agent **37** consists of a combination of moieties sensitive to both ^{1}H and ^{19}F MRI (Figure 12). Namely, its macrocyclic part based on a DO3A ligand and complexed with Gd(III) resembles the active structure described for Ca-sensitive probes suitable for T_1-weighted MRI (Section 3.1). In addition, the presence of a CF_3 group on the other terminus of the probe and coupled to the paramagnetic moiety through the EGTA-derived chelator ensures the Ca-dependent change in the Gd···F distance. In the absence of Ca(II), the distance is longer and hence the ^{19}F T_1 and T_2 relaxation times are long as well. In contrast, interaction with Ca(II) results in a shorter distance between Gd(III) and the ^{19}F nuclei, which shortens the ^{19}F T_1 and T_2 relaxation times. Importantly, this effect is not exclusive to Gd(III). Other paramagnetic cations, such as Dy(III), can be chelated (probe **38**), also affecting the ^{19}F T_1 and T_2 relaxation times to a different extent [51]. The PRE effect depends on $1/r^6$, where r is the distance between the paramagnetic center and the observed nuclei [52]. Thus, relatively small changes in r cause remarkable variations in the relaxation times.

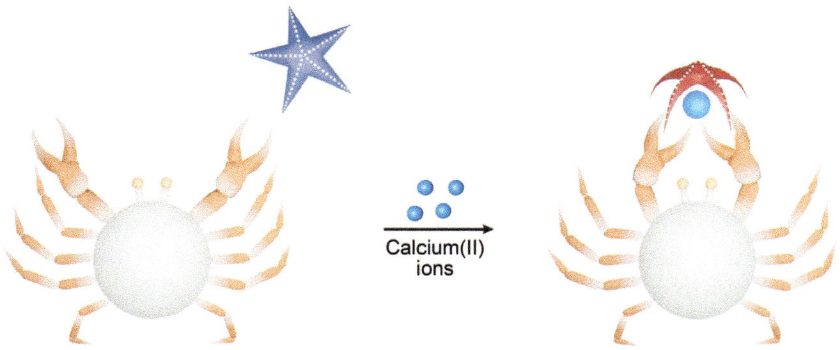

FIGURE 11 Ca-induced changes in ^{19}F MRI using the PRE effect: interaction between Ca(II)- and ^{19}F-containing probe causes the change in distance between fluorine and the paramagnetic metal ion, which results in change of ^{19}F T_1 and T_2 of the probe.

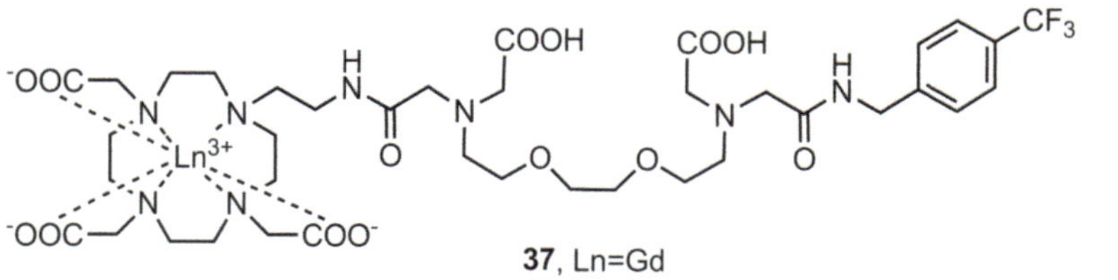

37, Ln=Gd
38, Ln=Dy

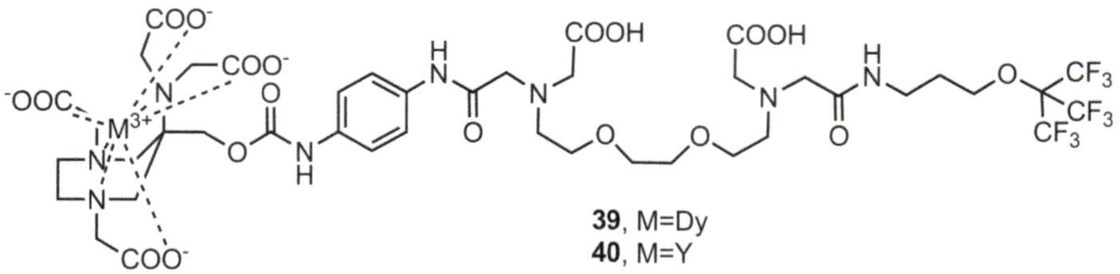

39, M=Dy
40, M=Y

FIGURE 12 Structures of the sensor probes **37–40** suitable for ^{19}F MRI.

Following the same mechanism, an improved version of the ^{19}F MRI sensor for Ca(II) was developed. Probes **39–40** bear three times more fluorine atoms than the analogues **37–38** (9 vs. 3 atoms per molecule). Moreover, the metal cation chelator is replaced by an AAZTA-based ligand that does not display conformational isomers in the NMR spectrum, resulting in a single ^{19}F resonance that maximizes the MR signal from the probe. Interaction of **39** with Ca(II) produces greater changes in ^{19}F NMR than those observed for **37–38**, which in combination with diamagnetic probe **40** can be used for the development of a ratiometric methodology based on ^{19}F chemical shift imaging to quantify the amount of ^{19}F nuclei in the sample [53].

The third possibility to affect the ^{19}F MRI signal involves formation of nanoribbons from fluorinated peptide amphiphiles. The mechanism and probe leading to the Ca-sensitive effect is discussed in Section 3.6 regarding nanosized probes.

3.5 Hyperpolarized Ca-Responsive Probes

Similarly to ^{19}F, other nuclei like ^{13}C or ^{15}N can change their frequencies notably in the presence of an analyte like Ca(II), albeit to a somewhat lesser extent (up to a few ppm). The design principle is almost identical. The Ca-chelating (receptor) moiety should be coupled to the reporting moiety (i.e., the NMR-active heteronucleus), usually in its close vicinity. Since the natural abundance and sensitivity of ^{13}C or ^{15}N is much lower than that of ^{19}F, a couple of specific approaches are applied when designing a Ca-sensitive probe. Low abundance of the NMR-active nucleus can easily be addressed by using an isotope-enriched molecule. On the other hand, hyperpolarization is the perfect approach to overcome the low sensitivity issue. The Ca-sensors discussed in this section are both enriched with ^{13}C or ^{15}N isotopes, and hyperpolarized in order to amplify the changes generated by the interaction of Ca(II) with the probe, which is then followed by means of ^{13}C or ^{15}N NMR.

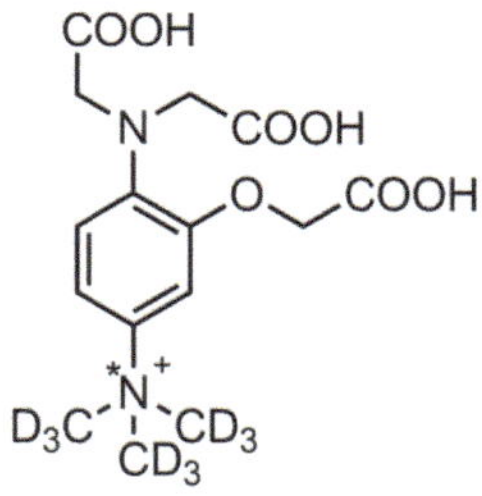

41

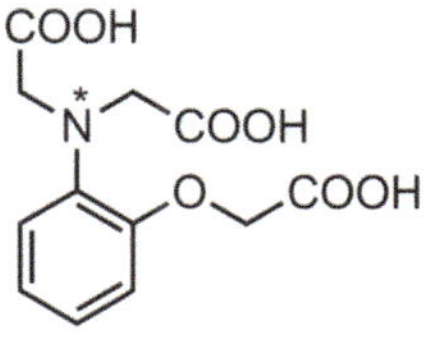

42

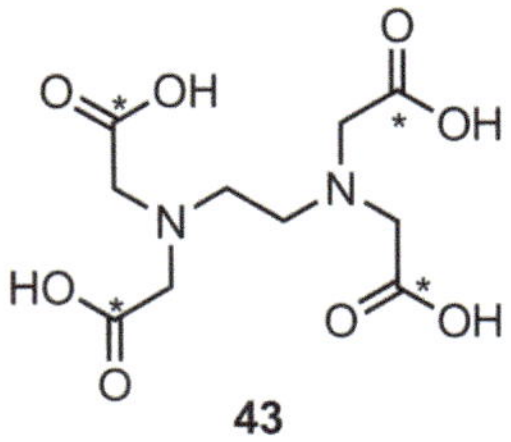

43

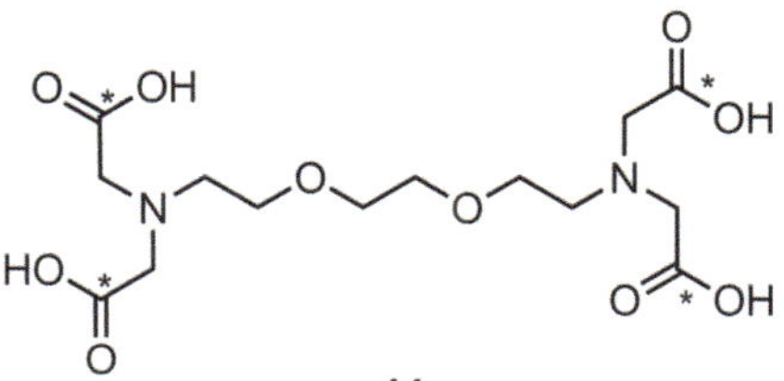

44

FIGURE 13 Structures of the hyperpolarized sensor probes **41–44**. Asterisks in the structures indicate isotope labeled positions with ^{15}N (**41–42**) or ^{13}C (**43–44**).

The first example from this class is the ^{15}N-enriched APTRA-derived probe **41** (Figure 13). The interaction with Ca(II) induces a frequency shift of slightly over 1 ppm, which showcases the major principle of action of the hyperpolarized sensor probes [54]. The follow-up probe **42** is a structural analog that now has the ^{15}N nuclei coordinating with Ca(II), which results in a larger frequency shift of 5 ppm [55].

Conversely, the Ca(II) chelates derived from EDTA (**43**) and EGTA (**44**) and enriched with ^{13}C at the coordinating carboxylates can be used for sensing this ion. Hyperpolarized probes **43–44** show ~5 and 10 ppm frequency shifts upon binding to Ca(II), respectively. Using this modern methodology, they can be used to quantify the amounts of Ca(II) by means of NMR or MRS, as well as to detect the presence and differentiate between various biologically important or toxic metal cations [56].

3.6 Nanosized Ca-Responsive Probes

Besides the above-mentioned Ca-sensitive probes of different types that can be classified as the 'small-size' probes (MW up to 1,000–1,500 Da), a separate class of so-called nanosized and Ca-responsive probes (MW > 3 kDa) was developed and reported. These probes contain a nanosized carrier molecule, such as the dendrimer, liposome, or another type of nanoparticle that is appended with the small-size responsive probes (Figure 14). There are a few reasons that led to the development of these probes. The low sensitivity of MRI can partially be circumvented by using multimeric MRI contrast agents, which are expected to provide sufficient MR contrast enhancement by taking a high payload of MR-active units to the target tissue. Next, the larger size of the agent may slow down the diffusion of the probe and thus increase its retention time in the tissue. Some of the nanocarriers can also improve

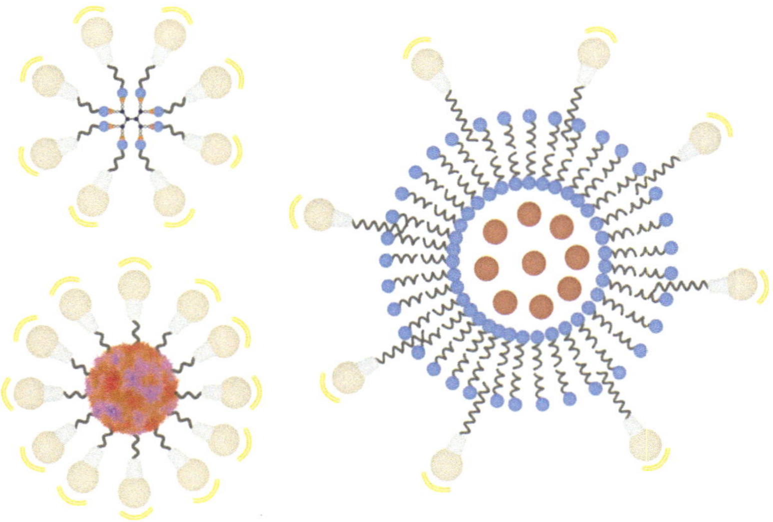

FIGURE 14 Nanosized and multimeric Ca-responsive probes. Different nanocarriers can be used for coupling to the MRI-active monometic units, e.g., dendrimers (illustration top left), nanoparticles (illustration bottom left), or liposomes and related lipid-based nanosized particles (right). The inner space of latter can be used for incorporation/entrapment of other functional molecules that may aid quantification of the probe (see the example of PFCE-containing lipid nanoparticles combined with **46**).

the biocompatibility of the entire nanosized responsive system. Finally, certain phenomena can only take place and the associated MRI changes can be triggered only by using nanosized systems.

The first example of the nanosized and Ca-responsive MR probe was based on the monomacrocyclic sensor **22**, whose aromatic nitro group was converted to an isothiocyanate and coupled to the amine groups on the surface of a G1 PAMAM dendrimer. This probe remained active toward Ca(II), albeit with a slightly weaker r_1 response compared to its small-sized precursor **22** [35]. When a series of G0–G2 dendrimers with the same precursor were studied at a low magnetic field (at 20 MHz), it was clearly demonstrated that the size of the probe not only affected the starting r_1 value (larger size → higher r_1), but also the amplitude of the Ca-induced relaxivity changes was more pronounced as the MW and size of the probe increased [57].

Dendrimeric agents were also used to develop a method for monitoring the Ca-flux by means of the rapid bSSFP imaging protocol. Here the G4 PAMAM dendrimer was conjugated with the monomeric probe **22**, while the changes in Ca(II) concentration were captured by following both ^{1}H T_1 and T_2 relaxation times. Due to the formation of the complex between the G4 dendrimer conjugate and Ca(II), a change in the nanoparticle diameter takes place. This affects the T_2 relaxation time much more than T_1; subsequently, a T_2/T_1-sensitive bSSFP sequence exhibits contrast-to-noise ratio changes that are a few folds higher than those obtained with conventional T_1- or T_2-weighted imaging protocols [58]. The follow-up study on a series of PAMAM dendrimeric probes appended with different types of monomeric MR units indicated

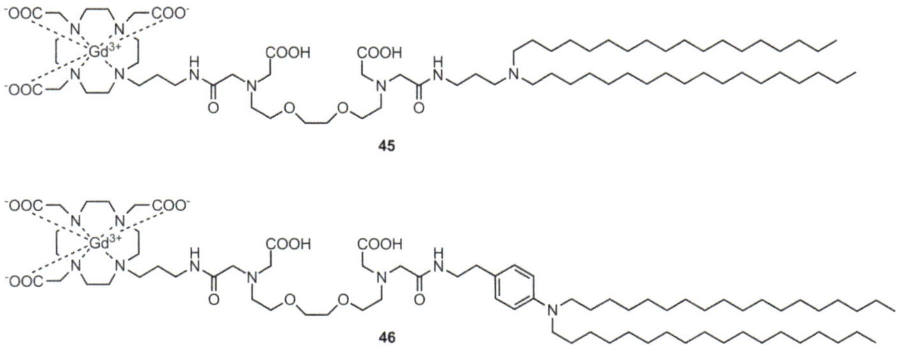

FIGURE 15 Structures of the amphiphilic monomacrocyclic probes **45** and **46**.

that a synergy of the physicochemical properties, including the change in hydration state, size of the probe, and its rigidity leads to changes that are more pronounced for r_2 rather than r_1 relaxivity [59].

A similar type of nanosized and Ca-responsive probe from the monomer **22** was prepared when using a siloxane-based nanoparticle carrier. The resulting ultrasmall and rigid particle has only a few nm in diameter and has the desired r_1 response toward Ca(II) at a high magnetic field (7 T) [60]. Due to its preferred biocompatible features, MRI *in vivo* tests were performed with this nanosized probe (Section 3.7). On the other hand, a trimeric agent that bears an RGD (Arg-Gly-Asp) peptide sequence known to target protein integrin has been prepared using a solid-phase synthesis protocol in an attempt to develop a responsive and target-specific MR agent that can accumulate in the desired tissue before performing possible functional imaging experiments [61]. The targeting features of this probe were also assessed *in vivo* (Section 3.7).

Examples of lipid-based and responsive nanoparticle probes were also prepared and reported. Using a liposome formulation, the amphiphilic monomeric probe **45** (Figure 15) was incorporated into the liposome bilayer formulation, resulting in particles with ~140 nm diameter. When they are studied at a low magnetic field (0.5 T), the overall r_1 change exceeds 400% upon the saturation with Ca(II). This dramatic r_1 enhancement is mainly caused by a combination of q changes and restricted local rotational motion of the chelated Gd(III). This effect is pronounced in the studied field; otherwise, the effect of the hydration would be the most influential parameter in higher magnetic fields [62]. If another formulation is used to prepare the amphiphilic monomeric probe **46** together with the PFCE, responsive and perfluorinated lipid nanoparticles can be prepared to have a dual purpose. The sensor part based on paramagnetic Gd(III) can report on Ca(II) concentration changes at the ^1H frequency, while the high amount of fluorine atoms incorporated into the lipid nanoparticle can be used to determine the amount of the probe based on the quantitative ^{19}F NMR signal [63]. Subsequently, this nanosized probe was used for quantification of ^{19}F and hence the Gd(III) amount *in vivo* (Section 3.7).

Finally, the fluorinated nanoparticles were also used to prepare Ca-responsive probes suitable for ^{19}F MRI. Highly fluorinated peptide amphiphiles were prepared by conjugating three different peptides with tridecafluoroheptanoyl chloride using a

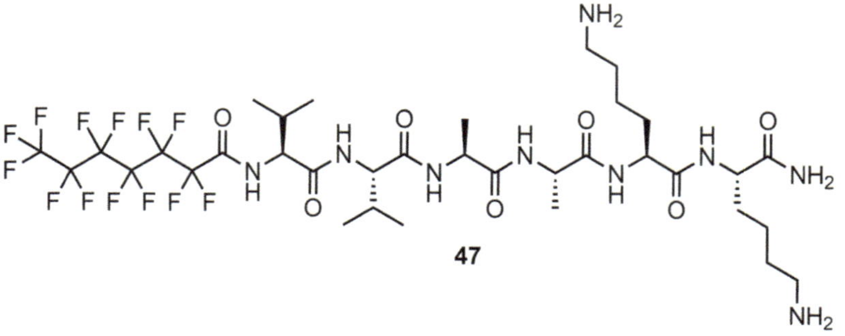

47

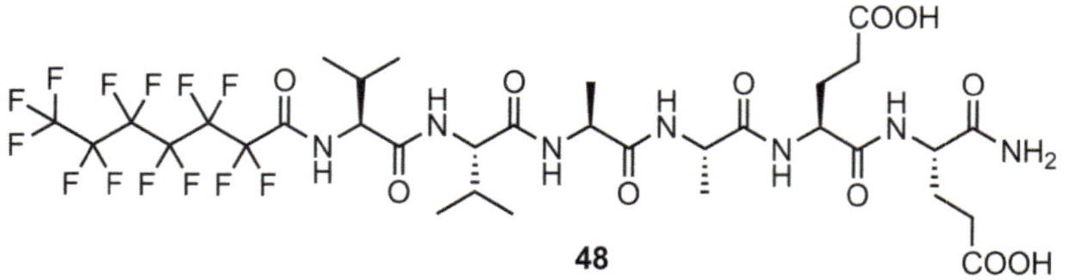

48

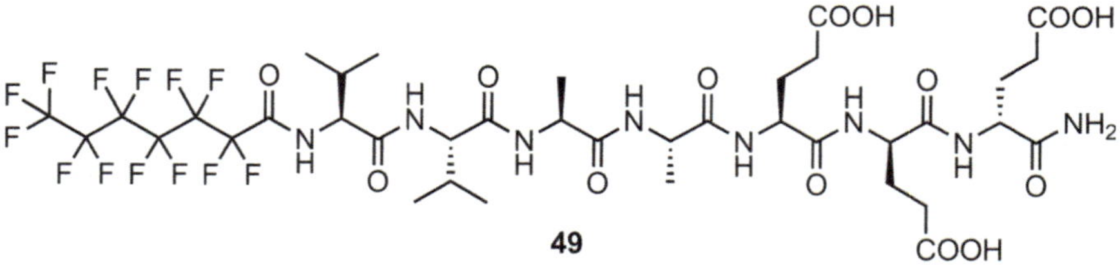

49

FIGURE 16 Structures of the peptide amphiphiles **47–49**.

solid-phase peptide synthesis protocol. The polypeptides **47–49** consisted of a Val-Val-Ala-Ala sequence combined with charged Lys-Lys, Glu-Glu, or Glu-Glu-Glu groups to improve the solubility (Figure 16). The peptide amphiphiles self-assembled in the presence of Ca(II), making the nanoribbons. Formation of these nanostructures significantly reduces the local rotational dynamics of the fluorine atoms, which shortens the ^{19}F T_2 relaxation times of the fluorine nuclei in **47–49** and results in a reduction of ^{19}F NMR signal intensity [64].

3.7 *In Vivo* MRI with Ca-Responsive Probes

The production of MR probes responsive to Ca(II) was typically followed with an extensive physicochemical characterization and, to a lesser extent, a successful validation *in vivo* of selected examples discussed in the previous sections.

The preliminary tests *in vivo* were performed with the aminobisphosphonate sensor **28**. The probe was administered in the rat cerebral cortex, where it displayed slow diffusion, possibly due to the Ca-mediated interaction of the aminobisphosphonate group with cell membranes. However, the probe showed good biocompatibility and did not affect the neuronal function, which was assessed by means of electrophysiology [65].

Many more studies have been performed with the mono- or bismacrocyclic probes that contain a EGTA-derived Ca-chelator as either small- or nanosized probes. The monomacrocyclic sensor **22** conjugated to a G1 PAMAM dendrimer exhibited slower diffusion in the rat cerebral cortex compared to the monomeric probe **22**. This property can be useful for performing functional experiments, as the agent has a longer retention time in the tissue along with a more stable MR signal [35]. When the same monomeric probe is grafted onto the siloxane nanoparticle and administered intravenously, this nanosized probe is rapidly eliminated through the renal route due to its ultrasmall size (~3 nm). In the proof-of-concept and the first successful MRI *in vivo* study, it was shown that the MR signal in kidneys increases twice: first, within a few minutes due to the probe's rapid renal excretion, and later when the ionic Ca(II) is administered as the chloride salt, which causes the r_1 enhancement of the MR probe. The laser-induced breakdown spectroscopy experiments *ex vivo* proved that the MR probe accumulates in kidneys and chelates Ca(II), suggesting that MR signal enhancement is a consequence of Ca(II) binding to the nanosized and responsive probe [60].

A couple more studies with the nanosized probes that contain monomacrocyclic agents showed additional affirmative features of the multifunctional nanosized probes. An RGD-containing multimeric bifunctional (responsive and target-specific) probe exhibited a longer retention time in living tissue, which was a direct cause of the RGD moiety [61]. In another approach, the responsive and perfluorinated lipid nanoparticles were used for quantification of Gd(III) content in rat brain. Following the administration of a suspension with this nanosized probe in the rat somatosensory cortex, different types of ^{1}H and ^{19}F MRI protocols were applied to obtain the MR signal generated by the probe. The ^{19}F MRI signal intensity was compared to the signal recorded from a reference fluorinated sample to quantify the amount of fluorine injected *in vivo*. In turn, knowing the ratio of Gd(III) and fluorine atoms in the prepared nanosized probe, the amount of fluorine was used to conveniently estimate the amount of Gd(III) in the tissue [63]. This methodology can be an important asset for functional imaging experiments, as it may allow easier correlation of the obtained ^{1}H MRI signal with the amount of Gd(III), aiding the determination of its r_1 relaxivity and indirectly the concentration of Ca(II).

The greatest progress in performing successful functional MRI experiments was made by using the bismacrocyclic class of Ca(II) sensors. In a preliminary feasibility study on different types of cell cultures, it was shown that administration of probe **9** did not affect cellular signaling, as the cells remained functional and responded to an ATP-induced stimulation. Besides, the changes in T_1 relaxation time were observed in the experimental setup when applying **9** in a 3D cellular model, while the estimations suggested that detection *in vivo* should be possible [66].

The ultimate validation of this approach was made in a study that focused on monitoring cerebral ischemia by applying the tMCAO experimental protocol. The probe **9** was continuously infused into the rat somatosensory cortex intracranially, while the ischemic stroke induction and tissue reperfusion by means of tMCAo were triggered remotely outside the MR scanner. The induction of ischemia is typically followed by a rapid decrease in extracellular Ca(II) [67], which could be observed

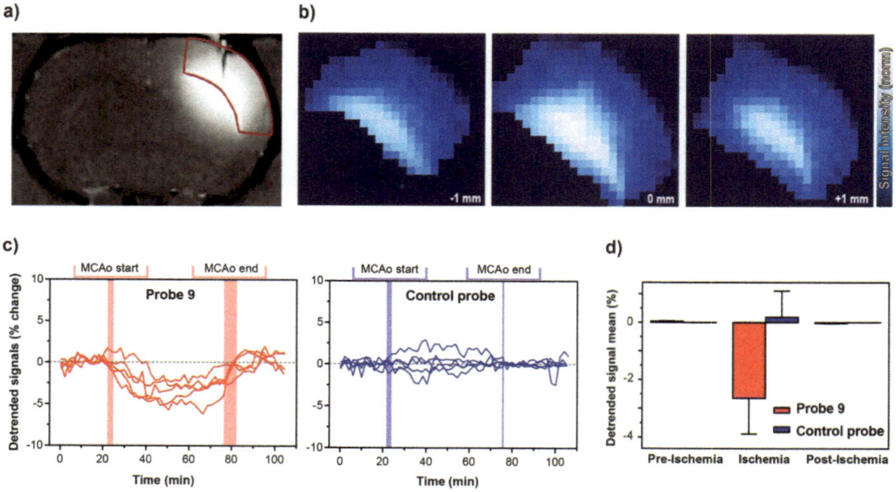

FIGURE 17 Monitoring ischemic stroke by means of a functional MRI. (a) Ca-responsive probe **9** was infused in the rat somatosensory cortex; (b) The probe enhanced MRI signal in a few MRI slices, which were then analyzed as a function of time, i.e., relative to the MCAo onset and the tissue reperfusion; (c) Detrended signals of the probe **9** (left) and the control probe (right) showing that only **9** causes the MRI signal change along with the start and end of MCAo; (d) Means of detrended signals for the probe **9** and control probe before, during, and after ischemia. (Reproduced with permission from Ref. [68]. © 2019, National Academy of Sciences.)

through the drop in MR signal with high spatial and temporal resolution. With the tissue reperfusion and the re-establishment of the resting-state Ca(II) concentration, the MR signal recovers as well (Figure 17). The control and non-responsive MR probe did not show the same behavior [68]. This methodology demonstrated the ability of the Ca-sensitive MR probe to directly detect and visualize in real-time transient cerebral ischemia, while showing great potential to investigate many other physiological processes that involve Ca(II) fluctuations.

Another approach for sensing extracellular Ca(II) levels was presented with the use of protein-based T_2-weighted probes (Section 3.2). Here a combination of lipid-coated iron-oxide nanoparticles and the calcium-binding domains of the synaptotagmin proteins reversibly responds to the appropriate concentration of Ca(II) (0.2–1.0 mM). Upon the infusion of the responsive MR probes, the neurons were activated using a few pharmacological or electrical stimuli (Figure 18). When potassium ions or glutamate were infused, the induced Ca(II) concentration changes can be followed by means of a T_2-weighted imaging sequence, whereas the iron nanoparticle probes without the calcium-binding domains did not show MR signal change. The activity induced by the MFB electrical stimulation could also be reliably followed using the responsive protein-based MRI probes with a temporal resolution in the order of several seconds [45].

While both of the above-presented approaches targeted the extracellular Ca(II), the responsive probes based on the paramagnetic Mn(III) complex described above

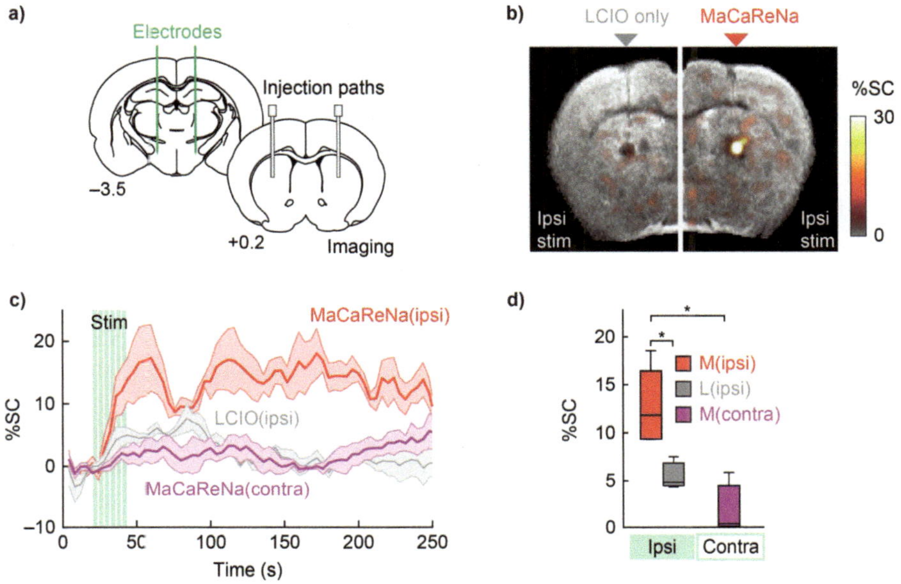

FIGURE 18 Molecular fMRI with the iron nanoparticle (T_2-weighted) Ca-sensitive probe. (a) Injection paths of the sensor (MaCaReNa) and the control (LCIO) probes and positioning of the stimulation electrodes; (b) The ipsilateral (ipsi) MFB stimulation caused MRI signal changes when the sensor probe was infused (right) but did not change when the control probe was infused; (c) Average responses of the sensor probe after the MFB stimulation (green vertical lines) is applied ipsi (red) and contralaterally (purple), or of the control probe when applied ipsi (gray); (d) Mean signal changes after the MFB stimulation for all the conditions is described under (c). (Reproduced with permission from Ref. [45]. © 2018, Springer Nature.)

(Section 3.1) were used to sense the changes inside the cell. Here the probe **30** was administered into the rat brain. It shows the persistence of the MR signal over longer periods of time due to the internalization into the cells, unlike most of the probes discussed above that remain in the extracellular space and have faster washout. Upon the ester hydrolysis and conversion of **30** to **31**, the activated probe responds to Ca(II) changes that are caused by potassium-induced neural depolarization [41]. This methodology expanded the scope of the entire molecular fMRI approach, showing a capability to detect both intra- and extracellular Ca(II) fluctuations and hence monitor a wide variety of essential molecular and cellular processes.

4 PRACTICAL CHALLENGES FOR USING CALCIUM-SENSITIVE MRI PROBES

Development of the functional MRI methodology to track Ca(II) (or other ionic or molecular targets) is quite a challenging task that goes beyond the preparation of the MRI sensor probe. Indeed, several other practical aspects should be considered while designing a new probe. These can be either technical issues related to the physics of MRI as a method or related to the physicochemical properties of MR probes.

The OI techniques, mainly those based on fluorescence emission, are currently dominant in performing functional imaging studies of Ca(II) dynamics in biological systems [8, 69]. While MRI can substantially expand the knowledge on the role of various biological targets including Ca(II), one should note that this technique is based on entirely different physical phenomena. Due to the physical principles used to generate the MR signal, this technique is much less sensitive than the former. This creates several consequences that limit the scope of MRI in functional studies. First, the Ca-induced change of the essential physicochemical properties of the probes discussed in Sections 3.1–3.3 (T_1 or T_2 relaxation times or CEST effect) often results in small changes in the MR signal. This cannot be compared to the orders of magnitude changes often observed in OI studies. The exception to this could be the use of hyperpolarized probes that can change the MRI signal dramatically. Conversely, hyperpolarization is limited by the short-lived state, which requires the experiments to be performed quickly. Often the administration of the hyperpolarized probe cannot be executed before the hyperpolarized signal relaxes. The design of hyperpolarized probes with particularly long relaxation times would be very beneficial to overcome these limitations.

A couple of additional issues reflect the poorer sensitivity of MRI. The first is the necessity to apply high quantities of the probes for the MR signal to be detected. Since the design of MRI probes usually requires a much larger size of the molecules than for the OI probes, this often limits the delivery of the sensor probes in the target tissue. A typical example is delivery in the brain and the inability of probes to cross the blood–brain barrier; currently such studies are being done by performing intracranial infusion of probes. The same is valid for delivery of the amounts required to obtain a response inside the cells. In addition, these probes are often hydrophilic to improve their own biocompatibility, making their cellular internalization extremely challenging. Second, the buildup of the MR signal requires more time, hence the acquisition of the MRI experiment lasts longer. This also affects the temporal resolution of MRI [70], which often cannot be compared with the one achieved by OI techniques. However, one should note that the large field of view and the ability to penetrate the tissue are the aspects making MRI the preferred methodology to study biological phenomena on a large scale.

Another hurdle that often appears in the use of MR sensor probes is the ambiguity of the MR signal changes. Namely, the probes themselves generate signal in the steady state in most cases (for example, T_1- and T_2-weighted agents); hence, it may be quite challenging to differentiate the MR signal alteration caused by the execution of the triggering event (preferred scenario) or changes in the local concentration of the probe (undesired but expected event). There are several methodological approaches that try to address this problem, mainly through an attempt to develop the so-called ratiometric methods, or those that can detect MR signal changes independent of the probe's concentration [71].

Finally, the MR probe itself should have the appropriate selectivity and sensitivity. The selectivity toward Ca(II) can successfully be handled by the choice of chelator or binding protein. On the other hand, the affinity for Ca(II) is directing the possible application of the MR probe: high-affinity probes (roughly $K_d \leq 1\ \mu M$) are more suitable for intracellular studies, provided that an appropriate strategy for their delivery into the cells exists. If the probe displays affinity in the high μM to low mM range, it should be suitable for targeting extracellular Ca(II). The majority of the probes discussed in this chapter (except for **1** and **31**) belong to the latter class, which may have an advantage in terms of easier delivery to the target tissue, as well as easier MR signal detection due to the larger quantities of the probes that are used in this case.

5 CONCLUSIONS

Great progress has been made in the development of MR probes sensitive to Ca(II), especially in the past decade. A large number of sensors were prepared and investigated for application with the conventional ^1H MRI techniques, although several examples of emerging methodologies, such as hyperpolarization, have also been recently reported. The development of the probes was followed by a few remarkable and exciting *in vivo* studies that validated the entire approach, investigating phenomena related to both extracellular and intracellular Ca(II) fluctuations. These studies showed the great power of molecular fMRI in the visualization of the physiology and pathology of living organisms. Since Ca(II) is regarded as an essential ion in numerous cell-signaling processes, much more functional studies should be expected in the near future considering all the forthcoming methodological advantages. The use of Ca(II) probes for MRI can be a great asset to modern functional molecular techniques in improving understanding and better controlling essential biological processes.

ACKNOWLEDGMENTS

The author thanks Ms. Gao Ya for help in preparing graphics and Prof. Carlos Platas-Iglesias for useful comments during the preparation of the manuscript. The financial support of the Shanghai Municipal Science and Technology Major Project (Grant No. 2019SHZDZX02) is gratefully acknowledged.

ABBREVIATIONS AND DEFINITIONS

AAZTA	6-amino-6-methylperhydro-1,4-diazepine tetraacetate
AM	acetomethoxyl
APTRA	*o*-Aminophenol-*N*,*N*,*O*-triacetate
BAPTA	1,2-bis(*o*-aminophenoxy)ethane-*N*,*N*,*N'*,*N'*-tetraacetic acid
bSSFP	balanced steady-state free precession
CEST	chemical exchange saturation transfer
DO3A	1,4,7,10-tetraazacyclododecane-1,4,7-triacetate
DTPA	diethylenetriaminepentaacetic acid
EDTA	ethylenediaminetetraacetic acid
EGTA	ethylene glycol-bis(2-aminoethylether)-*N*,*N*,*N'*,*N'*-tetraacetic acid
MFB	medial forebrain bundle
MRI	magnetic resonance imaging

MRS	magnetic resonance spectroscopy
NMR	nuclear magnetic resonance
OI	optical imaging
PAMAM	polyamidoamine
PFCE	perfluoro-15-crown-5-ether
PRE	paramagnetic relaxation enhancement
RF	radiofrequency
SNR	signal-to-noise ratio
tMCAo	transient middle cerebral artery occlusion

REFERENCES

1. I. Bertini, H. B. Gray, S. J. Lippard, J. S. Valentine, Bioinorganic Chemistry, University Science Books, Mill Valley, CA, **1994**.
2. K. P. Carter, A. M. Young, A. E. Palmer, *Chem. Rev.* **2014**, *114*, 4564–4601.
3. R. Y. Tsien, *Biochemistry* **1980**, *19*, 2396–2404.
4. G. Grynkiewicz, M. Poenie, R. Y. Tsien, *J. Biol. Chem.* **1985**, *260*, 3440–3450.
5. M. Schäferling, *Angew. Chem. Int. Ed.* **2012**, *51*, 3532–3554.
6. A. Miyawaki, J. Llopis, R. Heim, J. M. McCaffery, J. A. Adams, M. Ikura, R. Y. Tsien, *Nature* **1997**, *388*, 882–887.
7. V. Pérez Koldenkova, T. Nagai, *Biochim. Biophys. Acta Mol. Cell. Res.* **2013**, *1833*, 1787–1797.
8. M. Z. Lin, M. J. Schnitzer, *Nat. Neurosci.* **2016**, *19*, 1142–1153.
9. A. E. Merbach, L. Helm, É. Tóth, *The Chemistry of Contrast Agents in Medical Magnetic Resonance Imaging*, 2nd ed., Wiley, Chichester, **2013**.
10. E. Terreno, D. D. Castelli, A. Viale, S. Aime, *Chem. Rev.* **2010**, *110*, 3019–3042.
11. J. Wahsner, E. M. Gale, A. Rodríguez-Rodríguez, P. Caravan, *Chem. Rev.* **2019**, *119*, 957–1057.
12. S. Lacerda, D. Ndiaye, É. Tóth, in *Metal Ions in Bio-Imaging Techniques*, Eds.: A. Sigel, E. Freisinger, R. K. O. Sigel, Vol. 22, De Gruyter, **2021**, pp. 71–100.
13. C. F. G. C. Geraldes, M. Hélène Delville, in *Metal Ions in Bio-Imaging Techniques*, Eds.: A. Sigel, E. Freisinger, R. K. O. Sigel, Vol. 22, De Gruyter, **2021**, pp. 271–298.
14. S. Ogawa, T. M. Lee, A. R. Kay, D. W. Tank, *Proc. Natl. Acad. Sci. U. S. A.* **1990**, *87*, 9868–9872.
15. N. K. Logothetis, *Nature* **2008**, *453*, 869–878.
16. P. C. M. van Zijl, N. N. Yadav, *Magn. Reson. Med.* **2011**, *65*, 927–948.
17. A. Rodríguez-Rodríguez, M. Zaiss, D. Esteban-Gómez, G. Angelovski, C. Platas-Iglesias, in *Metal Ions in Bio-Imaging Techniques* Eds.: A. Sigel, E. Freisinger, R. K. O. Sigel, Vol. 22, De Gruyter, **2021**, pp. 101–136.
18. J.-X. Yu, R. R. Hallac, S. Chiguru, R. P. Mason, *Prog. Nucl. Magn. Reson. Spectrosc.* **2013**, *70*, 25–49.
19. J. Ruiz-Cabello, B. P. Barnett, P. A. Bottomley, J. W. M. Bulte, *NMR Biomed.* **2011**, *24*, 114–129.
20. I. Tirotta, V. Dichiarante, C. Pigliacelli, G. Cavallo, G. Terraneo, F. B. Bombelli, P. Metrangolo, G. Resnati, *Chem. Rev.* **2015**, *115*, 1106–1129.
21. P. Nikolaou, B. M. Goodson, E. Y. Chekmenev, *Chem. Eur. J.* **2015**, *21*, 3156–3166.
22. G. A. Smith, R. T. Hesketh, J. C. Metcalfe, J. Feeney, P. G. Morris, *Proc. Natl. Acad. Sci. U. S. A.* **1983**, *80*, 7178–7182.
23. W. H. Li, S. E. Fraser, T. J. Meade, *J. Am. Chem. Soc.* **1999**, *121*, 1413–1414.
24. W. H. Li, G. Parigi, M. Fragai, C. Luchinat, T. J. Meade, *Inorg. Chem.* **2002**, *41*, 4018–4024.

25. K. W. MacRenaris, Z. Ma, R. L. Krueger, C. E. Carney, T. J. Meade, *Bioconjugate Chem.* **2016**, *27*, 465–473.
26. E. L. Que, C. J. Chang, *Chem. Soc. Rev.* **2010**, *39*, 51–60.
27. K. Dhingra, P. Fouskova, G. Angelovski, M. E. Maier, N. K. Logothetis, E. Toth, *J. Biol. Inorg. Chem.* **2008**, *13*, 35–46.
28. A. Mishra, P. Fouskova, G. Angelovski, E. Balogh, A. K. Mishra, N. K. Logothetis, E. Toth, *Inorg. Chem.* **2008**, *47*, 1370–1381.
29. G. Angelovski, P. Fouskova, I. Mamedov, S. Canals, E. Toth, N. K. Logothetis, *ChemBioChem* **2008**, *9*, 1729–1734.
30. L. Connah, R. Joshi, S. Vibhute, G. Gambino, J. D. G. Correia, G. Angelovski, *Org. Lett.* **2019**, *21*, 5378–5382.
31. I. Mamedov, N. K. Logothetis, G. Angelovski, *Org. Biomol. Chem.* **2011**, *9*, 5816–5824.
32. A. Mishra, N. K. Logothetis, D. Parker, *Chem. Eur. J.* **2011**, *17*, 1529–1537.
33. K. Dhingra, M. E. Maier, M. Beyerlein, G. Angelovski, N. K. Logothetis, *Chem. Commun.* **2008**, 3444–3446.
34. K. D. Verma, A. Forgács, H. Uh, M. Beyerlein, M. E. Maier, S. Petoud, M. Botta, N. K. Logothetis, *Chem. Eur. J.* **2013**, *19*, 18011–18026.
35. S. Gündüz, N. Nitta, S. Vibhute, S. Shibata, M. E. Maier, N. K. Logothetis, I. Aoki, G. Angelovski, *Chem. Commun.* **2015**, *51*, 2782–2785.
36. L. Connah, V. Truffault, C. Platas-Iglesias, G. Angelovski, *Dalton Trans.* **2019**, *48*, 13546–13554.
37. J. Henig, I. Mamedov, P. Fouskova, E. Tóth, N. K. Logothetis, G. Angelovski, H. A. Mayer, *Inorg. Chem.* **2011**, *50*, 6472–6481.
38. V. Kubicek, T. Vitha, J. Kotek, P. Hermann, L. Vander Elst, R. N. Muller, I. Lukes, J. A. Peters, *Contrast Media Mol. Imag.* **2010**, *5*, 294–296.
39. P. Caravan, C. T. Farrar, L. Frullano, R. Uppal, *Contrast Media Mol. Imag.* **2009**, *4*, 89–100.
40. A. Barandov, B. B. Bartelle, B. A. Gonzalez, W. L. White, S. J. Lippard, A. Jasanoff, *J. Am. Chem. Soc.* **2016**, *138*, 5483–5486.
41. A. Barandov, B. B. Bartelle, C. G. Williamson, E. S. Loucks, S. J. Lippard, A. Jasanoff, *Nat. Commun.* **2019**, *10*, 897.
42. A. Bar-Shir, L. Avram, S. Yariv-Shoushan, D. Anaby, S. Cohen, N. Segev-Amzaleg, D. Frenkel, O. Sadan, D. Offen, Y. Cohen, *NMR Biomed.* **2014**, *27*, 774–783.
43. T. Atanasijevic, M. Shusteff, P. Fam, A. Jasanoff, *Proc. Natl. Acad. Sci. U. S. A.* **2006**, *103*, 14707–14712.
44. E. Rodriguez, V. S. Lelyveld, T. Atanasijevic, S. Okada, A. Jasanoff, *Chem. Commun.* **2014**, *50*, 3595–3598.
45. S. Okada, B. B. Bartelle, N. Li, V. Breton-Provencher, J. J. Lee, E. Rodriguez, J. Melican, M. Sur, A. Jasanoff, *Nat. Nanotechnol.* **2018**, *13*, 473–477.
46. H. F. Ozbakir, A. D. C. Miller, K. B. Fishman, A. F. Martins, T. E. Kippin, A. Mukherjee, *ACS Sens.* **2021**, *6*, 3163–3169.
47. G. Angelovski, T. Chauvin, R. Pohmann, N. K. Logothetis, É. Tóth, *Bioorg. Med. Chem.* **2011**, *19*, 1097–1105.
48. K. Du, A. E. Thorarinsdottir, T. D. Harris, *J. Am. Chem. Soc.* **2019**, *141*, 7163–7172.
49. A. Bar-Shir, A. A. Gilad, K. W. Y. Chan, G. S. Liu, P. C. M. van Zijl, J. W. M. Bulte, M. T. McMahon, *J. Am. Chem. Soc.* **2013**, *135*, 12164–12167.
50. A. Bar-Shir, N. N. Yadav, A. A. Gilad, P. C. M. van Zijl, M. T. McMahon, J. W. M. Bulte, *J. Am. Chem. Soc.* **2015**, *137*, 78–81.
51. P. Kadjane, C. Platas-Iglesias, P. Boehm-Sturm, V. Truffault, G. E. Hagberg, M. Hoehn, N. K. Logothetis, G. Angelovski, *Chem. Eur. J.* **2014**, *20*, 7351–7362.
52. J. A. Peters, K. Djanashvili, C. F. G. C. Geraldes, C. Platas-Iglesias, *Coord. Chem. Rev.* **2020**, *406*, 213146.

53. G. Gambino, T. Gambino, R. Pohmann, G. Angelovski, *Chem. Commun.* **2020**, *56*, 3492–3495.
54. H. Nonaka, R. Hata, T. Doura, T. Nishihara, K. Kumagai, M. Akakabe, M. Tsuda, K. Ichikawa, S. Sando, *Nat. Commun.* **2013**, *4*, 2441.
55. R. Hata, H. Nonaka, Y. Takakusagi, K. Ichikawa, S. Sando, *Chem. Commun.* **2015**, *51*, 12290–12292.
56. A. Mishra, G. Pariani, T. Oerther, M. Schwaiger, G. G. Westmeyer, *Anal. Chem.* **2016**, *88*, 10790–10794.
57. F. Garello, S. Gündüz, S. Vibhute, G. Angelovski, E. Terreno, *J. Mater. Chem. B* **2020**, *8*, 969–979.
58. S. Gündüz, T. Savić, R. Pohmann, N. K. Logothetis, K. Scheffler, G. Angelovski, *ACS Sens.* **2016**, *1*, 483–487.
59. L. Connah, G. Angelovski, *Biomacromolecules* **2018**, *19*, 4668–4676.
60. A. Moussaron, S. Vibhute, A. Bianchi, S. Gündüz, S. Kotb, L. Sancey, V. Motto-Ros, S. Rizzitelli, Y. Crémillieux, F. Lux, N. K. Logothetis, O. Tillement, G. Angelovski, *Small* **2015**, *11*, 4900–4909.
61. G. Gambino, T. Gambino, L. Connah, F. La Cava, H. Evrard, G. Angelovski, *J. Med. Chem.* **2021**, *64*, 7565–7574.
62. F. Garello, S. Vibhute, S. Gündüz, N. K. Logothetis, E. Terreno, G. Angelovski, *Biomacromolecules* **2016**, *17*, 1303–1311.
63. G. Gambino, T. Gambino, G. Angelovski, *Chem. Commun.* **2020**, *56*, 9433–9436.
64. A. T. Preslar, L. M. Lilley, K. Sato, S. R. Zhang, Z. K. Chia, S. I. Stupp, T. J. Meade, *ACS Appl. Mater. Interfaces* **2017**, *9*, 39890–39894.
65. I. Mamedov, S. Canals, J. Henig, M. Beyerlein, Y. Murayama, H. A. Mayer, N. K. Logothetis, G. Angelovski, *ACS Chem. Neurosci.* **2010**, *1*, 819–828.
66. G. Angelovski, S. Gottschalk, M. Milošević, J. Engelmann, G. E. Hagberg, P. Kadjane, P. Andjus, N. K. Logothetis, *ACS Chem. Neurosci.* **2014**, *5*, 360–369.
67. T. Kristián, B. K. Siesjö, *Stroke* **1998**, *29*, 705–718.
68. T. Savić, G. Gambino, V. S. Bokharaie, H. R. Noori, N. K. Logothetis, G. Angelovski, *Proc. Natl. Acad. Sci. U. S. A.* **2019**, *116*, 20666–20671.
69. C. Grienberger, A. Konnerth, *Neuron* **2012**, *73*, 862–885.
70. M. G. Shapiro, T. Atanasijevic, H. Faas, G. G. Westmeyer, A. Jasanoff, *Magn. Reson. Imag.* **2006**, *24*, 449–462.
71. L. A. Ekanger, M. J. Allen, *Metallomics* **2015**, *7*, 405–421.

3 Sensing Calcium Dynamics and Calcium Signaling

Li Tian, Xiaonan Deng, and Susan Kleinfelter
Department of Chemistry, Center for Diagnostics and Therapeutics and Advanced Translational Imaging Facility, Georgia State University, Atlanta, GA 30303, USA, Georgia State University

*Michael Kirberger**
School of Science and Technology, Georgia Gwinnett College, Lawrenceville, GA 30043, USA

*Jenny Yang**
Department of Chemistry, Center for Diagnostics and Therapeutics and Advanced Translational Imaging Facility, Georgia State University, Atlanta, GA 30303, USA

CONTENTS

* Corresponding authors.

DOI: 10.1201/9781003229971-3

Abstract

As a first and second messenger, calcium (Ca^{2+}) regulates numerous biological processes. Calcium homeostasis is modulated by intracellular Ca^{2+} stores, membrane channels, pumps, and/or receptors. Numerous diseases have been reported to be related to mishandling of calcium. Calcium signaling and calcium dynamics are mediated by major calcium stores and extracellular receptors, especially by calcium-sensing receptor (CaSR). In this chapter, we first introduce common properties for calcium-binding sites in proteins. We will then review efforts in addressing a pressing need to integrate Ca^{2+} signaling intracellularly and extracellularly, using a novel class of genetically encoded Ca^{2+} indicators (GECIs) with rapid kinetics. We describe a general strategy to rationally design a new class of GECIs, with a single Ca^{2+}-binding site and fast kinetics, by tuning rapid protein dynamics to modulate the chromophore conformational ensemble and alter the electrostatics around the chromophore. We further review recent progress in structure and function of CaSR and the molecular mechanism of extracellular calcium signaling. Endogenous ambient amino acids, and different allosteric modulators/drugs as well as disease mutations, further regulate functional cooperativity of the receptor to sense calcium and calcium signaling. These studies represent a milestone for visualizing and integrating Ca^{2+} dynamics, and facilitating drug discovery related to endoplasmic reticulum, CaSR, and Ca^{2+} dysfunction.

KEYWORDS

Binding; Calcium; EF-Hand; Genetically Encoded; Protein; Receptor; Sensing; Signaling

1 INTRODUCTION

1.1 Calcium Signaling and Spatial–Temporal Ca^{2+} Dynamics

Ionic calcium (Ca^{2+}) plays an integral and extensive role in nearly all aspects of cellular activity (Figure 1). In the Ca^{2+}-signaling network, Ca^{2+} acts as both a first and second messenger, through direct interactions with proteins and through interactions of Ca^{2+}-activated proteins with other biomolecules. To accomplish these many functions, Ca^{2+} concentration levels must be tightly regulated in both the intra- and extracellular environments [1–3].

Intracellular Ca^{2+} concentrations in the basal state can be as low as ~100 nM, increasing to ~10 μM during excited states. Conversely, extracellular concentrations typically range from 1.1 to 1.8 mM, while concentrations in bone are as high as 40 mM. The concentration of Ca^{2+} within cells is regulated by multiple processes, including Ca^{2+} channels and pumps that regulate the flow of ions in and out of the cell, and intracellular stores like the endoplasmic/sarcoplasmic reticulum (ER/SR), which can sequester or release Ca^{2+} during regulation of cellular activity.

Calcium signal transduction involves spatial–temporal Ca^{2+} dynamics associated with Ca^{2+} concentration gradients across subcellular membranes [4]. Changes in the gradients produce signal cascades that regulate numerous functions, including muscle contraction, fertilization, and gene regulation. The time scale for Ca^{2+}-induced cytosolic calcium oscillation can vary from ms to days (Figure 1). Ca^{2+} enters the cells from the extracellular environment as a result of changes in membrane potential. These fluctuations activate voltage-gated Ca^{2+} channels, facilitating Ca^{2+} transport across the membrane to induce Ca^{2+} signaling. During muscle contraction, the protein dihydropyridine receptor senses changes in the T-tubule membrane potential and activates the ryanodine receptor (RyR) on the SR resulting in millisecond release of Ca^{2+} [5]. Following the release of Ca^{2+}, Ca^{2+} sparks are generated in the cytosol that peak at 10 ms and exhibit half-decay rates of 20 ms. This is extremely fast when compared to transient receptor potential channels and Orai on the membrane, where peaks are observed between 1 and 10 s. Ca^{2+} dynamics are also on a much longer time scale for IP_3R (~10–60 s) and SERCA (~0.05–5 s). Mitochondrial membrane Ca^{2+} uniporter (MCU) exhibits moderate Ca^{2+} kinetics.

1.2 Major Ca^{2+} Stores, Channels, and Related Diseases

Intracellular Ca^{2+} transients are controlled by the ER, a major Ca^{2+} store, which also regulates important cellular tasks including protein synthesis, folding, post-translational modification, and protein trafficking. Dyshomeostasis of ER Ca^{2+} levels can result in misfolded proteins, triggering ER stress. Several high-capacity Ca^{2+}-binding proteins (CaBPs) in the ER/SR include calsequestrin, calnexin, and calreticulin. Calnexin (CNX) and calreticulin (CRT) are molecular chaperones involved in folding and subunit assembly of glycoproteins that pass through the ER. In the SR, the Ca^{2+} buffering protein calsequestrin stores Ca^{2+} for release through RyR to facilitate muscle contraction, and further acts as a Ca^{2+} channel sensor [7].

Although mitochondria in cells are known primarily for production of ATP, they also play a significant role in intracellular Ca^{2+} signaling and Ca^{2+} homeostasis by preventing intracellular overload from cytosolic Ca^{2+} influx. Resting Ca^{2+} levels in mitochondria are similar to cytosolic Ca^{2+} concentrations (<100 nM). Studies have also reported that mitochondria are active in fine-tuning neuronal activity [8] based on their ability to modulate localized instances of high Ca^{2+} increases [9] through the Ca^{2+} selective mitochondria uniporter [10].

ER channels that regulate Ca^{2+} mobilization include RyR, IP_3R, and SERCA (Figure 1). Ryanodine receptor (RyR) has three human isoforms: RyR1, RyR2, and RyR3. RyR1 is primarily found in skeletal muscle, cerebellum, hippocampus, extraocular muscle (EOM), and the diaphragm [11–13]. RyR_2 is expressed in cardiac muscle, cerebellum, and hippocampus, and has been identified in pregnant human myometrial tissue [11, 14, 15]. The third isoform, RyR3, is expressed in the liver, kidney, brain, placenta, and skeletal muscle. Diseases associated with mutations in RyR include catecholaminergic polymorphic ventricular tachycardia (CPVT) and malignant hyperthermia (MH) [16].

The membrane glycoprotein complex inositol trisphosphate receptor (IP_3R) is a Ca^{2+} channel, activated by inositol trisphosphate (IP_3), which mediates the release of

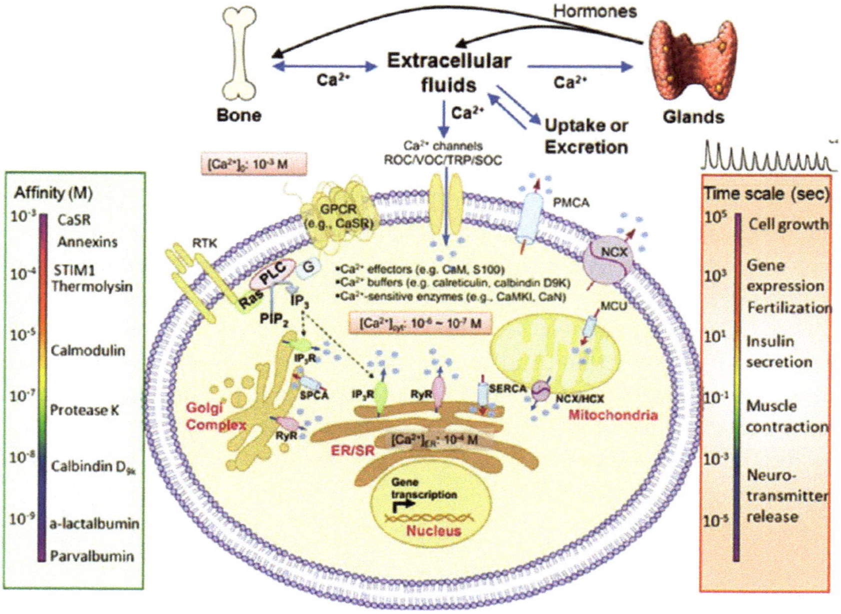

FIGURE 1 Schematic of Ca^{2+} signaling and time scale of intracellular calcium oscillation in different biological processes orchestrated by calcium-binding proteins with different affinities at different cellular environments. Extracellular Ca^{2+} homeostasis, mediated by CaSRs, involves the coordinated actions of hormones, bone cells, and the balanced uptake and excretion of Ca^{2+} in the intestine and kidney. Internal Ca^{2+} homeostasis is regulated through the actions of Ca^{2+}-signaling toolkits [6].

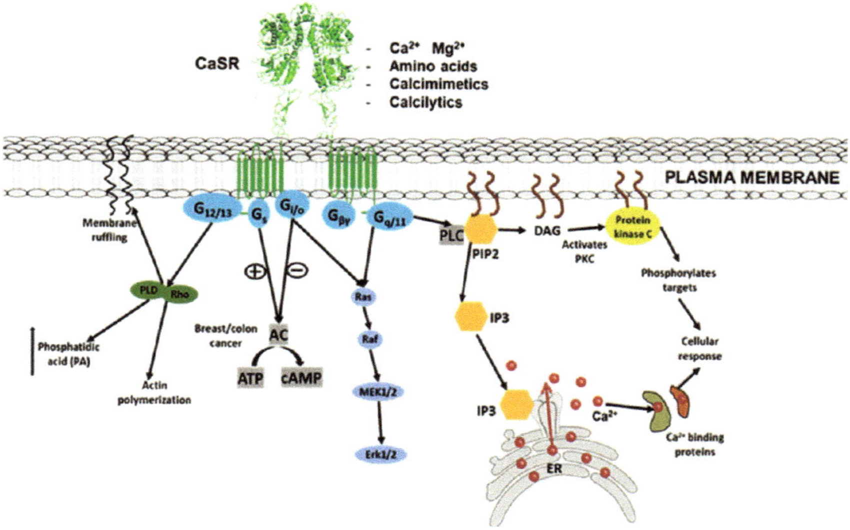

FIGURE 2 CaSR is a pleiotropic receptor for G protein-mediated intracellular signaling pathways.

Ca^{2+} from the ER [17]. The resulting Ca^{2+} signal interacts with multiple cytoplasmic targets, and regulates oxidative metabolism and cell survival through mitochondrial activity [18, 19]. IP_3, which functions both as an activator and second messenger, is generated following the activation of G protein-coupled receptors (GPCRs). A number of neurological disorders may result from mutations in IP_3Rs, including spinocerebellar ataxia 15 (SCA15), spinocerebellar ataxia 29 (SCA29) [19–21], Gillespie syndrome [22, 23], Alzheimer's disease (AD), Huntington's disease (HD), and Parkinson's disease (PD) [24–27].

Sarcoendoplasmic reticulum (SR) calcium transport ATPase (SERCA) pumps, which control cytosolic Ca^{2+} by transporting Ca^{2+} from the cytoplasm into the SR, are P-type ATPases expressed from three distinct genes (SERCA1, SERCA2, and SERCA3), resulting in ten different SERCA protein isoforms [28]. SERCA1 is mainly expressed in fast-twitch skeletal muscle fibers. SERCA2 gene encodes isoform SERCA2a, expressed predominantly in cardiac and slow-twitch skeletal muscle [29]. Protein isoforms encoded by SERCA3 are expressed primarily in non-muscle cells, but are also found at the mRNA level in different tissues [30–32]. Mutations on SERCAs are related to diseases, such as Brody's disease, which is characterized by muscle cramping and stiffness [32]. In addition, dysfunction of MCU causes abnormal mitochondrial Ca^{2+} dynamics, resulting in cell death and neurodegenerative diseases [33].

1.3 Extracellular Calcium-Signaling and Calcium-Sensing Receptor

The role of Ca^{2+} as a first messenger was established based on the discovery of calcium-sensing receptor (CaSR) by Dr. Edward Brown et al. in 1993 [34, 35]. CaSR is a member of the family C G protein-coupled receptors (cGPCRs), which includes γ-aminobutyric acid receptors (GABAs), metabotropic glutamate receptors (mGluR), and taste receptors (T1Rs), among others [36]. CaSRs have been identified in multiple species in tissues associated with extracellular Ca^{2+} and Mg^{2+} homeostasis (e.g., parathyroid, thyroid, kidney, bone), as well as in brain, skin, and other non-homeostatic tissues [37, 38].

Structurally, CaSR is comprised of a large extracellular domain (ECD), a seven-helix transmembrane domain (TMD), and a long, largely disordered intracellular tail. The ECD is further subdivided into a two-lobed Venus flytrap (VFT) domain, a rigid cysteine-rich domain (CRD), and a linker that joins the CRD to the TMD at the cell surface (Figure 2). CaSR is an obligate homodimer. Changes in conformation of the ECD in response to agonists and modulators trigger the movement of the two TMD protomers from an inactive to active position (from TMD protomers distant from each other to TMD protomers in close contact). Activation of the TMD in turn results in binding to the hetero-trimeric G protein and intracellular signaling. There are multiple Ca^{2+}-binding sites in the VFT domain, with different intrinsic affinities for Ca^{2+}. Research into other possible CaSR activating ligands determined that CaSRs are capable of integrating multiple metabolic signals from the local metabolic environment, in addition to Ca^{2+}, including polypeptides, polyvalent cations, amino acids, pH, and ionic strength [39]. Small changes in extracellular concentrations thus can produce a strong homotropic cooperative response in the ECD. CaSR activation is essential for several functions, including intracellular Ca^{2+} signaling, stimulation of calcitonin secretion in C-cells in both calcitropic and non-calcitropic systems, and inhibition of parathyroid hormone (PTH) release in parathyroid cells [40–45]. PTH and calcitonin regulate Ca^{2+} homeostasis through CaSR [36]. CaSR in the parathyroid glands reduces PTH secretion in response to an increase in serum Ca^{2+}. This results in reduction of Ca^{2+} reabsorption and reduced secretion of $1{,}25(OH)_2D$ in the kidney, reduced absorption of Ca^{2+} in the gut, and reduced skeletal resorption of Ca^{2+} [46]. The resulting Ca^{2+} increase produces a negative feedback loop involving an integrated hormonal response, which restores serum Ca^{2+} and closes the negative feedback loop.

Multiple mutations and polymorphisms have been identified in the extracellular, transmembrane, and carboxyl-terminal domains of the receptor, resulting in disorders of Ca^{2+} homeostasis [47]. Pathologies associated with inactivating human mutations include familial hypocalciuric hypercalcemia (FHH) and neonatal severe hyperparathyroidism (NSHPT), which can reduce CaSR's sensitivity to extracellular Ca^{2+}, whereas enhanced sensitivity to extracellular Ca^{2+} and Mg^{2+} ions results from activating mutations associated with autosomal dominant hypocalcemia (ADH) [48, 49]. Modulators that alter CaSR's affinity for calcium, enhance or reduce response to other regulators, or enhance or reduce conformational change of the ECD or TMD, are all potential drug candidates.

In the chapter, we will first introduce common structural properties for calcium-binding sites in proteins to reveal key determinants for calcium-binding affinities. We will then describe efforts to create GECIs to address a pressing need to integrate Ca^{2+} signaling intracellularly and extracellularly. We further review recent progress in determining the structural and functional roles of CaSR and the molecular mechanism of extracellular calcium signaling. Endogenous ambient amino acids and different allosteric modulators/drugs further regulate the functional cooperativity of the receptor to sense calcium and calcium signaling. These studies represent recent milestones for visualizing and integrating Ca^{2+} dynamics, and will facilitate drug discovery related to ER, CaSR, and Ca^{2+} dysfunction.

2 Ca^{2+}-BINDING PROTEINS AND METAL-BINDING COORDINATION IN PROTEINS

Calcium acts as a messenger involved in nearly all aspects of cellular function [50]. This role is inextricably linked to the interaction between the Ca^{2+} ion and the proteins that bind Ca^{2+} through different coordination geometries. Thus, an important aspect of studying Ca^{2+} interactions requires structural information on the protein and calcium-binding sites.

Modern structural studies of CaBPs can be traced back to the crystallization and structural elucidation of the calcium-loaded form of carp muscle parvalbumin by Kretsigner and Nockolds in 1973 [51]. This highly cited work described what would become known as the canonical EF-hand Ca^{2+}-binding motif, consisting of a 29-residue helix-loop-helix structure capable of binding a single Ca^{2+} ion in either an octahedral or a pentagonal-bipyramidal geometry (Figure 3A) coordinated by six oxygen ligands [52]. Subsequent studies of EF-hand proteins, including parvalbumin and calmodulin (CaM), confirmed that the Ca^{2+}-binding ligands were distributed across a 12-residue sequence within the canonical EF-hand motif [53], with ligands occupying relative positions 1, 3, 5, 7, 9 and 12 (Figure 3B). The residue in position 12 is generally Asp or Glu. The carboxyl group contributes two ligand oxygen atoms in a bidentate mode, which likely serves as an anchor during localized conformational restructuring to chelate Ca^{2+} [54]. CaM has been shown to interact with various proteins including channels and receptors, such as CaSR, to regulate numerous biological processes [55].

We have since gained a much deeper understanding of the variety and complexity of structures that can be classified as Ca^{2+}-binding sites in proteins and enzymes, as well as the different roles that Ca^{2+} plays in signaling pathways and intracellular regulation. Similar to the EF-hand motif, pseudo EF-hands were identified in the S100 proteins [56, 57]. The pseudo EF-hand motif is comprised of 14 residues in the loop, and has been found to coordinate the Ca^{2+} ion predominantly with main-chain carbonyl oxygen atoms, as opposed to side-chain oxygen atoms that are typical of the EF-hand motif.

By 1998, at least 45 distinct subfamilies of EF-hand proteins, including pseudo EF-hand proteins, had been identified and classified [57]. For these proteins, the coordinating ligands are still close in proximity to each other within the sequence. These conserved patterns could be utilized to identify previously unknown Ca^{2+}-binding sites in proteins, including those in bacteria [58].

In subsequent years, however, it had become increasingly apparent that many CaBP's did not conform to these earlier classifications as structures for new non-EF-hand proteins were identified [59] utilizing a broader range of coordination numbers (4–12), and adopting coordination geometries that exhibited less symmetry [60].

Accordingly, new models of classification evolved to accommodate the increasing variety of binding site types in CaBPs. A new model based on the proximity

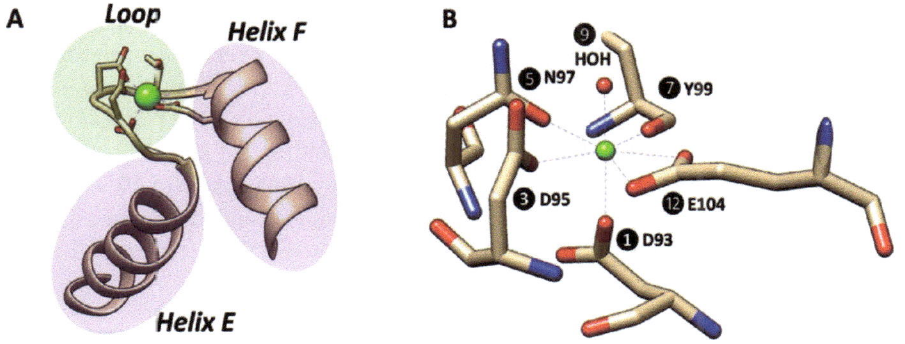

FIGURE 3 CaM EF-III Ca^{2+}-binding site. (a) Helix-Loop-Helix formation around Ca^{2+} ion. (b) Coordinating residues in relative positions 1, 3, 5, 7, 9, and 12. Pentagonal-bipyramidal geometry is holospheric, surrounding the Ca^{2+} ion.

of coordinating ligands in the sequences divided Ca^{2+} sites into Classes I–III [61] (Figure 4). Class I sites include those where all of the ligands are close within the sequence, and include canonical EF-hand, pseudo EF-hand and other non-canonical EF-hand motifs, such as calpain. In Class II sites, most of the coordinating ligands are close in the sequence, with some ligands distant in the sequence but close in the three-dimensional structure, such as hcv helicase. Class III sites, the smallest subgroup, includes coordinating ligands that are mostly separated in the sequence. The C2 domain of protein kinase C (PKC) would be an example of Class III.

Another classification model, described as a Hull property, was proposed to address the structure of the binding sites based on the extent to which the coordinating ligands surrounded the metal ion. Holospheric binding sites, where oxygen atoms surround the Ca^{2+} ion on all sides, are typically observed with higher coordination numbers (CN > 4) and include both the pentagonal-bipyramidal and octahedral geometries. Hemispheric binding sites more closely resemble a bowl shape, as observed with the PKC C2 domain. In some instances, these sites appear to be incomplete holospheric sites, which may reflect further evolutionary change over time [62]. The case of the C2 domain appears more complicated, where binding ligands are shared or "bridge" multiple Ca^{2+} ions, which has been observed with transition metal complexes. In addition to holospheric and hemispheric models, a lesser number of sites have been observed exhibiting irregular planar binding sites. Proteinase K is an example of a Class III, planar site coordinated by four ligand protein atoms (Figure 4).

Due to high off-rates associated with weak CaBPs ($K_d > 0.1$ mM), often observed with channels, pumps, and receptors, Ca^{2+} ions in these binding sites are often "invisible". An additional challenge is that calcium-binding sites are frequently located at flexible regions with conformational ensembles that are difficult to be determined. To predict or identify Ca^{2+}-binding sites (or other ions) in 3D structures, computer algorithms have been developed to capture potential calcium-binding coordination geometries by ligand clustering based graph theories and statistical analysis of calcium-binding sites in proteins [63–65]. These developed algorithms have been applied to address the problem of identifying Ca^{2+}-binding sites in transmembrane proteins. Prediction algorithms developed by Yang et al. have facilitated visualization of calcium-binding sites family C of GPCR, including the 'CaSR and mGluR' as well as CaBPs in bacteria and virus genomes [66–70]. Recent advancements in structural biology, especially Cryo-EM technology, are making it possible to visualize cell membrane proteins with increasingly refined resolution. This is transforming the capabilities of structural biology, and rapidly expanding the number of resolved protein structures available in the Electron Microscopy Data Bank (EMDB) and the Protein Data Bank (PDB) [71, 72]. This increasing number and variety of proteins, such as RyR [73], channels, and receptors, with experimental structures, will make accurate prediction of calcium-binding sites more feasible [74, 75].

3 SENSING CALCIUM-BINDING GECIs

Specific tools are essential for capturing calcium dynamics and probing the irregularities in Ca^{2+} signaling that lead to the aforementioned diseases. Significant advances in the understanding of intracellular calcium signaling can be attributed to progress

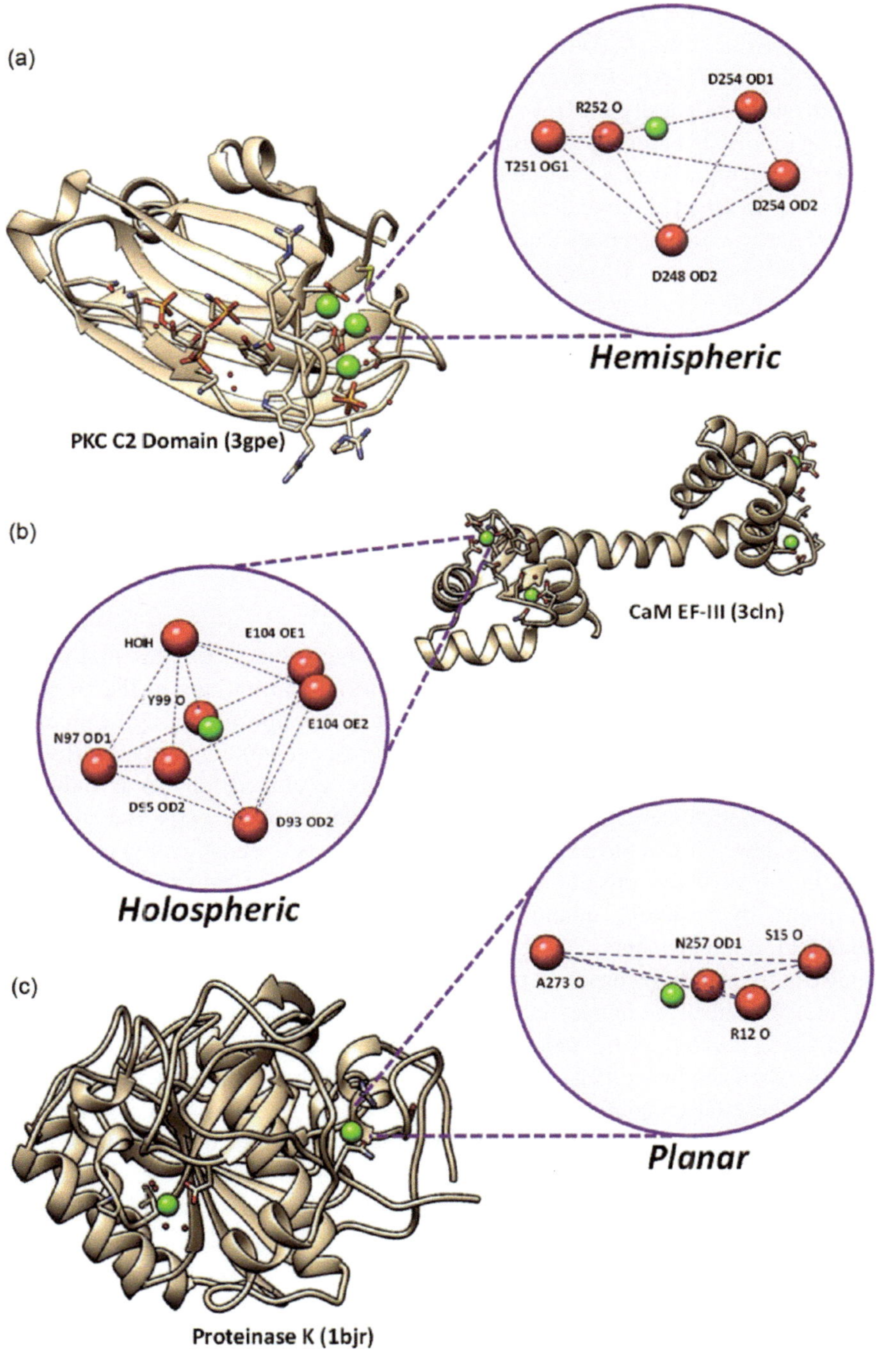

FIGURE 4 Classes of Ca^{2+}-binding sites. (a) Class II, hemispheric or "bowl" Ca^{2+}-binding site. (b) Class I, holospheric or pentagonal-bipyramidal Ca^{2+}-binding site. (c) Class III, planar (distorted) Ca^{2+}-binding site.

made in the creation of synthetic calcium dyes [76]. MagFura-2 and Fluo-5N have dissociation constants (K_d) of 25 and 90 μM, respectively, and are frequently used to study ER Ca^{2+} [77–80]. To overcome the limitations associated with the compartmentalization propensity of dyes in non-specific organelles [76], the discovery and creation of GECIs based on fluorescent proteins with different affinities and kinetic responses has significantly extended our ability to monitor calcium transients *in situ* and *in vivo* in targeted intracellular locations [81] (Table 1).

Several criteria are extremely important in the development of functional and successful Ca^{2+} indicators. (i) Indicators must function without interference from other metals or physiological molecules, such as Mg^{2+} and Zn^{2+}. (ii) Indicators require Ca^{2+} binding affinity values appropriate to specific subcellular organelles. (iii) Optical signal changes, fluorescence changes, color changes, or SNRs resulting from Ca^{2+} binding must be stable and sufficient to detect very small Ca^{2+} fluctuations, particularly at physiological temperatures. (iv) Indicators cannot induce toxicity, or otherwise perturb cellular activity. (v) Indicators require fast Ca^{2+} association and disassociation kinetics to detect fast Ca^{2+} dynamics. (vi) Indicators should form 1:1 binding complexes with Ca^{2+} ions to quantitatively interpret the signal.

3.1 Classes of GECIs

Aequorin is a 22 kDa bioluminescent Ca^{2+}-sensing protein complex first discovered in 1962 from the jellyfish Aequorea victoria. Limitations in the oxidation reaction associated with the complex made it less than optimal for real-time imaging applications. Subsequently, numerous fluorescent proteins (FPs) with chromophores producing different colors have been developed for use as indicators of activity, and for enhanced microscopic visualization in cells and in tissues [82–85]. These GECIs have found broad use in efforts to elucidate the role of Ca^{2+} signaling in different physiological mechanisms, and for different pathological conditions, by providing quantitative analyses of Ca^{2+} fluctuations in different subcellular compartments.

GECIs frequently incorporate the Ca^{2+}-binding domains from CaM or Troponin C (TnC) for sensitive changes in Ca^{2+} concentration. Following cooperative binding of Ca^{2+} ions in the EF-hand pairs, both proteins exhibit further conformational changes when binding other proteins or peptides. When coupled with fluorescent proteins, these GECI conformational changes can reduce the distance between the proteins, allowing overlap of the emission wavelength from one protein with the excitation wavelength of the second protein. This Forster resonance energy transfer (FRET) thus acts as an indicator of Ca^{2+} binding. Early efforts to develop Ca^{2+} indicators were pioneered by Anthony Persechini [86] and Roger Tsien [87], leading to the engineering of green fluorescent protein (GFP) [88] as a molecular template for the development of numerous fluorescent protein applications. These early models were subsequently refined to reduce interference with intracellular CaM or CaM targeting proteins by either modifying the CaM-binding site [89] or by enhancing the specificity of TnC for its target protein TnI [90].

The second class of calcium indicator GCaMP was created by a fusion of GFP, CaM, and the M13 peptide sequence from myosin light-chain kinase (MLCK)

TABLE 1
Summary of Ca^{2+} Indicators and Dyes

Name	Source	Binding Site Residues/Mutations	Fluorescence Dynamic Range	K_d	Kinetics
Genetically Encoded Calcium Indicators					
G-CatchER	EGFP	S147E, S202D, Q204E, F223E, and T225E	$\Delta F/F = 1.89 \pm 0.03$	0.18 ± 0.02 mM	$K_{on} \approx 3.89 \times 10^6 M^{-1}s^{-1}$ $K_{off} \approx 700\ s^{-1}$
G-CatchER+	G-CatchER	S175G, S30R, and Y39N		1.2 ± 0.2 mM	K_{obs} (association) $= 62\ s^{-1}$ K_{obs} (dissociation) $= 63\ s^{-1}$
R-CatchER	mApple	A145E, E147, D196, K198D, R216E, and T225	$\Delta F/F = 4.22 \pm 0.04$	0.35 ± 0.03 mM	$K_{on} \geq 7 \times 10^6\ M^{-1}s^{-1}$ $K_{off} > 2 \times 10^3\ s^{-1}$
GCaMP-1	M13-cp-EGFP-CaM		$\Delta F/F = 4.5$	235 nM	$\tau_d \approx 200$ ms
GCaMP-2	GCaMP-1	V76A, S88G, D93Y, A119K, V251I on cp-EGFP	$\Delta F/F = 5$	840 nM	rise $T_{1/2} = 95$ ms decay $T_{1/2} = 480$ ms (in neurons)
GCaMP-3	GCaMP-2	T116V, M66K, N363D on cp-EGFP	$\Delta F/F = 13.5$	660 nM	$K_{off} \approx 4.6\ s^{-1}$ rise $T_{1/2} = 95$ ms decay $T_{1/2} = 650$ ms (in neurons)
R-GECO1	M13-cp-mApple-CaM	T47A, L60P, E61V, S62V, E63S, R80G, K82R, Y130C, M158L, N165D, V233A, S293P, I366F, K380N, S404G, N414D, E430V on cp-mApple	$\Delta F/F = 16$	150 nM	$K_{on} = 9.52 \times 10^9 M^{-n}s^{-1}$ $K_{off} \approx 0.752\ s^{-1}$
f-RGECO1	R-GECO1	Q306D, M339F, D323A on cp-mApple		$1{,}213 \pm 15$ nM	$K_{off} = 34 \pm 1\ s^{-1}$
f-RGECO2	R-GECO1	Q306D, M339F, W44Y, D432A on cp-mApple		$1{,}261 \pm 11$ nM	$K_{off} = 26 \pm 1\ s^{-1}$
Twitch-2B	TnC(mCerulean3/cpVenus)		$\Delta F/F = 8$	200 nM	decay $T_{1/2} = 2.1$ s (in neurons)
Synthezied Calcium Dye					
BAPTA					$K_{on} = 500 \times \mu M^{-1}s^{-1}$
OGB-1	BAPTA			240 nM	$K_{off} \approx 2.6\ s^{-1}$
Fura-2					$K_{on} = 6.02 \times 10^8 M^{-1}s^{-1}$ $K_{off} \approx 96.7\ s^{-1}$
Fluo-4					$K_{on} = 6.0 \times 10^8 M^{-1}s^{-1}$ $K_{off} \approx 210\ s^{-1}$

(Figure 5) [91]. Tremendous efforts have been directed toward improving the dynamic range and SNR of GCaMPs, resulting in 5- to 10-fold increases in dynamic range occurring approximately every 5 years [4].

Currently, GCaMPs exhibit binding affinities that range from sub-μM to hundreds of μM. GCaMP indicators have been successfully utilized across multiple species for both *in vitro* and *in vivo* imaging, and have been further modified to produce different colors [92–105]. Similarly, recent developments with near-IR fluorescent Ca^{2+} indicators have enabled deep tissue imaging and multicolor Ca^{2+} imaging that does not involve spectral crosstalk [106, 107].

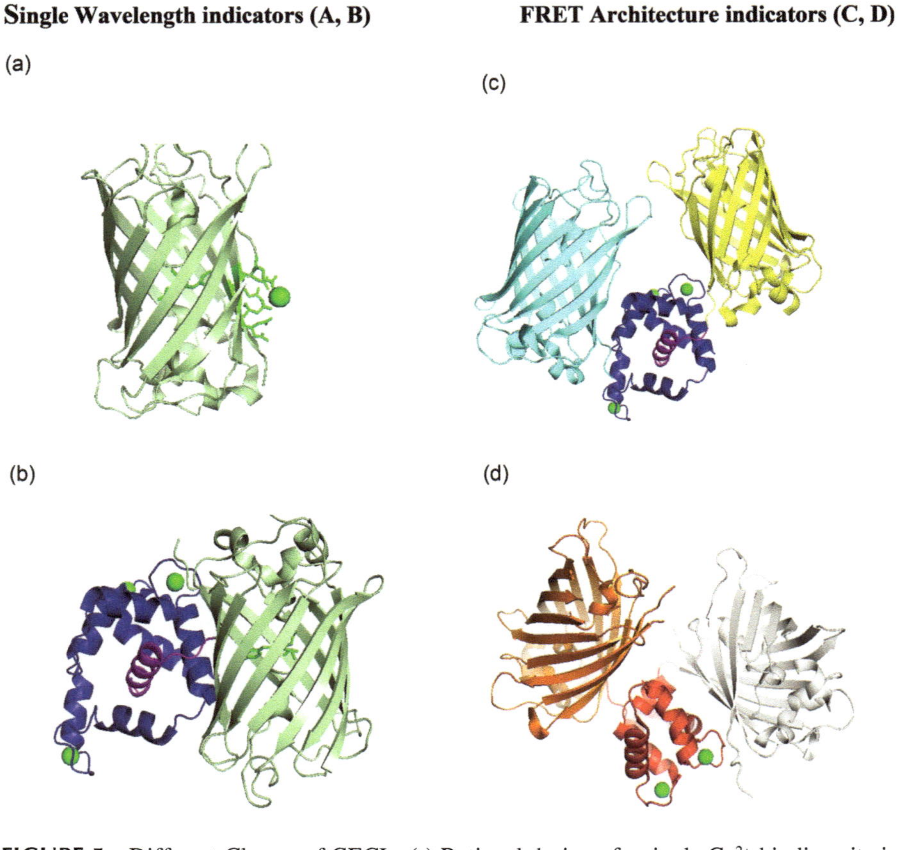

FIGURE 5 Different Classes of GECIs. (a) Rational design of a single Ca^{2+}-binding site in CatchER series. A calcium-binding site is created directly on EGFP (G-CatchER, PDB: 4L1I). (b) CaM-based single wavelength indicators. A fusion protein composed of the calmodulin-binding peptide M13, a circularly permuted fluorescent protein cp-EGFP, and calmodulin (GCaMP2, PDB: 3EK4). (c) CaM-based FRET indicators. A fusion protein that consists of a cyan fluorescent protein *ECFP*, calmodulin, the calmodulin-binding peptide M13, and a yellow fluorescent protein *EYFP* (Cameleon YC2.0). (d) Troponin C-based FRET indicators. A fusion protein composed of two fluorescent proteins (mCerulean3 and cpVenus) separated by calcium-binding protein Troponin C (Twitch2B, PDB: 6GEL). Yellow spheres represent Ca^{2+}. EGFP in green, ECFP in cyan and EYFP in bright yellow, mCerulean3 in orange, cpVenus in gray, M13 in purple, and CaM is shown in dark blue, respectively.

3.2 Kinetics Limitation of GECIs

To date, the upper limit rate for endogenous Ca^{2+} signaling is not known. It is also not clear whether current GECI kinetic responses are adequate to reliably report cellular Ca^{2+} signaling in excitable cells during electrophysiological stimulation. To study this, GECIs were evaluated in conjunction with the small molecule Ca^{2+} indicator Oregon-Green-BAPTA-1 (OGB-1), known for its fast kinetics, in presynaptic motoneuron boutons of transgenic *Drosophila* larvae [108]. It was hypothesized that OGB-1 could monitor Ca^{2+} spikes up to 10 Hz given its K_{on} rate of $4.3 \times 10^8 s^{-1}M^{-1}$, and a K_{off} rate of $10^3 s^{-1}$. However, this approach resulted in a smooth curve at 10 Hz stimulation rather than distinguishable peaks [108]. Other GCaMPs reporting increased sensitivity and improved kinetics, as well as the ability to overcome autofluorescence problems, and include G-CaMP3 and G-CaMP5 family Ca^{2+} sensors capable of *in vivo* monitoring of neuronal activity in different organisms [93, 94].

Subsequent efforts to generate more responsive GCaMPs capable of overcoming slow response kinetics and the stepwise Ca^{2+} response of CaM/TnC-based Ca^{2+} indicators, particularly for the quantification of high Ca^{2+} concentrations in organelles, such as the ER, have resulted in the development of Fast-GCaMPs, including Fast-GCaMP6f-RS06 and Fast-GCaMP6f-RS09 [109]. The Fast-GCaMPs exhibited faster off-rates (up to four-fold) than previous GCaMPs, which resulted from a design approach that protected the GFP chromophore by destabilizing the association of the hydrophobic pocket of calcium-bound CaM with the RS20-binding domain [109]. More recently, newer versions also include improvements in optical properties. GCaMPs exhibiting rapid kinetics have been designed based on red fluorescent proteins (RFPs), including mApple and mRuby. These variants have demonstrated that they are useful for neuronal imaging, and include variants f-RGECO1, f-RGECO2, f-RCaMP1, and f-RCaMP2 that exhibit Ca^{2+} dissociation (K_{off}) rates as much as ten times faster than the original versions [110]. The improved kinetics (increased Ca^{2+} dissociation constants and improved K_{off} rates) were the result of deactivating one of the paired Ca^{2+}-binding sites through mutation and reducing the affinity between Ca^{2+}-loaded CaM and RS20 targeting binding peptide. The K_{on} and K_{off} rates of the indicators also exhibited 4- to 10-fold increases at physiological temperature (37°C), compared with 20°C.

Troponin C-based GECIs with reduced Ca^{2+} binding sites have also been designed to reduce the Hill coefficient to 1, producing a linear quantitative response [111, 112]. Like CaM, Troponin C comprises N- and C-domains, where each domain includes two EF-hand Ca^{2+}-binding motifs. N-terminal EF-hand sites exhibit weaker binding affinities to Ca^{2+} compared with C-terminal sites. Deletion of the TnC N-terminal domain produced a novel sensor with strong binding affinity and a low Hill coefficient. Metal selectivity of this novel sensor for Ca^{2+} over Mg^{2+} was further improved by site-directed mutation of Ca^{2+} binding residues Asn15 and Asp17 to Asp15 and Asn17. This new class of Twitch GECIs exhibited a stronger binding affinity for Ca^{2+} (K_d: 150–450 nM), making it suitable to detect cytosolic Ca^{2+} signaling [113]. Single EF-hand-based GECIs created from Twitch exhibited FRET responses acceptable for cellular Ca^{2+} imaging and had a wide range of Ca^{2+} binding affinities from 2.6

to 257 μM. This made them viable for measuring the high concentrations of Ca^{2+} in Ca^{2+} stores, such as in the endoplasmic reticulum (ER).

In a specific application, Thy1-TwitchER transgenic mice, which express ER-targeted Twitch-2B 54S+ [113], were generated to directly monitor axonal ER Ca^{2+} depletion induced using either thapsigargin or caffeine. The result was a response range comparable to other highly sensitive ER sensors [111]. Similarly, the GECI-based fluorescent sensor CaMPARI, in combination with permanent post hoc staining of immediate early genes (IEGs) was applied to detect elevated intracellular Ca^{2+} in neuronal cells based on the irreversible green-to-red conversion of CaMPARI [114, 115].

Finally, another notable example is the development of improved variants of yellow cameleon-Nano (YC-Nano), YC2.60, and YC3.60, which exhibited high Ca^{2+} affinities (K_d = 15–140 nM) and large signal changes (1,450%), and were capable of visualizing intracellular signaling dynamics and neuronal activity in zebrafish embryos [116].

3.3 Rational Design of Calcium Sensors

Yang lab has pioneered the development of Ca^{2+} sensors and indicators by rational design based on the integration of data from statistical and structural analyses, computational design, and experimental methods to alter protein sequences and/or synthesize new constructs incorporating multiple proteins or peptides [117–120]. The rational design of calcium sensors surmounts the limitations associated with coupled calcium-binding sites in natural CaBPs with relatively slow responses to calcium dynamics.

One example is the green Ca^{2+} indicator, "G-CatchER," which was created by incorporating a Ca^{2+} binding site into enhanced GFP (EGFP). A novel Ca^{2+}-binding site was created by site-directed mutagenesis of residues S147E, S202D, Q204E, F223E, and T225E near the chromophore and near residues H148, T203, and E222 to form a hand-like site that may be responsible for its fast kinetic properties [119]. Introducing acidic residues near the chromophore altered the electrostatics, changing the chromophore state from anionic to neutral. These modifications resulted in a new absorption maximum at 398 nm, but at the expense of signal loss at 490-nm. Binding of Ca^{2+} to G-CatchER then produced an increased absorbance at 490 nm and decreased absorbance at 398 nm. The G-CatchER:Ca^{2+} stoichiometry was observed to form a 1:1 complex, which allowed the sensor to monitor Ca^{2+} dynamics in high Ca^{2+} concentration cellular compartments like the ER. G-CatchER exhibited strong binding affinity, with an apparent K_d of 0.18 ± 0.02 mM and an off-rate estimated at approximately 700 s^{-1} [121].

Three additional mutations, S175G, S30R and Y39N, were later incorporated into G-CatchER to improve intensity and thermostability at 37°C. The new variant, G-CatchER$^+$, exhibited a K_d value for Ca^{2+} of 1.2 ± 0.2 mM. G-CatchER$^+$ was specifically capable of reporting rapid local ER Ca^{2+} dynamics, with increased fluorescence lifetime and enhanced Ca^{2+}-dependent fluorescence. Calcium association and release values were calculated as K_{obs} = 62 s^{-1}, and K_{obs} = 63 s^{-1}. G-CatchER$^+$ was further able to monitor agonist/antagonist triggered Ca^{2+} dynamics orchestrated by IP_3Rs, RyRs, and SERCAs in multiple cell types, including primary neurons that are orchestrated by IP_3Rs, RyRs, and SERCAs, as well as in *Drosophila* [118]. Upon localization of G-CatchER$^+$ to the lumen of the RyR channel (G-CatchER$^+$-JP45), we report a rapid local Ca^{2+} release that is likely due to ER membrane-associated calsequestrin. Transgenic expression of G-CatchER$^+$ in *Drosophila* muscle demonstrates its utility as an *in vivo* reporter of stimulus-evoked SR local Ca^{2+} dynamics.

The Yang lab also recently reported a general strategy for the rational design of GECIs with improved signal intensity and fast kinetics. This design strategy involved the inclusion of a single Ca^{2+}-binding site, and residue modification to modulate the local geometry and electrostatics near the chromophore to enhance protein dynamics. Another Ca^{2+} indicator, R-CatchER, was developed based on the RFP, mApple. The new sensor R-CatchER exhibited superior *in vitro* kinetics in multiple cell types and was able to monitor spatiotemporal ER Ca^{2+} dynamics in neurons and dendritic branchpoints. R-CatchER was able both to achieve the first reporting of ER Ca^{2+} oscillations mediated by CaSR, and to reveal Ca^{2+}-based functional cooperativity of CaSR [117].

4 STRUCTURE AND FUNCTION OF CaSR

Our understanding of the essential physiological process of extracellular signaling mediated by CaSR is greatly hampered by the lack of knowledge of key molecular determinants in the ECD involved with the regulation and functional cooperativity of CaSR-mediated calcium responses. There are also critical challenges associated with membrane proteins, glycosylation, and weak binding processes of calcium-binding sites.

A 2007 study by Huang et al. reported a modeled structure of the ligand-binding domain of the extracellular region that was created based on the X-ray structures of mGluR1 (PDB IDs 1EWT and 1ISR), which share 27% sequence homology with the CaSR ECD [67]. This modeled structure has a VFT scaffold that is common in other family C of GPCRs [42, 122, 123]. Based on results derived from Ca^{2+}-binding prediction algorithms [65, 124–126], five Ca^{2+}-binding sites were predicted in each CaSR ECD monomer as follows: Site 1 (S147, S170, D190, Y218, and E297) and Site 2 (D215, L242, S244, D248, and Q253) located at the hinge region; Site 3 (E224, E228, E229, E231, and E232) located at the dimer interface of the lobe 2; Site 4 (E350, E353, E354, N386, and S388); and Site 5 (E378, E379, T396, D398, and E399) at the long flexible loop [66, 67, 127]. Predicted Ca^{2+}-binding sites 1, 3, and 5 were subsequently grafted into a non-CaBP CD2 scaffold with a flexible linker to determine their intrinsic Ca^{2+}-binding affinities [67]. Sub-domains representing wild-type or mutated Ca^{2+}-binding sites were also generated to study functional cooperativity using multiple techniques, including Tb-FRET binding and Ca^{2+} competition assays, Trp fluorescence, and 1D^{1}H NMR [66]. Results of these studies suggested that Predicted Site 1 likely plays an important role in positive cooperativity between multiple CaSR ECD Ca^{2+}-binding sites [66].

In 2016, Yang group reported the first high-resolution X-ray structure that included a Mg^{2+}-bound form of native CaSR ECD dimer (1–540) at pH 7.0 and at a resolution of 2.1 Å (PDB ID 5FBK) and a Gd^{3+}-loaded form at 2.7 Å (PDB ID 5FBH) [70]. The two putative Mg^{2+}/Ca^{2+} hinge binding site 1 (D190,S170, Y218, and E297), and hinge site 2 (D216, D275, and S272) of the CaSR ECD were reported, and are very similar to predicted sites 1–2 reported by Huang [66] (Figure 6). Also, a single metal-binding site was identified at the homodimer interface of lobe 2, comprised of negatively charged residues E228, E231, S240, and E241. Soaking the crystal in the presence of Gd^{3+} resulted in a rearrangement of coordination ligands (E228, E229, and E232) that moved the binding site geometry closer to Huang's original Predicted Site 3. Moreover, CaSR activity could be significantly reduced following mutations E228I/E229I, suggesting the important role of calcium binding to this highly negatively charged dimer interface (acidic patch) [70]. A Mg^{2+}/Ca^{2+} new binding site (LB1.site1) composed of backbone carbonyl groups (I81, S84, L87, and L88), was revealed at lobe 1 of the ECD. Consistent with earlier docking and biochemical studies, a Trp derivative, L-1,2,3,4-tetrahydronorharman-3-carboxylic acid (TNCA), as a co-agonist was revealed and located at the CaSR hinge adjacent to the hinge metal-binding site 1 to contribute functional cooperativity of the receptor [128]. Later, Liu et al. reported use of an innovative cell-free Förster resonance energy transfer (FRET)-based conformational CaSR bio-sensor to further validate the important role of calcium binding to these two reported putative Ca^{2+}-binding sites at the hinge region ((S147, S170, D190, Y218, and E297) and (D216, D275, and S272)), which appeared to be important for stabilizing active conformation of the receptor [129].

A subsequent study by Geng et al. reported structure determination of both apo- and holo- forms of the CaSR ECD (PDB IDs 5K5S and 5K5T) [130]. These structures included mutations in glycosylation sites and the Cys-rich domain under different pH conditions and different concentrations of Ca^{2+}, phosphate, and sulfate [130].

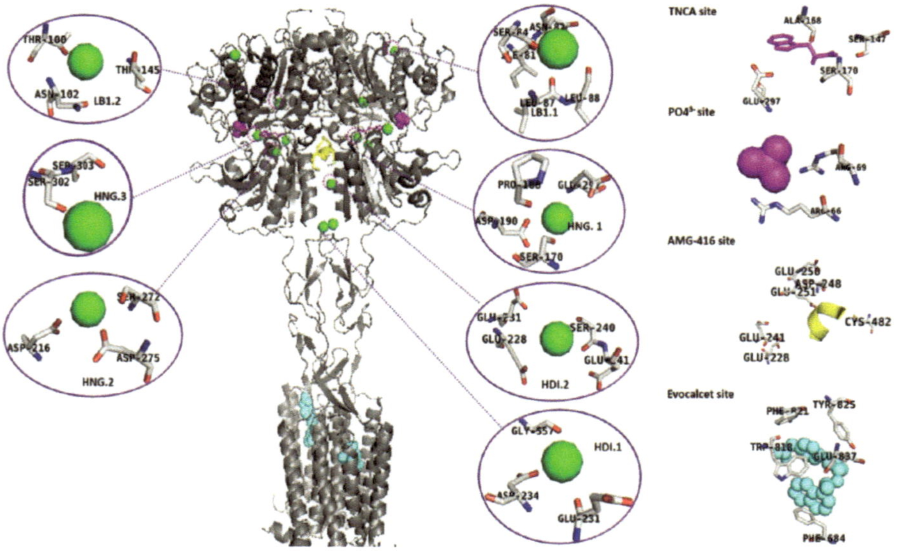

FIGURE 6 Modeled CaSR structure with summary of all reported Ca^{2+}-binding sites, TNCA/Trp-binding site, anion-binding site, and PAM/NAM-binding sites. Green spheres represent Ca^{2+}. L-Trp is indicated by magenta sticks. PO_4^{3-} ions are shown in magenta spheres. AMG-416 are indicated as blue ribbon, and evocalcet are shown as cyan spheres.

Comparison of VFT domain structures between the apo and holo forms found them to be similar (RMSD = 4.4 Å). The holo-ECD structure (PDB ID 5K5S) with mutations (N386Q and S402N, and/or N468Q) was determined at 2.6 Å resolution in 1.6 M NaH_2PO_4, 0.4 M K_2HPO_4, 100 mM Na_2HPO_4/citric acid, 10 mM $CaCl_2$, and 10 mM L-Trp at pH 4.2. Four Ca^{2+} ions were identified in each monomer. The apo-ECD structure (PDB ID 5K5T) was determined at 3.1 Å in 1.5 M Li_2SO_4, 100 mM Tris, 2 mM $CaCl_2$ at pH 8.5, with only one Ca^{2+} ion in each monomer. The new site located at lobe 1 identified by Zhang et al. was also identified in the 5K5T structure, and later, by Park et al. (PDB ID 7SIL) and Gao et al. (PDB ID 7M3G) [131, 132]. A metal-binding site (coordinated by D234, E231, and G557) in the same dimer interface of lobe was also reported but located at a slightly lower position compared to the dimer site reported by Zhang. However, Ca^{2+}-binding sites located at lobe 1 (T100, N102, and T145) and the hinge region (S302 and S303) were reported in the 5K5T structure.

Cryo-EM has also emerged as a potentially helpful tool to study the role of Ca^{2+} in CaSR, and recent advances have led to a rapid increase in the number of experimental protein structures. In 2021 and early 2022, 16 experimental structures of CaSR, including both the ECD and TM domains, were published by five different labs, bringing the total number of experimental structures available for study from four up to 20 (Table 2). This has made it possible to corroborate several earlier predictions about calcium binding in the ECD.

There are several main calcium-binding sites in the CaSR ECD which have been reported in studies from multiple labs (Figure 6). VFT Lobe 1 site 1 (LB1.1) comprises residues I81, S84, L87, and L88 (backbone O in all cases) and is occupied by calcium in six structures (5K5S, 7M3E, 7M3F, 7M3G, 7SIL, and 7SIM) [130–132] and by magnesium in two structures that were purified without calcium (5FBH, 5FBK) [70]. Hinge sites 1 and 2 were first revealed by Zhang et al. [70] and later experimentally verified [66, 70, 129]. Hinge site 1, with the addition of residue P188 (backbone O), was also reported by Chen et al. from the Geng group, in the cryo-electron microscopy structures of full-length CaSR (PDB ID 7E6T), while Chen et al reported a Ca^{2+}-binding site in the same location comprising residues S170, D190, E297, Y489, and P188 [129] (Figure 6).

Homodimer interface site (HDI.1), which falls between the two protomers just where LB2 of the VFT meets the CRD, is comprised of residues D234 (atom OD1 and/or OD2), and G557 (in *H. sapiens*, G556 in *G. gallus*) of the other protomer (backbone O). This site is occupied by calcium in 12 models with Cys-rich domain (5K5S, 7DD5, 7DD6, 7DD7, 7DTT, 7DTV, 7E6T, 7M3E, 7M3F, 7M3G, 7SIL, and 7SIM) [130–135].

This site partially overlaps with the homodimer interface site 2 (HDI.2) at residue S240 (atoms O and OG), occupied by magnesium in the early pair of structures without Cys-rich domain (5FBH, 5FBK) [70]. Soaking the crystal in the presence of Gd^{3+} resulted in a rearrangement of coordination ligands (E228, E229, and E232) due to a highly negatively charged acidic patch.

VFT Lobe 1 site 2 (LB1.2) is comprised of residues T100 and T145 (atom OG1), and A144 (backbone O), and is occupied by calcium in eight structures (5K5S, 5K5T, 7DD5, 7DD6, 7DD7, 7E6T, 7SIL, and 7SIM) [130–132]. This site shows density for an ion in two other structures where it was identified as chloride (5FBH, 5FBK). Hinge region site 3 (HNG.3) is comprised of S302 (atom OG), near but not binding

TABLE 2
Binding Sites in CaSR Structures

Site Type	Site Location	Key Residues	Ligand in Site	References
Ca^{2+} actual or putative	ECD lobe 1, top site 1	**I81, S84, L87, N82, L88**	Mg^{2+}/Ca^{2+}	Zhang et al. (2016), *Science Advances*, 2(5), e600241.
			Ca^{2+}	Geng et al. (2016), *Elife*, 5, e13662.
			Ca^{2+}	Park et al. (2021), *Proceedings of the National Academy of Sciences*, 118(51).
			Ca^{2+}	Gao et al. (2021), *Nature*, 595(7867), 455–459.
	ECD lobe 1 site 2	T100, N102, T145	Cl^-	Zhang et al. (2016), *Science Advances*, 2(5), e600241.
			Ca^{2+}	Geng et al. (2016), *Elife*, 5, e13662.
			Ca^{2+}	Park et al. (2021), *Proceedings of the National Academy of Sciences*, 118(51).
			Ca^{2+}	Chen et al. (2021), *eLife*, 10, e68578
			Ca^{2+}	Wen et al. (2021), *Science Advances*, 7(23), eagb1483.
	Hinge region site 1	D190, S170, E297, **P188**	$Mg^{2+}/Ca^{2+}/H_2O$	Zhang et al. (2016), *Science Advances*, 2(5), e600241.
			Ca^{2+}	Chen et al. (2021), *eLife*, 10, e68578
	Hinge region site 2	D216, D275, S272	$Mg^{2+}/Ca^{2+}/H_2O$	Zhang et al. (2016), *Science Advances*, 2(5), e600241.
	Hinge region site 3	S303, S302	Cl^-	Zhang et al. (2016), *Science Advances*, 2(5), e600241.
			Ca^{2+}	Geng et al. (2016), *eLife*, 5, e13662
			Ca^{2+}	Park et al. (2021), *Proceedings of the National Academy of Sciences*, 118(51).
			Ca^{2+}	Wen et al. (2021), *Science Advances*, 7(23), eagb1483.
			Ca^{2+}	Ling et al. (2021), *Cell Research*, 31(4), 383–394.
	Homodimer interface site 1 (Slightly lower than site 2)	D234, **E231**, G557	Ca^{2+}	Geng et al. (2016), *eLife*, 5, e13662.
			Ca^{2+}	Park et al. (2021), *Proceedings of the National Academy of Sciences*, 118(51).
			Ca^{2+}	Chen et al. (2021), *eLife*, 10, e68578
			Ca^{2+}	Gao et al. (2021), *Nature*, 595(7867), 455–459.
			Ca^{2+}	Wen et al. (2021), *Science Advances*, 7(23), eagb1483.
			Ca^{2+}	Ling et al. (2021), *Cell Research*, 31(4), 383–394.

(Continued)

TABLE 2 (*Continued*)
Binding Sites in CaSR Structures

Site Type	Site Location	Key Residues	Ligand in Site	References
	Homodimer interface site2	E228, **E231**, S240, E241, (E228, E229, E232)	Mg^{2+}/Ca^{2+} (Gd^{3+})	Zhang et al. (2016), *Science Advances*, 2(5), e600241.
TNCA/L-Trp	Hinge region	S147, S170, E297, A168	TNCA	Zhang et al. (2016), *Science Advances*, 2(5), e600241.
			L-Trp	Gent et al. (2016), *eLife*, 5, e13662.
			TNCA	Park et al. (2021), *Proceedings of the National Academy of Sciences*, 118(51).
			TNCA	Chen et al. (2021), *eLife*, 10, e68578
			L-Trp	Gao et al. (2021), *Nature*, 595(7867), 455–459.
			L-Trp	Wen et al. (2021), *Science Advances*, 7(23), eagb1483.
			L-Trp	Ling et al. (2021), *Cell Research*, 31(4), 383–394.
Anion sites	Hinge region, near TNCA/L-trp-binding site	R66, R69	CO_2^{2-}	Zhang et al. (2016), *Science Advances*, 2(5), e600241.
			PO_4^{3-}, SO_4^{3-}	Geng et al. (2016), *eLife*, 5, e13662.
			PO_4^{3-}	Park et al. (2021), *Proceedings of the National Academy of Sciences*, 118(51).
			PO_4^{3-}	Chen et al. (2021), *eLife*, 10, e68578
			PO_4^{3-}	Gao et al. (2021), *Nature*, 595(7867), 455–459.
			Cl^-	Wen et al. (2021), *Science Advances*, 7(23), eagb1483.
NAM	7 TM	Y825, F821, W818, F684, E837	NSP-2143	Park et al. (2021), *Proceedings of the National Academy of Sciences*, 118(51).
			NSP-2143	Gao et al. (2021), *Nature*, 595(7867), 455–459.
			NSP-2143	Wen et al. (2021), *Science Advances*, 7(23), eagb1483.
PAM			Evocalcet, Cinacalcet	Gao et al. (2021), *Nature*, 595(7867), 455–459.
			Evocalcet	Wen et al. (2021), *Science Advances*, 7(23), eagb1483.
	Lobe 2 interface	E251, C482, E228, E241, D248, E250	Etecalcetide (AMG-416)	Gao et al. (2021), *Nature*, 595(7867), 455–459.

S303 and G273, and is occupied by calcium in eight structures (5K5S, 7DD5, 7DD6, 7DD7, 7DTT, 7DTV, 7SIL, and 7SIM) [130–132, 135]. This site shows density for a small particle in one additional structure where it was interpreted as chloride (5FBK) [70] or water due to a lack of negatively charged ligand residues (5FBH).

A third hinge region site (HNG.3), consisting of residues R66 and R69 (atoms NH1, NH2, and/or NE), shows density for a ligand in five models. Given the positive charge of the ligating atoms, calcium in this location is unlikely.

4.1 Mechanistic Insights into Heterotropic Cooperativity Orchestrated by Allosteric Modulator Binding at ECD

Research suggests a role for L-amino acid on Ca^{2+} activation of CaSR by heterotropic cooperativity. Consistent with that, an L-Phe binding pocket was identified within the hinge region of CaSR adjacent to the ECD predicted Ca^{2+}-binding site using structure modeling, molecular docking, and functional assays. A study by Zhang et al. using saturation transfer difference NMR and purified CaSR ECD reported the determination of the binding affinity of L-Phe to CaSR ECD, which was enhanced in the presence of Ca^{2+} [69]. It is notable that multiple disease-related mutations are located either within or in close proximity to both the Ca^{2+} and amino acid-binding sites, which may interfere with Ca^{2+} binding and CaSR activation by disrupting cooperativity [68, 69, 136]. These studies led to a proposed mechanism involving co-activation of the CaSR by both Ca^{2+} and the amino acid at the ECD hinge region [68, 69].

Additionally, a tryptophan derivative, TNCA, was also identified at the hinge region in the 5FBK crystal structure bound by S147, S170, D190, Y218, and E297 (Figure 6). Conversely, Geng et al. proposed that this same region in the 5K5S structure was occupied by L-Trp, rather than TNCA. They further concluded that L-Trp binding was essential for receptor activation, but that Ca^{2+} binding played a secondary and non-activating role. This conclusion was inconsistent with previous functional studies [66–70] that suggested Ca^{2+} is the principal agonist, whereas L-Trp, other L-amino acids, and TNCA are co-agonists, facilitating Ca^{2+} binding but lacking the capability to activate CaSR by themselves. Liu et al. similarly suggested that Ca^{2+} alone can fully stabilize the active conformation of CaSR, while the L-amino acids function as positive allosteric modulators [129].

Interestingly in 2021, five publications validated this amino acid-binding site in their Cryo-EM structures of CaSR: three of these indicated occupation of L-Trp, while the other two showed occupation of TNCA, including Chen et al. from Geng's group [131–135]. Analysis of their sample preparation conditions strongly suggests that both compounds can bind to the same location, and occupation is dependent on the compound (TNCA or L-Trp) and its concentration.

There is also evidence that CaSR is modulated by γ-glutamyl peptides. These share the same pocket as L-Phe/TNCA based on observations that mutations T145A and S170T can diminish CaSR activation in the presence of both γ-glutamyl and L-Phe [137]. Moreover, results of structural modeling studies further suggested that glutathione and γ-glutamyl peptides likely bind at the same hinge region of the CaSR ECD [41, 68].

A study by Jiang et al. using a Ca^{2+}-binding site prediction algorithm and the mGluR1 structure identified a Ca^{2+}-binding site adjacent to the L-Glu-binding site in the hinge region of the mGluR1 ECD domain, which was analogous to that observed within the hinge region of CaSR ECD [138]. Results of this study suggested that either Ca^{2+} or L-Glu could individually activate the receptor, but that binding of both ligands together produced a synergistic effect with increased activation [138].

Similarly, a recent crystal structure of taste receptor type 1 revealed a binding site for different amino acids in the cleft between LB1 and LB2 using residues S142 and S165, which are conserved in cGPCRs [139]. Binding of agonists in these hinge regions results in ligand-induced closing motions, which appear to be critical to the function of CaSR, T1Rs, mGluR type1 (mGluR1), and $GABA_B$ receptors [140, 141]. For the $GABA_B$ receptor, agonist binding appeared to produce motion between the open and closed conformation only with GBR1b but not with GBR2.

4.2 Negative Regulation by Anion Binding and pH Effect

A strong sensitivity to pH in CaSR has also been reported. CaSR can be activated by an increase in pH in the physiological pH range 6.0–8.0, whereas CaSR is inactivated following a decrease in pH. Interestingly, CaSR becomes more sensitive to both Ca^{2+} and agonists at $pH<5.5$ [142]. Additionally, small changes occur in extracellular pH from 7.4 ± 0.2 without corresponding intracellular pH changes, and can rapidly inhibit or activate the receptor, which is associated with the production of parathyroid hormone PTH [143]. Similar phenomena have been observed in mGluRs that exhibited subtype-dependent pH sensitivity toward glutamate-induced activation, where mGluR4 responded to pH-dependent agonist activation, while mGluR1, mGluR5, and mGluR8 did not [144]. Similar behavior toward protons was also reported in a recent work by Campion et al. [143]. Results of this study refuted speculation that the observed nonlinear pH sensitivity was due to the presence of His residues in the CaSR ECD, because pH sensitivity was not altered following mutations of the His residues [143].

A bicarbonate-binding site was identified by Zhang et al. in close proximity to both the TNCA site and the metal-binding sites in the hinge region of the crystal structure of CaSR ECD (PDB ID 5FBK). The electron density of the bound ligand observed in this structure at 2.1 Å was tentatively identified as a bicarbonate anion due to its flat triangular shape, and because of its location in a positively charged pocket consisting of conserved residues R66, R69, W70, R415, I416, and S417. A sequence alignment of CaSR from different species indicated that these residues – R66, R69, W70, R415, I416, and S417 – are highly conserved across species except for fish, which is likely because they occupy an environment with different pH and salinity conditions. Abnormal changes in bicarbonate levels (normally 22–29 mM) are known to play an important role in cardiovascular damage and chronic kidney disease by changing serum $[Ca^{2+}]$, which negatively impairs the glomerular filtration rate [145]. Interestingly, the L-type Ca^{2+} channel also contains a carboxylate cluster that is responsible for Ca^{2+} selectivity and can sense pH changes [146]. Due to the pH sensitivity, CaSR should have different EC_{50} values for Ca^{2+} activation arising at different environmental pH values in the organs where it is located. It is possible that CaSR function may be altered by pH effects resulting from diet. Results of one study suggested that this pH modulation effect can be reduced following mutations of glutamate residues [142].

It is interesting to note that the bicarbonate-binding site identified by Zhang et al. in the 5FBH structure was instead identified as a phosphate-binding site in the 5K5S structure in the presence of very high concentrations of phosphate during crystallization [130]. Additionally, a later publication by Wen et al. identified chloride in this site, whereas the other Cryo-EM CaSR structures from the same year indicated the binding of phosphate ions [131–134]. Parathyroid mineral metabolism can be affected by a very narrow concentration range of PO_4^{3-} (~0.8–1.5 mM) coupled with the regulation of fibroblast growth factor 23 (FGF23) [147, 148]. FGF23 and CaSR are both associated with Ca^{2+}, Mg^{2+}, and phosphate homeostasis, and are related to dysfunction in FHH, ADH, and chronic kidney disease [149, 150]. Recent work by Centeno et al. directly characterized the mechanism of phosphate-induced PTH secretion. Additionally, mutation R62A was found to abolish inhibition of CaSR function by phosphate, although this effect may result from the impairment of the salt bridge between the upper and lower lobes [151].

4.3 Structural Implications for Drug Development and Regulation

A new CaSR agonist, AMG-416, was proposed to bind to the hinge region of CaSR ECD through the formation of a mixed disulfide bond between the agonist and C482 of the CaSR ECD. Gao et al. [132] revealed two identical copies of AMG-416 bound to C482 of each promoter in the 2.5Å structure of active-state CaSR (PDB:7M3G, Figure 6 and Table 2). The binding site is located at the ECD lobe2 interface, and comprises negatively charged residues [132]. However, C482 appears to play a non-essential role in normal CaSR function [152]. Zhang et al. identified three disulfide bonds in each of the two monomers of CaSR ECD [36]. The inter-monomer disulfide bonds between C129 and C131 in loop 2 were not identified because the loop was not present in one of the monomers. Because the loop 2 regions from the two monomers are anti-parallel, and due to the proximity of the two Cys residues, Zhang suggested that two inter-monomer disulfide bonds were formed between C129(A) and C131(B). Subsequent mutation studies indicated that mutation of either C129 or C131 or both failed to completely inhibit dimerization, indicating that the inter-monomer disulfide bond was not the only contributor to dimerization. Additionally, salt bridges between E456-R54, R172-D215, and R227-E231 in the CaSR ECD were identified as essential for agonist-induced homodimerization of CaSR. Similar interactions have also been reported in the T1R [139]. However, recent studies have reported increasing heterodimerization of CaSR with $GABA_{B1}R$ in patients with primary or secondary hyperparathyroidism. This dimerization resulted from hydrophobic interaction between the CaSR and $GABA_{B1}R$ monomer rather than through disulfide bond formation, suggesting a new mechanism for drug development to prevent PTH secretion [153].

Multiple attempts have been made to uncover molecular mechanisms of positive and negative allosteric modulation, specifically targeting the 7TM. Leach et al. combined site-directed mutagenesis and Ca^{2+} mobilization assays to elucidate the shared pocket at 7TM for NPS R568, cinacalcet, and AC265347, regardless of minor differences in binding residues and cooperativity [154]. Interestingly, they also identified a potential Ca^{2+}-binding site that overlaps with the allosteric binding site in the 7TM [154, 155]. However, they later determined that negative allosteric modulators share both overlapping and distinct binding sites at 7TM [156]. Furthermore, Bräuner-Osborn et al. reported that in order to completely block inhibition, two allosteric sites need to be prevented from binding negative allosteric modulators. As long as one allosteric site is bound per CaSR dimer by allosteric modulators, it is sufficient for achieving a positive allosteric effect [157]. With the utilization of advanced cryo-electron microscopy technology in protein structure determination, negative allosteric modulator NSP-2143 was identified in a binding site (Y825, F821, W 818, F684, and E837) within the 7TM domain of several CaSR structures (PDB: 7M3E, 7SIN, and 7DD5). Importantly, positive allosteric modulators like evocalcet and cinacalcet have also been shown to bind at the same site (Figure 6) [131, 132, 134].

We examined the conformation of the VFT domains in each protomer of all experimental CaSR structures currently available in the PDB (39 VFT domains total) (Figure 7). Models were superimposed on the dimer interface portion of LB1 (residues 101–188). Pairwise RMSD-Cα for the superimposed region was in all cases less than 1 Å (mean = 0.414 Å, std = 0.163 Å), except in comparisons with structure 7E6U (RMSD-Cα between residues 101–188 of 7E6U and all other protomers: mean = 4.372 Å, std = 0.378 Å). 7E6U has a sizeable protein ligand between the protomers at the VFT domain that significantly affects the entire ECD conformation. The RMSD-Cα of LB2 residues 189–382 was calculated for every combination of VFT domains after superimposing that pair on LB1 residues 101–188. Thirty-three of protomers show the VFT in a consistent closed conformation between LB1 and LB2 (max pairwise RMSD-Cα of LB2 mobile region 1.87 Å, mean = 0.88 Å, std = 0.34 Å). The other six protomers (5K5T chain A, 7E6U chains A and C, 7MJ3 chain B, 7SIN chains A and B) have a greater degree of opening between LB1 and LB2 (Figure 7).

The open VFT domains are distinctly different from the closed conformation (minimum RMSD-Cα of mobile region between an open and a closed conformation 6.64 Å) and more varied among themselves (mean RMSD-Cα of mobile region between two open VFTs 3.99 Å when 7E6U is excluded, 7.46 Å when 7E6U is included). In structure 7M3J, chain A shows the typical closed conformation, while chain B is open (RMSD-Cα of the mobile portion 8.02 Å between chains A and B). This asymmetrical arrangement has been described as "open-closed" [132]. In all other structures, the VFT domains of the two protomers were not significantly different (max pairwise RMSD-Cα of mobile portion between protomers in a single structure 1.27 Å).

4.4 CaSR Induced Ca^{2+} Oscillation and Interactome

CaSR activation triggers intracellular Ca^{2+} oscillation, which was determined to be regulated by protein kinase C (PKC) and amino acid in early 2000 [40]. In a similar study, it was reported that activation of PKC inhibited extracellular Ca^{2+}-elicited increases in induced Ca^{2+} oscillations as a result of phosphorylation of T888 in CaSR, indicating that PKC negatively modulates the coupling of the CaR to intracellular signaling systems [158]. A later study by Young et al. reported that human CaSR in HEK-293 cells responded to the addition of both L-Trp and

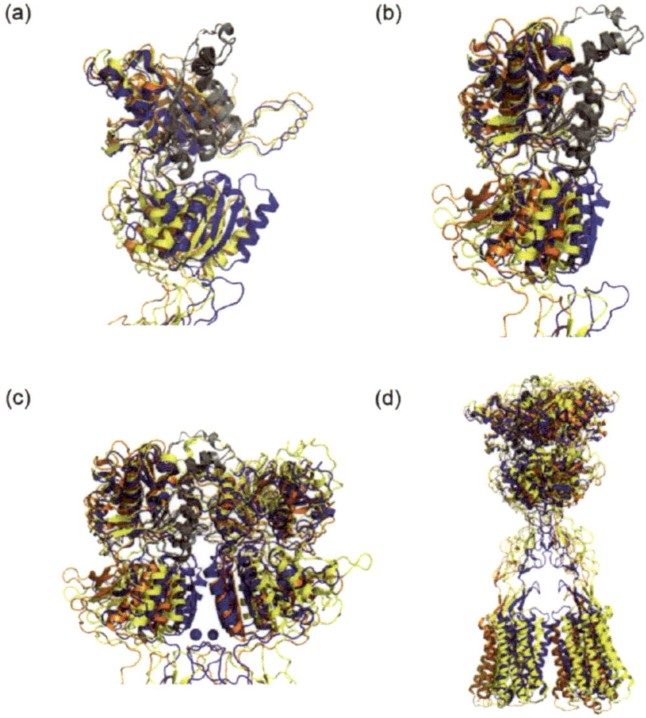

FIGURE 7 CaSR closed and open conformations of ECD and transmembrane regions. (a) VFT domain of PDB structures 7SIL (blue), 7SIN (yellow), and 7M3J (orange) superimposed at residues 101–188 of LB1 chain B (gray). 7SIL (blue) shows typical closed VFT conformation. LB2 helices (bottom lobe) are rotated upward, closer to LB1 (top lobe). 7SIN (yellow) and 7M3J (orange) show varying degrees of openness. (b) Panel A rotated so VFT dimer interface (residues 101–188, gray) is to the side. LB2 in closed conformation (blue) is both raised closer to LB1 (Panel A) and rotated closer to VFT dimer interface (Panel B). (c) Chains B (left) and A (right). LB2 helices in dimer interface are brought together in closed conformation (7SIL, blue), but not open (7SIN, yellow) or open-closed (7M3J, orange, open VFT on left and closed VFT on right). (d) CRD and TMD. Helix TM6 of chains A and B are brought into contact in closed position (7SIL, blue), but not in open (7SIN, yellow) or open-closed (7M3J, orange, open VFT on left and closed VFT on right).

L-Phe, in addition to extracellular $Ca2^{2+}$, with significant induced Ca^{2+} oscillations [159]. However, extracellular Ca^{2+}-induced oscillations occurred at a much higher frequency (~4 min^{-1}) than that produced by amino acid stimulation (~1 min^{-1}). Additionally, induced Ca^{2+} oscillations exhibited a sinusoidal pattern, but oscillations induced by the aromatic amino acids exhibited a transient pattern. These results indicated the patterns and frequencies of Ca^{2+} oscillations for both agonist types were bound to the CaSR, but the different patterns also provided a means to differentiate between the two [159].

It is known that CaSR-mediated crosstalk between extra- and intracellular Ca^{2+} signaling is critically important to protein biosynthesis and trafficking [47, 160]. Following prolonged exposure to extracellular Ca^{2+}, nascent CaSR from the ER [161], Golgi, and ER-Golgi intermediate compartments (ERGIC) are transported to the plasma membrane [47]. CaSR can then be transported for degradation in the lysosome or proteasome [47]. Physiological disorders capable of altering subcellular Ca^{2+}-mediated signaling are associated with disruption of protein biosynthesis and trafficking, such as in the ER [62]. Although research has established the role of CaSR as a regulator of systemic calcium (Ca^{2+}) homeostasis, the mechanisms through which extracellular Ca^{2+} and CaSR mediate networks of intracellular Ca^{2+}-signaling remain unclear. Recently, Gorkhali et al. reported the first CaSR protein-protein interactome with 94 novel putative and eight previously published interactors using proteomics [162]. The study further reported that extracellular Ca^{2+} has been found to promote enrichment in 66% of the identified CaSR interactors involving functions related to Ca^{2+} dynamics, endocytosis, degradation, trafficking, and protein processing in the endoplasmic reticulum (ER). These enhanced ER-related processes are governed by extracellular Ca^{2+}-activated CaSR and subsequent modulation of Ca^{2+} in the ER.

The first direct observations of ER Ca^{2+} oscillations were acquired using a recently developed novel ER-targeted Ca^{2+}-sensor, R-CatchER [117]. R-CatchER evolved to address the limitations of G-CatchER, a rationally designed fluorescent Ca^{2+} sensor created by inserting a single Ca^{2+}-binding site directly into EGFP to alter the electrostatic environment near the chromophore in response to binding of Ca^{2+} [121, 124]. G-CatchER was subsequently based on the mApple scaffold, an RFP modified by site-directed mutagenesis to create an enhanced Ca^{2+}-binding site on the surface of the protein. R-CatchER was able to capture spatiotemporal ER Ca^{2+} dynamics in neurons and hotspots at dendritic branchpoints, thereby providing the first confirmation of CaSR mediation of ER Ca^{2+} oscillations, and further revealing ER Ca^{2+} -based functional cooperativity of CaSR [45, 117]. R-CatchER also enabled detection of extracellular Ca^{2+} and agonists, such as L-Phe and TNCA, and cooperative tuning of ER Ca^{2+} oscillations mediated by CaSR. Moreover, this enabled for the first time the opportunity to observe how disease mutations can alter ER Ca^{2+} responses, oscillation frequency, and cooperativity. R-CatchER is expected to be invaluable in further providing a molecular mechanism for CaSR mediation of Ca^{2+} signaling and extending our capability to visualize Ca^{2+} dynamics at various subcellular compartments, cell types, and organs, for applications in drug discovery related to ER dysfunction and Ca^{2+} mishandling.

The role of CaSR in extracellular mediated Ca^{2+} interactions has also been explored with regulatory proteins vesicle-associated membrane protein-associated A (VAPA) (involved in trafficking from the ER to the Golgi), and 78-kDa glucose-regulated protein (GRP78/Bip/HSPA5), involved in ER protein processing [162]. Results of live cell imaging have also suggested that CaSR and VAPA are inter-dependent during Ca^{2+}-induced enhancement of near-cell membrane expression. Analysis of the CaSR interactome has further identified new potential biomolecules and signaling pathways that should further our understanding of the complex role of CaSR in Ca^{2+}_{ER} dynamics, as well as agonist mediated ER protein processing.

5 CONCLUSIONS

Our knowledge about calcium-binding sites in proteins has increased largely due to the rapid development of structural biology tools, such as Cryo-EM. The exciting computational capability in predicting protein structures largely extends our knowledge of protein and molecular recognition, especially for CaSR, which plays major roles in the integration of extracellular signaling to intracellular signaling and calcium dynamics. Further determination of the complex form of G protein with CaSR will provide a comprehensive view of regulation of CaSR-mediated signaling. Improved visualization of the role of calcium and calcium-signaling pathways will also require further development of prediction algorithms to visualize various types of calcium-binding sites in complex proteins and "invisible and dynamic conformational ensembles." Moreover, further development of calcium sensors with desired optical, metal binding, and kinetic properties will largely facilitate rapid progress in capturing calcium dynamics in many biological and pathological processes, integration of calcium signals, and development of therapeutics against human diseases.

ACKNOWLEDGMENTS

We would like to express our sincere appreciation to Dr. Edward Brown, Jian Hu, Donald Hamelberg, Kelley Moremen, and Guangtao Chen for their years of insightful guidance and collaborative research with our laboratory. We also want to acknowledge contributions to research cited in this work from previous lab members and collaborators. We thank Shen Tang, You Zhuo, Cassandra L. Miller, Florence Reddish, and Rakshya Gorkhali for their helpful discussion and contribution. This work was supported in part by an AHA grant to J.J.Y., CDT fellowship to X.D., and B&B fellowship to L.T.

ABBREVIATIONS AND DEFINITIONS

AD	Alzheimer's disease
ADH	autosomal dominant hypocalcemia
CaBP	calcium-binding protein
CaM	calmodulin
CaSR (CSR)	calcium sensing receptor
CN	coordination number
CNX	calnexin
CPVT	catecholaminergic polymorphic ventricular tachycardia
CRT	calreticulin
DHPR	dihydropyridine receptor
ECD	extracellular domain
EGFP	enhanced green fluorescent protein
EMDB	Electron Microscopy Data Bank
EOM	extraocular muscle
ER	endoplasmic reticulum
FHH	familial hypocalciuric hypercalcemia
FRET	Forster resonance energy transfer

GABA	γ-aminobutyric acid receptors
GECI	genetically encoded Ca^{2+} indicator
GFP	green fluorescent protein
GPCR	G-protein coupled receptors
HD	Huntington's disease
IP3R	inositol trisphosphate receptor
MCU	mitochondrial calcium uniporter
mGluR	metabotropic glutamate receptors
MH	malignant hyperthermia
MLCK	myosin light-chain kinase
NMR	Nuclear Magnetic Resonance
NSHPT	neonatal severe hyperparathyroidism
PD	Parkinson's disease
PDB	Protein Data Bank
PKC	protein kinase C
PTH	parathyroid hormone
RFP	red fluorescent protein
RMSD	root-mean-square deviation
RyR	ryanodine receptor
SERCA	sarco/endoplasmic reticulum Ca^{2+}-ATPase
SR	sarcoplasmic reticulum
TMD	transmembrane domain
TnC	troponin C
TNCA	L-1,2,3,4-tetrahydronorharman-3-carboxylic acid
TRP	transient receptor potential
VFT	Venus flytrap domain
VGCC	voltage-gated Ca^{2+} channel

REFERENCES

1. D. E. Clapham, *Cell* **2007**, *131* (6), 1047–1058.
2. D. K. Atchison, W. H. Beierwaltes, *Pflugers Arch.* **2013**, *465* (1), 59–69.
3. S. Smajilovic, J. Tfelt-Hansen, *Cardiovasc. Res.* **2007**, *75* (3), 457–467.
4. S. Tang, F. Reddish, Y. Zhuo, J. J. Yang, *Curr. Opin. Chem. Biol.* **2015**, *27*, 90–97.
5. E. Polakova, A. Zahradnikova, Jr., J. Pavelkova, I. Zahradnik, A. Zahradnikova, *J. Physiol.* **2008**, *586* (16), 3839–3854.
6. Y. Zhou, S. Xue, J. J. Yang, *Metallomics* **2013**, *5* (1), 29–42.
7. N. A. Beard, L. Wei, A. F. Dulhunty, *Clin. Exp. Pharmacol. Physiol.* **2009**, *36* (3), 340–345.
8. O. Kann, R. Kovacs, *Am. J. Physiol. Cell Physiol.* **2007**, *292* (2), C641–C657.
9. M. R. Duchen, *J. Physiol.* **2000**, *529* (Pt 1), 57–68.
10. Y. Komori, M. Tanaka, M. Kuba, M. Ishii, M. Abe, N. Kitamura, A. Verkhratsky, I. Shibuya, G. Dayanithi, *Cell Calcium* **2010**, *48* (6), 324–332.
11. C. Martin, K. E. Chapman, J. R. Seckl, R. H. Ashley, *Neuroscience* **1998**, *85* (1), 205–216.
12. A. Sonnleitner, A. Conti, F. Bertocchini, H. Schindler, V. J. T. E. J. Sorrentino, *EMBO J.* **1998**, *17* (10), 2790–2798.

13. M. Sekulic-Jablanovic, A. Palmowski-Wolfe, F. Zorzato, S. Treves, *Biochem. J.* **2015**, *466* (1), 29–36.
14. S. O. Marx, S. Reiken, Y. Hisamatsu, T. Jayaraman, D. Burkhoff, N. Rosemblit, A. R. Marks, *Cell* **2000**, *101* (4), 365–376.
15. S. S. Awad, H. K. Lamb, J. M. Morgan, W. Dunlop, J. I. Gillespie, *Biochem. J.* **1997**, *322* (Pt 3), 777–783.
16. R. Robinson, D. Carpenter, M. A. Shaw, J. Halsall, P. Hopkins, *Hum. Mutat.* **2006**, *27* (10), 977–989.
17. A. Bartok, D. Weaver, T. Golenar, Z. Nichtova, M. Katona, S. Bansaghi, K. J. Alzayady, V. K. Thomas, H. Ando, K. Mikoshiba, S. K. Joseph, D. I. Yule, G. Csordas, G. Hajnoczky, *Nat. Commun.* **2019**, *10* (1), 3726.
18. M. R. Duchen, *Pflugers Arch.* **2012**, *464* (1), 111–121.
19. K. Mikoshiba, *Adv. Biol. Regul.* **2015**, *57*, 217–227.
20. H. Ivanova, T. Vervliet, L. Missiaen, J.B. Parys, H. De Smedt, G. Bultynck, *Biochim. Biophys. Acta.* **2014**, *1843* (10), 2164–2183.
21. M. Tada, M. Nishizawa, O. Onodera, *Neurochem. Int.* **2016**, *94*, 1–8.
22. S. Gerber, K. J. Alzayady, L. Burglen, D. Bremond-Gignac, V. Marchesin, O. Roche, M. Rio, B. Funalot, R. Calmon, A. Durr, V. L. Gil-da-Silva-Lopes, M. F. Ribeiro Bittar, C. Orssaud, B. Heron, E. Ayoub, P. Berquin, N. Bahi-Buisson, C. Bole, C. Masson, A. Munnich, M. Simons, M. Delous, H. Dollfus, N. Boddaert, S. Lyonnet, J. Kaplan, P. Calvas, D. I. Yule, J. M. Rozet, L. Fares Taie, *Am J Hum. Genet.* **2016**, *98* (5), 971–980.
23. M. McEntagart, K. A. Williamson, J. K. Rainger, A. Wheeler, A. Seawright, E. De Baere, H. Verdin, L. T. Bergendahl, A. Quigley, J. Rainger, A. Dixit, A. Sarkar, E. Lopez Laso, R. Sanchez-Carpintero, J. Barrio, P. Bitoun, T. Prescott, R. Riise, S. McKee, J. Cook, L. McKie, B. Ceulemans, F. Meire, I. K. Temple, F. Prieur, J. Williams, P. Clouston, A. H. Nemeth, S. Banka, H. Bengani, M. Handley, E. Freyer, A. Ross, D. D. D. Study, V. van Heyningen, J. A. Marsh, F. Elmslie, D. R. FitzPatrick, *Am. J. Hum. Genet.* **2016**, *98* (5), 981–992.
24. I. Bezprozvanny, M. R. Hayden, *Biochem. Biophys. Res. Commun.* **2004**, *322* (4), 1310–1317.
25. T. Cali, D. Ottolini, M. Brini, *Cell Tissue Res.* **2014**, *357* (2), 439–454.
26. S. Magi, P. Castaldo, M.L. Macri, M. Maiolino, A. Matteucci, G. Bastioli, S. Gratteri, S. Amoroso, V. Lariccia, *Biomed. Res. Int.* **2016**, *2016*, 6701324.
27. T. Glaser, V. F. Arnaud Sampaio, C. Lameu, H. Ulrich, *Semin. Cell Dev. Biol.* **2019**, *95*, 25–33.
28. M. Periasamy, A. Kalyanasundaram, *Muscle Nerve* **2007**, *35* (4), 430–442.
29. A. Zarain-Herzberg, D. H. MacLennan, M. Periasamy, *J. Biol. Chem.* **1990**, *265* (8), 4670–4677.
30. A. Hovnanian, *Subcell. Biochem.* **2007**, *45*, 337–363.
31. F. Wuytack, B. Papp, H. Verboomen, L. Raeymaekers, L. Dode, R. Bobe, J. Enouf, S. Bokkala, K. S. Authi, R. Casteels, *J. Biol. Chem.* **1994**, *269* (2), 1410–1416.
32. E. R. Chemaly, L. Troncone, D. Lebeche, *Cell Calcium* **2018**, *69*, 46–61.
33. Y. Liao, Y. Dong, J. Cheng, *Int. J. Mol. Sci.* **2017**, *18*(2), 248–263.
34. A. M. Hofer, E. M. Brown, *Nat. Rev. Mol. Cell Biol.* **2003**, *4* (7), 530–538.
35. E. M. Brown, G. Gamba, D. Riccardi, M. Lombardi, R. Butters, O. Kifor, A. Sun, M. A. Hediger, J. Lytton, S. C. Hebert, *Nature* **1993**, *366* (6455), 575–580.
36. C. Zhang, C. L. Miller, R. Gorkhali, J. Zou, K. Huang, E. M. Brown, J. J. Yang, *Front. Physiol.* **2016**, *7*, 441.
37. M. Uhlen, L. Fagerberg, B. M. Hallstrom, C. Lindskog, P. Oksvold, A. Mardinoglu, A. Sivertsson, C. Kampf, E. Sjostedt, A. Asplund, I. Olsson, K. Edlund, E. Lundberg, S. Navani, C. A. Szigyarto, J. Odeberg, D. Djureinovic, J. O. Takanen, S. Hober, T. Alm,

P. H. Edqvist, H. Berling, H. Tegel, J. Mulder, J. Rockberg, P. Nilsson, J. M. Schwenk, M. Hamsten, K. von Feilitzen, M. Forsberg, L. Persson, F. Johansson, M. Zwahlen, G. von Heijne, J. Nielsen, F. Ponten, *Science* **2015**, *347* (6220), 1260419.

38. X. Zhao, B. Schindell, W. Li, L. Ni, S. Liu, C. U. B. Wijerathne, J. Gong, C. M. Nyachoti, K. O, C. Yang, *J. Anim. Sci.* **2019**, *97* (6), 2402–2413.
39. G. E. Breitwieser, S. U. Miedlich, M. Zhang, *Cell Calcium* **2004**, *35* (3), 209–216.
40. A. D. Conigrave, S. J. Quinn, E. M. Brown, *Proc. Natl. Acad. Sci. U. S. A.* **2000**, *97* (9), 4814–4819.
41. M. Wang, Y. Yao, D. Kuang, D. R. Hampson, *J. Biol. Chem.* **2006**, *281* (13), 8864–8870.
42. M. Bai, *Cell Calcium* **2004**, *35* (3), 197–207.
43. W. Chang, D. Shoback, *Cell Calcium* **2004**, *35* (3), 183–196.
44. A. D. Conigrave, H. C. Lok, *Clin. Exp. Pharmacol. Physiol.* **2004**, *31* (5–6), 368–371.
45. X. Deng, Y. Xin, C. L. Miller, D. Hamelberg, M. Kirberger, K.W. Moremen, J. Hu, J. J. Yang, *Curr. Opin. Physiol.* **2020**, *17*, 269–277.
46. M. Peacock, *Clin. J. Am. Soc. Nephrol.* **2010**, *5* (Suppl 1), S23–S30.
47. G. E. Breitwieser, *Mol. Endocrinol.* **2012**, *26* (9), 1482–1495.
48. R. V. Thakker, *Cell Calcium* **2004**, *35* (3), 275–282.
49. Y. Kinoshita, M. Hori, M. Taguchi, S. Watanabe, S. Fukumoto, *J. Clin. Endocrinol. Metab.* **2014**, *99* (2), E363–E368.
50. E. Carafoli, J. Krebs, *J. Biol. Chem.* **2016**, *291* (40), 20849–20857.
51. R. H. Kretsinger, C. E. Nockolds, *J. Biol. Chem.* **1973**, *248* (9), 3313–3326.
52. N. C. Strynadka, M. N. James, *Annu. Rev. Biochem.* **1989**, *58*, 951–998.
53. M. R. Nelson, W. J. Chazin, *The EF-Hand Calcium-Binding Proteins Data Library,* The Scripps Research Institute. 1997.
54. W. Yang, H. W. Lee, H. Hellinga, J. J. Yang, *Proteins* **2002**, *47* (3), 344–356.
55. Y. Huang, Y. Zhou, H. C. Wong, A. Castiblanco, Y. Chen, E. M. Brown, J. J. Yang, *J. Biol. Chem.* **2010**, *285* (46), 35919–35931.
56. K. L. Yap, J. Kim, K. Truong, M. Sherman, T. Yuan, M. Ikura, *J. Struct. Funct. Genom.* **2000**, *1* (1), 8–14.
57. H. Kawasaki, S. Nakayama, R. H. Kretsinger, *Biometals* **1998**, *11* (4), 277–295.
58. Y. Zhou, W. Yang, M. Kirberger, H. W. Lee, G. Ayalasomayajula, J. J. Yang, *Proteins* **2006**, *65* (3), 643–655.
59. E. Pidcock, G. R. Moore, *J. Biol. Inorg. Chem.* **2001**, *6*, 479–489.
60. M. Kirberger, X. Wang, H. Deng, W. Yang, G. Chen, J. J. Yang, *J. Biol. Inorg. Chem* **2008,** *13* (7), 1169–81.
61. Kirberger M., Yang J.J. (2013) Calcium-Binding Protein Site Types. In: Kretsinger R.H., Uversky V.N., Permyakov E.A. (eds) *Encyclopedia of Metalloproteins.* Springer, New York, NY. https://doi.org/10.1007/978-1-4614-1533-6_35
62. M. Brini, D. Ottolini, T. Cali, E. Carafoli, *Met. Ions Life Sci.* **2013**, *13*, 81–137.
63. X. Wang, M. Kirberger, F. Qiu, G. Chen, J. J. Yang, *Proteins* **2008**, *75* (4), 787–798.
64. X. Wang, K. Zhao, M. Kirberger, H. Wong, G. Chen, J. Yang, *Protein Sci.* **2010,** *19* (6), 1180–90.
65. K. Zhao, X. Wang, H. C. Wong, R. Wohlhueter, M. P. Kirberger, G. Chen, J. J. Yang, *Proteins* **2012**, *80* (12), 2666–2679.
66. Y. Huang, Y. Zhou, A. Castiblanco, W. Yang, E. M. Brown, J. J. Yang, *Biochemistry* **2009**, *48* (2), 388–398.
67. Y. Huang, Y. Zhou, W. Yang, R. Butters, H. W. Lee, S. Li, A. Castiblanco, E. M. Brown, J. J. Yang, *J. Biol. Chem.* **2007**, *282* (26), 19000–19010.
68. C. Zhang, N. Mulpuri, F. M. Hannan, M. A. Nesbit, R. V. Thakker, D. Hamelberg, E. M. Brown, J. J. Yang, *PLoS One* **2014**, *9* (11), e113622.
69. C. Zhang, Y. Zhuo, H. A. Moniz, S. Wang, K. W. Moremen, J. H. Prestegard, E. M. Brown, J. J. Yang, *J. Biol. Chem.* **2014**, *289* (48), 33529–33542.

70. C. Zhang, T. Zhang, J. Zou, C. L. Miller, R. Gorkhali, J. Y. Yang, A. Schilmiller, S. Wang, K. Huang, E. M. Brown, K. W. Moremen, J. Hu, J. J. Yang, *Sci. Adv.* **2016**, *2* (5), e1600241.
71. E. Callaway, *Nature* **2020**, *578* (7794), 201.
72. H. M. Berman, J. Westbrook, Z. Feng, G. Gilliland, T. N. Bhat, H. Weissig, I. N. Shindyalov, P. E. Bourne, *Nucleic Acids Res.* **2000**, *28* (1), 235–242.
73. V. Bauerova-Hlinkova, D. Hajduchova, J. A. Bauer, *Molecules* **2020**, *25* (18), 4040.
74. C. Andrews, Y. Xu, M. Kirberger, J. J. Yang, *Int. J. Mol. Sci.* **2020**, *22* (1) 308.
75. Y. Ye, H. W. Lee, W. Yang, S. Shealy, J. J. Yang, *J. Am. Chem. Soc.* **2005**, *127* (11), 3743–3750.
76. J. V. Gerasimenko, O. H. Petersen, O. V. Gerasimenko, *Membr. Transp. Signal* **2014**, *3* (3), 63–71.
77. A. M. Hofer, I. Schulz, *Cell Calcium* **1996**, *20* (3), 235–242.
78. A. P. Ziman, C. W. Ward, G. G. Rodney, W. J. Lederer, R. J. Bloch, *Biophys. J.* **2010**, *99* (8), 2705–2714.
79. R. M. Paredes, J. C. Etzler, L. T. Watts, W. Zheng, J. D. Lechleiter, *Methods* **2008**, *46* (3), 143–151.
80. B. Raju, E. Murphy, L. A. Levy, R. D. Hall, R. E. London, *Am. J. Physiol. Cell Physiol.* **1989**, *256* (3), C540–C548.
81. D. M. Chudakov, M. V. Matz, S. Lukyanov, K. A. Lukyanov, *Physiol. Rev.* **2010**, *90* (3), 1103–1163.
82. B. N. Giepmans, S. R. Adams, M. H. Ellisman, R. Y. Tsien, *Science* **2006**, *312* (5771), 217–224.
83. J. Lippincott-Schwartz, G. H. Patterson, *Science* **2003**, *300* (5616), 87–91.
84. R. Rizzuto, A. W. Simpson, M. Brini, T. Pozzan, *Nature* **1992**, *358* (6384), 325–327.
85. A. M. Hofer, B. Landolfi, L. Debellis, T. Pozzan, S. Curci, *EMBO J.* **1998**, *17* (7), 1986–1995.
86. A. Persechini, J. A. Lynch, V. A. Romoser, *Cell Calcium* **1997**, *22* (3), 209–216.
87. A. Miyawaki, J. Llopis, R. Heim, J. M. McCaffery, J. A. Adams, M. Ikura, R. Y. Tsien, *Nature* **1997**, *388* (6645), 882–887.
88. R. Heim, A. B. Cubitt, R. Y. Tsien, *Nature* **1995**, *373* (6516), 663–664.
89. A. E. Palmer, R. Y. Tsien, *Nat. Protoc.* **2006**, *1* (3), 1057–1065.
90. N. Heim, O. Griesbeck, *J. Biol. Chem.* **2004**, *279* (14), 14280–14286.
91. J. Nakai, M. Ohkura, K. Imoto, *Nat. Biotechnol.* **2001**, *19* (2), 137–141.
92. Y. N. Tallini, M. Ohkura, B. R. Choi, G. Ji, K. Imoto, R. Doran, J. Lee, P. Plan, J. Wilson, H. B. Xin, A. Sanbe, J. Gulick, J. Mathai, J. Robbins, G. Salama, J. Nakai, M. I. Kotlikoff, *Proc. Natl. Acad. Sci. U. S. A.* **2006**, *103* (12), 4753–4758.
93. L. Tian, S. A. Hires, T. Mao, D. Huber, M. E. Chiappe, S. H. Chalasani, L. Petreanu, J. Akerboom, S. A. McKinney, E. R. Schreiter, C. I. Bargmann, V. Jayaraman, K. Svoboda, L. L. Looger, *Nat. Methods* **2009**, *6* (12), 875–881.
94. J. Akerboom, T. W. Chen, T. J. Wardill, L. Tian, J. S. Marvin, S. Mutlu, N. C. Calderon, F. Esposti, B. G. Borghuis, X. R. Sun, A. Gordus, M. B. Orger, R. Portugues, F. Engert, J. J. Macklin, A. Filosa, A. Aggarwal, R. A. Kerr, R. Takagi, S. Kracun, E. Shigetomi, B. S. Khakh, H. Baier, L. Lagnado, S. S. Wang, C. I. Bargmann, B. E. Kimmel, V. Jayaraman, K. Svoboda, D. S. Kim, E. R. Schreiter, L. L. Looger, *J. Neurosci.* **2012**, *32* (40), 13819–13840.
95. T. W. Chen, T. J. Wardill, Y. Sun, S. R. Pulver, S. L. Renninger, A. Baohan, E. R. Schreiter, R. A. Kerr, M. B. Orger, V. Jayaraman, L. L. Looger, K. Svoboda, D. S. Kim, *Nature* **2013**, *499* (7458), 295–300.

96. J. H. Cho, C. J. Swanson, J. Chen, A. Li, L. G. Lippert, S. E. Boye, K. Rose, S. Sivaramakrishnan, C. M. Chuong, R. H. Chow, *ACS Chem. Biol.* **2017**, *12* (4), 1066–1074.
97. J. de Juan-Sanz, G. T. Holt, E. R. Schreiter, F. de Juan, D. S. Kim, T. A. Ryan, *Neuron* **2017**, *93* (4), 867–881.
98. Y. Zhao, S. Araki, J. Wu, T. Teramoto, Y. F. Chang, M. Nakano, A. S. Abdelfattah, M. Fujiwara, T. Ishihara, T. Nagai, R. E. Campbell, *Science* **2011**, *333* (6051), 1888–1891.
99. M. J. Henderson, H. A. Baldwin, C. A. Werley, S. Boccardo, L. R. Whitaker, X. Yan, G. T. Holt, E. R. Schreiter, L. L. Looger, A. E. Cohen, D. S. Kim, B. K. Harvey, *PLoS One* **2015**, *10* (10), e0139273.
100. J. Suzuki, K. Kanemaru, K. Ishii, M. Ohkura, Y. Okubo, M. Iino, *Nat. Commun.* **2014**, *5*, 4153.
101. N. Helassa, X. H. Zhang, I. Conte, J. Scaringi, E. Esposito, J. Bradley, T. Carter, D. Ogden, M. Morad, K. Torok, *Sci. Rep.* **2015**, *5*, 15978.
102. M. Ohkura, T. Sasaki, J. Sadakari, K. Gengyo-Ando, Y. Kagawa-Nagamura, C. Kobayashi, Y. Ikegaya, J. Nakai, *PLoS One* **2012**, *7* (12), e51286.
103. A. Muto, M. Ohkura, T. Kotani, S. Higashijima, J. Nakai, K. Kawakami, *Proc. Natl. Acad. Sci. U. S. A.* **2011**, *108* (13), 5425–5430.
104. J. Wu, L. Liu, T. Matsuda, Y. Zhao, A. Rebane, M. Drobizhev, Y. F. Chang, S. Araki, Y. Arai, K. March, T. E. Hughes, K. Sagou, T. Miyata, T. Nagai, W. H. Li, R. E. Campbell, *ACS Chem. Neurosci.* **2013**, *4* (6), 963–972.
105. H. Dana, Y. Sun, B. Mohar, B. K. Hulse, A. M. Kerlin, J. P. Hasseman, G. Tsegaye, A. Tsang, A. Wong, R. Patel, J. J. Macklin, Y. Chen, A. Konnerth, V. Jayaraman, L. L. Looger, E. R. Schreiter, K. Svoboda, D. S. Kim, *Nat. Methods* **2019**, *16* (7), 649–657.
106. A. A. Shemetov, M. V. Monakhov, Q. Zhang, J. E. Canton-Josh, M. Kumar, M. Chen, M. E. Matlashov, X. Li, W. Yang, L. Nie, D. M. Shcherbakova, Y. Kozorovitskiy, J. Yao, N. Ji, V. V. Verkhusha, *Nat. Biotechnol.* **2020**, *39*, 368–377.
107. Y. Qian, K. D. Piatkevich, B. Mc Larney, A. S. Abdelfattah, S. Mehta, M. H. Murdock, S. Gottschalk, R. S. Molina, W. Zhang, Y. Chen, J. Wu, M. Drobizhev, T. E. Hughes, J. Zhang, E. R. Schreiter, S. Shoham, D. Razansky, E. S. Boyden, R. E. Campbell, *Nat. Methods* **2019**. *16* (2), 171–174.
108. T. Hendel, M. Mank, B. Schnell, O. Griesbeck, A. Borst, D. F. Reiff, *J. Neurosci.* **2008**, *28* (29), 7399–7411.
109. A. Badura, X. R. Sun, A. Giovannucci, L. A. Lynch, S. S. Wang, *Neurophotonics* **2014**, *1* (2), 025008.
110. S. Kerruth, C. Coates, C. D. Durst, T. G. Oertner, K. Torok, *J. Biol. Chem.* **2019**, *294* (11), 3934–3946.
111. M. E. Witte, A. M. Schumacher, C. F. Mahler, J. P. Bewersdorf, J. Lehmitz, A. Scheiter, P. Sanchez, P. R. Williams, O. Griesbeck, R. Naumann, T. Misgeld, M. Kerschensteiner, *Neuron* **2019**, *101* (4), 615–624.
112. M. Mank, D. F. Reiff, N. Heim, M. W. Friedrich, A. Borst, O. Griesbeck, *Biophys. J.* **2006**, *90* (5), 1790–1796.
113. T. Thestrup, J. Litzlbauer, I. Bartholomaus, M. Mues, L. Russo, H. Dana, Y. Kovalchuk, Y. Liang, G. Kalamakis, Y. Laukat, S. Becker, G. Witte, A. Geiger, T. Allen, L. C. Rome, T. W. Chen, D. S. Kim, O. Garaschuk, C. Griesinger, O. Griesbeck, *Nat. Methods* **2014**, *11* (2), 175–182.
114. B. F. Fosque, Y. Sun, H. Dana, C. T. Yang, T. Ohyama, M. R. Tadross, R. Patel, M. Zlatic, D. S. Kim, M. B. Ahrens, V. Jayaraman, L. L. Looger, E. R. Schreiter, *Science* **2015**, *347* (6223), 755–760.

115. T. A. Zolnik, F. Sha, F. W. Johenning, E. R. Schreiter, L. L. Looger, M. E. Larkum, R. N. Sachdev, *J. Physiol.* **2017**, *595* (5), 1465–1477.
116. K. Horikawa, Y. Yamada, T. Matsuda, K. Kobayashi, M. Hashimoto, T. Matsu-ura, A. Miyawaki, T. Michikawa, K. Mikoshiba, T. Nagai, *Nat. Methods* **2010**, *7* (9), 729–732.
117. X. Deng, X. Q. Yao, K. Berglund, B. Dong, D. Ouedraogo, M. A. Ghane, Y. Zhuo, C. McBean, Z. Z. Wei, S. Gozem, S. P. Yu, L. Wei, N. Fang, A. M. Mabb, G. Gadda, D. Hamelberg, J. J. Yang, *Angew. Chem. Int. Ed. Engl.* **2021**, *60* (43), 23289–23298.
118. F. N. Reddish, C. L. Miller, X. Deng, B. Dong, A. A. Patel, M. A. Ghane, B. Mosca, C. McBean, S. Wu, K. M. Solntsev, Y. Zhuo, G. Gadda, N. Fang, D. N. Cox, A. M. Mabb, S. Treves, F. Zorzato, J. J. Yang, *iScience* **2021**, *24* (3), 102129.
119. Y. Zhang, F. Reddish, S. Tang, Y. Zhuo, Y. F. Wang, J. J. Yang, I. T. Weber, *Acta Crystallogr. D Biol. Crystallogr.* **2013**, *69* (Pt 12), 2309–2319.
120. J. Zou, A. M. Hofer, M. M. Lurtz, G. Gadda, A. L. Ellis, N. Chen, Y. Huang, A. Holder, Y. Ye, C. F. Louis, K. Welshhans, V. Rehder, J. J. Yang, *Biochemistry* **2007**, *46* (43), 12275–12288.
121. S. Tang, H. C. Wong, Z. M. Wang, Y. Huang, J. Zou, Y. Zhuo, A. Pennati, G. Gadda, O. Delbono, J. J. Yang, *Proc. Natl. Acad. Sci. U. S. A.* **2011**, *108* (39), 16265–16270.
122. J. Hu, A. M. Spiegel, *J. Cell Mol. Med.* **2007**, *11* (5), 908–922.
123. C. Silve, C. Petrel, C. Leroy, H. Bruel, E. Mallet, D. Rognan, M. Ruat, *J. Biol. Chem.* **2005**, *280* (45), 37917–37923.
124. H. Deng, G. Chen, W. Yang, J. J. Yang, *Proteins* **2006**, *64* (1), 34–42.
125. X. Wang, M. Kirberger, F. Qiu, G. Chen, J. J. Yang, *Proteins* **2009**, *75* (4), 787–798.
126. X. Wang, K. Zhao, M. Kirberger, H. Wong, G. Chen, J. J. Yang, *Protein Sci.* **2010**, *19* (6), 1180–1190.
127. Y. Kubo, T. Miyashita, Y. Murata, *Science* **1998**, *279* (5357), 1722–1725.
128. C. Zhang, Y. Huang, Y. Jiang, N. Mulpuri, L. Wei, D. Hamelberg, E. M. Brown, J. J. Yang, *J. Biol. Chem.* **2014**, *289* (8), 5296–5309.
129. H. Liu, P. Yi, W. Zhao, Y. Wu, F. Acher, J. P. Pin, J. Liu, P. Rondard, *Proc. Natl. Acad. Sci. U. S. A.* **2020**, *117* (35), 21711–21722.
130. Yong Geng, Lidia Mosyak, Igor Kurinov, Hao Zuo, Emmanuel Sturchler, Tat Cheung Cheng, Prakash Subramanyam, Alice P Brown, Sarah C Brennan, Hee-chang Mun, Martin Bush, Yan Chen, Trang X Nguyen, Baohua Cao, Donald D Chang, Matthias Quick, Arthur D Conigrave, Henry M Colecraft, Patricia McDonald, Qing R Fan (2016) Structural mechanism of ligand activation in human calcium-sensing receptor eLife 5:e13662 https://doi.org/10.7554/eLife.13662
131. J. Park, H. Zuo, A. Frangaj, Z. Fu, L. Y. Yen, Z. Zhang, L. Mosyak, V. N. Slavkovich, J. Liu, K. M. Ray, B. Cao, F. Vallese, Y. Geng, S. Chen, R. Grassucci, V. P. Dandey, Y. Z. Tan, E. Eng, Y. Lee, B. Kloss, Z. Liu, W. A. Hendrickson, C. S. Potter, B. Carragher, J. Graziano, A. D. Conigrave, J. Frank, O. B. Clarke, Q. R. Fan, *Proc. Natl. Acad. Sci. U. S. A.* **2021**, *118* (51).
132. Y. Gao, M. J. Robertson, S. N. Rahman, A. B. Seven, C. Zhang, J. G. Meyerowitz, O. Panova, F. M. Hannan, R. V. Thakker, H. Brauner-Osborne, J. M. Mathiesen, G. Skiniotis, *Nature* **2021**, *595* (7867), 455–459.
133. X. Chen, L. Wang, Q. Cui, Z. Ding, L. Han, Y. Kou, W. Zhang, H. Wang, X. Jia, M. Dai, Z. Shi, Y. Li, X. Li, Y. Geng, *eLife* **2021**, *10*, e68578.
134. T. Wen, Z. Wang, X. Chen, Y. Ren, X. Lu, Y. Xing, J. Lu, S. Chang, X. Zhang, Y. Shen, X. Yang, *Sci. Adv.* **2021**, *7* (23), eabg1483.

135. S. Ling, P. Shi, S. Liu, X. Meng, Y. Zhou, W. Sun, S. Chang, X. Zhang, L. Zhang, C. Shi, D. Sun, L. Liu, C. Tian, *Cell Res.* **2021**, *31* (4), 383–394.
136. F. M. Hannan, M. A. Nesbit, C. Zhang, T. Cranston, A.J. Curley, B. Harding, C. Fratter, N. Rust, P. T. Christie, J. J. Turner, M. C. Lemos, M. R. Bowl, R. Bouillon, C. Brain, N. Bridges, C. Burren, J.M. Connell, H. Jung, E. Marks, D. McCredie, Z. Mughal, C. Rodda, S. Tollefsen, E. M. Brown, J. J. Yang, R. V. Thakker, *Hum. Mol. Genet.* **2012**, *21* (12), 2768–2778.
137. G. K. Broadhead, H. C. Mun, V. A. Avlani, O. Jourdon, W. B. Church, A. Christopoulos, L. Delbridge, A. D. Conigrave, *J. Biol. Chem.* **2011**, *286* (11), 8786–8797.
138. Y. Jiang, Y. Huang, H. C. Wong, Y. Zhou, X. Wang, J. Yang, R. A. Hall, E. M. Brown, J. J. Yang, *J. Biol. Chem.* **2010**, *285* (43), 33463–33474.
139. N. Nuemket, N. Yasui, Y. Kusakabe, Y. Nomura, N. Atsumi, S. Akiyama, E. Nango, Y. Kato, M. K. Kaneko, J. Takagi, M. Hosotani, A. Yamashita, *Nat. Commun.* **2017**, *8*, 15530.
140. Y. Geng, M. Bush, L. Mosyak, F. Wang, Q. R. Fan, *Nature* **2013**, *504* (7479), 254–259.
141. N. Kunishima, Y. Shimada, Y. Tsuji, T. Sato, M. Yamamoto, T. Kumasaka, S. Nakanishi, H. Jingami, K. Morikawa, *Nature* **2000**, *407* (6807), 971–977.
142. S. J. Quinn, M. Bai, E. M. Brown, *J. Biol. Chem.* **2004**, *279* (36), 37241–37249.
143. K. L. Campion, W. D. McCormick, J. Warwicker, M. E. Khayat, R. Atkinson-Dell, M. C. Steward, L. W. Delbridge, H. C. Mun, A. D. Conigrave, D. T. Ward, *J. Am. Soc. Nephrol.* **2015**, *26* (9), 2163–2171.
144. C. Levinthal, L. Barkdull, P. Jacobson, L. Storjohann, B. C. Van Wagenen, T. M. Stormann, L. G. Hammerland, *Pharmacology* **2009**, *83* (2), 88–94.
145. C. Voiculet, O. Zara, C. Bogeanu, I. Vacaroiu, G. Aron, *J. Med. Life* **2016**, *9* (4), 449–454.
146. B. Prod'hom, D. Pietrobon, P. Hess, *J. Gen. Physiol.* **1989**, *94* (1), 23–42.
147. M. G. Penido, U. S. Alon, *Pediatr. Nephrol.* **2012**, *27* (11), 2039–2048.
148. M. Mizobuchi, T. Suzuki, *Clin. Calcium* **2012**, *22* (10), 1543–1549.
149. R. Tyler Miller, *Best Pract. Res. Clin. Endocrinol. Metab.* **2013**, *27* (3), 345–358.
150. V. N. Babinsky, F. M. Hannan, S. C. Youhanna, C. Marechal, M. Jadoul, O. Devuyst, R. V. Thakker, *PLoS One* **2015**, *10* (3), e0119459.
151. P. P. Centeno, A. Herberger, H. C. Mun, C. Tu, E. F. Nemeth, W. Chang, A. D. Conigrave, D. T. Ward, *Nat. Commun.* **2019**, *10* (1), 4693.
152. S. T. Alexander, T. Hunter, S. Walter, J. Dong, D. Maclean, A. Baruch, R. Subramanian, J. E. Tomlinson, *Mol. Pharmacol.* **2015**, *88* (5), 853–865.
153. W. H. Chang, C. L. Tu, F. G. Jean-Alphonse, A. Herberger, Z. Q. Cheng, J. Hwong, H. Ho, A. Li, D. W. Wang, H. D. Liu, A. D. White, I. Suh, W. Shen, Q. Y. Duh, E. Khanafshar, D. M. Shoback, K. H. Xiao, J. P. Vilardaga, *Nat. Metab.* **2020**, *2* (3), 243–+.
154. K. Leach, K. J. Gregory, I. Kufareva, E. Khajehali, A. E. Cook, R. Abagyan, A. D. Conigrave, P. M. Sexton, A. Christopoulos, *Cell Res.* **2016**, *26* (5), 574–592.
155. A. N. Keller, I. Kufareva, T. M. Josephs, J. Diao, V. T. Mai, A. D. Conigrave, A. Christopoulos, K. J. Gregory, K. Leach, *Mol. Pharmacol.* **2018**, *93* (6), 619–630.
156. T. M. Josephs, A. N. Keller, E. Khajehali, A. DeBono, C. J. Langmead, A. D. Conigrave, B. Capuano, I. Kufareva, K. J. Gregory, K. Leach, *Br J Pharmacol*, **2020**, *177* (8), 1917–1930.
157. S. E. Jacobsen, U. Gether, H. Brauner-Osborne, *Sci. Rep.* **2017**, *7*, 46355.
158. M. Bai, S. Trivedi, C. R. Lane, Y. Yang, S. J. Quinn, E. M. Brown, *J. Biol. Chem.* **1998**, *273* (33), 21267–21275.

159. S. H. Young, E. Rozengurt, *Am. J. Physiol. Cell Physiol.* **2002**, *282* (6), C1414–C1422.
160. G. E. Breitwieser, *Best Pract. Res. Clin. Endocrinol. Metab.* **2013**, *27* (3), 303–313.
161. S. Pidasheva, M. Grant, L. Canaff, O. Ercan, U. Kumar, G. N. Hendy, *Hum. Mol. Genet.* **2006**, *15* (14), 2200–2209.
162. R. Gorkhali, L. Tian, B. Dong, P. Bagchi, X. Deng, S. Pawar, D. Duong, N. Fang, N. Seyfried, J. Yang, *Sci. Rep.* **2021**, *11* (1), 20576.

4 Fluorescent Bio-Sensors for Manganese(II) and Iron(II)

Jennifer Park, Jiansong Xu*, and Joseph A. Cotruvo, Jr.†*
Department of Chemistry, The Pennsylvania State University, University Park, PA 16802, USA

CONTENTS

* These authors contributed equally to this work.
† Corresponding author.

DOI: 10.1201/9781003229971-4

Abstract

The ubiquitous biological roles of manganese and iron motivate the development of chemical biology tools to interrogate the concentrations, localization, and dynamics of these metal ions within cells in real time. However, because Mn(II) and Fe(II) are the lowest metal ions in the Irving–Williams series and tend to bind weakly to both biological and synthetic molecules alike, designing molecules that are able to selectively bind them to enable their detection and quantification is an intrinsically difficult chemical problem. Here we review small-molecule synthetic and genetically encoded fluorescent sensors for these metals. While recent years have brought some success in selective detection of Fe(II), particularly in terms of reaction-based synthetic probes and riboswitch-based genetically encoded sensors, selective Mn(II) detection continues to be elusive. We discuss promising future directions for addressing these challenges.

KEYWORDS

Fluorophores; Genetically Encoded Sensors; Irving-Williams Series; Labile Metal Pools; Metal Selectivity; Reaction-Based Probes; Recognition-Based Sensors

1 THE IMPORTANCE AND CHALLENGES OF SELECTIVE DETECTION OF Mn(II) AND Fe(II)

Nearly all organisms require iron for essential cellular functions; the few that do not require iron require manganese instead [1]. Iron is necessary for respiration, electron transfer, O_2 activation, Lewis acid catalysis, and many other important cellular processes. In contrast to the widespread essentiality of iron in enzymatic chemistry, iron can also be toxic because of the reaction of Fe(II) with O_2 to form destructive reactive oxygen species [2]. This reactivity is implicated in a recently discovered form of iron-dependent cell death (ferroptosis) [3], which exhibits links to cancer [4] and neurodegeneration [5]. Restriction of iron availability has been the traditional focus of investigations into the immune response to invading pathogens [6], but recent evidence has suggested that iron overload may also be encountered in many bacteria, including human pathogens [7–9]. The physiological causes and necessary concentrations for iron overload are not completely understood, but appear to be linked to oxidative stress under aerobic conditions [10], and may be important under anaerobic conditions as well [11,12].

In prokaryotes, manganese assumes increased importance under iron-limited conditions and oxidative stress, conditions encountered by pathogens invading a host [13–17]. Conversely, the regulation and trafficking of manganese in humans is not well understood. However, manganese levels are known to be developmentally regulated during neuronal differentiation [18], mouse models of Huntington's disease are manganese-depleted [19,20], and hyperaccumulation of manganese (manganism) is associated with parkinsonian neurological effects [21–23]. Recent identification of Mn(II) transporters associated with inherited neuropathies [24–28] has further increased interest in human manganese homeostasis [29, 30].

As a result, methods to probe the pools of Fe(II) and Mn(II), within both prokaryotic and eukaryotic cells and in real time, are needed. Such approaches can help make sense of the mechanisms associated with the normal and disease physiologies mentioned above, including intracellular metal concentration, localization, and dynamics, and identification and characterization of proteins and small molecules involved in metal uptake, trafficking, and efflux. Analyte-selective fluorescent probes (in which analyte recognition leads to an irreversible reaction that leads to a fluorescence response: "reaction-based probes") and sensors (in which analyte recognition is reversible: "recognition-based sensors") offer an attractive approach to provide this information [31–35]. However, Mn(II) and Fe(II) are particularly challenging analytes to detect selectively due to their tendency to bind the most weakly of all first-row transition metal ions to biological ligands, according to the Irving–Williams series [36]. It is not a coincidence that the largest number and among the most biologically successful metal sensors developed to date are for relatively tight-binding metal ions of biological interest: Ca(II) [37–41], Zn(II) [42–47], and Cu(I) [48–52].

In this chapter, we consider the approaches used for the development of selective fluorescent sensors for Mn(II) and Fe(II), which both involve small-molecule and genetically encoded systems. We note that, as of 2014, the state of the field of fluorescent bio-sensors for metal ions was comprehensively reviewed by Palmer and co-workers [33]. Here our focus is on more recent work, though we mention key earlier developments as well. Small-molecule probes for iron have blossomed in the last decade, largely as a result of the identification of several approaches for reaction-based detection of Fe(II) that can be applied in cells. By contrast, reliable detection of Mn(II) using synthetic sensors is limited to only a handful of examples, each with significant disadvantages. We also address very recent examples of genetically encoded sensors for both metal ions. These relatively recent developments presage exciting opportunities for the broader application of probes and sensors for these metals in the near future and will contribute to the understanding of the physiology of these metals.

2 INTERROGATING TOTAL, CHELATABLE, AND LABILE METAL POOLS

Before discussing probes and sensors for Mn(II) and Fe(II) more specifically, it is important to consider what these molecules are detecting. A fundamental, general challenge in the selective sensing of metals in biological systems is that they exist in multiple "pools," usually with poorly defined speciation. Depending on the investigation, analysis of different pools may be desirable, but here we will consider three broad categories: total cellular metal, "chelatable" metal ions, and the "labile" (or free or bioavailable) pools.

Total metal content in a cell or organism is the most straightforward, and it can be quantified by a variety of methods, including inductively coupled plasma mass spectrometry (ICP-MS) and atomic absorption spectrometry. While these methods obliterate spatial information, other techniques, such as laser ablation ICP-MS (LA-ICP-MS) or synchrotron X-ray fluorescence (XFM) nanoimaging, can visualize metal distribution at a tissue, cellular, or even subcellular level. The application of these and other methods for imaging metals in cells has been recently comprehensively reviewed [53]. These methods are destructive, low-throughput, and require specialized instrumentation, but they provide valuable information about overall metal accumulation or deficiency in the system of interest.

The next level of distinction is between the metals that are essentially irreversibly bound on the timescale of an experiment to ligands (e.g., metal ions in an enzyme active site or in storage), and those that are reversibly bound. The latter may be termed the "chelatable" metal pool: everything bound more weakly to ligands than the affinity of the chelator being used as a probe. A small subset of this pool is the "labile" pool. Cells sense deficiency, sufficiency, or excess of metal ions not through the total quantity in the cell, but through the labile concentrations of those metals [54]. The labile pool is the result of buffering of the metal ions by relatively weak protein, nucleic acid ligands, in addition to abundant small-molecule ligands. For example, glutathione is proposed to be an important buffer for Cu(I) [55] and possibly Fe(II) [56], whereas phosphates are likely important for buffering Mn(II) [57]. In a classic study, Outten and O'Halloran demonstrated that, although *Escherichia coli* grown in standard minimal medium contain ~100 μM total zinc, the labile Zn(II) concentration is sensed by metalloregulatory transcription factors that respond in the femtomolar range [58]. These setpoints are important because they help to prevent the mismetalation of enzymes (and aberrant responses of metalloregulators) [59]. Therefore, it is arguably most essential in probing the "metal state" of a cell to know where in the buffered range the cell is at a given time, i.e., interrogating the labile pool.

In the cases of manganese and iron, the +II oxidation state is the most labile form and the form in which the metals are predominantly trafficked in the cell. The buffered concentration setpoints for Mn(II) and Fe(II) are relatively high – in the high nanomolar to low-mid micromolar range, depending on the organism [54, 56, 60] – reflecting their inherently weak binding to most ligands. Because the essence of sensing the labile pool is tuning a sensor to be able to exchange metal with the ligands buffering that pool without perturbing it significantly, the micromolar buffering range for Mn(II) and Fe(II) crystallizes the problem of how to sense these metal ions selectively.

The methods considered in the present chapter probe the chelatable and labile pools, which can interact reversibly with molecular probes/sensors for their detection and quantification. For a detailed discussion of the mechanisms of modulation of fluorescence signals, general classes of sensors, and important considerations for fluorescence imaging, we refer the reader to the excellent review by Palmer and coworkers [33]. Here we will simply summarize the basic information necessary to evaluate the molecules presented below.

The most basic type of sensor is recognition-based: binding of an analyte to the sensor leads to a change in signal that can be monitored. Often this is a change in signal intensity, either an increase (turn-on) or decrease (turn-off), although in certain cases the excitation or emission wavelength can also shift. Metal ions, especially paramagnetic ones, such as Mn(II) and Fe(II), often act as fluorescence quenchers (turn-off). Turn-off sensors are generally less desirable than turn-on sensors because a loss of signal associated with analyte binding corresponds to a loss of spatial information. However, a turn-off sensor can be turned into a ratiometric sensor by appending a fluorophore that is unresponsive to the analyte, yielding two signals the ratio of which can be reported. This strategy is also helpful in allowing normalization for different probe concentrations (which is usually difficult to independently verify) in different cell types or experiments. In certain cases, fluorescence quenching can be overcome using fluorophores and receptors that take advantage of photo-induced electron transfer (PET), or charge transfer processes in order to produce a fluorescence turn-on upon metal ion binding [61, 62]. Finally, another approach to overcome challenges of selectivity based on recognition alone, or of fluorescence quenching associated with metal binding, is to take advantage of unique reactivities of particular metal ions. In this strategy, binding of a metal ion to a reactive trigger leads to uncaging of a fluorophore and a fluorescence increase [34, 35]. The discussion below will provide examples of all of these types of probe/sensor designs, with the exception of reaction-based Mn(II) probes, which have not been reported to date.

3 BIO-SENSORS FOR Mn(II)

The development of selective, fluorescent sensors for Mn(II) is challenging because Mn(II) is the lowest ion in the Irving–Williams series, ranking relative stabilities of complexes of divalent first-row transition metal ions with ligands [36]. As summarized below, this challenge is still largely unresolved [63], although turn-off, turn-on, displacement-based, nanoparticle-based, and genetically encoded sensors have been reported.

3.1 Traditional Spectroscopic Methods

Besides the general methods for quantifying metals described in the previous section, magnetic resonance methods have been particularly informative for Mn(II). The distinctive electron paramagnetic resonance (EPR) signal of Mn(II) has allowed for quantification of total and labile Mn(II) concentrations through early whole-cell studies [64]. It has also enabled detailed analyses of Mn(II) coordination environments and speciation through advanced EPR methods, such as electron nuclear double resonance spectroscopy, in *Deinococcus radiodurans*, *E. coli*, and yeast [57, 65, 66]. Magnetic resonance imaging methods have also been used to study manganese accumulation in the brains of occupationally exposed human patients [67, 68].

3.2 Quantification of Chelatable Mn(II) Using Turn-Off Fluorescent Sensors

While the methods mentioned above are useful in quantification of total cellular manganese and provide some information about subcellular distribution and ligation, they are limited because they are destructive and/or require specialized instrumentation. To overcome these disadvantages, Bowman and co-workers have developed two fluorescence quenching-based assays for *ex-vivo* quantification of cellular Mn(II). First, the cellular Fura-2 Mn(II) extraction assay (CFMEA) was established to quantify cellular

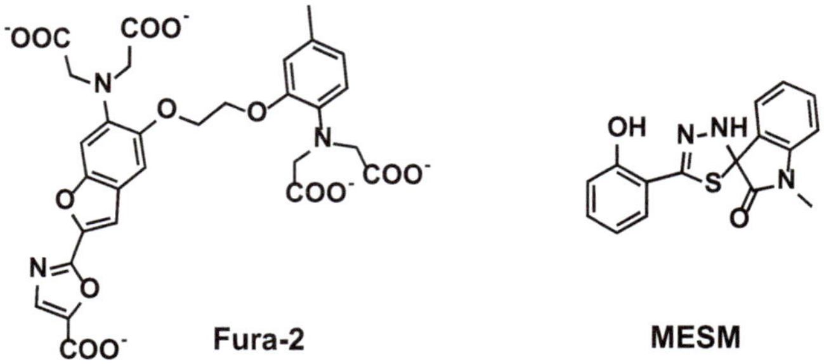

FIGURE 1 Structures of (left) Fura-2 and (right) MESM.

Mn(II) levels following cell lysis by Mn(II)-induced fluorescence quenching of the non-specific Ca(II) indicator **Fura-2** (Figure 1) [69]. Building on this approach, the Bowman group sought to extract cellular Mn(II) without lysing cells. Through screening molecules that affect total manganese levels [70], they identified one of those molecules as acting as a "manganese-extracting small molecule" (MESM, Figure 1) [71]. MESM is an ionophore that promotes Mn(II) efflux from the cell within ~2 minutes, while it does not cause significant efflux of other metals, including Ca(II). This selectivity is important because the extracted Mn(II) is then quantified using Fura-2, as in the CFMEA, and Ca(II) and other metals would interfere with this downstream assay. Validation of this method, termed MESMER, with ICP-MS yielded manganese quantification results that were not significantly different when cells were exposed to high Mn(II) concentrations; however, basal manganese was underestimated by the assay. This result suggests that MESM accesses some or all of the chelatable Mn(II) pool, but it is unable to define subcellular Mn(II) localization or exactly which pools are accessed. Furthermore, because MESM shows low cytotoxicity, cell samples can be assayed repeatedly to assay Mn(II) levels over time, although at the expense of rapid dynamic information. While this approach does simplify Mn quantification significantly by avoiding the need for ICP-MS analysis, it does not replace the need for sensors that can directly detect Mn(II) within cells in real time.

3.3 Turn-On Fluorescent Sensors for Mn(II)

Despite the paramagnetism of Mn(II), two promising designs for turn-on sensors for Mn(II) have been developed. These sensors operate by a PET or charge transfer mechanism, in which the fluorescence of the unbound fluorophore is quenched by electron transfer from the receptor in the excited state of the fluorophore; metal complexation inhibits the fluorescence quenching, resulting in a fluorescence turn-on [61, 62]. This general strategy has been widely used to develop turn-on sensors for many metal ions. First, Canary and co-workers [72] took advantage of coordination similarities between Ca(II) and Mn(II) [73] to modify the well-known Ca(II) chelator, BAPTA (Figure 2a) [37]. BAPTA itself features a slight preference for Mn(II) over Ca(II) (Table 1). In accordance with hard–soft acid–base principles, the substitution of one of the hard carboxylate ligands of BAPTA with the softer pyridine nitrogen donor to yield chelator **1** increases the selectivity for Mn(II) over Ca(II) by

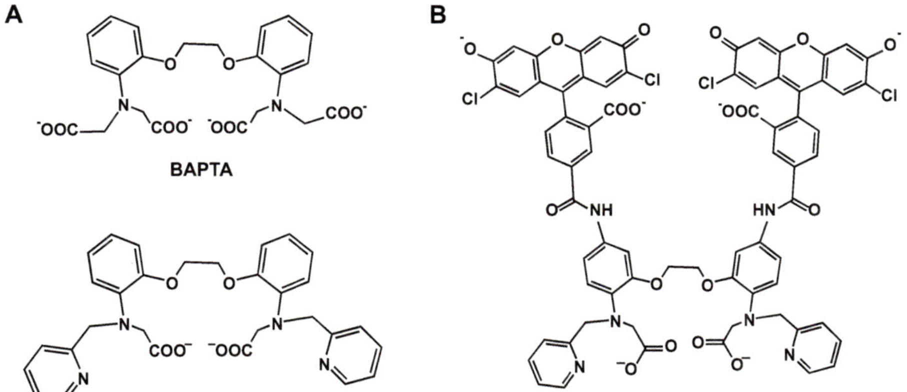

FIGURE 2 Canary's BAPTA-derived turn-on sensors for Mn(II). (a) Comparison of the structures of BAPTA and ligand **1**. (b) Sensor **3**.

~1000-fold, while retaining high affinity for Mn(II) (Figure 2a, Table 1). Combining that chelator as a receptor with a fluorophore similar to calcium green-2 yielded a turn-on sensor, **3**, with four-fold response to Mn(II) (Figure 2b). While this sensor solved the problem of Ca(II) and Mg(II) selectivity, other metal ions – in particular, Fe(II) and Zn(II) – also cause a similar fluorescence response that would be problematic for cellular applications. Still, the sensor showed a response to Mn(II) in HeLa cells supplemented with 200 μM Mn(II), which could be reversed by treatment with *N,N,N′,N′*-tetrakis(2-pyridylmethyl)ethylenediamine [74] (TPEN). Therefore, this work represents a promising approach but will require decreasing affinity for Fe(II) and Zn(II) for selective sensing of Mn(II).

Second, Datta and co-workers reported another promising recognition-based sensor, **M1**, consisting of a boron dipyrromethene (BODIPY) fluorophore connected to a penta-aza macrocycle with pendant methyl ester arms, and meant to achieve pentagonal bipyramidal coordination geometry in complex with Mn(II) (Figure 3) [75]. The PET mechanism is mediated by the dimethylaniline moiety connecting the fluorophore and the macrocycle. M1 exhibited green fluorescence ($\lambda_{ex}=480$, $\lambda_{em}=508$ nm) and a 60-fold turn-on to Mn(II), with good selectivity over Ca(II), Fe(II), and other ions, although it did show a lesser response to Zn(II) and Cu(II). However, this *in vitro* characterization was carried out in acetonitrile, as the sensor is

TABLE 1

Log Stability Constants (log *K*) for BAPTA and Canary Mn(II) sensors [72]

Metal Ions	BAPTA	1	3
Mn(II)	9.1	8.6	8.0
Ca(II)	6.9	3.8	3.3
Zn(II)	9.4	9.2	7.1
Mg(II)	1.7	0.8	0.9

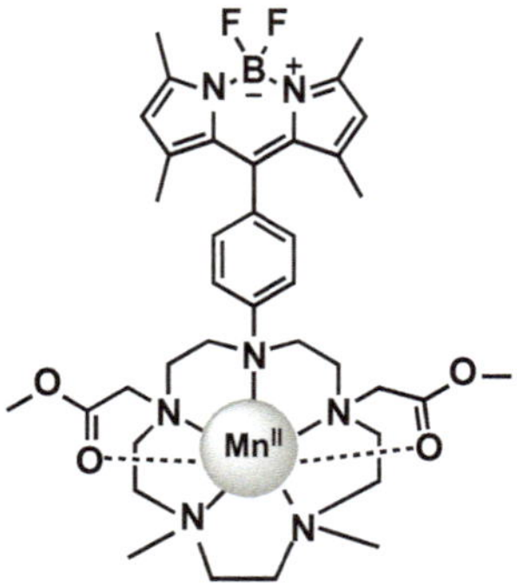

FIGURE 3 Datta's BODIPY-based turn-on sensor, **M1**.

poorly soluble in water. In acetonitrile, the K_d of Mn(II)-**M1** is 4 μM. A study utilizing this receptor for a ^{19}F-magnetic resonance probe for Mn(II) sensing implies that in aqueous environments the K_d of the receptor for Mn(II) is much weaker [76], outside of the biologically relevant range. In line with the hydrophobicity of the sensor *in vitro*, in-cell application of the sensor leads to perinuclear punctate localization, and co-localization with Nile Red suggests staining of lipid-rich areas of the cell [75]. It was later revealed that M1 partitions to the Golgi in mammalian cells [77], which also happens to be a major site of manganese accumulation [78, 79]. M1 fluorescence does increase upon cellular exposure to higher concentrations of Mn(II), and it qualitatively reports on Mn(II) in the manganese-accumulating bacterium *D. radiodurans* [80]. The lipophilicity is undesirable, and unfortunately work to render the sensor more hydrophilic by substitution of the BODIPY fluorophore with hemicyanine yielded a sensor that responded to Hg(II) rather than to Mn(II) [81]. Despite apparently successful imaging studies showing Mn(II) responsiveness, the weak K_d and lipophilicity of the sensor suggests that M1 is not responding reversibly to labile Mn(II) in cellular studies, although an irreversible mechanism is possible, where Mn(II)-bound sensor partitions to lipids where the Mn(II) can no longer interact with the labile pool. A stronger binding and more selective receptor and less hydrophobic fluorophore will be necessary for further pursuit of this strategy.

Finally, two fluorescent probes for Mn(II) were developed by coupling of rhodamine and fluorescein hydrazides with 2,6-diformyl-*p*-cresol, leading to two orange-emitting fluorophores (λ_{ex} = 520 nm, λ_{em} = 558 and 605 nm, respectively) with good selectivity to Mn(II) over Ca(II), although Fe(II) was not tested [82]. Unfortunately, these sensors exhibit low aqueous solubility, having been characterized *in vitro* in >95% v/v acetonitrile or DMSO and calling into question the relevance of the reported in-cell response to Mn(II).

3.4 Displacement-Based, Polymer-Based, and Nanoparticle-Based Sensors for Mn(II)

Several additional modes of Mn(II) detection have been reported. However, they have minimal application in biological systems to date, and some are not suitable for in-cell analysis. The first method, developed by Canary and co-workers, involves a two-chelator system in which competition between Mn(II) and another metal ion (e.g., Zn(II) or Cd(II)) for the stronger chelator displaces the other metal, leading to a colorimetric [83] or fluorescent [84] response of the weaker chelator to that displaced metal. In the case of fluorescence detection, the researchers utilized a two-dye system, calcein blue (CB) and a weaker chelator, FluoZin-1 (Fz1) (Figure 4) [84]. In the absence of Mn(II), Cd(II)-bound CB shows strong blue fluorescence ($\lambda_{ex}=350$ nm, $\lambda_{em}=433$ nm) and the unbound Fz1 exhibits weak green fluorescence ($\lambda_{ex}=493$ nm, $\lambda_{em}=518$ nm). Upon Mn(II) addition, Mn(II) competes with Cd(II) for CB binding (K_d of Mn(II)-CB is 170 nM versus 14 nM for Cd(II)-CB [84]), quenching its fluorescence, and the displaced Cd(II) binds to Fz1, increasing its fluorescence. This strategy was applied in HEK cells overexpressing the divalent metal transporter DMT1. Cells were preloaded with the acetoxymethyl (AM) ester forms of the sensors [85] as the free carboxylates cannot cross the membrane; the ester linkages are cleaved by intracellular esterases. The cells were also preloaded with 10 μM Cd(II) to saturate the CB-binding site. Uptake of extracellular Mn(II) could be followed with time by a decrease in CB fluorescence and an increase in Fz1 fluorescence. Of course, the disadvantage of this approach is the requirement for addition of Cd(II), which has no physiological role in most organisms and may interfere with homeostasis of other metals.

Several types of nanoparticle-based sensors for Mn(II) have also been developed. The most successful is that by Dhenadhayalan et al., who linked silicon quantum dots to various crown ethers and showed that Mn(II) binding to the aza-18-crown-6 derivative (Figure 5) effectively suppresses the PET-based fluorescence quenching, leading to a turn-on response [86]. Mn(II) gave the largest fluorescence activation, although the sensor also responded significantly to similar concentrations of K^+ and Zn(II) out of the other biologically relevant ions tested. The response to K^+ may be

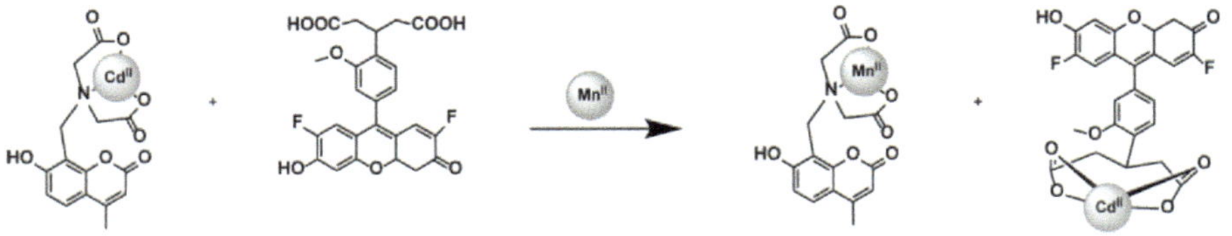

FIGURE 4 Displacement-based sensing of Mn(II) using calcein blue and FluoZin-1. The coordination environments of each metal are only indicated schematically.

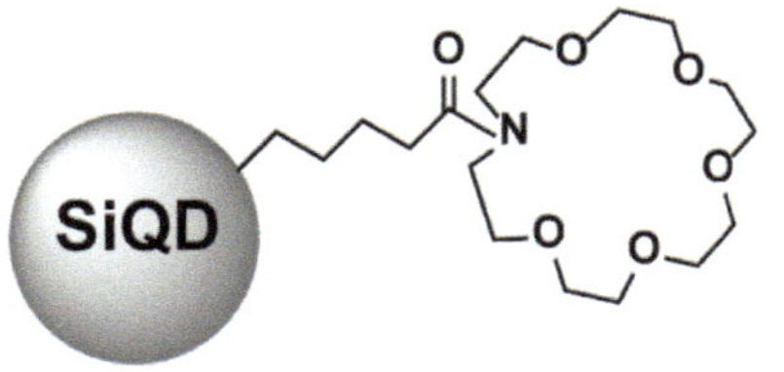

FIGURE 5 Aza-18-crown-6 receptor linked to silicon quantum dot (SiQD) detects Mn(II) by suppressing PET upon Mn(II) binding.

a particular challenge for cellular applications. Nevertheless, a slight fluorescence response was observed in HeLa cells loaded with the sensor upon the addition of Mn(II), although the sensor exhibited punctate distribution that was suggested to stem from the hydrophobicity of the aza-crown ethers. Another strategy using 1-thio-glycerol-capped CdTe quantum dots has also been reported [87], exhibiting selective response to Mn(II) at pH 11 but not at physiological pH. Other strategies featuring selective colorimetric responses to Mn(II) have also been reported [88, 89], but they are not fluorescent and have not been applied in cellular systems.

An additional PET-based approach for Mn(II) sensing uses graphene nanosheets coupled with 1,2-bis-(2-pyren-1-ylmethylamino-ethoxy) ethane (NPEY) as a metal ion receptor [90]. Mn(II) binding to NPEY was found to increase sensor fluorescence seven-fold through a PET quenching mechanism, presumably through π–π stacking between NPEY and the graphene. This material exhibits good selectivity against ions, such as Ca(II) and Zn(II), but it was not tested with Fe(II) and detection was only tested in the range of 10–100 μM Mn(II). The material shows a fluorescence response in HeLa cells exposed to 1 mM Mn(II).

Finally, various sensors have also been developed that give a colorimetric response to Mn(II) but no fluorescence response or quenching, whereas other (diamagnetic) ions give fluorescence activation. Several of these types of sensors utilize a Schiff base coupled to a hydroxyjulolidine- [91, 92], phenol- [93] or hydroxynaphthalene-based [94] chromophore. However, these sensors would not be suitable for cellular applications due to a lack of selectivity, fluorescence quenching, and poor solubility in an aqueous solution.

The preceding discussion demonstrates that the development of small-molecule or nanoparticle-based sensors for Mn(II) is still in its infancy. PET-based sensors show the most promise in overcoming the paramagnetism of Mn(II), but a suitable selective receptor has not yet been identified – a testament to the Irving–Williams series. This lack of success motivates the exploration of other approaches, particularly genetically encoded strategies for imaging Mn(II) in cells.

3.5 Genetically Encoded Sensors for Mn(II)

Genetically encoded sensors offer the potential to combine selectivity evolved by biological systems with desirable properties lacking in available synthetic Mn(II) sensors, including a K_d in the physiologically relevant range, real-time and reversible detection, aqueous solubility, ease of ratiometric analyte detection, and predictable localization. Most genetically encoded fluorescent sensors couple a fluorescence readout with a conformational change induced by analyte binding to a receptor, usually a protein [32, 33] or, more recently, a gene-regulatory RNA structure (riboswitch) [95–97]. Protein-based sensors have been developed for Ca(II) (based on calmodulin and related proteins) [38–40], Zn(II) (based on zinc fingers), [44, 46] Cu(I) (based on a Cu regulatory protein), [50] and Mg(II) [98]. Many of these sensors rely on a Förster resonance energy transfer (FRET) mechanism in which the protein's metal-mediated conformational change alters the distance and orientation between donor and acceptor fluorescent proteins, yielding a ratiometric fluorescence response.

In the case of Mn(II), utilization of a biological scaffold may be particularly beneficial as one of the primary challenges of synthetic sensor development has been finding a suitable, selective receptor. One of the few well-characterized proteins that can selectively bind Mn(II) (and therefore can serve as inspiration for the design of a fluorescent sensor for Mn(II)) is the *Bacillus subtilis* transcription factor, MntR [14, 99]. Unfortunately, the conformational change induced in MntR by Mn(II) binding appears to be subtle, such that a FRET-based sensor would be challenging to engineer. However, the pentagonal bipyramidal coordination geometry enforced by MntR is similar to that of Ca(II) in the EF-hand motifs of Ca(II)-binding proteins, such as calmodulin [100], which have been used extensively in the generation of protein-based calcium sensors [38, 40, 101]. Mn(II) binds to calmodulin with micromolar affinity, but it fails to induce the same conformational change [102]. Together, these ideas motivated us to re-engineer EF hands to respond to Mn(II).

We recently succeeded in designing the first two genetically encoded sensors for Mn(II), MnLaMP1 and MnLaMP2 (Figure 6) [103 (J. Park, J.A. Cotruvo, Jr., et al., submitted)]. These sensors were created through rational mutagenesis of the metal-binding sites (EF hands) of lanmodulin, a natural, high-affinity lanthanide-binding protein our group discovered in 2018 [104]. Lanmodulin undergoes a large conformational change in response to lanthanide ion binding. We had previously exploited this in the development of a selective FRET-based sensor for lanthanides by flanking the lanmodulin domain using the commonly used FRET pair, enhanced cyan fluorescent protein and citrine [105]. The apparent K_d values for the Mn(II) complexes of MnLaMP1 and MnLaMP2 are 29 and 7 μM, respectively, with robust selectivity over crucial competitors Fe(II), Ca(II), and Mg(II), as well as other biologically relevant divalent metal ions. Both sensors enable real-time quantification of labile Mn(II) levels in *E. coli*, a first for any Mn(II) sensor. However, the sensor design will require improvement for mammalian studies because labile Mn(II) concentrations in eukaryotic systems appear to be in the sub-micromolar range [64, 106].

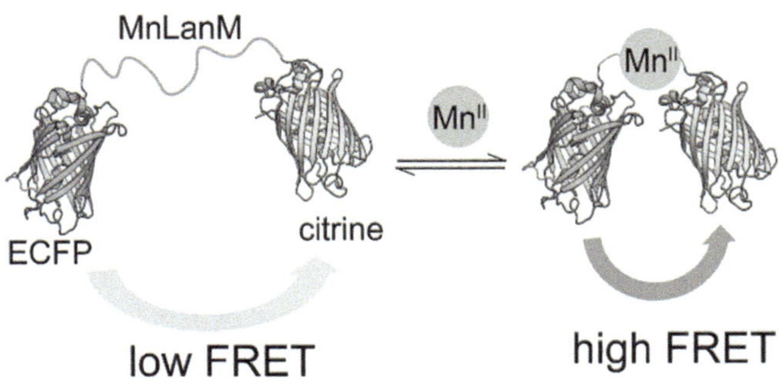

FIGURE 6 Schematic representation of the genetically encoded FRET-based sensors for Mn(II), MnLaMP1, and MnLaMP2. The lanmodulin variant engineered to respond to Mn(II) is MnLanM.

4 METHODS FOR QUANTIFYING Fe(II)

While being one spot higher in the Irving–Williams series than Mn(II), Fe(II) still suffers from the same difficulties of selective, recognition-based detection. However, the more facile conversion of the prevalent oxidation state in the cell, Fe(II), to Fe(III) in the presence of a probe ligand offers unique strategies for selective detection that have been exploited by researchers.

4.1 Traditional Approaches for Quantifying Total and Chelatable Fe(II)

In addition to standard methods for total cellular iron concentration, such as ICP-MS and atomic absorption spectrometry, total iron in cellular samples can be quantified using less specialized instrumentation using the sensitive Fe(II)-chelating dye, ferrozine [107], ($\varepsilon_{564\ nm} = 27.9\ mM^{-1}cm^{-1}$ for the Fe(II)-tris(ferrozine) complex), which has been adapted for cellular samples [108]. The total cellular concentration of iron can be estimated by approximating the total cellular volume; however, the volume determination may be challenging and may be impacted by iron availability, as is the case for *E. coli* [109, 110].

Given the limitations of total-metal methods, researchers have sought to develop new methods to study the dynamic nature of iron homeostasis. An early method to quantify labile iron pools in cells was developed by Keyer and Imlay and uses whole-cell EPR spectroscopy [111]. This method utilizes desferrioxamine B (DFO), a high-affinity Fe(III) chelator (siderophore) that also binds Fe(II) and facilitates the oxidation of Fe(II) to Fe(III). The resulting Fe(III)-DFO complex exhibits an EPR signal at $g = 4.3$, the concentration of which can be determined using external Fe(III)-DFO standards. The intracellular Fe(III)-DFO concentration can be estimated after correction for cell packing density in the EPR tube. This method was initially applied to investigate the iron levels of *E. coli* strains with disruptions in iron homeostasis and oxidative defense: Δ*fur* (lacking the Fe(II)-sensing transcription factor), Δ*sodAB* (lacking the Mn- and Fe-dependent superoxide dismutases), and a strain lacking all three genes [111]. These experiments gave signals that quantified to 10, 70, 80, and 160 μM iron, for wild-type, Δ*fur*, Δ*sodAB*, and Δ*fur* Δ*sodAB* strains, respectively. However, this method features a few notable caveats. First, because it is an endpoint assay requiring DFO addition and freezing of cells, samples cannot be measured at multiple timepoints. Second, and perhaps more importantly, because it depends on the essentially irreversible conversion of Fe(II) to Fe(III), cells may compensate for low labile Fe(II) through import or release from other pools, such as storage. Therefore, the concentration of labile Fe(II) is most likely overestimated, and in fact this method more likely determines the size of the chelatable pool of Fe(II) (i.e., all the Fe(II) that is bound more weakly than the affinity of DFO for Fe(II)), rather than the labile iron concentration that is being sensed by regulators [59].

4.2 Recognition-Based Sensors for Fe(II)

In contrast to the above methods, a fluorescent sensor would enable imaging of Fe(II) localization and dynamics in real time in an intact system. Key properties of both recognition-based sensors and reaction-based probes (discussed in Section 4.3) for

Fe(II) are summarized in Table 2. The major challenges of the recognition-based sensors are: (i) the low placement of Fe(II) in the Irving–Williams series, making selectivity difficult to achieve, and (ii) Fe(II), like Mn(II), readily quenches fluorescence, making the design of turn-on sensors more challenging. The classic sensor utilizing the recognition strategy is **calcein** (Figure 7, Left) [112]. Calcein consists of fluorescein derivatized with two iminodiacetic acid (IDA) moieties as receptors, and metal binding results in fluorescence quenching. At 0.5 μM calcein and 2 μM Fe(II), 46% quenching was observed. The complexity of two binding sites per calcein and how each binding event affects fluorescence response, in addition to the ability of IDA to form 2:1 complexes with Fe(II), makes precise K_d determination complex [112]. While Ca(II) and Mg(II) do not significantly affect fluorescence *in vitro*, Mn(II), Co(II), Ni(II), Cu(II), Zn(II), and Fe(III) also quench to differing extents, not surprisingly given the stability constants of IDA itself [113]. Because the carboxylate moieties hinder cellular uptake, the AM-ester calcein derivative was used for in-cell studies in human erythroleukemia K-562 cells. The sensor reversibly responded to Fe(II) but also to Mn(II) as well as to Co(II) and Ni(II) (the latter being less relevant to mammalian systems) [112]. Although widely used as an Fe(II) sensor, these considerations make clear that calcein's turn-off, promiscuous response to multiple biologically relevant transition metals makes it an imperfect tool with which to probe Fe(II) in cells.

TABLE 2
Properties of Fluorescent Probes and Sensors for Fe(II) Discussed Herein

Probe Name	Type	Excitation (nm)	Emission (nm)	F/F_0	Ref.
Calcein	Recognition	486	517	0.54[a]	[112]
Phen green SK	Recognition	507	532	0.15[a]	[114, 115]
BDP-Cy-Tpy	Recognition	485, 569 (free) 485, 596 (bound)	507, 635	7	[116]
DansSQ	Recognition	352	425	15	[117]
FlCFe1	Recognition	405	490/530	5	[118]
IP1	Reaction	470	508	6	[119]
LCy7	Reaction	560	690	5	[120]
RhoNox	Reaction	492 (free) 555 (bound)	575	30	[121]
HMRhoNox-M/LysoRhoNox	Reaction	550	575	40	[122]
FluNox-1	Reaction	450	530	10	[123]
CoNox-1	Reaction	295	495	8	[123]
SiRhoNox-1	Reaction	575	660	45	[123]
FIP-1	Reaction	495/545	515/556	2.25	[124]
Probe 1	Reaction	337	465	10	[125]
YTP	Reaction	374	560	7	[126]

[a] Turn-off sensor.

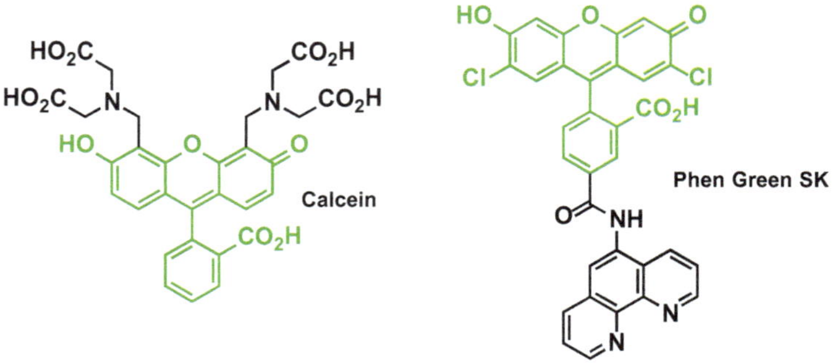

FIGURE 7 Classic recognition-based Fe(II) sensors, calcein and Phen Green SK.

Another approach for imaging Fe(II) by Rauen and co-workers used **Phen Green SK**, a sensor in which the fluorophore was linked to a 1,10-phenanthroline receptor (Figure 7, right) [114, 115]. Using a multi-step *ex-situ* calibration procedure, the chelatable iron concentration in hepatocytes was estimated at 4–10 μM [115]. This sensor performed better than calcein in a direct comparison, although it is not clear whether it was due to more efficient fluorophore quenching or because 1,10-phenanthroline is a higher-affinity Fe(II) [and Fe(III)] chelator. While the stability constant for the 1:1 Fe(II):phenanthroline complex is ~6, the stability constant for the ML_3 (3:1) complex is ~21 [127]. Therefore, this sensor would most likely aid in the quantification of the total chelatable iron pool, much like the DFO EPR method. The 1,10-phenanthroline ligand is also an excellent chelator for Cu(I). Overall, Phen Green SK did not solve the fundamental shortcomings of calcein as an Fe(II) sensor.

Calcein and Phen Green SK exhibit fluorescence quenching upon Fe(II) binding, an undesirable outcome due to loss of spatial resolution. Tang and co-workers designed the sensor **BDP-Cy-Tpy** to include an internal standard, thereby converting what would otherwise be a turn-off sensor into a ratiometric sensor (Figure 8, left) [116]. BDP-Cy-Tpy contains two fluorophores, a BODIPY (orange, emission at 569 nm, in Figure 8) and cyanine (red, emission at 635 nm, in Figure 8), where

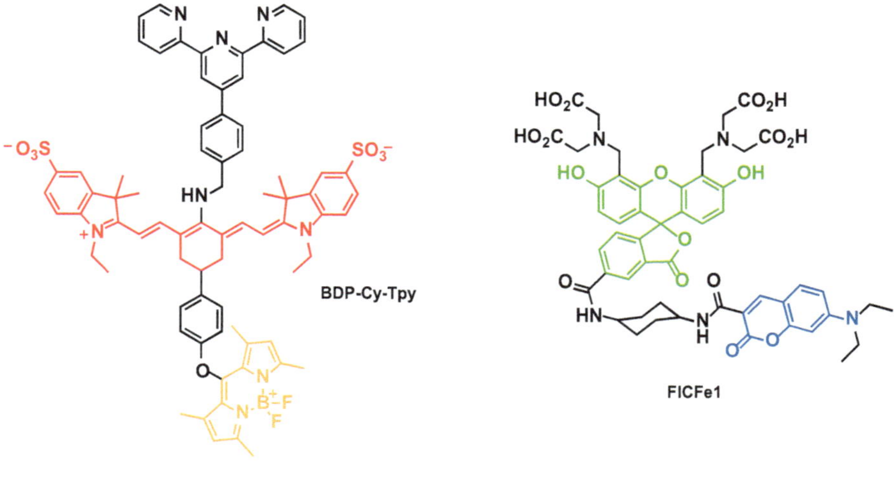

FIGURE 8 Ratiometric recognition-based Fe(II) sensors, BDP-Cy-Tpy and FlCFe1.

TABLE 3
Stability Constants for tpy [128]

Metal Ions	Log K_1	Log K_2
Mn(II)	4.4	N.A.
Fe(II)	7.1	13.8
Co(II)	8.4	9.9
Ni(II)	10.7	11.1
Zn(II)	6.7	5.2

N.A., not applicable.

the BODIPY acts as the internal standard and cyanine is connected to the Fe(II)-responsive group, terpyridine (tpy). As summarized in Table 3, the stability constants for Fe(II)-tpy complexes suggest reasonable selectivity for Fe(II), largely through formation of the Fe(II)-tpy$_2$ complex. Binding of Fe(II) results in nearly complete quenching of the cyanine fluorescence, whereas BODIPY fluorescence is minimally affected. Overall, the maximum change in ratio between the two emission signals is approximately seven-fold. The apparent K_d reported for BDP-Cy-Tpy is 2.5 μM [116]. The sensor reversibly detected changes in Fe(II) levels in human liver cells (HL-7702) with the addition of Fe(II) and TPEN.

Given their electronic properties and stability constants, other metals like Co(II), Ni(II), and Cu(II) surprisingly did not quench BDP-Cy-Tpy fluorescence on their own, and the presence of those ions did not interfere appreciably with the formation of the complex with Fe(II) in competition assays. Furthermore, TPEN could only partially recover the fluorescence of iron-bound sensor, suggesting a kinetic issue or possibly oxidation of the iron. The authors of this study did not mention whether the *in vitro* characterization experiments were done anaerobically, a requirement because of the rapid reactivity of Fe(II) with O_2 to form Fe(III), which could also quench fluorescence and preclude metal exchange. It is therefore possible that some of these unusual properties stem from aerobic characterization experiments.

More recently, another approach to a ratiometric Fe(II) sensor design was taken by New and co-workers, who conjugated calcein to 7-diethylaminocoumarin using a cyclohexanediamine linker to yield **FlCFe1** [118]. This sensor exhibited a five-fold ratiometric response and was responsive to iron supplementation and chelation in DLD-1 cells. Whereas this sensor is an improvement on calcein, the issue of low metal selectivity remains.

Reddy, Varma, and co-workers used DANSYL–styrylquinoline conjugate as the basis for a rare turn-on sensor for Fe(II), **DansSQ** (Figure 9) [117]. *In vitro* characterization experiments were carried out in 9:1 acetonitrile:water, showing that the sensor responded selectively to Fe(II) over Mn(II), Zn(II), and Cu(II) *in vitro*. However, Cu(I) was not tested. Because of this sensor's high hydrophobicity, it has not found an in-cell application.

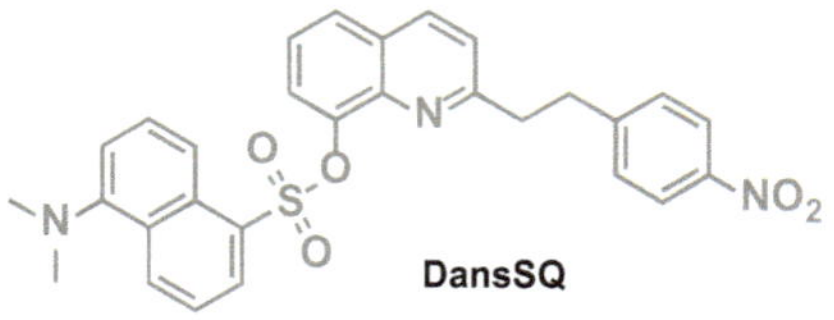

FIGURE 9 Recognition-based turn-on Fe(II) sensor, DansSQ.

Finally, the selective nickel sensor **NS1** [129], which was developed by Chang and co-workers and features a mixed N/S/carboxylate receptor, exhibits a very weak response to 100 μM Fe(II). This type of scaffold, utilized extensively for sensors for a number of metal ions including Cu(I), Pb(II), and Hg(II) [34], is attractive because unlike the recognition-based sensors described above, metal binding leads to a turn-on response. It may be possible to tune the metal-binding receptor in this sensor to more tightly and selectively bind Fe(II).

4.3 Reaction-Based Probes for Fe(II)

4.3.1 IP1-Type N_4O Receptor-Based Probes

Because weak Fe(II) binding, selectivity, and turn-off responses are challenges for binding-based sensors, the reactivity of Fe(II) with physiological oxidants, in particular O_2, has been utilized to develop reaction-based probes for Fe(II) [34]. Building on the work of Taki and co-workers for reaction-based detection of Cu(I) using a tris[(2-pyridyl)methyl]amine receptor-capped fluorescein [130], Chang and co-workers utilized a receptor containing two (2-pyridyl)methyl, one hydroxymethyl, and an amine functionality for recognition of Co(II) and, upon reaction with O_2, the release of the pendant Tokyo Green fluorophore [131]. This probe undergoes an 18-fold turn-on selectively in the presence of Co(II) with only very weak response to Fe(II) and Cu(I). This approach was then extended with the development of iron probe 1 (**IP1**), which features a pentadentate N_4O ligand set with three pyridine, one amine, and one carboxylate donor ligands, modeled after the 2-histidine-1-carboxylate coordination commonly observed in non-heme iron O_2-activating enzymes (Figure 10) [119]. Upon Fe(II) binding, the open coordination site can bind dioxygen and oxidatively cleave the ether linkage to release the fluorophore (Figure 10). This reaction leads to up to six-fold fluorescence enhancement after one hour of incubation with Fe(II). The probe is also reactive with Co(II) and O_2, although free Co(II) concentrations in cells are very low and this cross-reactivity is unlikely to cause interference in cells. Studies in HepG2/C3A cells yielded a ~30% increase in fluorescence upon exposure to 100 μM Fe(II) as ferrous ammonium sulfate, and a ~30% decrease when labile Fe(II) is chelated using DFO. The probe also showed that adding hepcidin or ascorbate increased fluorescence response inside the cell [119].

This general principle was extended to other detection modalities, such as the bioluminescent probe **FP-1**, in which Fe(II) binding to the same N_4O receptor linked to a caged luciferin, followed by reaction with O_2, liberates luciferin [132]. FP-1 was applied in FVB-luc+ mice expressing luciferase in order to study iron accumulation and Fe(II) localization during sepsis, inspired by prior studies utilizing the

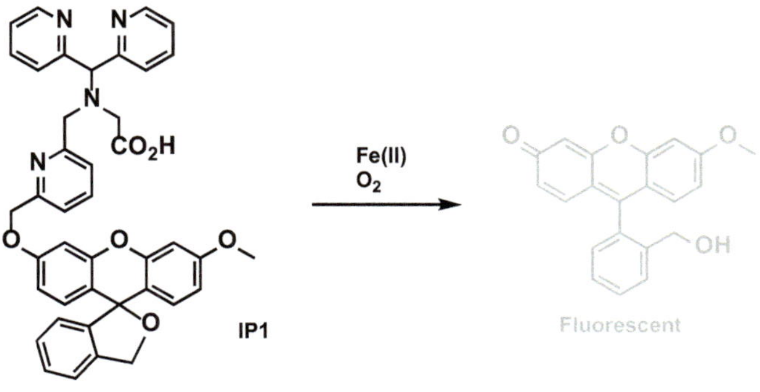

FIGURE 10 Reaction scheme of IP1-type reaction-based Fe(II) probes.

endoperoxide-based luminescent probe **ICL-1** described below [133]. Finally, the probe **LCy7** utilized this receptor linked to the cyanine dye QCy7 to enable near-infrared imaging of Fe(II) in hepatocytes (*in vitro*) and in mouse livers (*in vivo*) in acetaminophen-induced ER stress [120].

4.3.2 RhoNox-Based Probes

Another class of reaction-based probes for Fe(II) is "**RhoNox**," initially developed in 2013 by Hirayama and co-workers. These probes, the first being based on rhodamine B, feature *N*-oxide moieties that render the fluorophore only weakly fluorescent, primarily using a twisted intramolecular charge transfer (TICT) mechanism instead of a PET process [121]. Selectively in the presence of Fe(II), the *N*-oxide is reduced to an amine, restoring fluorescence (Figure 11). The original RhoNox probe exhibits an excellent, 30-fold fluorescence response (Table 2). This basic probe architecture has been functionalized in several ways. Initial studies focused on the optimization of reaction kinetics by changing the diaminoethyl moieties to diaminomethyl. Alteration of the spirolactone to a hydroxymethyl ether reduced the fluorescence background yielding **HMRhoNox-M**, which exhibits a 40-fold response [122]. Later work revealed that this probe is localized to the lysosome [134]. Different colors of RhoNox probes were also developed, replacing the rhodamine fluorophore with rhodol (FluNox-1122), coumarin (CoNox-1123), and Si-rhodamine B (SiRhoNox-1123). While FluNox-1, like RhoNox itself, localized to the cytosol, CoNox-1 and SiRhoNox-1 localized to the endoplasmic reticulum. Of all of these derivatives, HMRhoNox-M had the best performance *in vitro*, but it and the original RhoNox probe had similar

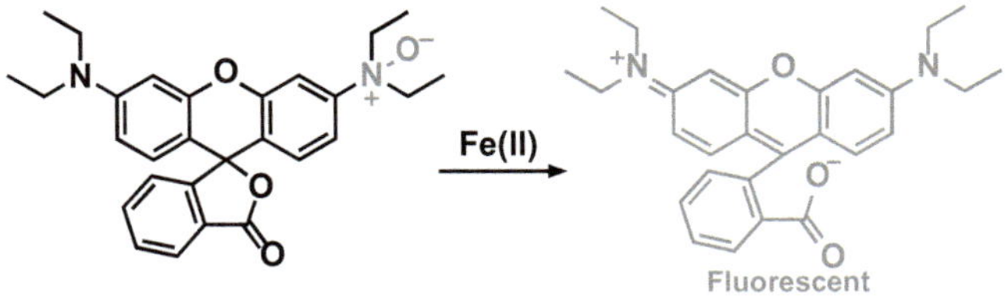

FIGURE 11 Fluorescence activation of RhoNox.

responses in HepG2 cells. When the cells were supplemented with 100 μM ferrous ammonium sulfate, the probes displayed an approximately two-fold increased fluorescence [121, 122] and could also monitor transferrin-mediated iron uptake [122]. Later work investigated a number of cyclic amine *N*-oxides and showed that using *N*-Boc piperazine *N*-oxide (**RhoNox-4**) enhanced *in vitro* fluorescence response to 160-fold and in-cell response to approximately three-fold [135]. However, because these probes are intensitometric, it is unknown how much the chelatable iron pool was increased under these experimental conditions.

The RhoNox probes have also been functionalized to target individual cellular organelles using well-established strategies. **Mem-RhoNox** was developed with palmitoyl chains conjugated to RhoNox, enabling anchoring of the probe to the cell membrane [136]. This probe was used to visualize the endosomal reduction of Fe(III) to Fe(II) during the endocytic iron uptake pathway involving Fe(III)-transferrin-bound transferrin receptor. **Ac-MtFluNox** features a triphenylphosphonium moiety conjugated to FluNox for localization of the probe to mitochondria [137]. The addition of a myristoyl chain enables the localization of SiRhoNox-1 to the Golgi (**Gol-SiRhoNox**) [134]; as mentioned above, CoNox-1 and SiRhoNox-1 localized to the ER [123], and HMRhoNox-M (renamed LysoRhoNox) localized to the lysosome [134] without addition of targeting groups. The fluorophores can be swapped and the probes used in combination, allowing simultaneous multi-color and multi-organelle imaging. However, as with all organelle-targeted, reaction-based probes, activation of the probe prior to accumulation in its targeted organelle may not be easily distinguished from a probe that is activated in the organelle itself. Whether this is a concern depends in part on the reaction kinetics of the probe. The original RhoNox probe reacts with Fe(II) rather rapidly, with full fluorescence activation within 1 h *in vitro*, and HMRhoNox-M is even faster, within 20 minutes [121]. The localization of certain probes to organelles without the addition of targeting groups may be beneficial, but it also highlights the occasional unpredictability of small-molecule probes.

Finally, the probe **Rh-T** was designed using a related approach, but using a nitroxide instead of the *N*-oxide of the RhoNox-type probes (Figure 12) [138]. Rh-T was designed to be used as a fluorescent probe as well as an EPR probe. Upon one-electron reduction of the nitroxide to the hydroxylamine by Fe(II), fluorescence was activated by 2.5-fold *in vitro*, and the EPR signal associated with the nitroxide disappeared. Although the probe was tested against a panel of biologically relevant metal ions *in vitro*, Cu(I) and other radicals like superoxide [139], which could also conceivably reduce the nitroxide, were not included. Rh-T responded to added Fe(II)

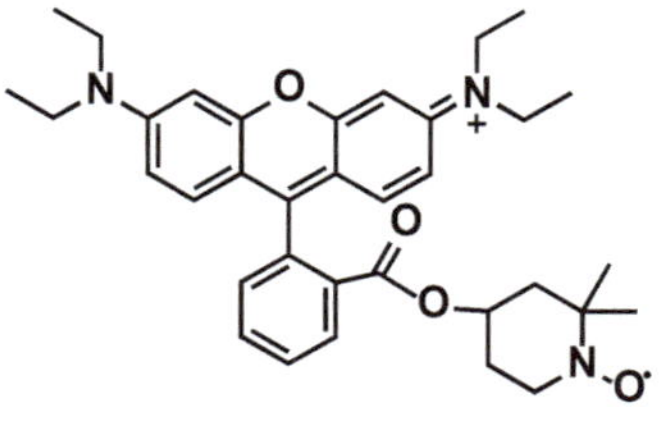

FIGURE 12 Reaction-based probe, **Rh-T**. Note the nitroxide rather than an *N*-oxide.

in human wsl fibroblasts, and yielded a punctate distribution in cells that suggested mitochondrial localization by comparison with MitoTracker dye. Reduction of the nitroxide to hydroxylamine was also confirmed by whole-cell EPR [138]. Given that other cellularly relevant radicals might also react with the probe, and their concentration might also be dependent on Fe(II) levels in cells, it is not clear whether the probe's response in cells was solely to Fe(II). Nevertheless, this probe represents an interesting approach for Fe(II) detection in cells.

4.3.3 Endoperoxide-Based Probes

A third major class of reaction-based probes for Fe(II) are based on the adamantyl endoperoxide group, inspired by the anti-malarial natural product artemisinin. An important aspect of the mechanism of action of artemisinin appears to originate from reactivity of the 1,2,4-trioxane with heme iron released from hemoglobin by the malaria parasite. In order to gain mechanistic information about this reaction, Vennerstrom and co-workers modeled the natural product by the compound in Figure 13 ($R_1 = R_2 = H$). They examined its reactivity with Fe(II), which in the presence of an additional one-electron donor cleaved the endoperoxide to the indicated products [140]. An advantage of this approach for reaction-based Fe(II) detection over the O_2-dependent triggers described above is the reduction in molecularity for the reaction from three to two. Two groups envisioned this model reaction as a starting point for the construction of Fe(II) probes that were reported in rapid succession.

Renslo and co-workers [125] reported "Probe **1**," in which a DANSYL fluorophore was conjugated to the adamantyl moiety and a DABSYL to the piperidyl moiety, as a quencher of DANSYL fluorescence. Upon the addition of Fe(II), cleavage of the endoperoxide leads to dequenching of DANSYL fluorescence and an approximately ten-fold turn-on response. This probe exhibited a highly selective response to Fe(II), with only Cu(I) as a minor interference with a two-fold response. Unfortunately, however, the probe could not be taken up into cells. Therefore, the investigators developed "Probe **3**," which was used to deliver puromycin in a caged, carbamoylated form into cells and allowed Fe(II) detection through immunofluorescence [125].

Chang and co-workers adopted a similar strategy in designing **FIP-1** [124]. Rather than constructing an intensity-based probe like **1**, they instead utilized a FRET donor-acceptor pair, 5-aminomethylfluorescein (5-AMF) and Cy3 (Figure 13), thereby generating a ratiometric probe. In the absence of Fe(II), the probe will show a FRET signal from the 5-AMF to Cy3. After reaction with Fe(II), reduction and cleavage of the endoperoxide group dissociates the two fluorophores, abolishing FRET. This

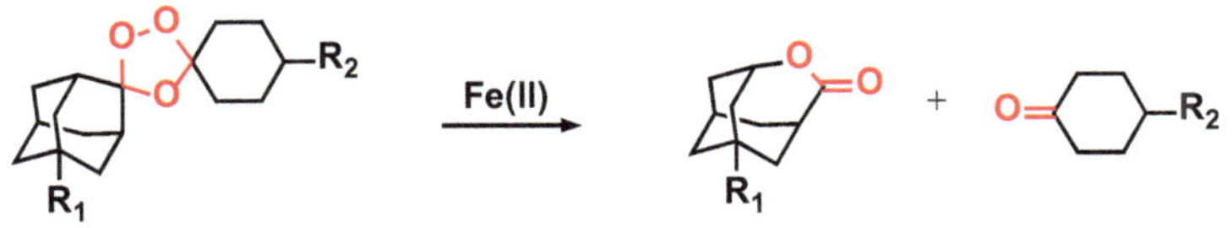

FIGURE 13 Reaction scheme for endoperoxide-based probes. In initial studies by Vennerstrom and co-workers, $R_1 = R_2 = H$. In **FIP-1**, R_1 and R_2 are 5-aminomethylfluorescein and Cy3, respectively. In **ICL-1**, $R_1 = H$ and R_2 is a carbamoylated D-aminoluciferin.

probe exhibits a ~2.5-fold FRET change and excellent selectivity against other metal ions, with super-physiological concentrations of Cu(I) again being a minor interference. It is likely that this probe would also be reactive with heme iron, but the levels of labile heme are in the low nanomolar range and therefore unlikely to compete with labile iron [141]. Application of FIP-1 in HEK293T cells showed responsiveness to iron supplementation and chelation, and the probe also allowed interrogation of labile iron in cancer cell lines and in a model of ferroptosis [124]. This class of probe was further developed to enable imaging of labile iron in live animals using bioluminescence (delivery of caged luciferin using **ICL-1** [133]) and using ^{18}F for positron emission tomography studies (**18F-TRX**) [142].

4.3.4 Quinoline-Derived Probes

The most recent example of a new class of Fe(II)-reactive probes is **YTP**. YTP uses a dimethylaminoquinoline fluorophore and utilizes the formation of a benzoxazole ring in the presence of Fe(II), leading to an approximately seven-fold fluorescence turn-on *in vitro* and in HeLa cells (Figure 14) [126]. The fluorescence was not evenly distributed throughout the cell, however, which might suggest lipophilicity or localization beyond the cytosol and deserves further study.

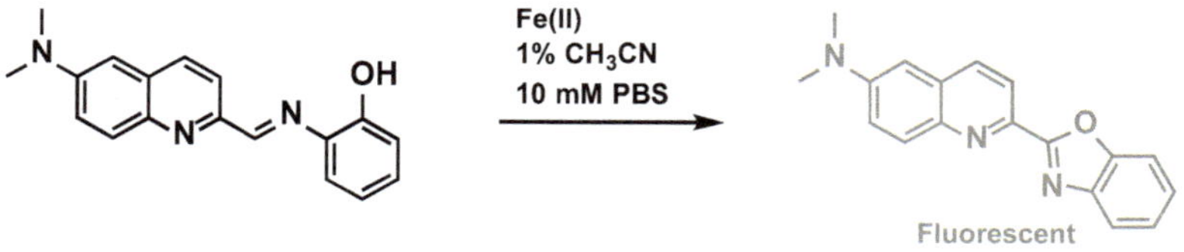

FIGURE 14 Reaction scheme for YTP.

4.3.5 Summary of Reaction-Based Fe(II) Probes

Small-molecule reaction-based probes overcome some of the main drawbacks of recognition-based sensors. They are highly selective for Fe(II) (despite some cross-reactivity with Co(II) [119] or Cu(I) [124], depending on the reactive trigger) and lead to fluorescence activation in the presence of Fe(II). The endoperoxide and *N*-oxide-based probes have the advantage that an exogenous oxidant is not required. Reaction-based probes are also able to magnify small changes in analyte concentration more effectively than recognition-based sensors. However, these advantages come at the expense of being reversible to be able to monitor metal dynamics. The desire to improve selectivity while retaining reversible binding, as well as to utilize a turn-on response, motivates the exploration of selective biological ligands for Fe(II) for the development of genetically encoded sensors, described below.

4.4 Genetically Encoded Methods for Detecting Fe(II)

As is the case for Mn(II), there are few obvious candidate proteins to act as recognition elements for a protein-based Fe(II) sensor, as such sensors typically exploit a large conformational change occurring upon analyte binding. The Fe(II)-responsive transcription factors like Fur, DtxR, and IdeR would seem to be starting points, but the conformational changes induced by metal binding to metalloregulatory proteins are generally subtle and poorly understood in most systems [99, 143, 144]. Nevertheless, a handful of reporters and sensors based on these and other metalloregulatory machinery have been reported recently.

4.4.1 Fur-Based Reporter

Although utilization of the Fe(II)-dependent conformational change of the ferric uptake regulator Fur directly for a FRET-based fluorescent sensor is a challenge, Silver and co-workers took advantage of the Fur-mediated transcriptional response to create a genetic reporter system for intracellular iron [145]. The investigators placed GFP under control of the *E. coli fiu* promoter, which possesses several Fur-binding sites. In the presence of Fe(II), dimerization of Fur will repress GFP expression. Therefore, this represents a "turn-off" reporter: in the presence of limiting iron, GFP fluorescence is maximal. Additionally, because GFP chromophore maturation is rather slow and GFP fluorescence will persist until protein is degraded, temporal resolution is limited. Studies using this reporter in *E. coli* estimated a basal free iron concentration of ~20 µM and suggested that the addition of millimolar extracellular iron could lead to intracellular free iron concentrations above 100 µM. The K_d of Fe(II)-Fur is ~1.2 µM [146], and therefore 20 µM is far above the saturation point of the Fur-mediated transcriptional response [59, 147]. Instead, these numbers correspond roughly to the free iron levels as quantified in wild-type and oxidatively challenged cells using the whole-cell EPR method [111]. Therefore, we suggest that these concentrations derived from the reporter studies may reflect the total chelatable iron pool, rather than true labile iron concentrations. Although this reporter was an important starting point, a turn-on genetically encoded sensor would be preferable over the reporter's turn-off response to Fe(II).

4.4.2 Riboswitch-Based Sensors and Reporters

In 2015, Breaker, Winkler, and co-workers reported the discovery and characterization of the NiCo (or *czcD*) class of riboswitches [148]. Along with two parallel papers characterizing manganese-sensing riboswitches [149, 150], this study introduced a new mechanism of RNA-based transition metal sensing in bacteria. The NiCo riboswitches were reported to bind 3 or 4 Co(II) and Ni(II) ions cooperatively and with low-micromolar affinity associated with a conformational change that is crucial to the gene-regulatory activity of these RNA elements. The riboswitches were claimed to not be Fe(II) responsive, but the studies were carried out aerobically, where Fe(II) would easily oxidize to insoluble Fe(III).

Our group suspected that the NiCo should also exhibit some response to Fe(II) *in vitro*, and that Fe(II) might even be a more likely ligand than Ni(II) and Co(II) under physiological conditions. However, typical methods of characterization of RNA conformational changes are difficult to adapt to anaerobic conditions. We sought to interrogate the metal selectivity of the NiCo riboswitch by designing a fluorescent-based sensor. Using the strategy developed by the Hammond [96, 151] and Jaffrey [97] groups to convert small-molecule-responsive riboswitches into fluorescent sensors, we designed a series of sensors based on the *Erysipelotrichaceae bacterium* (*Eba*) *czcD* riboswitch. In this approach (Figure 15a), a sensor is constructed by fusing the

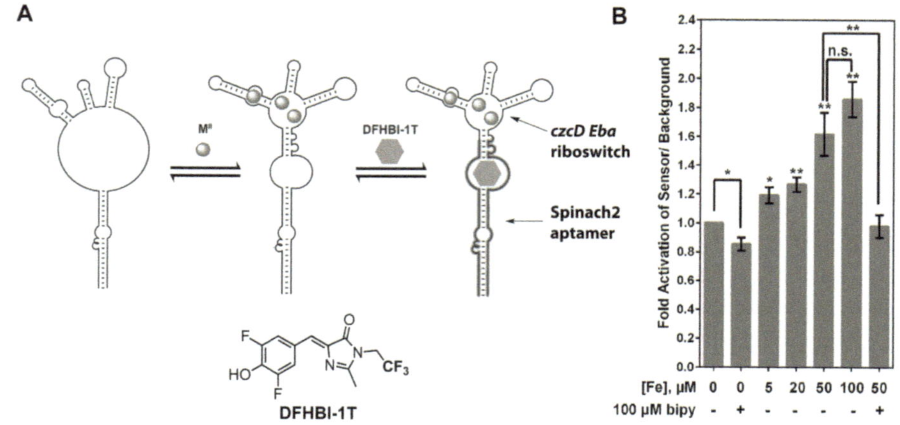

FIGURE 15 (a) Schematic representation of *czcD* riboswitch-based sensors. The *czcD* riboswitch was fused using the P1 stem to the Spinach2 aptamer. Upon Fe(II) (or other divalent metal ions) binding, the fluorophore-binding site in Spinach2 is formed. Binding of DFHBI-1T enhances its fluorescence, resulting in metal-dependent fluorescence increase overall. (b) Response of **czcD-2** to Fe(II) in *E. coli*, measured by flow cytometry. Cells were grown in M9 minimal medium with or without the indicated concentrations of added ferric citrate (1 hour). For selected samples, bipyridyl (bipy, at 100 µM) was added prior to analysis to illustrate the reversibility of the sensor's response in cells. *$p<0.05$; **$p<0.01$. n.s.: not significant. Copyright 2020, American Chemical Society; reproduced with permission.

RNA aptamer, Spinach2 [152], to one of the stems of the riboswitch. Binding of the cognate ligand to the riboswitch induces a conformational change, which organizes the transducer stem and forms the binding site in Spinach2 for the GFP chromophore analog DFHBI-1T [153] – (5*Z*)-5-[(3,5-difluoro-4-hydroxyphenyl)methylene]-3,5-dihydro-2-methyl-3-(2,2,2-trifluoroethyl)-4*H*-imidazol-4-one. In solution, DFHBI-1T fluorescence is weak due to bond rotation, but the rigid environment of the aptamer leads to a large increase in fluorescence. We found that in the case of the *Eba czcD* riboswitch, fusion of Spinach2 to the riboswitch's P1 stem at three different points led to a large turn-on response upon binding of metal to the riboswitch (Table 4). Characterization of these three sensors revealed that the riboswitch responds not

TABLE 4
Riboswitch-Based Sensors for Fe(II)

Probe Name	Excitation (nm)	Emission (nm)	$K_{d,\,app}$ (µM)	F/F_0	Ref.
czcD-1	475	505	4.9	11	154
czcD-2	475	505	0.40	7.7	154
czcD-3	475	505	0.20	2.5	154
Lmo-1	475	505	3.1	12	12
Lmo-2	475	505	0.27	4.1	12
Lmo-3	475	505	2.7	13	12

TABLE 5
Metal Selectivity and Fluorescence Response (F/F_0) of Two Riboswitch-Based Fluorescent Sensors, czcD-2 and Lmo-1

Metal	czcD-2[a]		Lmo-1[b]	
	$K_{d,app}$ (μM)	F/F_0	$K_{d,app}$ (μM)	F/F_0
Mn(II)	11	9.0	NR[c]	NR
Fe(II)	0.40	7.7	3.1	12.3
Co(II)	0.12	9.0	0.58	8.0
Ni(II)	0.061	8.2	0.17	5.3
Zn(II)	0.093	7.4	NR	NR

[a] Data from Ref. [154].
[b] Data from Ref. [12].
[c] NR, no significant response; $F/F_0 < 2$ at 50 μM metal ion.

only to Co(II) and Ni(II) but also to Fe(II), Zn(II), and (more weakly) to Mn(II) *in vitro* (Table 5).

The K_d of the **czcD-2** sensor for Fe(II) was best suited for in-cell application (Table 4). These studies utilized the *E. coli* BL21 Star (DE3) strain, which has a mutation in the gene encoding RNase E to improve mRNA stability; RNA-based sensors generally require this strain to be used for studies in *E. coli* [96]. Applying **czcD-2** and analyzing cellular fluorescence using flow cytometry, we demonstrated a dose-dependent fluorescence response of the sensor inside the cell, with a significant response to iron added at concentrations as low as 5 μM (Figure 15b). The ability to decrease fluorescence both at basal levels and high iron levels following treatment with bipyridyl demonstrates both the reversibility of the sensor and that the sensor is tuned to the physiological labile Fe(II) concentrations [154]. **czcD-2** did respond to the addition of 20 μM other metals, but it is not clear whether this is the result of high intracellular concentrations of those metals displacing Fe(II) from native binding sites or response of the riboswitch itself to the metals. Perhaps more likely, the BL21 Star (DE3) strain, derived from *E. coli* B, is defective in sensing cobalt and nickel [155], which also may have artificially increased the intracellular free concentrations of these metals and led to these responses. Normally both metals are regulated at orders of magnitude lower concentrations than Fe(II), levels that would be predicted to be inaccessible to the sensor [59, 60]. Regardless, a sensor with similar K_ds for Co(II) and Ni(II) as for Fe(II) would still be selective under most *in vivo* conditions. Therefore, our riboswitch-based sensor is, to the best of our knowledge, the first reversible genetically encoded sensor for iron.

More recently, we used an analogous strategy focusing again on fusions to the P1 stem to develop sensors based on the *Listeria monocytogenes czcD* riboswitch (**Lmo-1**, **Lmo-2**, and **Lmo-3**) [12]. Interestingly, the best-performing sensors of this set, **Lmo-1** and **Lmo-3**, only minimally responded to Mn(II) and Zn(II) even at 50 μM, above likely physiological levels for both metals (Table 5). Therefore, these sensors have the potential to more selectively sense Fe(II) than **czcD-2** for in-cell applications. This work also demonstrates that P1 stem fusions are a general strategy to create iron sensors based on this class of riboswitches.

While these genetically encoded iron sensors are promising first steps, there is also significant room for improvement. Riboswitch-based sensors are generally unstable in cells, and they require high expression levels that may alter analyte dynamics. They also cannot currently be extended to mammalian systems. Furthermore, the intensitometric nature of the **czcD-2** sensor necessitated a cumbersome calibration protocol for in-cell quantification. However, the recent development of ratiometric RNA-based sensor platforms [156] and a FRET system using the RNA aptamers Spinach and Mango [157] promises that the advantages of ratiometric protein-based sensors can also be extended to future riboswitch-based sensors. More specifically to sensing Fe(II), it is also important to note that the Fe(II)-**Lmo-1** K_d of 3.1 μM is higher than the Fe(II)-Fur K_d (1.2 μM), suggesting that the sensor may not be in its optimal response range under aerobic conditions. We have proposed that the *czcD* riboswitches may be tuned for response to iron overload under anaerobic conditions [12], where higher Fe(II) concentrations seem to be tolerated [158]. Perhaps fortuitously, an advantage of RNA-based sensors over typical fluorescent protein-based sensors, which require O_2 for fluorophore maturation, is that they can be applied under anaerobic conditions [159]. Therefore, these riboswitch-based iron sensors may be ideally suited for investigation of iron homeostasis in anaerobiosis.

Finally, Wang et al. have also used the *czcD* riboswitch in a genetic reporter system, by placing it upstream of a gene for expression of the red fluorescent protein, mCherry [160]. Because metal binding to this family of riboswitches induces transcription anti-termination, metal binding increases the expression of the downstream gene [148] – a turn-on fluorescence response unlike the Fur-based reporter described in 4.4.1. Three different riboswitches were evaluated for this *czcD*-based reporter, with the *Eba* riboswitch – the same as used for the **czcD-2** sensor – giving the best results. In *E. coli* grown in rich medium (LB), an increase in fluorescence was visible in response to >50 μM added Co(II), and >1 mM added Ni(II), after 2–4 hours. The sensor did not respond to added Fe(II) and Fe(III). However, under these growth conditions, the cells are likely iron sufficient. It is possible that the significant basal fluorescence observed with this reporter system could be from iron, given our data indicating that this riboswitch is also Fe(II)-responsive *in vitro* [12]. Therefore, we propose that this reporter would also be usable as an iron probe. Like the Fur-based reporter described above, a major disadvantage of this reporter system is that mCherry signal build ups over time. Thus, it would not be very responsive to a decrease in labile iron. Destabilizing the fluorescent protein by targeting it for degradation may enhance temporal resolution.

5 CONCLUSIONS AND PERSPECTIVE

The development of sensors for metal ions such as Ca(II), Mg(II), Cu(I), and Zn(II) is a relatively mature field of activity, and these sensors have been applied to discover new biology of these metals. By contrast, there are still significant challenges in designing and implementing broadly useful sensors for the lowest metals in the

Irving–Williams series, Mn(II) and Fe(II). Nevertheless, important strides have been made in the last decade on this inherently difficult problem. The greatest advances have been made in Fe(II) sensing, particularly with the development of multiple types of reaction-based turn-on probes that have been used to help uncover new aspects of iron physiology. Despite the high selectivity imparted by the reactive trigger, the downside of these tools is their irreversibility and the difficulty of quantification of iron concentrations. The field of Mn(II) sensing has been hindered by the paucity of receptors identified for binding with requisite selectivity over Fe(II), Ca(II), Zn(II), and Mg(II) for in-cell application. Although it has some disadvantages, the MESMER method is perhaps the best fluorescence-based method for quantification of total cellular Mn(II). No reaction-based schemes for probing Mn(II) have been developed to date, although this would be an intriguing direction for further exploration.

Recent work has shown that genetically encoded sensors for Mn(II) and Fe(II) can be constructed, although there is still significant room for improvement. Our work has yielded a genetically encoded FRET-based sensor for Mn(II) that performs well in bacteria but is not yet applicable in eukaryotic systems because its $K_{d,\text{app}}$ is slightly too high. Significantly more work is required for the development of genetically encoded iron sensors. Genetic reporters have limited temporal resolution, whereas the **czcD-2** sensor is relatively rapidly reversible but can only be implemented in strains where RNAs can be stably expressed. A protein-based sensor for Fe(II) would be more versatile than the riboswitch-based sensor, but an appropriate protein starting point would have to be identified. Nevertheless, the developments of the last few years in both synthetic and genetically encoded systems provide a clear path forward to better illuminate the biology of these important metal ions.

ACKNOWLEDGMENTS

The authors acknowledge an NIGMS grant GM138308 (to J.A.C.) for supporting this work and our laboratory's efforts to construct genetically encoded sensors for Mn(II) and Fe(II). J.A.C. also thanks a Louis Martarano Career Development Professorship for support. The authors declare no competing financial interest.

ABBREVIATIONS AND DEFINITIONS

ICP-MS	inductively coupled plasma mass spectrometry
LA-ICP-MS	laser ablation inductively coupled plasma mass spectrometry
XFM	X-ray fluorescence microscopy
PET	photo-induced electron transfer
EPR	electron paramagnetic resonance
ENDOR	electron nuclear double resonance
CFMEA	cellular Fura-2 Mn(II) extraction assay
MESM	manganese-extracting small molecule
BAPTA	1,2-bis(*o*-aminophenoxy)ethane-*N,N,N′,N′*-tetraacetic acid
TPEN	*N,N,N′,N′*-tetrakis(2-pyridylmethyl)ethylenediamine
BODIPY	boron dipyrromethene (4,4-difluoro-4-bora-3a,4a-diaza-s-indacene)
DMSO	dimethyl sulfoxide

CB	calcein blue
Fz1	FluoZin-1
HEK	human embryonic kidney
DMT1	divalent metal transporter 1
AM	acetoxymethyl
NPEY	1,2-bis-(2-pyren-1-ylmethylamino-ethoxy)ethane
SiQD	silicon quantum dot
FRET	Förster resonance energy transfer
B. subtilis	*Bacillus subtilis*
MntR	manganese transport regulator
MnLaMP	genetically encoded sensor for Mn(II) engineered from lanmodulin
ECFP	enhanced cyan fluorescent protein
E. coli	*Escherichia coli*
DFO	desferrioxamine B
Fur	ferric uptake regulator
K_d	dissociation constant
IDA	iminodiacetic acid
tpy	terpyridine
IP1	iron probe 1
TICT	twisted intramolecular charge transfer
RhoNox	rhodamine *N*-oxide-based iron probe
DANSYL	5-(dimethylamino)naphthalene-1-sulfonyl
DABSYL	4-(dimethylamino)azobenzene-4'-sulfonyl
5-AMF	5-aminomethylfluorescein
GFP	green fluorescent protein
Eba	*Erysipelotrichaceae bacterium*
DFHBI-1T	(5*Z*)-5-[(3,5-difluoro-4-hydroxyphenyl)methylene]-3,5-dihydro-2-methyl-3-(2,2,2-trifluoroethyl)-4*H*-imidazol-4-one
bipy	2,2'-bipyridine
Lmo	*Listeria monocytogenes*

REFERENCES

1. J. E. Posey, F. C. Gherardini, *Science* **2000**, *288*, 1651–1653.
2. K. Keyer, J. A. Imlay, *Proc. Natl. Acad. Sci. U. S. A.* **1996**, *93*, 13635–13640.
3. S. J. Dixon, K. M. Lemberg, M. R. Lamprecht; R. Skouta; E. M. Zaitsev; C. E. Gleason; D. N. Patel, A. J. Bauer, A. M. Cantley, W. S. Yang, B. Morrison III, B. R. Stockwell, *Cell* **2012**, *149* (5), 1060–1072.
4. S. J. Dixon, B. R. Stockwell, *Annu. Rev. Cancer Biol.* **2019**, *3* (1), 35–54.
5. S. Masaldan, A. I. Bush, D. Devos, A. S. Rolland, C. Moreau, *Free Rad. Biol. Med.* **2019**, *133*, 221–233.
6. M. I. Hood, E. P. Skaar, *Nat. Rev. Microbiol.* **2012**, *10*, 525–537.
7. G. Guan, A. Pinochet-Barros, A. Gaballa, S. J. Patel, J. M. Argüello, J. D. Helmann, *Mol. Microbiol.* **2015**, *98*, 787–803.
8. H. Pi, S. J. Patel, J. M. Argüello, J. D. Helmann, *Mol. Microbiol.* **2016**, *100*, 1066–1079.
9. H. Pi, J. D. Helmann, *Metallomics* **2017**, *9*, 840–851.
10. A. Pinochet-Barros, J. D. Helmann, *J. Bacteriol.* **2020**, *202*, e00697-19.

11. R. Guth-Metzler, M. S. Bray, M. Frenkel-Pinter, S. Suttapitugsakul, C. Montllor-Albalate, J. C. Bowman, R. Wu, A. R. Reddi, C. D. Okafor, J. B. Glass, L. D. Williams, *Nucl. Acids Res.* **2020**, *48*, 8663–8674.
12. J. Xu, J. A. Cotruvo, Jr., *ACS Bio Med Chem Au* **2022**, in revision doi: 10.1021/acsbiomedchemau.1c00069.
13. J. D. Aguirre, V. C. Culotta, *J. Biol. Chem.* **2012**, *287* (17), 13541–13548.
14. J. D. Helmann, *J. Biol. Chem.* **2014**, *289*, 28112–28120.
15. J. A. Imlay, *J. Biol. Chem.* **2014**, *289*, 28121–28128.
16. J. A. Cotruvo, Jr., J. Stubbe, *Metallomics* **2012**, *4*, 1020–1036.
17. M. B. Brophy, E. M. Nolan, *ACS Chem. Biol.* **2015**, *10*, 641–651.
18. K. K. Kumar, E. W. J. Lowe, A. A. Aboud, M. D. Neely, R. Redha, J. A. Bauer, M. Odak, C. D. Weaver, J. Meiler, M. Aschner, A. B. Bowman, *Sci. Rep.* **2014**, *4*, DOI: 10.1038/srep06801.
19. K. K. Kumar, C. R. Goodwin, M. A. Uhouse, J. Bornhorst, T. Schwerdtle, M. Aschner, J. A. McLean, A. B. Bowman, *Metallomics* **2015**, *7*, 363–370.
20. B. B. Williams, G. F. Kwakye, M. Wegrzynowicz, D. Li, M. Aschner, K. M. Erikson, A. B. Bowman, *Toxicol. Sci.* **2010**, *117* (1), 169–179.
21. R. G. Lucchini, E. Albini, L. Benedetti, S. Borghesi, R. Coccaglio, E. C. Malara, G. Parrinello, S. Garattini, S. Resola, L. Alessio, *Am. J. Ind. Med.* **2007**, *50* (11), 788–800.
22. D. Milatovic, S. Zaja-Milatovic, R. C. Gupta, Y. C. Yu, M. Aschner, *Toxicol. Appl. Pharm.* **2009**, *240* (2), 219–225.
23. A. B. Bowman, G. F. Kwakye, E. H. Hernandez, M. Aschner, *J. Trace Elem. Med. Bio.* **2011**, *25* (4), 191–203.
24. M. Stamelou, K. Tuschl, W. K. Chong, A. K. Burroughs, P. B. Mills, K. P. Bhatia, P. T. Clayton, *Movement Disord.* **2012**, *27*, 1317–1322.
25. K. Mukhtiar, S. Ibrahim, K. Tuschl, P. Mills, *Brain Develop.* **2016**, *38* (9), 862–865.
26. K. Tuschl, E. Meyer, L. E. Valdivia, N. Zhao, C. Dadswell, A. Abdul-Sada, C. Y. Hung, M. A. Simpson, W. K. Chong, T. S. Jacques, R. L. Woltjer, S. Eaton, A. Gregory, L. Sanford, E. Kara, H. Houlden, S. M. Cuno, H. Prokisch, L. Valletta, V. Tiranti, R. Younis, E. R. Maher, J. Spencer, A. Straatman-Iwanowska, P. Gissen, L. A. Selim, G. Pintos-Morell, W. Coroleu-Lletget, S. S. Mohammad, S. Yoganathan, R. C. Dale, M. Thomas, J. Rihel, O. A. Bodamer, C. A. Enns, S. J. Hayflick, P. T. Clayton, P. B. Mills, M. A. Kurian, S. W. Wilson, *Nat. Commun.* **2016**, *7*, 11601.
27. D. Leyva-Illades, P. Chen, C. E. Zogzas, S. Hutchens, J. M. Mercado, C. D. Swaim, R. A. Morrisett, A. B. Bowman, M. Aschner, S. Mukhopadhyay, *J. Neurosc.* **2014**, *34*, 14079–14095.
28. S. Jenkitkasemwong, A. Akinyode, E. Paulus, R. Weiskirchen, Hojyo, S.; Fukada, T.; Giraldo, G.; Schrier, J.; Garcia, A.; Janus, C.; Giasson, B.; Knutson, M. D., *Proc. Natl. Acad. Sci. U. S. A.* **2018**, *115* (8), e1769–e1778.
29. S. Mukhopadhyay, *Neurotoxicology* **2018**, *64*, 278–283.
30. C. Liu, T. Jursa, M. Aschner, D. R. Smith, S. Mukhopadhyay, *Proc. Natl. Acad. Sci. U. S. A.* **2021**, *118* (35) e2107673118.
31. E. L. Que, D. W. Domaille, C. J. Chang, *Chem. Rev.* **2008**, *108* (5), 1517–1549.
32. E. C. Greenwald, S. Mehta, J. Zhang, *Chem. Rev.* **2018**, *118* (24), 11707–11794.
33. K. P. Carter, A. M. Young, A. E. Palmer, *Chem. Rev.* **2014**, *114* (8), 4564–4601.
34. A. T. Aron, K. M. Ramos-Torres, J. A. Cotruvo, C. J. Chang, *Acc. Chem. Res.* **2015**, *48* (8), 2434–2442.
35. D. A. Iovan, S. Jia, C. J. Chang, *Inorg. Chem.* **2019**, *58* (20), 13546–13560.
36. H. Irving, R. J. P. Williams, *Nature* **1948**, *162*, 746–747.
37. R. Y. Tsien, *Biochemistry* **1980**, *19* (11), 2396–23404.

38. A. E. Palmer, M. Giacomello, T. Kortemme, S. A. Hires, V. Lev-Ram, D. Baker, R. Y. Tsien, *Chem. Biol.* **2006**, *13* (5), 521–530.
39. T. Thestrup, J. Litzlbauer, I. Bartholomaus, M. Mues, L. Russo, H. Dana, Y. Kovalchuk, Y. Liang, G. Kalamakis, Y. Laukat, S. Becker, G. Witte, A. Geiger, T. Allen, L. C. Rome, T.-W. Chen, D. S. Kim, O. Garaschuk, C. Griesinger, O. Griesbeck, *Nat. Methods* **2014**, *11* (2), 175–182.
40. T. W. Chen, T. J. Wardill, Y. Sun, S. R. Pulver, S. L. Renninger, A. Baohan, E. R. Schreiter; R. A. Kerr, M. B. Orger, V. Jayaraman, L. L. Looger, K. Svoboda, D. S. Kim, *Nature* **2013**, *499*, 295–300.
41. X. Zhou, K. J. Belavek, E. W. Miller, *Biochemistry* **2021**, *60* (46), 3547–3554.
42. E. M. Nolan, S. J. Lippard, *Acc. Chem. Res.* **2009**, *42* (1), 193–203.
43. Y. Qin, P. J. Dittmer, J. G. Park, K. B. Jansen, A. E. Palmer, *Proc. Natl. Acad. Sci. U. S. A.* **2011**, *108* (18), 7351–7356.
44. P. J. Dittmer, J. G. Miranda, Gorski, J. A.; Palmer, A. E., *J. Biol. Chem.* **2009**, *284*, 16289–16297.
45. J. G. Miranda, A. L. Weaver, Y. Qin, J. G. Park, C. I. Stoddard, M. Z. Lin, A. E. Palmer, *PLoS One* **2012**, *7* (11), 16.
46. J. L. Vinkenborg, T. J. Nicolson, E. A. Bellomo, M. S. Koay, G. A. Rutter, M. Merkx, *Nat. Methods* **2009**, *6* (10), 737–740.
47. Y. Qin, D. W. Sammond, E. Braselmann, M. C. Carpenter, A. E. Palmer, *ACS Chem. Biol.* **2016**, *11* (10), 2744–2751.
48. L. Yang, R. McRae, M. M. Henary, R. Patel, B. Lai, S. Vogt, C. J. Fahrni, *Proc. Natl. Acad. Sci. U. S. A.* **2005**, *102* (32), 11179–11184.
49. L. Zeng, E. W. Miller, A. Pralle, E. Y. Isacoff, C. J. Chang, *J. Am. Chem. Soc.* **2006**, *128* (1), 10–11.
50. J. Liu, J. Karpus, S. V. Wegner, P. R. Chen, C. He, *J. Am. Chem. Soc.* **2013**, *135*, 3144–3149.
51. S. C. Dodani, A. Firl, J. Chan, C. I. Nam, A. T. Aron, C. S. Onak, K. M. Ramos-Torres, J. Paek, C. M. Webster, M. B. Feller, C. J. Chang, *Proc. Natl. Acad. Sci. U. S. A.* **2014**, *111*, 16280–16285.
52. M. T. Morgan, B. Yang, S. Harankhedkar, A. Nabatilan, D. Bourassa, A. M. McCallum, F. Sun, R. Wu, C. R. Forest, C. J. Fahrni, *Angew. Chem. Int. Ed. Engl.* **2018**, *57* (31), 9711–9715.
53. C. M. Ackerman, S. Lee, C. J. Chang, *Anal. Chem.* **2017**, *89* (1), 22–41.
54. A. W. Foster, T. R. Young, P. T. Chivers, N. J. Robinson, *Curr. Opin. Chem. Biol.* **2022**, *66*, 102095.
55. L. Banci, I. Bertini, S. Ciofi-Baffoni, T. Kozyreva, K. Zovo, P. Palumaa, *Nature* **2010**, *465* (7298), 645–648.
56. S. J. Patel, A. G. Frey, D. J. Palenchar, S. Achar, K. Z. Bullough, A. Vashisht, J. A. Wohlschlegel, C. C. Philpott, *Nat. Chem. Biol.* **2019**, *15* (9), 872–881.
57. R. L. McNaughton, A. R. Reddi, M. H. S. Clement, A. Sharma, K. Barnese, L. Rosenfeld, E. B. Gralla, J. S. Valentine, V. C. Culotta, B. M. Hoffman, *Natl. Acad. Sci. U. S. A.* **2010**, *107* (35), 15335–15339.
58. C. E. Outten, T. V. O'Halloran, *Science* **2001**, *292* (5526), 2488–2492.
59. D. Osman, M. A. Martini, A. W. Foster, J. Chen, A. J. P. Scott, R. J. Morton, J. W. Steed, E. Lurie-Luke, T. G. Huggins, A. D. Lawrence, E. Deery, M. J. Warren, P. T. Chivers, N. J. Robinson, *Nat. Chem. Biol.* **2019**, *15*, 241–249.
60. M. R. Jordan, J. Wang, D. A. Capdevila, D. P. Giedroc, *Curr. Opin. Microbiol.* **2020**, *55*, 17–25.
61. A. P. de Silva, H. Q. N. Gunaratne, T. Gunnlaugsson, A. J. M. Huxley, C. P. McCoy, J. T. Rademacher, T. E. Rice, *Chem. Rev.* **1997**, *97* (5), 1515–1566.
62. A. W. Czarnik, *Acc. Chem. Res.* **1994**, *27* (10), 302–308.

63. S. Das, K. Khatua, A. Rakshit, A. Carmona, A. Sarkar, S. Bakthavatsalam, R. Ortega, A. Datta, *Dalton Transac.* **2019**, *48* (21), 7047–7061.
64. D. E. Ash, V. L. Schramm, *J. Biol. Chem.* **1982**, *257*, 9261–9264.
65. A. Sharma, E. K. Gaidamakova, V. Y. Matrosova, B. Bennett, M. J. Daly, B. M. Hoffman, *Natl. Acad. Sci. U. S. A.* **2013**, *110* (15), 5945–5950.
66. L. C. Tabares, S. Un, *J. Biol. Chem.* **2013**, *288* (7), 5050–5055.
67. C. J. da Silva, A. N. J. da Rocha, M. F. Mendes, A. P. S. D. Braga, S. Jeronymo, *Arch. Neurol-Chicago* **2008**, *65* (7), 983–983.
68. Long, Z. Y.; Jiang, Y. M.; Li, X. R.; William, F.; Xu, J.; Yeh, C. L.; Long, L. L.; Luo, H. L.; Jaroslaw, H.; James, M.; Zheng, W.; Ulrike, D., *Neurotoxicology* **2014**, *45*, 285–292.
69. G. F. Kwakye, D. Li, O. A. Kabobel, A. B. Bowman, *Curr. Protoc. Toxicol.* **2011**, *48*, 12–18.
70. K. K. Kumar, E. W. Lowe, A. A. Aboud, M. D. Neely, R. Redha, J. A. Bauer, M. Odak, C. D. Weaver, J. Meiler, M. Aschner, A. B. Bowman, *Sci. Rep.* **2014**, *4*, DOI: 10.1038/srep06801.
71. K. J. Horning, P. Joshi, R. Nitin, R. C. Balachandran, F. M. Yanko, K. Kim, P. Christov, M. Aschner, G. A. Sulikowski, C. D. Weaver, A. B. Bowman, *J. Biol. Chem.* **2020**, *295* (12), 3875–3890.
72. J. Liang, J. W. Canary, *Angew. Chem. Int. Ed. Engl.* **2010**, *49* (42), 7710–3.
73. F. T. Senguen, Z. Grabarek, *Biochemistry* **2012**, *51* (31), 6182–6194.
74. P. Arslan, F. Di Virgilio, M. Beltrame, R. Y. Tsien, T. Pozzan, *J. Biol. Chem.* **1985**, *260* (5), 2719–2727.
75. S. Bakthavatsalam, A. Sarkar, A. Rakshit, S. Jain, A. Kumar, A. Datta, *Chem. Commun.* **2015**, *51* (13), 2605–2608.
76. A. Sarkar, I. E. Biton, M. Neeman, A. Datta, *Inorg. Chem. Commun.* **2017**, *78*, 21–24.
77. S. Das, A. Carmona, K. Khatua, F. Porcaro, A. Somogyi, R. Ortega, A. Datta, *Inorg. Chem.* **2019**, DOI: 10.1021/acs.inorgchem.9b01389.
78. A. Carmona, G. Deves, S. Roudeau, P. Cloetens, S. Bohic, R. Ortega, *ACS. Chem. Neurosci.* **2010**, *1* (3), 194–203.
79. A. Carmona, S. Roudeau, L. Perrin, G. Veronesi, R. Ortega, *Metallomics* **2014**, *6* (4), 822–832.
80. F. Borsetti, F. Dal Piaz, F. D'Alessio, A. Stefan, R. Brandimarti, A. Sarkar, A. Datta, A. Montón Silva; T. den Blaauwen; M. Alberto, E. Spisni, A. Hochkoeppler, *Microbiology* **2018**, *164* (10), 1266–1275.
81. S. Das, A. Sarkar, A. Rakshit, A. Datta, *Inorg. Chem.* **2018**, *57* (9), 5273–5281.
82. S. Adhikari, A. Ghosh, A. Sahana, S. Guria, D. Das, *Anal. Chem.* **2016**, *88* (2), 1106–1110.
83. Z. Dai, N. Khosla, J. Canary, *Supramol. Chem.* **2009**, *21* (3–4), 296–300.
84. F. Gruppi, J. Liang, B. B. Bartelle, M. Royzen, D. H. Turnbull, J. W. Canary, *Chem. Commun.* **2012**, *48* (87), 10778–10780.
85. R. Y. Tsien, *Nature* **1981**, *290* (5806), 527–528.
86. N. Dhenadhayalan, H. L. Lee, K. Yadav, K. C. Lin, Y. T. Lin, A. H. H. Chang, *ACS Appl. Mater. Inter.* **2016**, *8* (36), 23953–23962.
87. S. H. Xu, C. L. Wang, H. S. Zhang, Q. F. Sun, Z. Y. Wang, Y. P. Cui, *J. Mater. Chem.* **2012**, *22* (18), 9216–9221.
88. R. Hu, L. Zhang, H. B. Li, *N. J. Chem.* **2014**, *38* (6), 2237–2240.
89. K. B. Narayanan, S. S. Han, *Res. Chem. Intermediat.* **2017**, *43* (10), 5665–5674.
90. X. Mao, H. Su, D. Tian, H. Li, R. Yang, *ACS Appl. Mater. Interfaces* **2013**, *5* (3), 592–597.
91. K. B. Kim, G. J. Park, H. Kim, E. J. Song, J. M. Bae, C. Kim, *Inorg. Chem. Commun.* **2014**, *46*, 237–240.

92. Y. J. Lee, C. Lim, H. Suh, E. J. Song, C. Kim, *Sensor Actuat. B-Chem.* **2014**, *201*, 535–544.
93. P. S. Hariharan, S. P. Anthony, *Spectrochim. Acta A* **2015**, *136*, 1658–1665.
94. S. Swami, A. Agarwala, D. Behera, R. Shrivastava, *Sensor Actuat. B-Chem.* **2018**, *260*, 1012–1017.
95. P. J. McCown, K. A. Corbino, S. Stav, M. E. Sherlock, R. R. Breaker, *RNA* **2017**, *23*, 995–1011.
96. C. A. Kellenberger, S. C. Wilson, J. Sales-Lee, M. C. Hammond, *J. Am. Chem. Soc.* **2013**, *135*, 4906–4909.
97. M. You, J. L. Litke, S. R. Jaffrey, *Proc. Natl. Acad. Sci. U. S. A.* **2015**, *112*, E2756–E2765.
98. L. H. Lindenburg, J. L. Vinkenborg, J. Oortwijn, S. J. A. Aper, M. Merkx, *PLoS One* **2013**, *8*, e82009.
99. A. M. McGuire, B. J. Cuthbert, Z. Ma, K. D. Grauer-Gray, M. Brunjes Brophy, K. A. Spear, S. Soonsanga, J. I. Kliegman, S. L. Griner, J. D. Helmann, A. Glasfeld, *Biochemistry* **2013**, *52* (4), 701–713.
100. J. J. Chou, S. Li, C. B. Klee, A. Bax, *Nat. Struct. Biol.* **2001**, *8*, 990–996.
101. G. S. Baird, D. A. Zacharias, R. Y. Tsien, *Proc. Natl. Acad. Sci. U. S. A.* **1999**, *96* (20), 11241–11246.
102. F. T. Senguen, Z. Grabarek, *Biochemistry* **2012**, *51*, 6182–6194.
103. J. A. Cotruvo, Jr., J. A. Mattocks, *Metal-Binding Protein and Use Thereof*, Application PCT/US2019/049652, 2019.
104. J. A. Cotruvo, Jr., E. R. Featherston, J. A. Mattocks, J. V. Ho, T. N. Laremore, *J. Am. Chem. Soc.* **2018**, *140*, 15056–15061.
105. J. A. Mattocks, J. V. Ho, J. A. Cotruvo, Jr., *J. Am. Chem. Soc.* **2019**, *141*, 2857–2861.
106. N. J. Kuhn, S. Ward, W. S. Leong, *Euro. J. Biochem.* **1991**, *195* (1), 243–250.
107. L. L. Stookey, *Anal. Chem.* **1970**, *42* (7), 779–781.
108. J. Riemer, H. H. Hoepken, H. Czerwinska, S. R. Robinson, R. Dringen, *Anal. Biochem.* **2004**, *331* (2), 370–375.
109. T. Romero, M. Gong, M. Y. Liu, A. Brun-Zinkernagel, *J. Bacteriol.* **1993**, *175*, 4744–4755.
110. C. Pourciao, A. Pannuri, A. Potts, H. Yakhnin, P. Babitzke, T. Romero, *mBio* **2019**, *10*, e01034.
111. K. Keyer, J. A. Imlay, *Proc. Natl. Acad. Sci. U. S. A.* **1996**, *93*, 13635–13640.
112. W. Breuer, S. Epsztejn, P. Millgram, I. Z. Cabantchik, *Am. J. Physiol.* **1995**, *268*, C1354–C1361.
113. A. E. Martell, R. M. Smith, *Critical Stability Constants*, Plenum Press, New York, 1974; Vol. 1: Amino Acids.
114. F. Petrat, H. de Groot, U. Rauen, *Arch. Biochem. Biophys.* **2000**, *376* (1), 74–81.
115. F. Petrat, U. Rauen, H. de Groot, *Hepatology* **2003**, *29* (4), 1171–1179.
116. P. Li, L. Fang, H. Zhou, W. Zhang, X. Wang, N. Li, H. Zhong, B. Tang, *Chemistry* **2011**, *17* (38), 10520–10531.
117. L. Praveen, M. L. P. Reddy, R. L. Varma, *Tetrahedron Lett.* **2010**, *51* (50), 6626–6629.
118. I. J. Carney, J. L. Kolanowski, Z. Lim, B. Chekroun, A. G. Torrisi, T. W. Hambley, E. J. New, *Metallomics* **2018**, *10* (4), 553–556.
119. H. Y. Au-Yeung, J. Chan, T. Chantarojsiri, C. J. Chang, *J. Am. Chem. Soc.* **2013**, *135* (40), 15165–15173.
120. L. Wu, Q. Ding, X. Wang, P. Li, N. Fan, Y. Zhou, L. Tong, W. Zhang, W. Zhang, B. Tang, *Anal. Chem.* **2020**, *92* (1), 1245–1251.
121. T. Hirayama, K. Okuda, H. Nagasawa, *Chem. Sci.* **2013**, *4* (3), 1250–1256.
122. M. Niwa, T. Hirayama, K. Okuda, H. Nagasawa, *Org. Biomol. Chem.* **2014**, *12* (34), 6590–6597.

123. T. Hirayama, H. Tsuboi, M. Niwa, A. Miki, S. Kadota, Y. Ikeshita, K. Okuda, H. Nagasawa, *Chem. Sci.* **2017**, *8* (7), 4858–4866.
124. A. T. Aron, M. O. Loehr, J. Bogena, C. J. Chang, *J. Am. Chem. Soc.* **2016**, *138* (43), 14338–14346.
125. B. Spangler, C. W. Morgan, S. D. Fontaine, M. N. Vander Wal, C. J. Chang, J. A. Wells, A. R. Renslo, *Nat. Chem. Biol.* **2016**, *12* (9), 680–685.
126. C. H. Cai, H. L. Wang, R. J. Man, *Spectrochim. Acta. A Mol. Biomol. Spectrosc.* **2021**, *255*, 119729.
127. C. V. Banks, R. I. Bystroff, *J. Am. Chem. Soc.* **1959**, *81* (23), 6153–6158.
128. A. Winter, M. Gottschaldt, G. R. Newkome, U. S. Schubert, *Curr. Top. Med. Chem.* **2012**, *12*, 158–175.
129. S. C. Dodani, Q. He, C. J. Chang, *J. Am. Chem. Soc.* **2009**, *131* (50), 18020–18021.
130. M. Taki, S. Iyoshi, A. Ojida, I. Hamachi, Y. Yamamoto, *J. Am. Chem. Soc.* **2010**, *132* (17), 5938–5939.
131. H. Y. Au-Yeung, E. J. New, C. J. Chang, *Chem. Commun.* **2012**, *48* (43), 5268–5270.
132. P. Feng, L. Ma, F. Xu, X. Gou, L. Du, B. Ke, M. Li, *Talanta* **2019**, *203*, 29–33.
133. A. T. Aron, M. C. Heffern, Z. R. Lonergan, M. N. Vander Wal, B. R. Blank, B. Spangler, Y. Zhang, H. M. Park, A. Stahl, A. R. Renslo, E. P. Skaar, C. J. Chang, *Proc. Natl. Acad. Sci. U. S. A.* **2017**, *114* (48), 12669–12674.
134. T. Hirayama, M. Inden, H. Tsuboi, M. Niwa, Y. Uchida, Y. Naka, I. Hozumi, H. Nagasawa, *Chem. Sci.* **2019**, *10* (5), 1514–1521.
135. T. Hirayama, M. Niwa, S. Hirosawa, H. Nagasawa, *ACS Sens.* **2020**, *5* (9), 2950–2958.
136. M. Niwa, T. Hirayama, I. Oomoto, D. O. Wang, H. Nagasawa, *ACS Chem. Biol.* **2018**, *13* (7), 1853–1861.
137. T. Hirayama, A. Miki, H. Nagasawa, *Metallomics* **2019**, *11* (1), 111–117.
138. S. Maiti, Z. Aydin, Y. Zhang, M. Guo, *Dalton Trans.* **2015**, *44* (19), 8942–8949.
139. A. Samuni, C. M. Krishna, J. B. Mitchell, C. R. Collins, A. Russo, *Free Rad. Res. Commun.* **1990**, *9* (3–6), 241–249.
140. Y. Tang, Y. Dong, X. Wang, K. Sriraghavan, J. K. Wood, J. L. Vennerstrom, *J. Org. Chem.* **2005**, *70*, 5103–5110.
141. D. A. Hanna, R. M. Harvey, O. Martinez-Guzman, X. Yuan, B. Chandrasekharan, G. Raju, F. W. Outten, I. Hamza, A. R. Reddi, *Proc. Natl. Acad. Sci.* **2016**, *113* (27), 7539.
142. R. K. Muir, N. Zhao, J. Wei, Y. H. Wang, A. Moroz, Y. Huang, Y. C. Chen, R. Sriram, J. Kurhanewicz, D. Ruggero, A. R. Renslo, M. J. Evans, *ACS Cent. Sci.* **2019**, *5* (4), 727–736.
143. Z. Ma, F. E. Jacobsen, D. P. Giedroc, *Chem. Rev.* **2009**, *109*, 4644–4681.
144. D. A. Capdevila, F. Huerta, K. A. Edmonds, M. T. Le, H. Wu, D. P. Giedroc, *eLife* **2018**, *7*, e37268.
145. X. Liu, P. A. Lopez, T. W. Giessen, M. Giles, J. C. Way, P. A. Silver, *Sci Rep* **2016**, *6* (1), 38019.
146. S. A. Mills, M. A. Marletta, *Biochemistry* **2005**, *44*, 13553–13559.
147. T. R. Young, M. A. Martini, A. W. Foster, A. Glasfeld, D. Osman, R. J. Morton, E. Deery, M. J. Warren, N. J. Robinson, *Nat. Commun.* **2021**, *12*, 1195.
148. K. Furukawa, A. Ramesh, Z. Zhou, Z. Weinberg, T. Vallery, W. C. Winkler, R. R. Breaker, *Mol. Cell* **2015**, *57*, 1088–1098.
149. I. R. Price, A. Gaballa, F. Ding, J. D. Helmann, A. Ke, *Mol. Cell* **2015**, *57*, 1110–1123.
150. M. Dambach, M. Sandoval, T. B. Updegrove, V. Anantharaman, L. Aravind, L. S. Waters, G. Storz, *Mol. Cell* **2015**, *57*, 1099–1109.
151. C. A. Kellenberger, C. Chen, A. T. Whiteley, D. A. Portnoy, M. C. Hammond, *J. Am. Chem. Soc.* **2015**, *137*, 6432–6435.

152. J. S. Paige, T. Nguyen-Duc, W. Song, S. R. Jaffrey, *Science* **2012**, *335*, 1194.
153. W. Song, R. L. Strack, N. Svensen, S. R. Jaffrey, *J. Am. Chem. Soc.* **2014**, *136*, 1198–1201.
154. J. Xu, J. A. Cotruvo, Jr., *Biochemistry* **2020**, *59*, 1508–1516.
155. C. Pinske, M. Bönn, S. Krüger, U. Lindenstrauß, R. G. Sawers, *PLoS One* **2011**, *6* (8), e22830.
156. R. Wu, A. P. K. K. Karunanayake Mudiyanselage, K. Ren, Z. Sun, Q. Tian, B. Zhao, Y. Bagheri, D. Lutati, P. Keshri, M. You, *ACS Appl. Bio Mater.* **2020**, *3* (5), 2633–2642.
157. M. D. E. Jepsen, S. M. Sparvath, T. B. Nielsen, A. H. Langvad, G. Grossi, K. V. Gothelf, Andersen, E. S., *Nat. Commun.* **2018**, *9* (1), 18.
158. N. A. Beauchene, E. L. Mettert, L. J. Moore, S. Keleş, E. R. Willey, P. J. Kiley, *Proc. Natl. Acad. Sci. U. S. A.* **2017**, *114*, 12261–12266.
159. X. C. Wang, S. C. Wilson, M. C. Hammond, *Nucl. Acids Res.* **2016**, *44*, e139.
160. X. Wang, W. Wei, J. Zhao, *Front. Chem.* **2021**, *9* (13), DOI: 10.3389/fchem.2021.631909.

5 Fluorescent Probes for Zinc Ions and Their Applications in the Life Sciences

Tingwen Wei
State Key Laboratory of Materials-Oriented Chemical Engineering, College of Chemical Engineering, Jiangsu National Synergetic Innovation Center for Advanced Materials (SICAM), Nanjing Tech University, Nanjing 211816, People's Republic of China

*Nahyun Kwon and Juyoung Yoon**
Department of Chemistry and Nano Science, Ewha Womans University, Seoul 03760, South Korea

*Xiaoqiang Chen**
State Key Laboratory of Materials-Oriented Chemical Engineering, College of Chemical Engineering, Jiangsu National Synergetic Innovation Center for Advanced Materials (SICAM), Nanjing Tech University, Nanjing 211816, People's Republic of China

CONTENTS

* Corresponding author.

DOI: 10.1201/9781003229971-5

Abstract

Zinc ion is one of the essential trace elements of the human body. It is involved in the synthesis of over a hundred enzymes and is also associated with the activities of nearly two-hundred enzymes. Zinc ion is also widely involved in the metabolism of lipids, saccharides, nucleic acids, enzymes, proteins, and gene transcription. It has been implicated in physiological processes associated with many diseases, such as diabetes and Alzheimer's disease, and also affects human growth, development, appetite, and taste. Therefore, developing methods for efficient and rapid detection of zinc ions in vitro and in vivo are quite significant. Fluorescent probe can be an excellent method for detecting and imaging zinc ion because of its simple operation, fast response, high sensitivity, and good selectivity. In the past decades, a large number and variety of fluorescent probes for sensing zinc ions have been developed. In this chapter, the response mechanism, recognition group, sensing type, targeted imaging, and applications of Zn(II)-selective fluorescent probes are introduced separately.

KEYWORDS

Bioimaging; Fluorescent Probes; Recognition; Response Mechanism; Zinc Ions Sensing

1 INTRODUCTION

Zinc has many uses in the manufacturing industry, especially in the galvanization industry, where it is used to coat the surface of steel or iron to produce a protective film [1]. This transition metal forms excellent alloys with many non-ferrous metals, which are indispensable materials in modern industry. In addition, this metal is widely used in battery manufacturing, medicine, rubber, paint and other industries [2–5]. Unfortunately, excessive emissions of zinc in mining, smelting, mechanical manufacturing, galvanizing, instrument production, organic synthesis, and papermaking processes cause serious pollution of soil, water, and other environments [6]. The seriousness of this problem is elevated by the multiple roles zinc plays in biological processes. Zinc ion is one of the most important trace elements and the second most abundant transition metal ion in the human body [7], which for adults typically contains about 2–3 g [8] corresponding to intracellular concentrations of 10–100 μM [9, 10]. Zinc ions are important cofactors in many enzyme-catalyzed reactions, and they play important roles in biological processes, such as apoptosis, mammalian reproduction, signaling, enzymatic function, immune function, and gene expression [11, 12]. Zinc has also been implicated in physiological processes associated with many diseases, such as diabetes and Alzheimer's disease [13, 14], and it also affects human growth, development, appetite, and taste [15]. Its significance as an environmental pollutant and a biologically important metal has made zinc ions targets of studies aimed at developing methods for rapid and efficient detection in water, soil, and biological samples.

Zinc ion has an electronic configuration containing five doubly occupied 3d orbitals, and thus, it possesses no empty d orbitals and no unpaired electron. As such zinc ion does not have a magnetic moment [16] and, as a result, cannot be detected using electron paramagnetic resonance spectroscopy. In addition, other conventional detection methods, such as nuclear magnetic resonance and UV-visible spectroscopy, are not applicable to direct real-time monitoring of zinc ions *in vitro* and *in vivo* [17]. In contrast, fluorescent detection technologies that utilize specifically designed probes have become extremely important in methods for the real-time detection of zinc ions inside and outside living organisms with high selectivity and sensitivity [18]. Ideal zinc-ion fluorescent probes have several advantageous properties, including high water solubility and optical stability, low background fluorescence, large and fast fluorescence signal responses, high detection selectivity and chemical stability, fast targeted transmission, and inertness to pH fluctuations. In addition, in order to have strong tissue penetration and low phototoxicity in biosamples, the excitation wavelength of an ideal probe should be at least in the visible light region or, preferably, the excitation and emission wavelengths would be in the near-infrared range [19].

Zinc-ion fluorescent probes usually contain two functional moieties, including the metal ion recognition center and the signaling fluorophore. Because zinc-ion fluorescent probes respond to zinc ions typically by first forming complexes, their recognition moieties are ligands that can donate electron pairs, such as Schiff bases, pyridines, diacids, quinolines, acyclic and cyclic polyamines, and other heterocyclic groups [20]. The signaling group in these sensors is commonly composed of a fluorophore or a nascent fluorophore, which is linked to the recognition site through either a varying length chain or directly, in which case the recognition group is part of the entire fluorophore.

2 MECHANISMS FOR RESPONSES OF FLUORESCENT PROBES TO ZINC IONS

Effective probes that detect zinc ion (Zn^{2+}) sensitively have low background fluorescence and high metal ion selectivity. The most widely used ones that have these properties operate through photo-induced electron transfer (PET) and intramolecular charge transfer (ICT) mechanisms [21]. In addition, probes have been designed to detect Zn^{2+} based on excited state intramolecular proton transfer (ESIPT) and Förster (fluorescence) resonance energy transfer (FRET) pathways.

2.1 Zinc-Ion Fluorescent Probes that Utilize a PET Mechanism

PET is a photo-induced process in which single electron transfer (SET) occurs to excited states of substances [22]. Nitrogens possessing a nonbonded electron pair are excellent SET donors in PET processes. As a result, substances containing these types of nitrogens serve as quenchers of fluorescence arising from excited states of fluorophores. Also, owing to their excellent coordinating ability, nitrogen centers possessing electron pairs strongly coordinate Zn^{2+} and by so doing lose their ability to serve as SET donors. These phenomena serve as the basis for the design of probes whose fluorescence is quenched in the absence of Zn^{2+} and "turned on" when this metal ion is present.

The polyamine-containing substances **1** and **2** (Figure 1) are examples of fluorescent probes for Zn^{2+} that function through PET-based mechanisms [23]. Probe **1** only weakly fluoresces due to the occurrence of PET from the di-2-picolylamine (DPA) electron donor site to the excited state of the conjugated fluorophore. When Zn^{2+} coordinates to the amine centers in **1**, PET is inhibited and fluorescence is recovered. Likewise, probe **2**, generated by esterification of **1** with acetic anhydride, is almost completely nonfluorescent due to both the operation of PET process and the lack of extended π-conjugation caused by formation of the spirocyclic lactone. Unlike most zinc-ion probes that are based on coordination with the metal ion, **2** functions through both coordination and a chemical reaction promoted by Zn^{2+}. Specifically, complexation of zinc ions with **2** not only leads to coordination to the polyamine centers, it also induces ester bond hydrolysis and subsequent spirolactone ring opening. Both these events lead to a significantly enhanced fluorescence efficiency.

Hydrazone derivative **3** is a Zn^{2+}-sensitive fluorescent probe with aggregation-induced emission (AIE) characteristics [24]. As shown in Figure 2, **3** achieves specific recognition for Zn^{2+} through coordination of the phenolic hydroxyl, imine nitrogen atom benzothiazole nitrogen. As a result, PET inhibition occurs and fluorescence emission is significantly enhanced in the presence of its metal ion.

Additionally, some unique modes of bonding can be found in Zn^{2+} probes that operate through the PET mechanism. An interesting example is **4**, which contains electron donor groups that coordinate with Zn^{2+}. Thus, the fluorescence of **4** is "turned on" in the presence of this metal ion. Moreover, like other tetraphenylethylene (TPE) derivatives, **4** takes advantage of AIE (Figure 3) [25]. Specifically, upon coordination to Zn^{2+} to the $N(CH_2COO)_2$ groups, **4** exhibits strong emission at 485 nm. It must be highlighted that the coordinating bonds between Zn^{2+} and **4** are both intramolecular and intermolecular in nature.

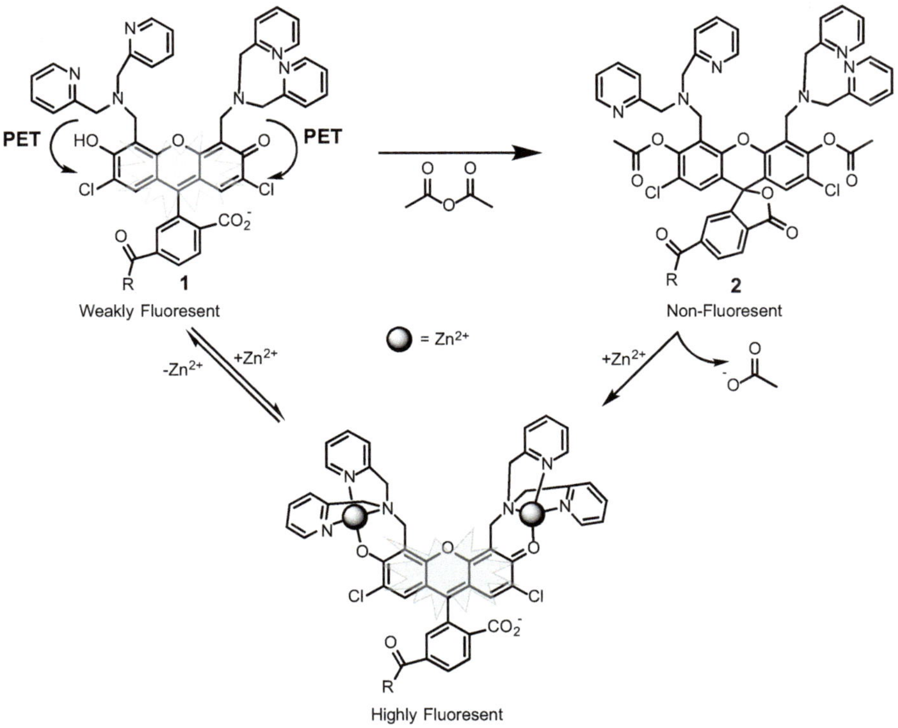

FIGURE 1 Schematic representation of zinc-induced fluorescence from probes **1** and **2**.

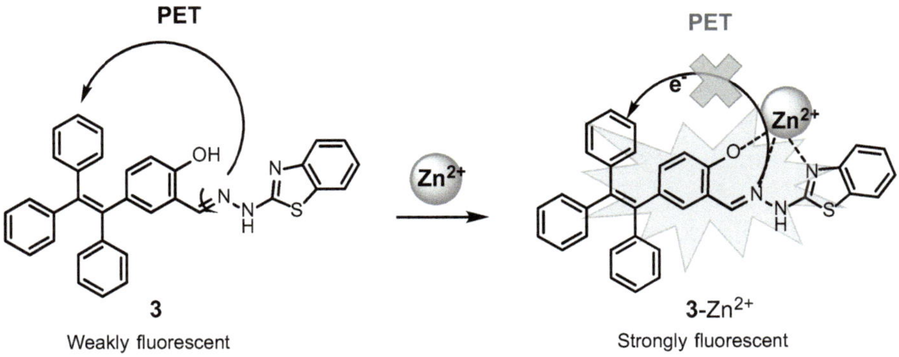

FIGURE 2 Possible sensing mechanism of probe **3** toward Zn^{2+}.

2.2 Zinc-Ion Fluorescent Probes that Operate Through an ICT Mechanism

Another important mechanism utilized in the design of Zn^{2+} sensors is ICT. In the electronically excited states of molecules, electrons tend to transfer from electron-rich terminal (donors) sites to electron-poor terminal (acceptors) sites in a process

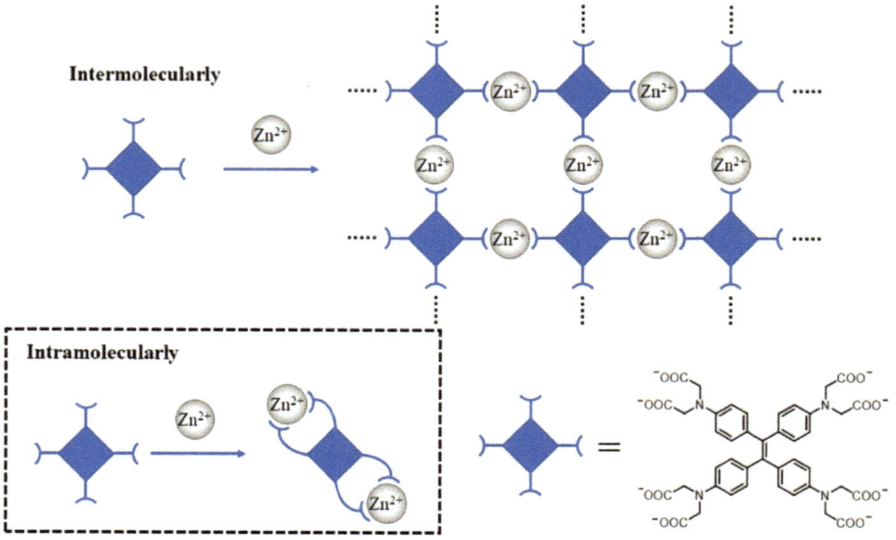

FIGURE 3 Coordination between **4** and Zn^{2+}. Adapted from Ref. [25] with permission. Copyright 2011, Organic Letters.

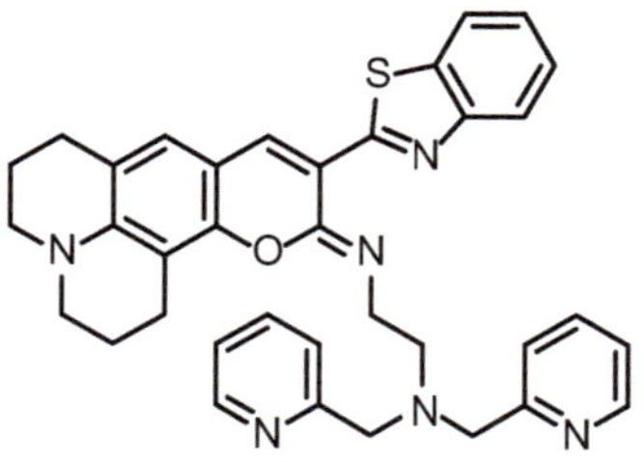

5

FIGURE 4 Structure of **5**.

called ICT [26]. Accompanying the alteration of the electron density, a remarkable red shift typically takes place, which helps avoid overlap of background and sensor fluorescent signals.

Some fluorophores, which themselves do not display a strong donor–acceptor (D-A) effect response in their excited states, can gain the ICT capability upon complexation with Zn^{2+}. An example is **5**, an iminocoumarin-based zinc sensor for ratiometric fluorescence imaging of neuronal Zn^{2+} (Figure 4) [27]. When Zn^{2+} interacts with **5**, the microenvironment of the fluorophore changes and causes a greater stabilization of the ICT excited state than the ground state. This leads to a red shift of the absorption and emission maxima of the probe. The behavior of **5** is not unique. The ratiometric fluorescent probe **6**, developed for imaging zinc ions in living cells (Figure 5) also functions in this manner [28]. This probe undergoes a red shift of 75 nm in the presence of Zn^{2+}, a large Stokes shift of the emission influenced by the ICT process.

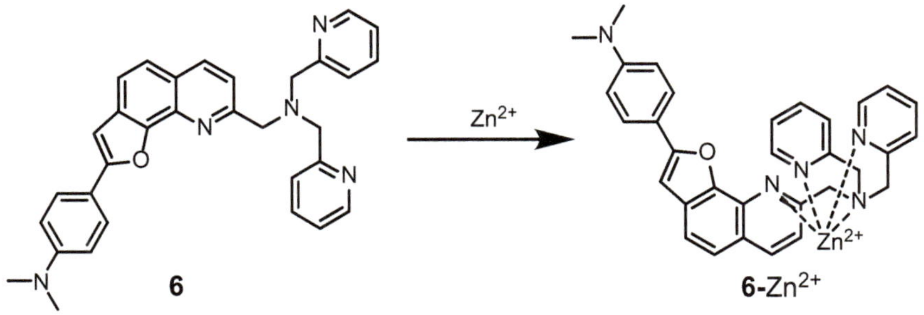

6

6-Zn2+

FIGURE 5 Structure of **6** and proposed mechanism for sensing Zn^{2+}.

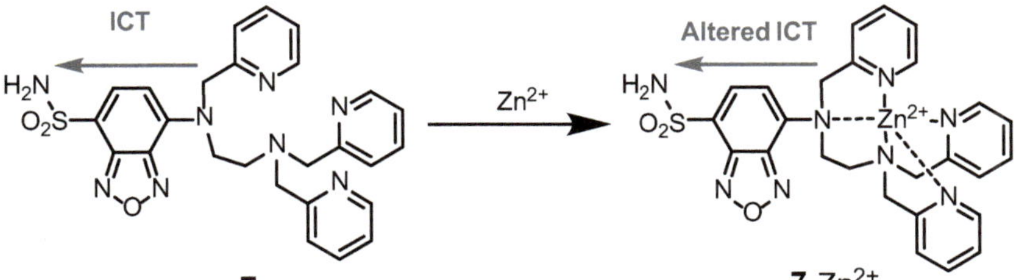

7

7-Zn2+

FIGURE 6 Structure of **7** and proposed mechanism for sensing Zn^{2+}.

Other fluorophores containing the potential for strong D-A interactions can display an ICT effect upon interaction with Zn^{2+}. For example, **7**, which has a large Stokes shift of 119 nm, has been developed as a visible light (460 nm) excitable ratiometric sensor for imaging Zn^{2+} *in vivo* (Figure 6) [29]. The probe has been applied to imaging zinc ions in zebrafish larvae. In the presence of zinc ions, the N atoms at the amine centers in the sensor coordinate to Zn^{2+}, which leads to a decrease in ICT. In addition, the increases in zinc-ion concentrations have little influence on the variation in the emission intensity, suggesting that a PET process not operative.

2.3 Zinc-Ion Fluorescent Probes Functioning through the FRET Mechanism

When light emission from one component (donor) of a system overlaps with the absorption of the other component (Acceptor), FRET can take place [30]. This energy exchange process results in a significant red shift in emissions. In probes that are based on FRET, the addition of Zn^{2+} leads to alteration of the properties of the donor, which thereby blocks the energy transfer and results in a change of the fluorescent signal.

The FRET-based sensor **8** (Figure 7), is constructed by integrating a two-photon excitable coumarin derivative as a donor and a 4-amine-7-sulfamoylbenzo[c][1,2,5]-oxadiazole (ASBD) as an acceptor [31]. Complexation of Zn^{2+} with the N,N,N′-tris(pyridin-2-ylmethyl)ethane-1,2-diamine (TPEA) moiety causes a large blue shift in the absorption of the ASBD moiety, which leads to a decrease in the

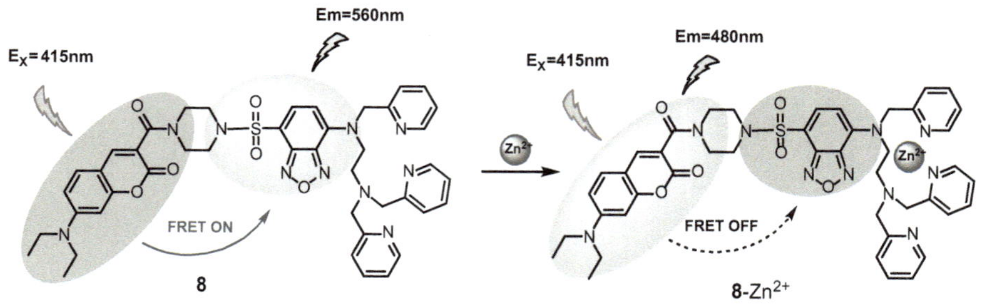

FIGURE 7 Structure of **8** and schematic illustration of its FRET-based sensing mechanism for Zn^{2+}.

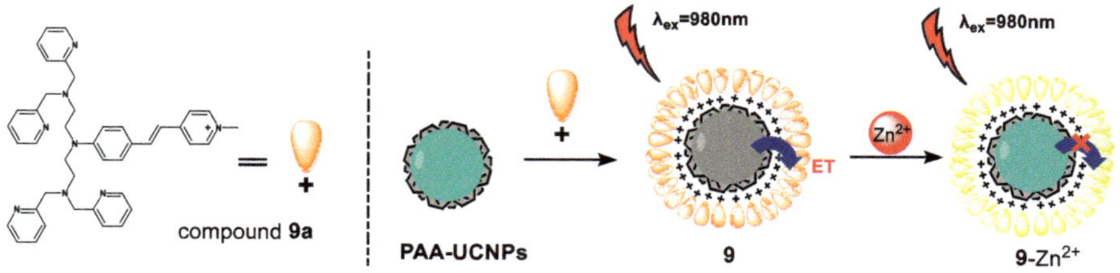

FIGURE 8 Structure of **9** and its response to Zn^{2+}. Adapted from Ref. [32] with permission. Copyright 2015, Journal of the American Chemical Society.

spectral overlap between donor emission and acceptor absorption. As a result, the FRET effect of sensor is diminished, leading to an increase in the $F_{480\,nm}/F_{560\,nm}$ ratio.

The nano-probe **9**, having a core-shell structure containing upconversion nanoparticles (UCNPs) (Figure 8) [32], emits visible fluorescence under the excitation of 980 nm light. As a result, it can be used for zinc-ion detection in biological tissues with a high penetration depth while avoiding the toxic side effects associated with the use of ultraviolet light. In the probe, NaYF4: Yb/Tm@NaYF4 (20/0.2 mol %) upconversion nanocrystals serve as the energy donor and Zn^{2+} responsive polyamine linked pyridyl-stilbene **9a** serves as the acceptor in a FRET process that leads to quenching of the upconversion luminescence. Added Zn^{2+} coordinates to **9a** in the complex, causing a blue shift from 475 to 360 nm, suppression of FRET from the UCNPs to **9a** and recovery of the upconversion luminescence signal at 475 nm.

2.4 Zinc-Ion Fluorescent Probes Based on the ESIPT Mechanism

ESIPT is a typical photophysical process that is accompanied by a large Stokes shift [33, 34]. This process is responsible for the specific fluorescence response of probe **10** (Figure 9) to Zn^{2+} in water (HEPES 10 mM, pH 7.4) [35]. Owing to the operation of ESIPT, the fluorescence spectrum of **10** displays a considerably large Stokes shift. However, in protic solvents, such as water, this process is inhibited and **10** has only weak emission at 382 nm. In the presence of Zn^{2+}, the ESIPT process returns in conjunction with a decrease in fluorescence at 382 nm and growth of a strong 440 nm fluorescence band, which can be seen using the naked eye.

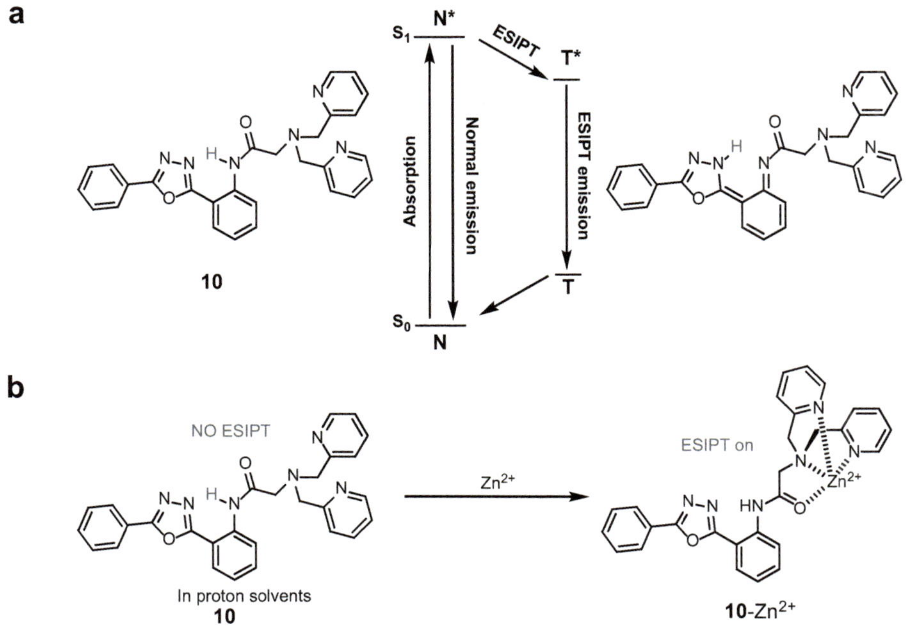

FIGURE 9 (a) ESIPT mechanism of sensor **10**; (b) Proposed mechanism for sensing Zn^{2+}.

3 Zn^{2+} RECOGNITION GROUP (LIGAND)

Owing to the presence of empty outer orbitals, zinc ion readily forms complexes with Lewis basic molecules and ions containing unshared electron pairs or π-electron systems. The coordination number of Zn^{2+} can be 3–6, with the most common coordination number being 4. Therefore, recognition groups for Zn^{2+} in fluorescent probes are most often hydroxyl, Schiff base, pyrazole, pyridine, diacetic acid, quinoline, and polyamine moieties [20, 21].

3.1 Schiff Base as Ligand Group

Schiff bases are a class of organic compounds containing azomethine or imine groups, usually formed readily by condensation of active carbonyl groups of aldehydes or ketones with primary amines [36]. The N atom of the Schiff base has an unshared pair of electrons that can participate in forming complexes with many metal ions through coordination bonds. Schiff bases are therefore a very important class of organic ligands [37]. By changing the connecting linkages and the positions of donor nitrogen, Schiff base ligands can form a variety of cyclic complexes through monodentate or multidentate coordination. Due to the simplicity of their synthesis on a large scale, Schiff bases are a rich source of raw materials used in the preparation of ion fluorescent probes [38] and specifically for zinc-ion fluorescence probes [39]. According to the substances used for their formation, Schiff bases can be further divided into several families depending on whether they derive from simple amines and carbonyl compounds, or they are part of hydrazones, imino acids, semicarbazones, guanidines, or heterocyclic ring systems. Another subclassification focuses on structure, including single Schiff bases [40], double Schiff bases [41], and macrocyclic Schiff bases [42], with double Schiff bases being divided into symmetric and asymmetric members [43, 44].

Single Schiff bases can be generated by condensation reactions of mono-aldehydes, and mono-ketones with monoamines. Representative examples of this group are those prepared from salicylic aldehyde, benzaldehyde, naphthalene formaldehyde, heterocyclic aldehydes, and ketones. Salicylaldehyde-derived Schiff base **11** is a ratiometric fluorescent probe for Zn^{2+} [45] (Figure 10). The operation of a twisted ICT (TICT) process from the donor (diethylamino, $–N(CH_3)_2$) to the acceptor (–CHO) occurs in **11** to give rise to an emission band at 487 nm. Upon the addition of Zn^{2+} to a solution of **11**, the intensity of the band at 487 nm decreases and a concomitant increase occurs in a new blue-shifted emission centered at 353 nm with excitation at 330 nm. Simultaneously, the fluorescence color of **11** changes from turquoise to blue upon addition of Zn^{2+}. The spectral changes are due to 1:2 complex formation through coordination of the imine and phenol electron donor centers in two probe molecules with a single zinc ion, which alters the electron-donating character and blocks TICT.

Probes **12–15** are single Schiff base, salicylic aldehyde derived zinc-ion fluorescent probes (Figure 11). In this group, **12** [46] and **13** [47], having respective emission wavelengths of 457 (excitation at 370 nm) and 458 nm (excitation at 370 nm), operate using PET and CHEF mechanisms in response to complexation with zinc ions. In contrast, **14** [48] and **15** [49] function through ESIPT and ICT mechanisms in response to zinc ions, with emission wavelengths of 510 nm (excitation at 390 nm) and 575 nm (excitation at 390 nm), respectively. The small cytotoxicities of probes **12–15** allow them to be employed as highly selective and sensitive bioimaging tools.

Double Schiff bases can be produced by condensation reactions of amines with bis-functionalized ketones or aldehydes. Compared to single Schiff base counterparts, double Schiff bases enable forming highly π-conjugated systems with metal ions by using multiple coordination sites. Double Schiff bases can be divided into symmetric and asymmetric double Schiff bases [43, 44].

As the name implies, the groups at the two Schiff base sites of asymmetric double Schiff bases are not identical. The simplest method for the synthesis of asymmetric

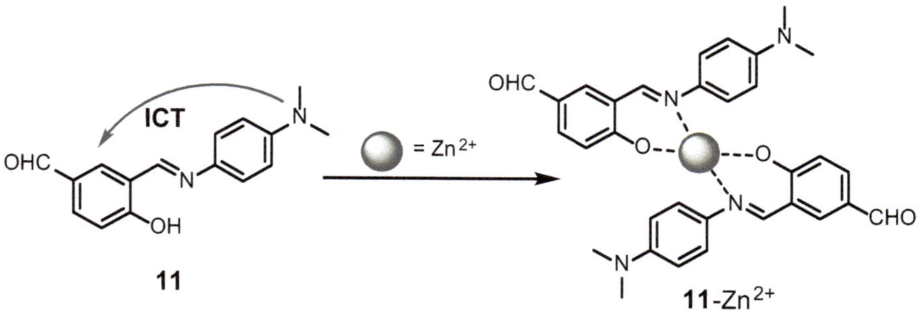

FIGURE 10 Structure of **11** and proposed mechanism for sensing Zn^{2+}.

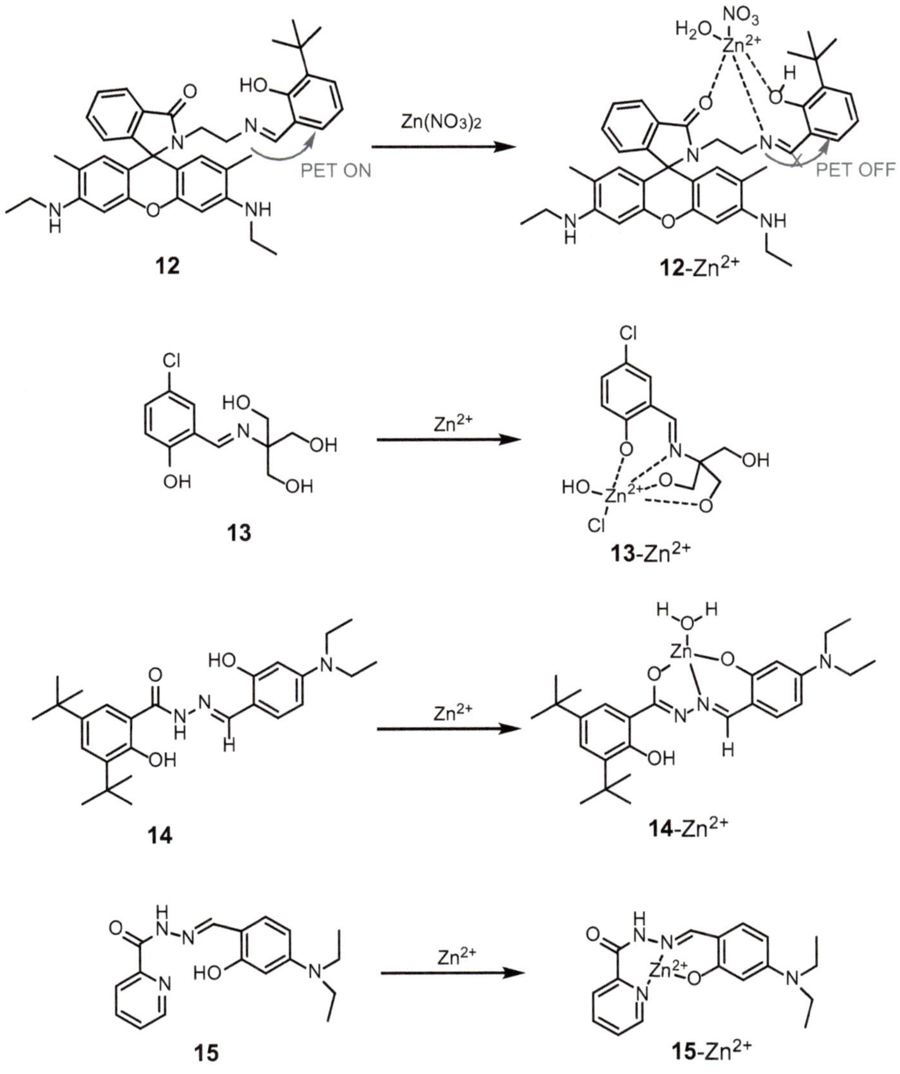

FIGURE 11 Structures of **12–15** and the proposed mechanism for sensing Zn^{2+}.

double Schiff bases uses a structurally asymmetric dicarbonyl compound. If the reaction monomer is symmetric, generation of the first Schiff base group can be accomplished by controlling the molar ratio of the carbonyl and amine components, or taking advantage of the use of selective protective groups. As shown in Figure 12, probe **16** is an asymmetric Schiff base zinc-ion fluorescence probe based on salicylic aldehyde [50]. X-ray crystallographic analysis shows that the Zn^{2+} complex with **16** is tetranuclear and that it crystallizes in a tetragonal space group, in which the symmetric unit is composed of four molecules of **16** and four zinc ions. That the complex of **16** with Zn^{2+} has 1:1 stoichiometry was demonstrated by utilizing Job's plot

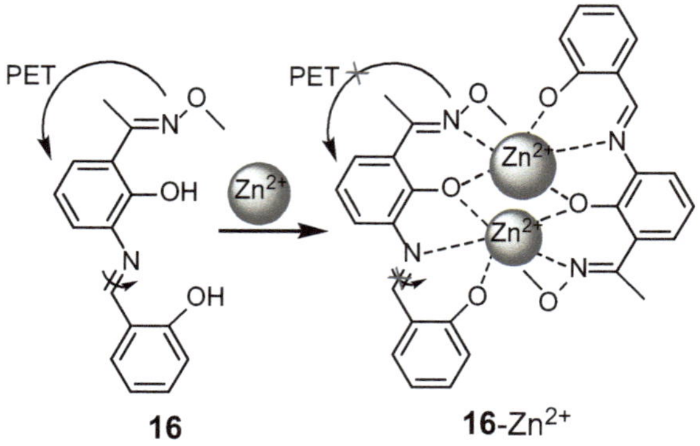

FIGURE 12 Structure of **16** and proposed mechanisms for sensing Zn^{2+}.

analysis. The binding mode and response mechanism of the probe to Zn^{2+} is shown in Figure 12. Probe **16** has weak fluorescence emission at 530 nm when excited at 430 nm due to the quenching by the PET effect. Binding of probe **16** to zinc ion causes a significant enhancement in the fluorescence intensity and a blue shift in the maximum from 530 to 516 nm, which is due to inhibition of the PET process and the occurrence of ICT. In addition, **16** has potential applications to the detection of trace amounts of Zn^{2+} in biological and environmental systems.

Probe **17**, containing a benzothiazole moiety, is another asymmetric double Schiff base type zinc-ion fluorescence probe (Figure 13) [51]. Owing to a consequence of ESIPT and AIE, probe **17** in DMF/H_2O (9:1, v/v) has orange fluorescence with an emission wavelength of 573 nm when excited at 411 nm. When Zn^{2+} is added, PET and ESIPT processes are inhibited and the fluorescence emission at 573 nm is enhanced and accompanied by a blue shift to 520 nm. This probe has a similar fluorescence response to Cd^{2+}, and can be employed to effectively detect and differentiate Zn^{2+} and Cd^{2+} by using cysteine as an auxiliary reagent. In addition, **17**

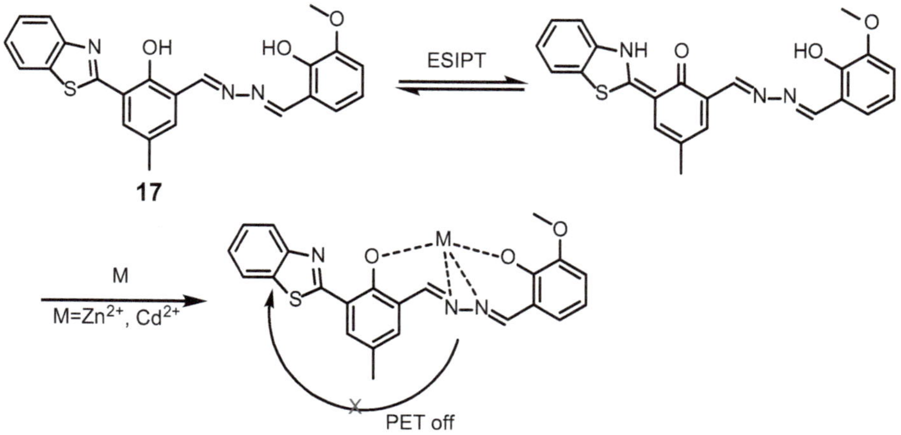

FIGURE 13 Structure of **17** and proposed mechanisms for sensing Zn^{2+} or Cd^{2+}.

can be used to detect Zn^{2+} and Cd^{2+} in live cells in real time. Finally, this probe can also be prepared in test paper form to facilitate rapid detection of Zn^{2+} and Cd^{2+} in practical applications.

The asymmetric double Schiff base derivative **18** is a near-infrared zinc-ion fluorescent probe (Figure 14) [52]. The probe displays essentially no fluorescence due to quenching using a C=N bond isomerization process. After coordination with Zn^{2+}, the fluorescence of the probe at 665 nm is enhanced by about 184-fold when excited at 440 nm. Furthermore, red fluorescence of the formed complex **18**-Zn^{2+} is transformed to dark red emission under 297 nm irradiation because of photochemical formation of the closed-ring isomer **18′**-Zn^{2+}, which can be reversibly converted to **18**-Zn^{2+} by visible light ($\lambda > 500$ nm) irradiation.

Symmetric double Schiff bases are prepared from monomers with symmetric structures and, as such, their synthesis is simpler than that of asymmetric double Schiff bases. These substances can be obtained by the reaction of a diamine with a reactive carbonyl compound or of a bis-carbonyl compound with an amine. Alternatively, it can be generated by the reaction of bis-carbonyl compounds with diamines by considering different reactivities of different functional groups, configuration inversion and steric hindrance. The symmetric double Schiff base has a symmetric structure, and the steric hindrance of it is smaller in cooperating with metals than the asymmetric configuration, so it is easier to coordinate with multiple metals. In addition, the symmetric double Schiff base complex is more stable and not easy to decompose. Symmetric double Schiff base coordinates with Zn^{2+} to produce symmetric complexes. As shown in Figure 15, probe **19**, a symmetric double Schiff base zinc-ion probe [53], is synthesized by the reaction of 2,6-diformyl-4-methylphenol with thiosemicarbazide. Job's plot analysis shows that **19** and Zn^{2+} form a 1:1 complex in which a hydroxy oxygen and an imine nitrogen bind to Zn^{2+} to form four stable hexagon heterocyclic structures, according to the results calculated by DFT. The rigidity, conjugation effects and stability of the complex are also greatly enhanced by the formation of a planar four-membered ring containing two Zn^{2+} and hydroxyl oxygen groups in the center of the complex. DFT calculations also indicate that strong

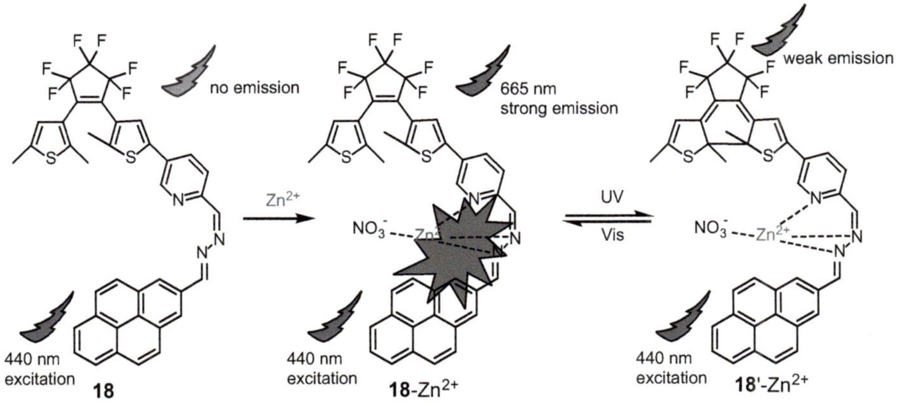

FIGURE 14 Structure of **18** and the proposed mechanism for sensing Zn^{2+}.

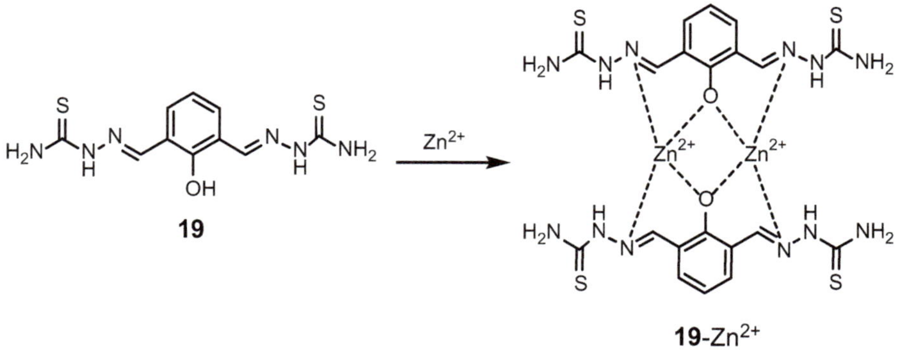

FIGURE 15 Structure of **19** and the proposed mechanism for sensing Zn^{2+}.

charge transfer occurs from the ligand to the metal (LMCT) in the complex. This is another reason for the fluorescence enhancement of **19** upon Zn^{2+} complexation. This probe has a similar response to Hg^{2+}, but the fluorescence emission at 580 nm (excitation at 390 nm) of the probe is enhanced by adding Hg^{2+}, while emission at 478 nm (excitation at 390 nm) is enhanced by adding Zn^{2+}. Probe **19** has been applied to the detection of Zn^{2+} and Hg^{2+} in living cells.

Probes **20–22** (Figure 16) are all comprised of symmetric double Schiff bases and function using ESIPT and PET mechanisms in their detection of Zn^{2+}. The rhodamine spirolactam ring in probe **20** does not open upon coordinating to Zn^{2+} [54]. Thus, when Zn^{2+} is added, the fluorescence emission of **20** at 468 nm (excitation

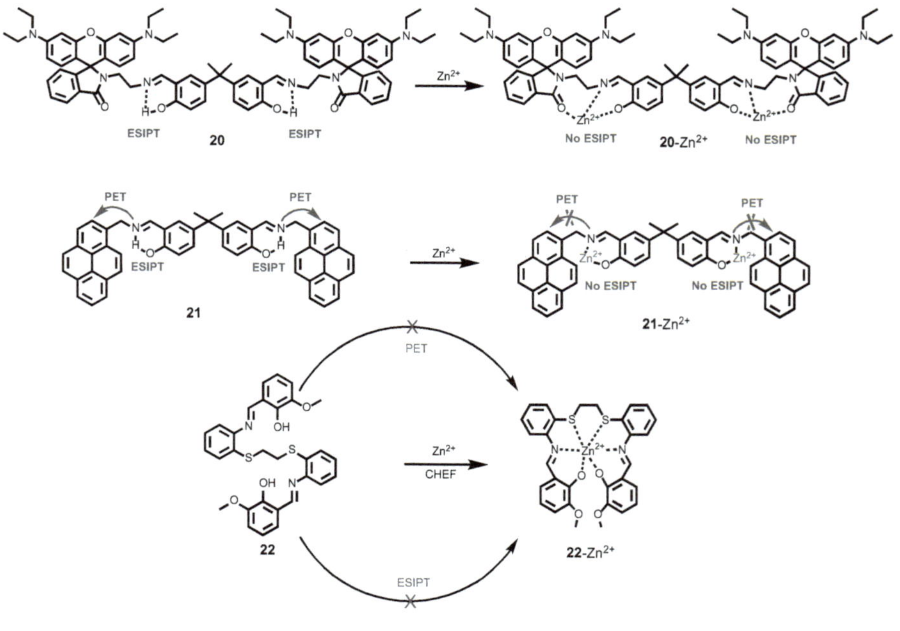

FIGURE 16 Structures of **20–22** and proposed mechanisms for sensing Zn^{2+}.

at 364 nm) is enhanced, while the fluorescence emission attributed to the rhodamine moiety does not change. Probe **21** is a zinc-ion fluorescent probe that contains pyrene fluorophores [55]. When complexed with Zn^{2+}, the blue emission of **21** centered at 472 nm (excitation at 365 nm), attributed to the pyrene fluorophore, is enhanced due to the inhibition of PET and ESIPT processes. Probe **22** is a fluorescent probe that possesses a phenol and vanillin moiety [56]. Upon binding to the zinc ions, the fluorescence of **22** is significantly enhanced and accompanied by a red shift from 411 to 555 nm. Interestingly, $H_2PO_4^-$ addition causes quenching of the fluorescence of the complex **22**-Zn^{2+}, but other anions do not promote this effect, which indicates that **22**-Zn^{2+} has the potential of being used to detect $H_2PO_4^-$. Probe **22** has practical applicability for imaging and sensing Zn^{2+} in physiological systems, including living cells.

Macrocyclic Schiff bases are cyclic polyfunctional compounds formed by reactions of polymolecule aldehydes, ketones, and amines. These substances generally contain four or more Schiff base moieties, which provide multiple coordination sites for metal ions. The macrocyclic substance **23** (Figure 17) containing two calix[4]arene head units and two *o,o*-bis(iminomethyl)phenol groups was developed to detect zinc ions [57]. Probe **23** sensitively and selectively responds to Zn^{2+} among other biologically related metal ions with a fluorescence enhancement at 484 nm (λ_{ex} = 409 nm) by a dual fluorescence enhancement/quenching effect. In the complex **23**-Zn-py$_2$, the adjacent calix[4]aromatic hydrocarbon unit self-assembles through intermolecular face-to-face CH···π interactions, and the zinc ion is hexacoordinated in the form of a twisted octahedron, consisting of two imino nitrogens and two phenolic oxygens from **23** and two *cis*-pyridine co-ligands. The free imine functional group in the complex points away from the center of the fluorophore, thus reducing the efficiency of PET quenching.

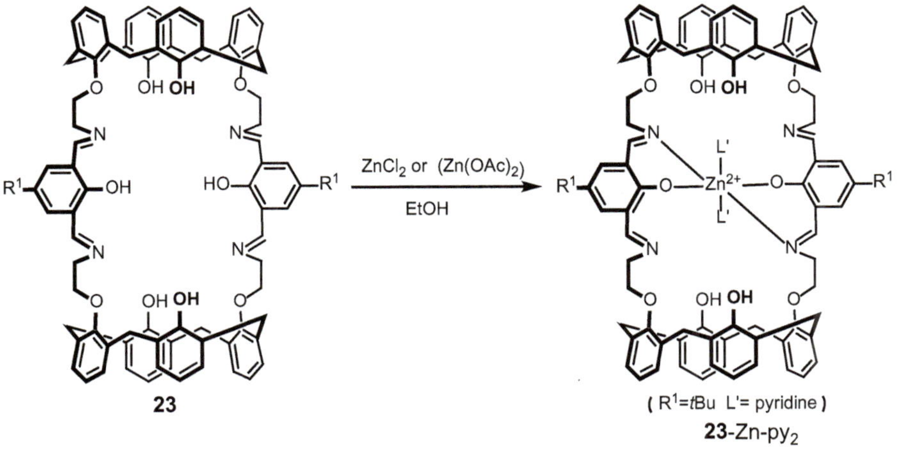

FIGURE 17 Structures of **23** and zinc complex (**23**-Zn-py$_2$).

3.2 DPA and Its Derivatives as a Zn^{2+} Ligand

DPA contains a secondary amine group (–NH–) and two pyridine groups (Figure 18). The pyridine nitrogen is alkaline because of the existence of a nonbonded electron pair containing an sp^2 hybrid orbit. Therefore, the N atoms on pyridine readily coordinate to metal ions, and pyridine is a good ligand group. The two pyridine groups of DPA act like a clamp to provide a total of three sites for metal ion chelation [58, 59], and thus DPA and its derivatives are commonly used in the design of zinc-ion fluorescent probes [60–63].

The secondary amine nitrogen center in DPA is a good reaction site for direct connection to various fluorophores. Probe **24** is a zinc-selective fluorescent sensor, which contains a 2-(2′-hydroxy-3′-naphthyl)benzoxazole (HNBO) chromophore and a DPA metal chelator (Figure 19a) [64]. The crystal structure shows that **24** binds to zinc ions in a pentagonal coordination mode through its DPA and HNBO moieties. The dissociation constant (K_d) of the complex was determined to be 12 pM. The fluorescent response of the probe to zinc ions is due to deprotonation of the HNBO group and the coordination by the DPA electron pairs, which inhibits PET quenching. The probe not only has cell permeability, but it also has 44-fold fluorescence enhancement in the presence of zinc ions under the physiological conditions (pH = 7). Therefore, **24** can be used for the detection and imaging of zinc ions in biological samples, such as cells and zebrafish (Figure 19b). In addition, **24** is capable of being utilized for visualization of endogenous zinc ions released from apoptotic cells.

Probes **25–27** are acetylated versions of PET-based probes that contain a variety of fluorophores and DPA groups (Figure 20) [65]. Binding of Zn^{2+} to these probes rapidly and selectively promotes hydrolytic cleavage of the acetate esters, causing a large fluorescence response. Probe **25** is spectroscopically silent at wavelengths >400 nm, and shows an absorption band centered at 505 nm while emits fluorescence at 525 nm $\Phi = 0.77$) light upon the addition of zinc ion. Also, **26** is nonfluorescent in the absence of zinc ion, while in its presence the absorption of this probe is red shifted ($\lambda_{max} = 354$ nm) and it fluoresces ($\lambda_{em} = 445$ nm, $\Phi = 0.84$). Finally, **27** displays weak emission at 626 nm, and its fluorescence intensity rapidly increases by 12.5-fold upon adding Zn^{2+}. Importantly, acetylation of these sensors significantly improves zinc detection sensitivity making it possible for them to be used at concentrations approximately ten times lower than non-acetylated analogs.

DPA-based probes show good selective "turn-on" or "ratiometric" fluorescence signals in response to Zn^{2+}. However, in contrast with the fact they have higher affinities for Zn^{2+} than alkali and alkaline earth metal ions, they often have poor binding selectivity for this metal ion over other transition and heavy metal (HTM) ions. This limitation has been removed by the design of probe **28**, which uses an amide containing DPA as the receptor and naphthalimide as a fluorophore (Figure 21) [66].

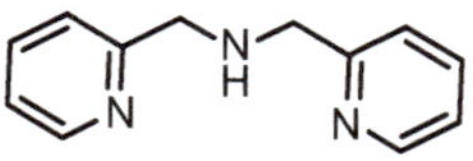

DPA

FIGURE 18 Structure of DPA.

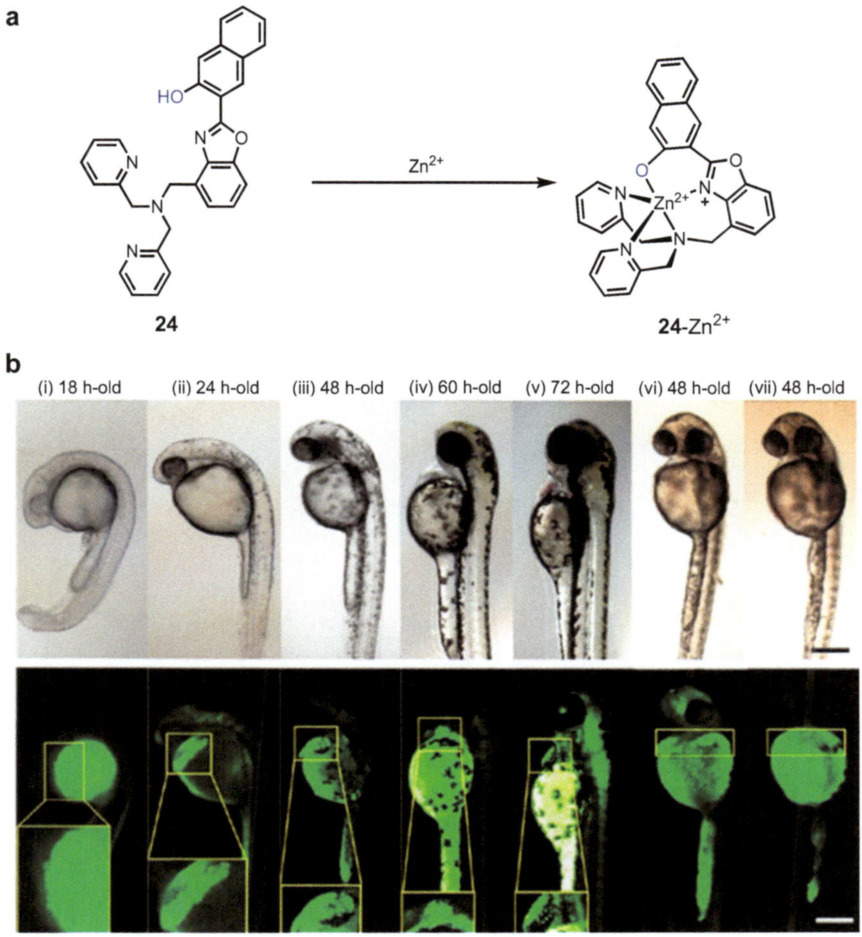

FIGURE 19 (a) Structure of probe **24** and the proposed mechanism for sensing Zn^{2+}. (b) Detection of endogenous zinc ions in zebrafish with **24**. Fluorescence microscope images of (i) 18, (ii) 24, (iii) 48, (iv) 60, and (v) 72 h-old zebrafish incubated with **24** for 1 h: (top panels) bright-field images; (bottom panels) fluorescence microscope images. Yellow boxes show magnified images. Fluorescence microscope images of 48 h-old zebrafish incubated with HNBO–DPA for 1 h in the absence (vi) and presence (vii) of 100 μM TPEN. Scale bar = 250 μm. Adapted from Ref. [64] with permission. Copyright 2012, Inorganic Chemistry.

Probe **28** strongly binds to Zn^{2+} in aqueous solutions (K_d = 5.7 nM) using an imidic acid tautomeric form of the amide-DPA unit while it binds to most other HTM ions in an amide tautomeric form (Figure 21). Probe **28** has excellent selectivity for Zn^{2+} over most other HTM ions due to this "receptor transformation" induced binding phenomenon. After binding to zinc ions, the fluorescence emission of probe **28** at 483 nm is significantly enhanced and red shifted to 514 nm. It is interesting that fluorescence of the probe is significantly enhanced and blue shifted from 483 to 446 nm

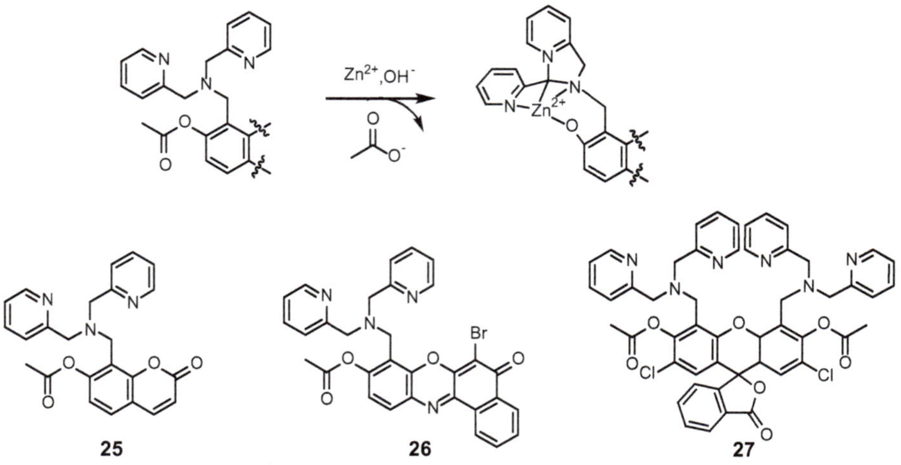

FIGURE 20 Structures of probes **25–27** and mechanisms for sensing Zn^{2+}.

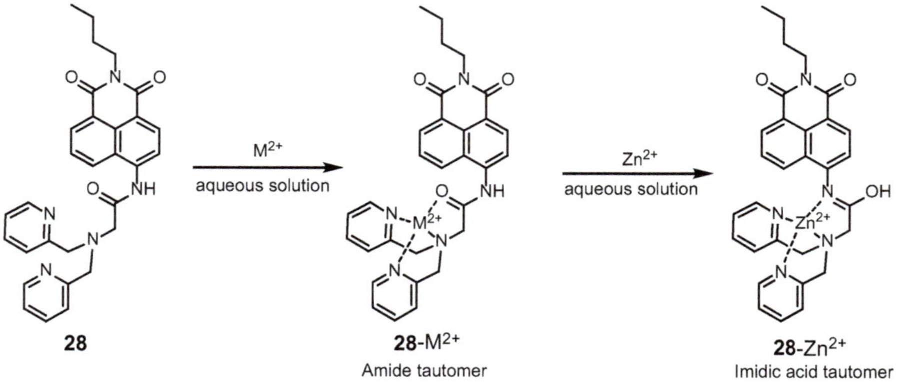

FIGURE 21 Different binding modes of probe **28** with HTM ions.

upon binding to Cd^{2+} (K_d=48.5 nM). Thus, **28** can be employed to discriminate Zn^{2+} and Cd^{2+} *in vitro* and *in vivo* by displaying green and blue fluorescence, respectively. Moreover, the probe can be used to detect Zn^{2+} during the development of living zebrafish embryos.

To provide more coordination sites and thus improve the affinity for metal ions, derivatives of DPA, such as N,N-bis(2-pyridinylmethyl)-1,2-ethanediamine (DPEN) and TPEA, have been utilized as recognition groups in zinc-ion probes (Figure 22) [21]. For example, probes **29** and **30** are both zinc-ion fluorescent probes that contain fluorescein as a fluorophore and DPEN as a ligand (Figure 23) [67]. The fluorescence quantum yields of **29** and **30** are 0.004 and 0.006, respectively, due to quenching by PET from the polyamine moiety to the xanthene fluorophore. As a result, they emit little fluorescence under physiological conditions. After adding Zn^{2+}, the fluorescence intensity of **29** increases 69-fold, while that of **30** increases 60-fold. The

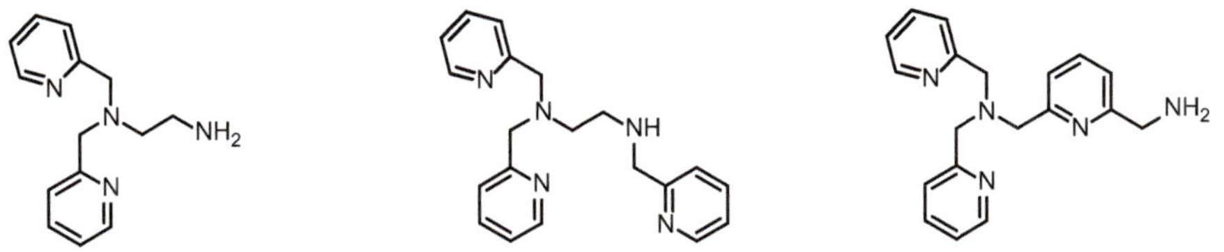

N,N-bis(2-pyridinylmethyl)-1, 2-ethanediamine

N,N,N′-tris(pyridin-2-ylmethyl)ethane-1, 2-diamine (TPEA)

1-(6-(aminomethyl)pyridin-2-yl)-*N*,*N*-bis(pyridin-2-ylmethyl)methanamine

FIGURE 22 Structure of common DPA derivatives

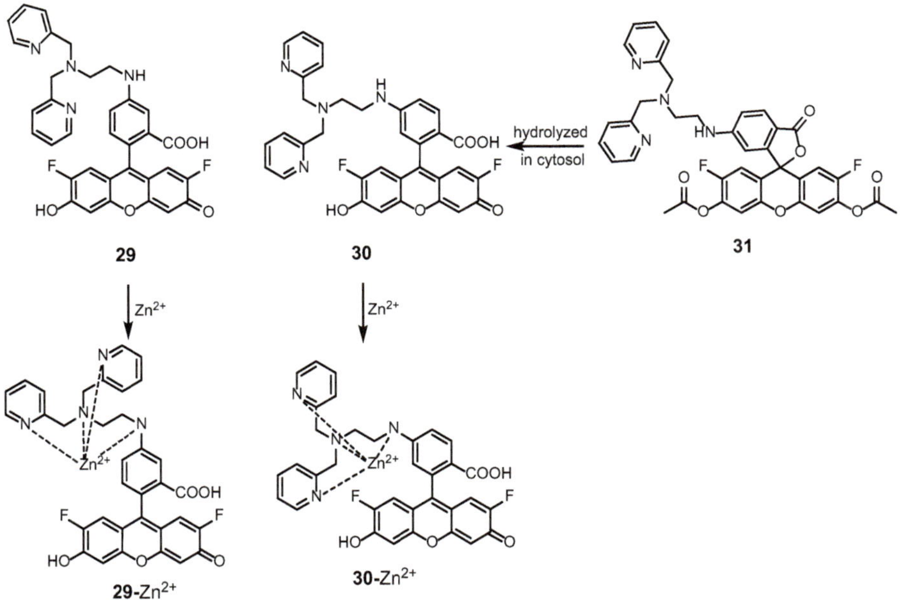

FIGURE 23 Structures of probe **29–31** and proposed mechanisms for sensing Zn^{2+}.

enhancement of the fluorescence intensity of these probes is likely due to the inhibition of PET by the coordination of Zn^{2+} with the nitrogen centers. In both cases, a 1:1 complex between the probe and Zn^{2+} is formed, and the binding rate constant of **29** is $3.5 \times 10^6\,M^{-1}\,s^{-1}$ and that of **30** is $3.2 \times 10^6\,M^{-1}\,s^{-1}$. Probes can be utilized to quantitatively measure Zn^{2+} concentrations in the range of 0.1–10 nM. Cell experiments showed that probe **30** is difficult to penetrate cell membranes. Thus, probe **31** was synthesized for intracellular zinc-ion detection. Probe **31** is able to cross the cell membrane. In the cytosol, it is hydrolyzed to produce **30**, which then responds to the intracellular Zn^{2+} concentration. Probe **30** can also be used to detect the concentration of extracellular Zn^{2+}. Its property of not penetrating the cell membrane is conducive to applications in neuroscience experiments.

The small molecule self-assembled nanofluorescent probe **32** was designed to selectively detect Zn^{2+} in human hair. The monomer of **32** is composed of a BODIPY derivative as the fluorophore and DPEN as a Zn^{2+} acceptor (Figure 24) [68]. The fluorescent dye **32** aggregates and self-assembles in the aqueous medium resulting in fluorescence quenching. In the presence of Zn^{2+}, the metal ion enters the nanoparticles (NPs) and combines with DPEN to "turn on" fluorescence. The fluorescent probe can be used to quantitatively measure Zn^{2+} in the concentration range of 0-10 μM and at pHs in the range of 5.0–8.0. In the presence of other metal ions, the NPs have good selectivity for Zn^{2+}, making it possible for them to be used to determine Zn^{2+} in complex mixtures. The detection limit for Zn^{2+} is approximately 61.3 nM. This method uses self-assembled nanoparticles to determine Zn^{2+} quantitatively and accurately in human hair samples, the accuracy of which was confirmed by using inductively coupled plasma emission spectrometry.

Another example is the colorimetric Zn^{2+} probe **33**, which utilizes tricarbocyanine as the fluorophore and tris(pyridin-2-yl-methyl)amine as the metal ion recognition site (Figure 25) [69]. When combined with Zn^{2+}, NH deprotonation of **33** occurs to shorten the conjugated π-electron system in the tricarbocyanine fluorophore, resulting in a blue shift of the absorption and emission maxima. The apparent dissociation constant (K_d) of the **33**-Zn^{2+} complex was determined to be 1.2 nM. The probe has high selectivity for Zn^{2+}, and a detection limit in the nanomolar range. In addition, **33** has been successfully used to detect Zn^{2+} in zebrafish embryonic neuroma.

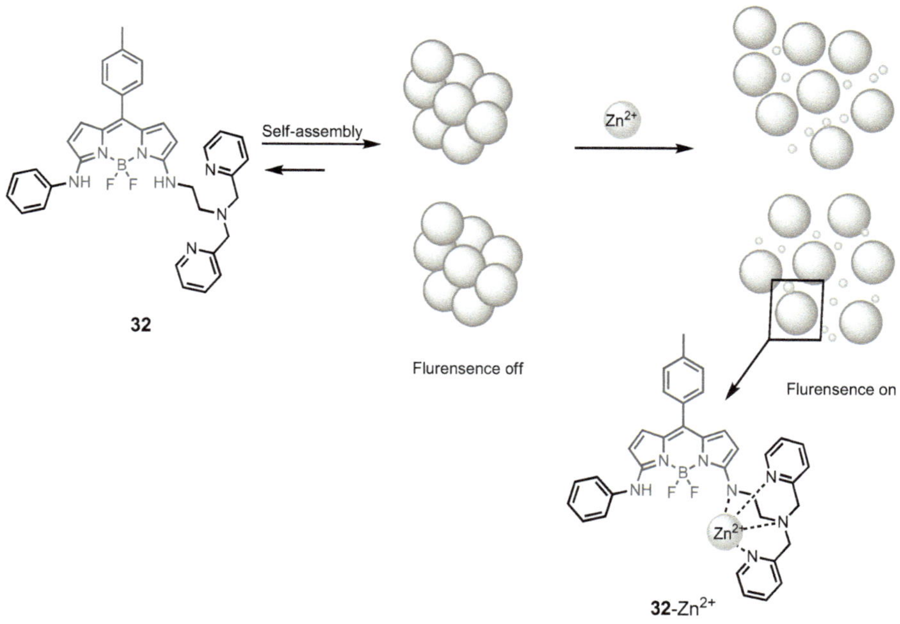

FIGURE 24 Structure of probe **32** and proposed mechanism of sensing Zn^{2+}.

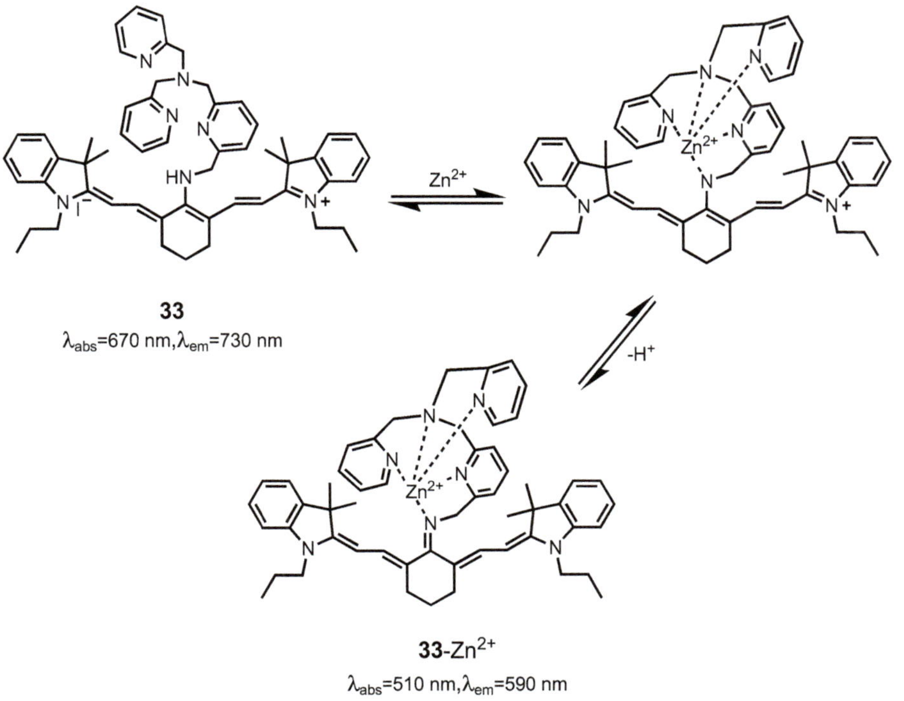

FIGURE 25 Structure of probe **33** and proposed binding mode to Zn^{2+}.

3.3 Bipyridines as Ligand Groups

2,2′-Bipyridines have high affinities for metal ions [70–72]. Interestingly, complexation with metal ions causes conformational restriction of the bipyridine group to take place that enables ICT to occur, which leads to significantly enhanced and red-shifted fluorescence. Therefore, bipyridines serve as both ligands and fluorescent signaling groups [73, 74]. For example, the bipyridine-containing probe **34**, which has donor-π-acceptor-π-donor properties, was developed for near-infrared ratiometric fluorescence detection of Zn^{2+} based (Figure 26) [75]. In this probe, pyrrole is the electron donor, and the 2,2′-bipyridine moiety is the electron acceptor and metal chelating center. The absorption maximum of the probe is 410 nm and its emission

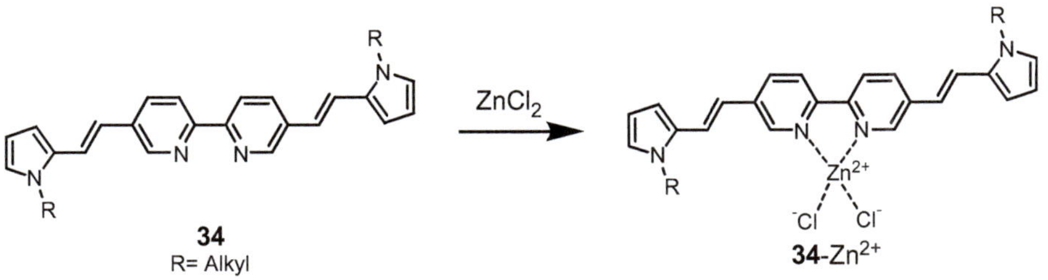

FIGURE 26 Structure of **34** and proposed mechanism for sensing Zn^{2+}.

maximum is 565 nm. After complexation with Zn^{2+}, absorption of the probe changes to 421 nm and the emission red shifts to 610 nm. The binding constant of **34** for Zn^{2+} is 3.6×10^5 M^{-1}. The probe has low toxicity and it was employed for detection of exogenous Zn^{2+} in C6 glioma cells. Moreover, it was also used for the imaging of endogenous Zn^{2+} in acute hippocampal slices, assessing the dynamics of endogenous Zn^{2+} in hippocampal slices under epileptic conditions, and detecting of Zn^{2+} and determining its dynamics in hippocampal slices by Raman spectroscopy.

In another example, the bipyridine derivative **35** was developed as a ratiometric two-photon fluorescence probe for Zn^{2+} (Figure 27a) [76]. One-photon excitation at 405 nm or two-photon excitation at 800 nm of a mixture of **35** and Zn^{2+} in methanol-HEPES (v/v = 1/1, 50 mM HEPES, pH 7.4) causes the emission peak at 465 nm to gradually decrease and a new emission peak to appear at 550 nm. The red shift and enhancement of fluorescence caused by an ICT process enables ratiometric determination of Zn^{2+}. Probe **35** has a high two-photon absorption (2PA) cross section (δ) of

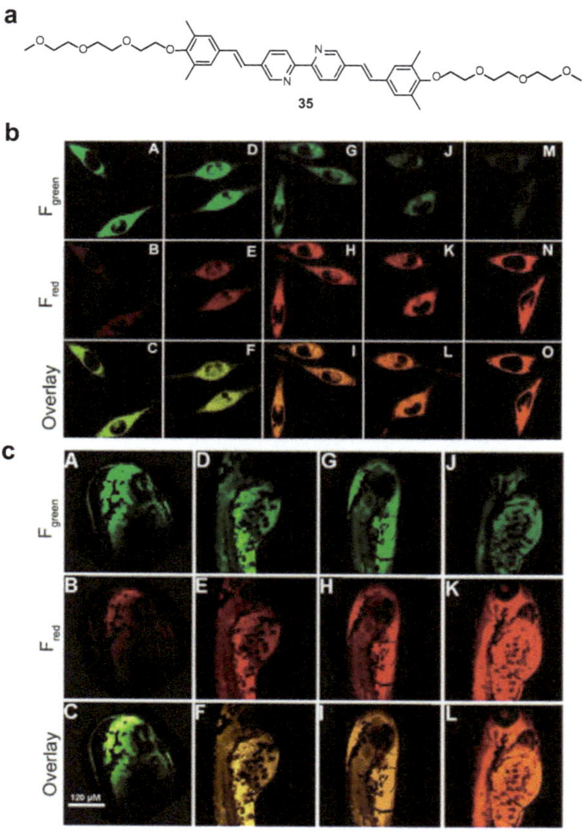

FIGURE 27 (a) Structure of **35**; (b) Confocal microscope images of live SHSY-5Y cells incubated with **35** in green channel, red channel, and overlay with increasing concentration of Zn^{2+}; (c) Confocal microscope images of zebrafish incubated with **35** upon adding increasing concentrations of Zn^{2+}. Adapted from Ref. [76] with permission. Copyright 2017, Analytical Chemistry.

516 ± 77 GM, which further increases to 958 ± 144 GM upon coordination with Zn^{2+}. Probe **35** was applied to imaging and biosensing of Zn^{2+} in live cells, brain tissues, and zebrafish (Figure 27b and c).

3.4 Quinolines as Ligand Groups

Quinolines, consisting of fused pyridine and benzene ring systems, are well-known ligands for transition metal ions [77, 78]. Quinoline and its derivatives, especially 8-hydroxyquinoline and 8-aminoquinoline that can chelate metal ions, are often used as fluorophores in the construction of fluorescent probes for metal ions [79]. The "N" atom in quinoline is susceptible to forming hydrogen bonds in protic polar solvents, which results in weak fluorescence. However, the coordination of metal ions to these N centers can induce ICT processes or inhibit PET processes. Thus, quinoline and its derivatives are often used as chelators and fluorophores in the construction of zinc-ion fluorescent probes [80–82].

For example, probe **36** containing covalently linked N-(8-quinolyl)-p-aminobenzene-sulfonamide (HQAS) and cyclodextrin (CD) moieties was designed for the detection of Zn^{2+} (Figure 28) [83]. The inclusion of the CD group effectively improves the water solubility of the probe, while HQAS is both the fluorophore and a zinc-ion chelating group. Upon adding Zn^{2+}, the fluorescence emission at 518 nm of the probe (excitation at 282 or 362 nm) is significantly enhanced. In addition, the probe was successfully applied to imaging living yeast cells.

Aminoquinoline derivatives **37** and **38** have been developed as fluorescent Zn^{2+} probes (Figure 29) [84, 85]. The 2-hydroxyethoxy groups in the probes not only provide enhanced water solubility, but also serve as coordination sites for metal ions. When Zn^{2+} is added, the fluorescence emission of **37** is significantly enhanced and red shifted to 515 nm and the fluorescence emission of **38** at 490 nm is enhanced. The respective association constants (K_a) of complexes **37**-Zn^{2+} and **38**-Zn^{2+} are 6.7×10^6 and $2.5 \times 10^5\ M^{-1}$. The probes show high selectivity for Zn^{2+}. In addition, both probes were successfully applied to imaging living yeast cells.

The dual emission nanofluorescent probe **39** was obtained by connecting the substance PEIQ, which contains polyethyleneimine (PEI) and 8-chloroacetylaminoquinoline moieties, to $[Ru(bpy)_3]^{2+}$-encapsulated silica nanoparticles (SiNPs)

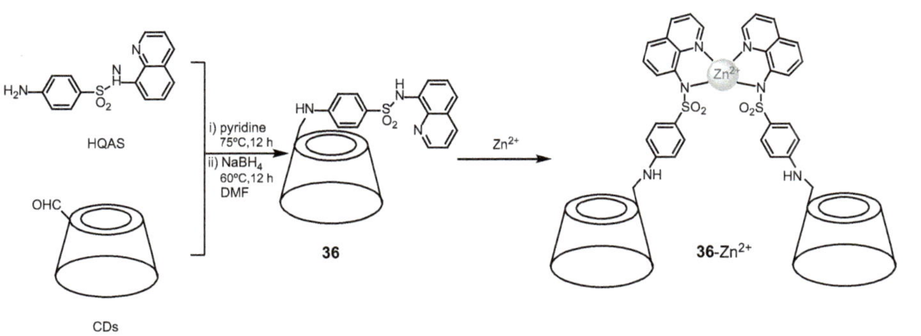

FIGURE 28 Synthesis of **36** and proposed mechanisms for sensing Zn^{2+}.

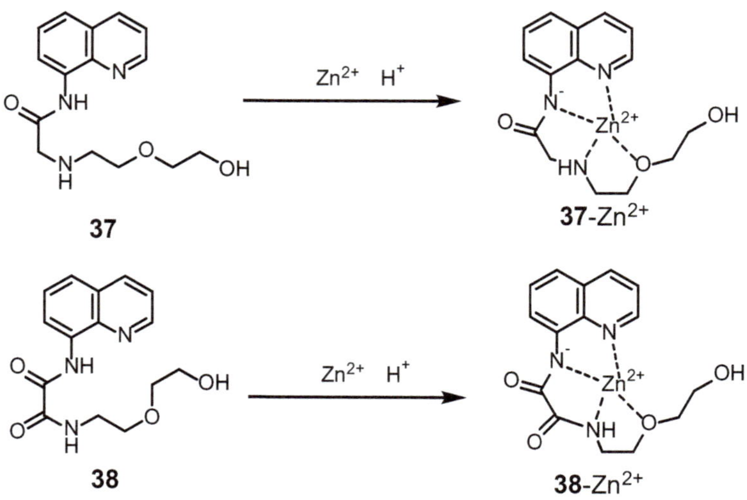

FIGURE 29 Structures of **37** and **38** and proposed mechanisms for sensing Zn^{2+}.

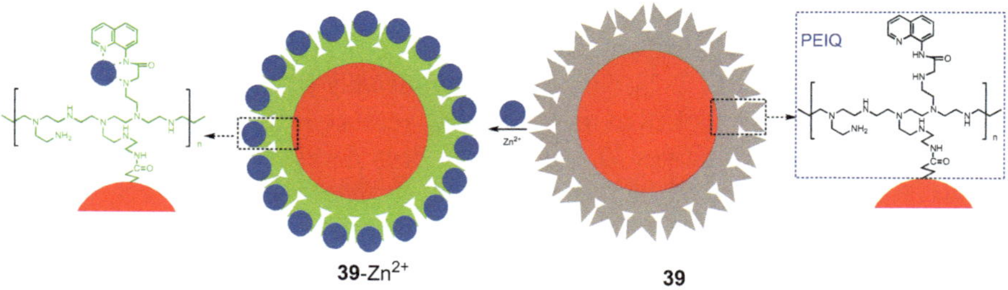

FIGURE 30 Structure of **39** and a schematic of Zn^{2+}sensing. Adapted from Ref. [86] with permission. Copyright 2014, Biosensors & Bioelectronics.

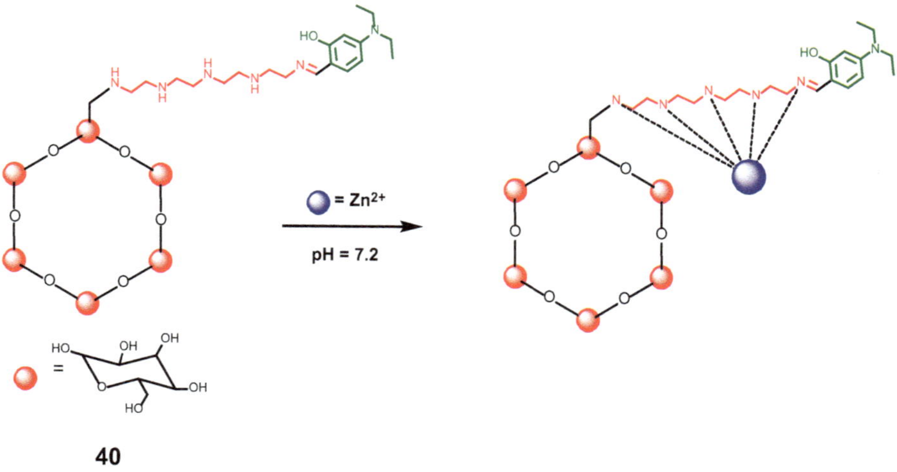

FIGURE 31 Structure of **40** and proposed mechanisms for sensing Zn^{2+}.

(Figure 30) [86]. Probe **39** itself displays strong fluorescence emission at 600 nm and weak fluorescence emission at 500 nm upon excitation at 365 nm. Upon continuous addition of Zn^{2+}, emission of the probe **39** at 600 nm remains almost unchanged, and the band at 500 nm increases continuously, corresponding to a transition from red to green emission that can be recognized by using the naked eye. In addition, the probe has excellent water solubility and biocompatibility, and high selectivity against interference from metal ions. The probe also shows high Zn^{2+} sensitivity and a detection limit as low as 0.5 μM. Finally, the probe has low biotoxicity and was successfully used for intracellular imaging of A549 cell line.

3.5 Polyamines as Ligand Groups

Aliphatic amines have a strong affinity for transition metal ions. Fluorescence probes constructed using electron-donating aliphatic amine groups as a ligand often operate using a PET mechanism. Once complexed with the transition metal, the amine is no longer an electron donor, and PET quenching is inhibited and fluorescence is significantly enhanced. The fluorescence enhancement effect brought about by complexation with zinc ions is usually large, because the full-shell electron configuration of zinc ions does not enable the operation of alternative electron or energy transfer mechanisms for excited state deactivation [21]. The zinc-ion fluorescent probe **40** [87] (Figure 31), containing a polyamine tethered phenol Schiff base linked to β-cyclodextrin derivative (β-CD), utilizes a PET mechanism for zinc-ion sensing. The polyamine is the site for binding zinc ions, and the β-CD moiety provides high water solubility. Upon the addition of Zn^{2+}, the fluorescence intensity at 410 nm (excitation at 330 nm) is significantly enhanced. The association constant between **40** and Zn^{2+} is 2.55×10^{13} M^{-1}. Moreover, **40** has low toxicity and excellent biocompatibility,

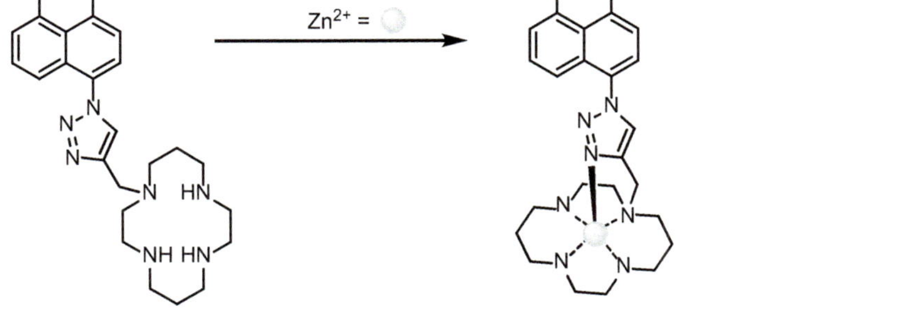

41

FIGURE 32 Structure of **41** and proposed mechanisms for sensing Zn^{2+}.

and as a result, it is suitable for use in the detection and quantification of zinc ions in cells or organisms.

Another example of a polyamine-containing probe is **41**, which was developed for imaging free Zn^{2+} in cancer cells. The probe, which contains a cyclam ligand and a naphthalimide fluorophore, is water soluble and has low toxicity and high selectivity (Figure 32) [88]. Upon interacting with zinc ions in 0.01 mM HEPES buffer, the fluorescence intensity of **41** at 412 nm (excitation at 346 nm) is significantly enhanced. The results of using **41** to image Zn^{2+} in breast cancer cells suggest that it selectively locates and permeates throughout the entire cancer cell, including the nucleus, which makes it useful for studying the cancer cells and the role Zn^{2+} plays in their development.

3.6 Triazoles as Ligand Groups

1,2,3-Triazoles can be obtained by Cu(I) catalyzed reactions of terminal alkynes and azides (click chemical reactions) [89, 90]. This chemistry is widely used for the construction of fluorescent probes and for connecting triazoles to fluorophores and receptors. The triazole moiety has a high affinity for zinc ions. An example of a probe of this type is the triazole/rhodamine derivative **42** (Figure 33), which serves as a chromogenic and fluorogenic sensor for Zn^{2+} [91]. Probe **42** selectively responds to Zn^{2+} in EtOH by undergoing a change from colorless to pink in association with the production of a complex with a binding constant of $7.4 \times 10^3\ M^{-1}$. It was demonstrated that the triazole groups and linker length in **42** are necessary for coordination of Zn^{2+} with nitrogens of the bis-triazolyl and the carbonyl group oxygen of the spirolactam ring, and for promotion of the key ring-opening process, which leads to a significant enhancement of fluorescence emission at 579 nm with excitation at 530 nm.

A coumarin and quinoline ring system was incorporated into the design of fluorescent probe **43** (Figure 34) for detection of Zn^{2+} [92]. The coumarin group acts as a fluorophore and electron donor in this probe, while the quinoline and triazole groups serve as electron acceptors and ligands for Zn^{2+}. In the absence of Zn^{2+}, **43** in buffer solution (CH_3CN: $H_2O = 1{:}3$, v/v, 10 mM HEPES, pH = 7.2) exhibits bright green fluorescence (excitation at 380 nm). After binding to zinc ions, the fluorescence of **43** shifts from green (516 nm) to yellow (556 nm) followed by the enhancement of ICT effect from the coumarin moiety to the quinoline and triazole groups. Moreover, **43** has a high sensitivity to zinc ions with a detection limit of 48.1 nM. Probe **43** was applied to detect Zn^{2+} in real samples, such as drinking, pond, and river water. The probe has strong cell membrane permeability and low biotoxicity and, as a result, it has been applied to fluorescence imaging of zinc ions in living MCF-7 cells.

3.7 Iminodiacetic Acid as a Ligand Group

Iminodiacetic acid and derivatives, such as 1,2-bis-(*o*-aminophenoxy)ethane-*N,N,N′,N′*-tetraacetic acid and ethylene glycolbis(2-aminoethylether)-*N,N,N′,N′*-tetraacetic acid, are often used to construct fluorescent probes for Ca^{2+} and Mg^{2+} [93, 94]. Actually, the concentration of Zn^{2+} can also be measured using these types of fluorescent probes that contain iminodiacetic acid derivatives. However, Ca^{2+} or

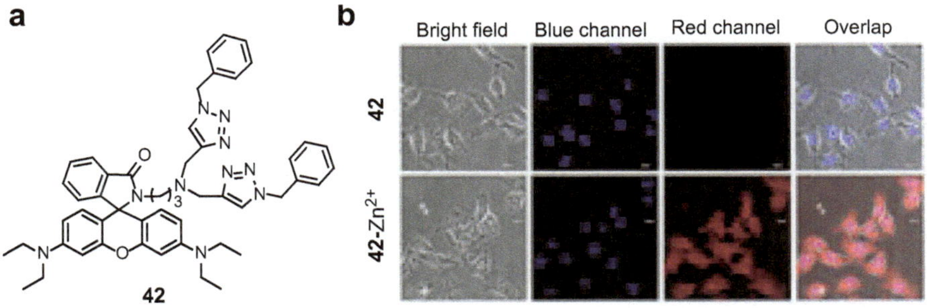

FIGURE 33 (a) Structure of **42**; (b) Confocal fluorescence microscope images of **42** in HeLa cells with and without Zn^{2+}. Adapted from Ref. [91] with permission. Copyright 2016, Sensors and Actuators B: Chemical.

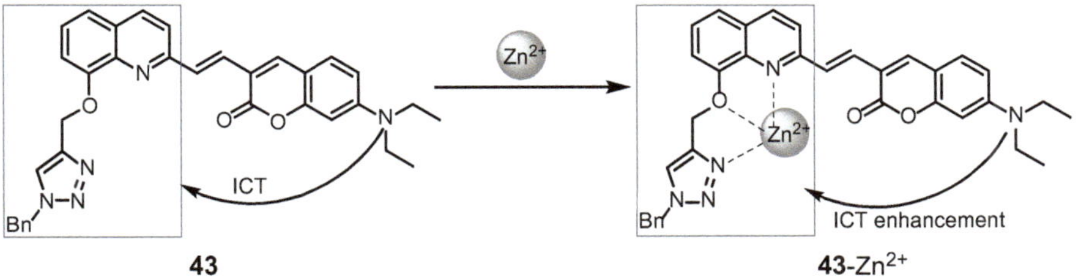

FIGURE 34 Structure of **43** and the proposed mechanism for sensing Zn^{2+}.

Mg^{2+} in the sample interferes with these zinc-ion determinations. To produce a probe that has a higher selectivity for Zn^{2+} over Ca^{2+} or Mg^{2+}, one or more of the chelating moieties can be removed [21]. For example, probe **44** contains a naphthalimide fluorophore and an iminodiacetic acid moiety as a Zn^{2+} receptor (Figure 35) [95]. Interaction of Zn^{2+} with the iminodiacetic acid center blocks PET quenching and results in a 34-fold increase in the intensity of green naphthalimide-derived fluorescence at 550 nm (excitation at 470 nm) in an aqueous solution. The probe has good selectivity for Zn^{2+}, while it has only very weak fluorescence responses to Ca^{2+} and Mg^{2+}. This probe was utilized to detect Zn^{2+} in biological systems.

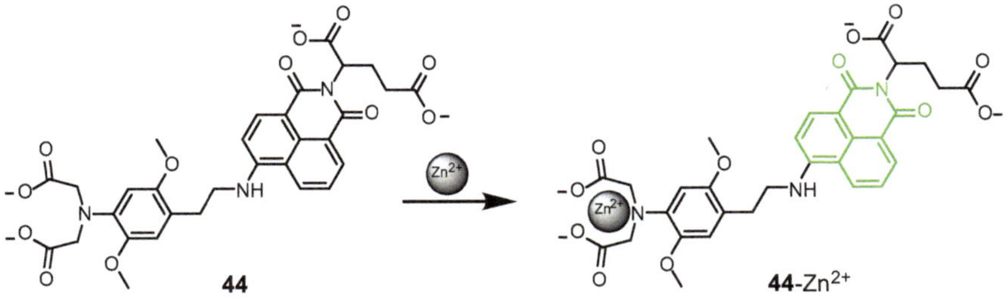

FIGURE 35 Structure of **44** and mechanism for sensing Zn^{2+}.

4 TYPE OF ZINC-ION FLUORESCENT PROBES

After decades of study, a large number and variety of zinc-ion probes have been developed. The probes can be divided into "turn off" and "turn on" types based on whether fluorescence emission arises or decreases in the presence of the metal ion or not. Moreover, the fluorophores in the probes can be classified into biosynthetic fluorophores and bioprotein fluorophores [96]. Also, the probes can be subdivided according to the number of signals (imaging channels) used to evaluate the response (single-channel imaging probes and multichannel imaging or ratiometric probes). Zinc-ion probes can be classified according to the excitation mode, including single- and two-photon excitation [97]. Lastly, ultraviolet, visible, and near-infrared classification is also possible based on the wavelength of fluorescence emission observed. In this section, we describe single- and multichannel fluorescent probes, near-infrared probes, two-photon fluorescent probes, and bioprotein fluorescent probes for zinc ions.

4.1 Single-Channel Fluorescent Probes for Zinc Ion

A single-channel zinc-ion fluorescent probe is one in which fluorescence changes in only one emission band occur without an accompanying emission wavelength shift. These types of probes can be divided into two types, including "turn on" and "turn off." Turn on fluorescent probes are often more suitable for biological imaging than turn off fluorescent probes. Furthermore, most of the single-channel zinc-ion fluorescent probes operate using the PET mechanism [17].

The first single-channel intracellular zinc-ion imaging sensor was the Zinquin-based probe **45** (Figure 36) [98] developed by Zalewski. In response to zinc ions, fluorescence emission from **45** at 490 nm is significantly enhanced with excitation at 370 nm. Probe **45** is a membrane-permeable ester that is hydrolyzed by esterases to form Zinquin acid that is retained inside the cell. The probe mainly displays punctate diffuse fluorescence in the extracellular nucleus. In the 1990s, Zinquin was one of the most used Zn^{2+} sensors for live imaging of cytoplasmic Zn^{2+}.

Like **45**, the first-generation zinc-ion probes are usually excited using UV light, which has problems associated with severe photodamage and photobleaching. Moreover, fluorescence from endogenous fluorophores, occurring when UV excitation

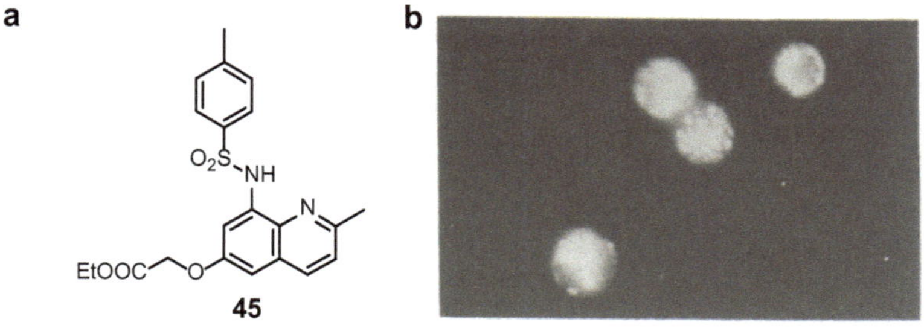

FIGURE 36 (a) Structure of **45**. (b) wide-field fluorescence images of CLL cells incubated with **45**. Adapted from Ref. [98] with permission. Copyright 1993, Biochemical Journal.

is employed, can interfere with single-channel imaging of zinc ions. These issues are detrimental to the application of probes in biological samples. To overcome the disadvantages of UV excitation, zinc-ion probes that are activated by visible light excitation were developed. These typically use fluorescein, naphthalimide, and rhodamine derivatives as fluorophores. For example, **46** and **47** (Figure 37) are fluorescein-based probes designed for detection of zinc ions [99]. In the presence of Zn^{2+}, the

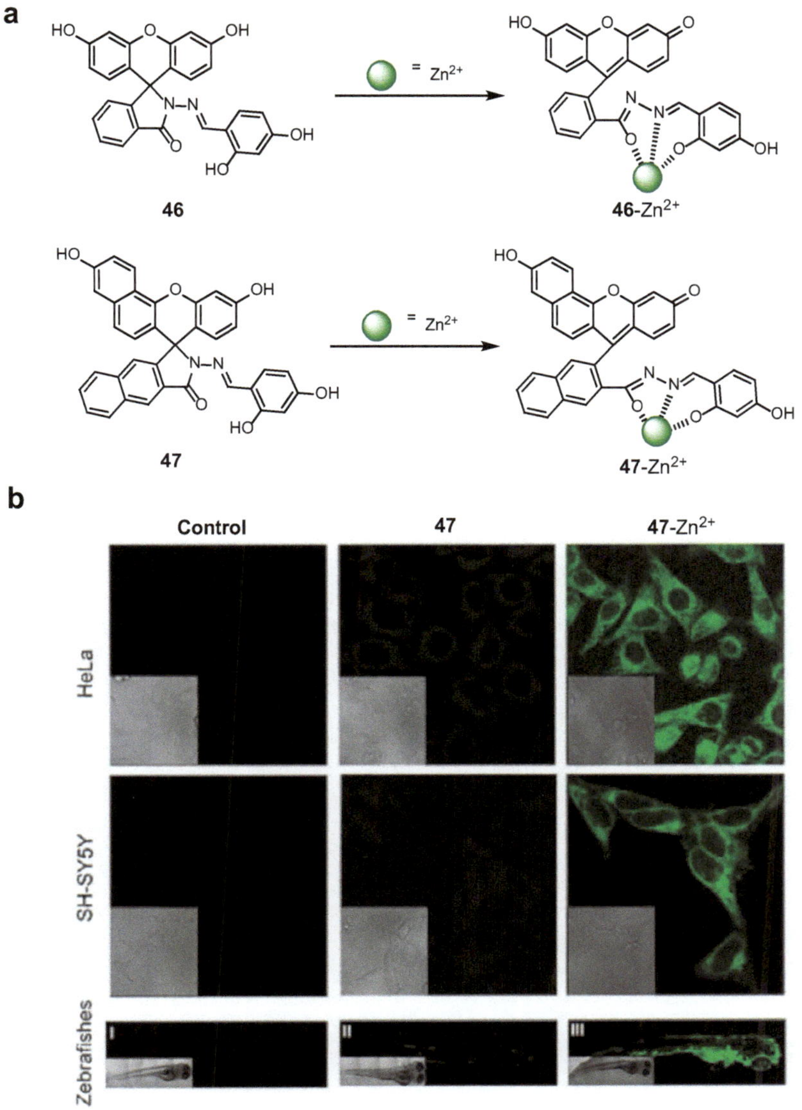

FIGURE 37 (a) Structures of **46**, **47** and mechanisms for sensing Zn^{2+}; (b) Confocal fluorescence microscope images of HeLa, SH-SY5Y cells, and zebrafishes incubated with **47** pretreatment of Zn^{2+}. Adapted from Ref. [99] with permission. Copyright 2020, Spectrochimica Acta. Part A, Molecular and Biomolecular Spectroscopy.

fluorescence emission of **46** at 520 nm (excitation at 480 nm) is significantly enhanced and the fluorescence emission of **47** at 575 nm (excitation at 425 nm) is significantly enhanced. The changes are attributed to the zinc ion-induced opening of the spirocyclic ring. Furthermore, probe **47** has good biocompatibility and low cytotoxicity, and it has been applied to endogenous/exogenous Zn^{2+} imaging in living cells and zebrafish.

4.2 Multichannel (Ratiometric) Fluorescent Probes for Zinc Ion

"Turn on" fluorescent probes based on a single-channel response are susceptible to problems associated with several factors, including instrument parameters, microenvironment, probe concentration, and photobleaching. Changes in intensities at two or more fluorescence emission wavelengths, like those occurring with ratiometric fluorescence probes, greatly improve the accuracy and sensitivity of the analytical results. By using a ratio of signals, probes of this type can effectively avoid errors

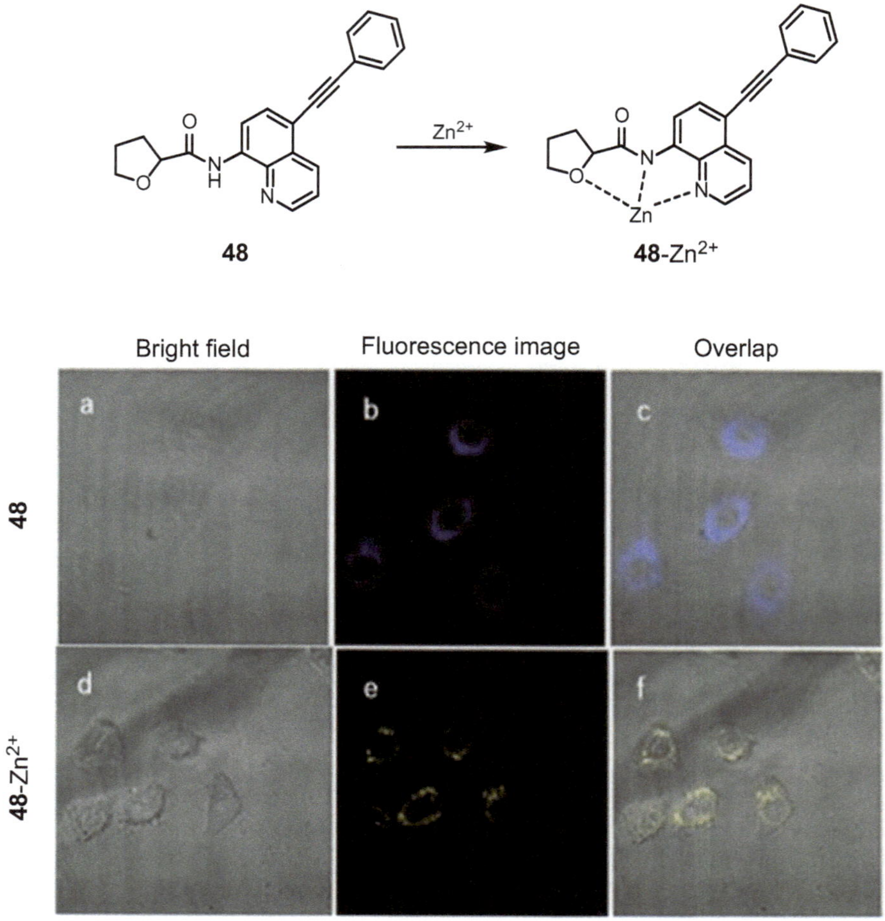

FIGURE 38 Structure of **48** and mechanism for sensing Zn^{2+}. And fluorescence and bright-field images of the probe in the HepG-2 cell line. Adapted from Ref. [101] with permission. Copyright 2017, RSC Advances.

caused by the instrument, environment, and methods, and can also eliminate the effect of probe concentrations. In addition, ratiometric probes have advantages of good optical stability, quantitative accuracy, low detection limit, wide linear range, and high selectivity and sensitivity [100]. Therefore, the design of ratiometric fluorescent probes that give ratios of responses to zinc ions has had important significance in the detection and imaging of zinc ions in biological samples. Ratiometric zinc-ion fluorescent probes mostly operate using ICT or FRET mechanisms.

One probe of this type developed for highly selective Zn^{2+} detection is **48** (Figure 38), which contains a tetrahydrofuran and quinoline system [101]. Adding Zn^{2+} to a DMSO/H_2O (v/v = 8/2) solution of **48** results in a significant reduction in the intensity of emission at 430 nm (excitation at 390 nm) and production of a new emission band at 525 nm, corresponding to a red shift of 95 nm. The presence of the phenylacetylene moiety at C-5 of the quinoline ring in this probe is responsible for the occurrence of an enhanced ICT process and a red shift in emission in the metal complex in which electron transfer occurs to Zn^{2+}. This sensor has been applied to cell imaging.

As shown in Figure 39, the ratiometric Zn^{2+} sensor **49** contains a coumarin aldehyde and a thiophene-2-carbohydrazide group [102]. Upon addition of Zn^{2+}, the color of a solution of **49** changes from green to faint yellow, emission is enhanced and the emission maximum shifts from 517 to 563 nm (excitation at 450 nm). The enhanced fluorescence is attributed to the formation of a complex in which Zn^{2+} is coordinated

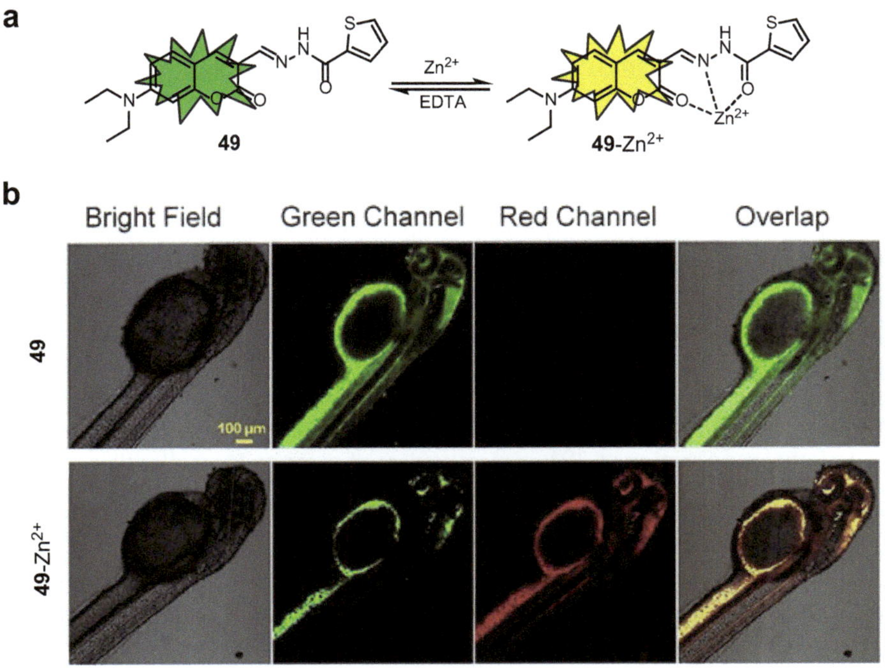

FIGURE 39 (a) Structure of **49** and mechanism for sensing Zn^{2+}; (b) Fluorescent microscope image of zebrafish stained with **49** and further incubated with Zn^{2+}. Adapted from Ref. [102] with permission. Copyright 2020, Dyes and Pigments.

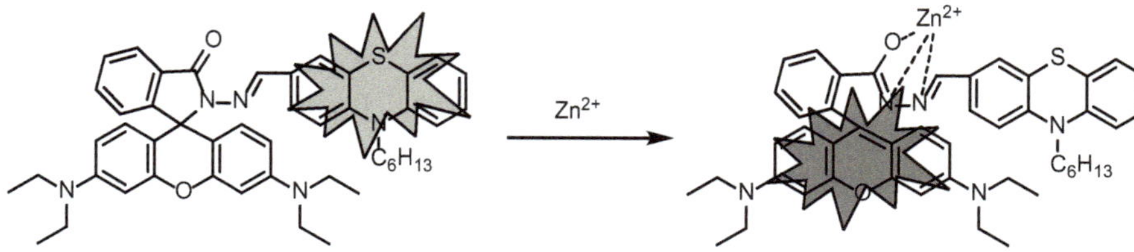

FIGURE 40 Structure of **50** and mechanism for sensing Zn^{2+}.

with the coumarin fluorophore. Owing to its high membrane permeability and low toxicity, **49** was successfully applied to detect Zn^{2+} in zebrafish.

Another example of a ratiometric Zn^{2+} probe is **50**, which contains a phenothiazine–rhodamine couple and operates using a FRET-based mechanism (Figure 40) [103]. This probe was successfully applied to detecting Zn^{2+} with high selectivity and sensitivity. Under excitation at 559 nm, **50** displays green fluorescence at 528 nm, which is characteristic of the phenothiazine fluorophore. Upon adding Zn^{2+}, a new emission band at 608 nm appears while the band at 528 nm decreases, in association with an isoemissive point at 575 nm. The above phenomena are attributed to Zn^{2+} promoted ring opening of the spirolactam ring system and introduction of the operation of FRET from phenothiazine to rhodamine.

4.3 Two-Photon Fluorescent Probes for Zinc Ion

Two-photon absorption (2PA) is a process in which long-wavelength absorption promotes short-wavelength emission [104]. The transition of organic molecules from the ground state to the excited state by simultaneously absorbing two photons through a virtual intermediate state can be promoted by using very high-intensity light. Figure 41 is a schematic of single-photon and 2PA processes, with S_0 representing the ground state, S_1 the first excited singlet excited state, and a dashed line for the virtual intermediate state of a substance. During two-photon excitation, organic molecules absorb two photons nearly simultaneously, with the sum of the energies of the two photons being equal to the difference between the excited and ground-state energies of the molecule. If the two photons used have the same frequency, the 2PA process is degenerate, while if the frequencies of the two photons differ, it is non-degenerate. The virtual level that serves as an intermediate in the 2PA process has a lifetime on the order of femtoseconds. In general, the cross section of 2PA is many orders of magnitude smaller than that for

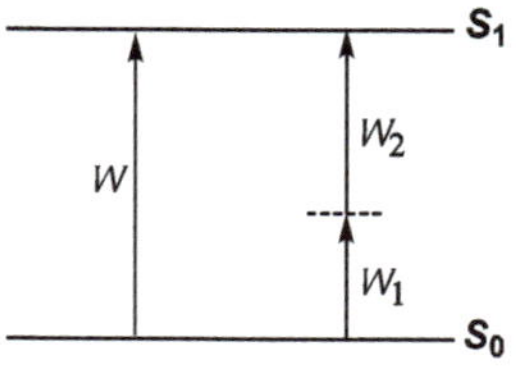

FIGURE 41 Schematic representation of single- and two-photon absorption. The virtual level is indicated by the dashed line.

single-photon absorption. As a result, carrying out 2PA requires an intense femtosecond pulse laser excitation.

As two-photon imaging instruments have become increasingly mature, two-photon fluorescent probes have also been more extensively studied. Relative to fluorescence imaging using conventional UV-visible light excitation, the excitation light for two-photon imaging is long-wavelength (mostly in the near-infrared region). Therefore, two-photon imaging has the advantages of deeper tissue penetration, lower tissue spontaneous fluorescence, and better 3D spatial positioning capabilities [105], properties that are excellent for cell and tissue imaging.

One example of a two-photon imaging probe for Zn^{2+} is **51** and its 6-carboxylic acid derivative **52**, which contain fluorescein fluorophores. The probes are membrane-free and have been used to observe Zn^{2+} at the synaptic level in adult mouse brain tissue and to monitor the release of Zn^{2+} from individual neurons using a two-photon laser microscope (Figure 42a) [106]. Both **51** and **52** mimic earlier developed ZinAlkylPyr (ZAP) probes, except they contain a pentafluorobenzyl group to reduce basicity. The presence of carboxylic acid on the bottom ring also significantly increases the absorption rate of the probe. Probe **51** was crystallized from methanol and its X-ray structure was determined, becoming the fourth metal-free Zinpyr probe characterized in this manner so far. However, the absorption of **51** is weak, which severely limits the brightness of its emission and limits its applicability in the study of biological processes. In contrast, **52** has a single photophysical response to Zn^{2+} and a stronger absorption than **51**. Moreover, **52** only stains Zn^{2+} in active recovery vesicles, showing that it is retained in the vesicles near the active region. The fact that these vesicles are more likely to be released under stimulation enables **52** to detect Zn^{2+} release more effectively. The use of **52** combined with two-photon imaging provides a unique tool for monitoring the release and dynamics of Zn^{2+} in living tissue at a single synaptic resolution. This two-photon Zn^{2+} probe can be employed to monitor presynaptic zinc-ion dynamics to improve understanding of the physiological role of mobile zinc ions in normal and abnormal neurological function (Figure 42b). Zn^{2+}

a

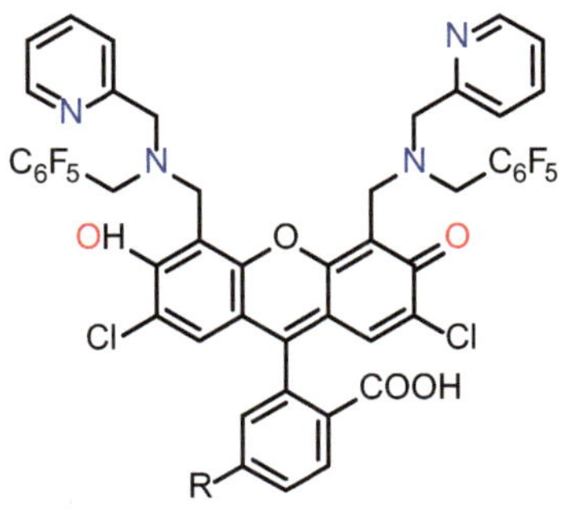

51 R = H
52 R = CO_2H

b

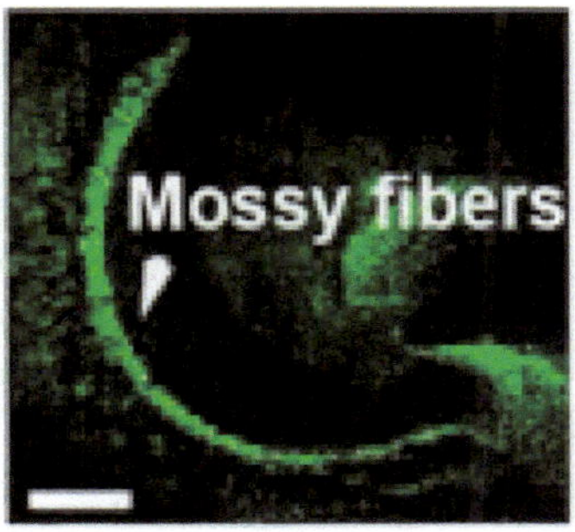

52

FIGURE 42 (a) Structures of **51** and **52**. (b) Representative images of Zn^{2+} stained mossy fiber regions in hippocampal slices. Acute adult mouse hippocampal slices were stained with **52** (5 μM). (Scale bars: 200 μm.) Adapted from Ref. [106] with permission. Copyright 2014, Proc. Natl. Acad. Sci. USA.

two-photon imaging with **52** has improved spatial and temporal resolution and thus can be used to determine the precise location of the associated synapses. Using the membrane-impermeable Zn^{2+} fluorescent probe **52** and a two-photon fluorescence microscope should provide an approach to studying the function of presynaptic Zn^{2+} in synaptic plasticity, toxicity, and Zn^{2+}-induced encephalopathy.

Probe **53** is a D-π-A-π-D type 2PA fluorescent probe for Zn^{2+} that contains a bipyridine moiety. This probe displays an unprecedented 2PA-promoted enhancement in response to Zn^{2+} (Figure 43a) [107] owing to the formation of an extended π-conjugated system upon complexation with zinc ion. Moreover, the ethylene oxide side chain in **53** benefits water solubility and cell permeability, which aid in imaging of free zinc ions in living cells. Based on the operation of an excited state charge transfer in the π-main chain of the Zn^{2+} complex, the fluorescence following 2P excitation is significantly enhanced, in association with a 13-fold increase in the 2PA cross section and a nine-fold increase in the fluorescence intensity at 620 nm compared to the probe in the absence of Zn^{2+}. When the 2PA section of 1433 GM and the two-photon excited fluorescence brightness of 860 GM were combined with Zn^{2+}, the 2P excited fluorescence changed well in the range of 517~620 nm, which was advantageous to monitor the ratio of free zinc ions in cells. The apparent dissociation constant (K_d^{TP}) of **53** was estimated to be 10 μM, and the detection limit is in the micromole range. The high cell permeability, and fluorescence quantum yield, coupled with a large 2PA cross section and high 2P fluorescence brightness make **53** useful in two-photon imaging of free zinc ions in living cells. The low cytotoxicity and high photostability of **53** allow its use in two-photon Zn^{2+} imaging of HeLa cells. Furthermore, use of the probe for tissue imaging and monitoring of free zinc ions in living cells was demonstrated by its application to two-photon imaging of zinc ions in living rat hepatocytes (Figure 43b).

Another simple Zn^{2+} fluorescent probe, **54**, which contains quinoline as the two-photon fluorophore and N,N-bis(pyridine-2-methyl)methylamine as the Zn^{2+} acceptor, operates by a ICT mechanism (Figure 44a) [108]. After **54** coordinates with Zn^{2+} in HEPES buffer, the emission peak at 455 nm decreases and the new emission peak at 515 nm increases upon two-photon excitation at 750 nm. The apparent dissociation constant (K_d) of **54** for Zn^{2+} was calculated to 2.94 ± 0.50 μM, and the detection range is 0–6.00 μM and a limit of detection for Zn^{2+} of 25 ± 5 nM. Probe **54** has significant selectivity, high water solubility, low cytotoxicity, and good biocompatibility, enabling it to be used for imaging endogenous Zn^{2+} released by NO-stimulation in neurons (Figure 44b). It can also be used for ratio imaging and biosensing of Zn^{2+} in zebrafish larvae.

4.4 NIR Fluorescent Probes for Zinc Ion

Early systems for fluorescence analysis focused on using light in the UV-visible region. However, the vast majority of mammalian tissues are opaque to light in the visible spectrum. Some endogenous molecules (e.g., hemoglobin, fat, and water) have significant scattering and absorption in the visible light region, which leads to signal attenuation as tissue depth increases. In addition, spontaneous fluorescence of endogenous molecules, including flavins, reticular proteins, and

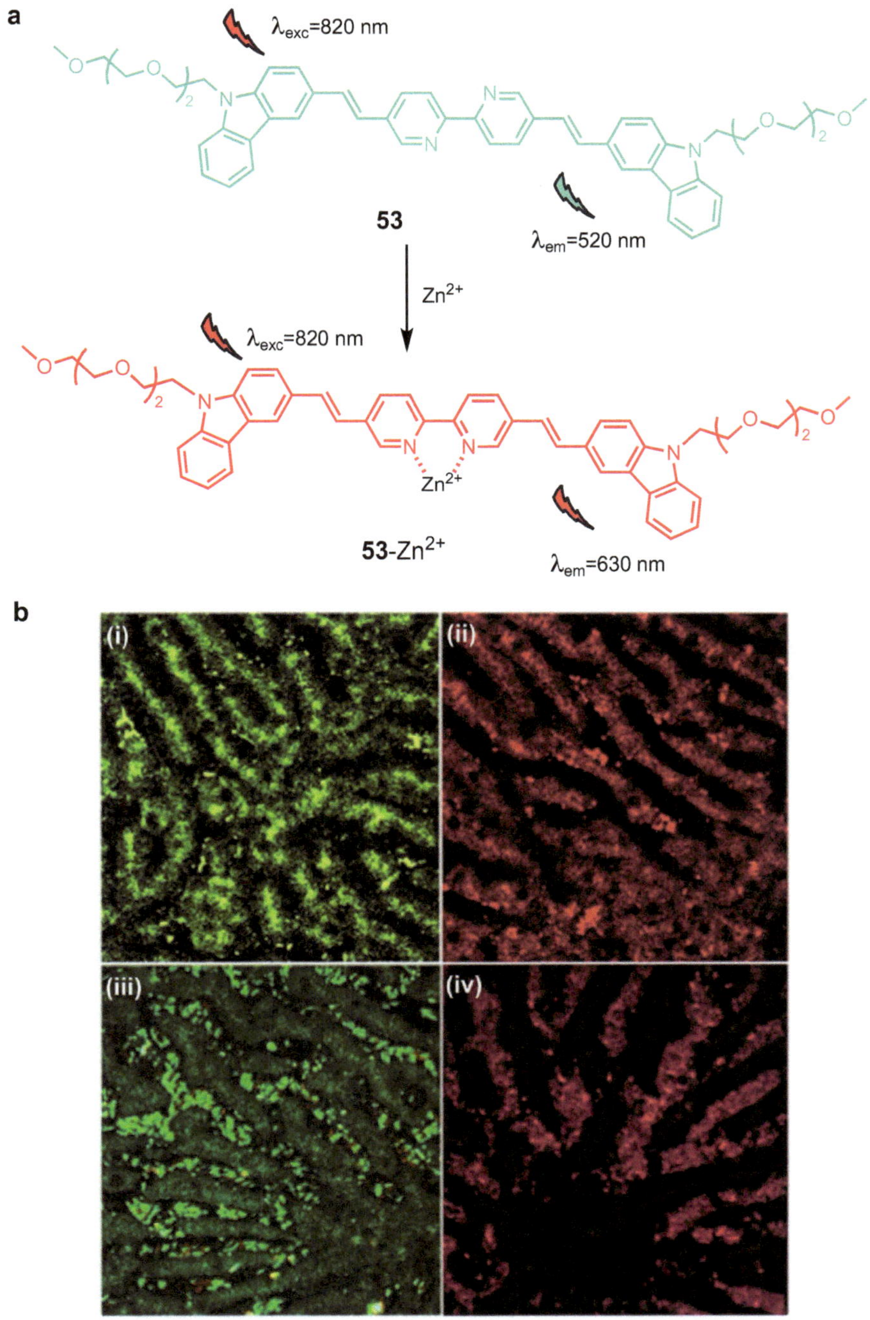

FIGURE 43 (a) Schematic representation of the two-photon response of **53** and **53**-Zn^{2+}. (b) *In vivo* 2P imaging using **53** and **53**-Zn^{2+}. 2P tissue images of hepatocytes in live rat after intravenous injection with (i) **53** (λ_{ex} = 820 nm, λ_{em} = 530 nm) and (ii) **53**-Zn^{2+} (λ_{ex} = 820 nm, λ_{em} = 650 nm). Corresponding multispectral images of tissues after intravenous injection with (iii) **53** (λ_{ex} = 820 nm, λ_{em} = 500–600 nm) and (iv) **53**-Zn^{2+} (λ_{em} = 600–700 nm). Adapted from Ref. [107] with permission. Copyright 2014, Chemical Science.

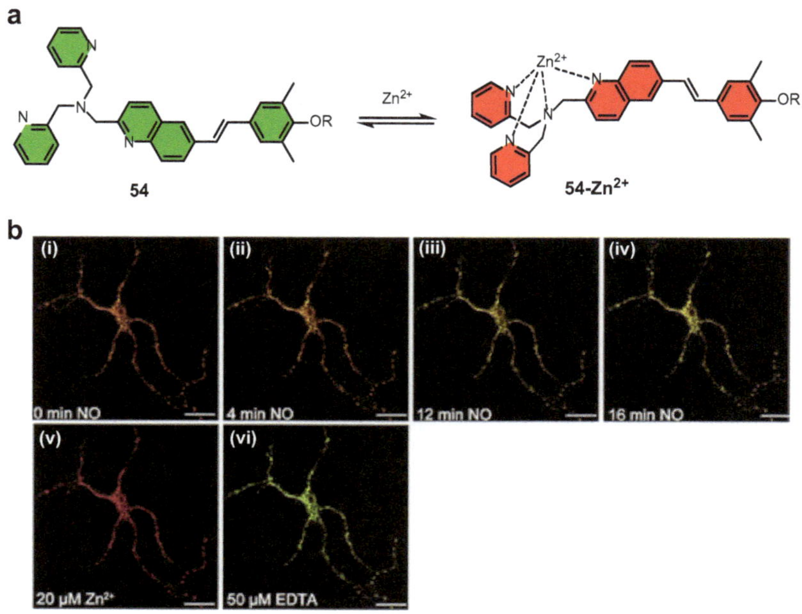

FIGURE 44 (a) Structure of **54** and illustration of the Zn^{2+} sensing process. (b) Two-photon confocal fluorescence microscope images of NO-induced neurins with 5 μM **54**. Fluorescence images of **54** before (i) and after being incubated with NO for (ii) 4 minutes, (iii) 12 minutes, (iv) 16 minutes, (v) then treated with 10 μM Zn^{2+}/pyrithione (1:2), and (vi) followed 20 μM EDTA. Adapted from Ref. [108] with permission. Copyright 2020, Biosensors & Bioelectronics.

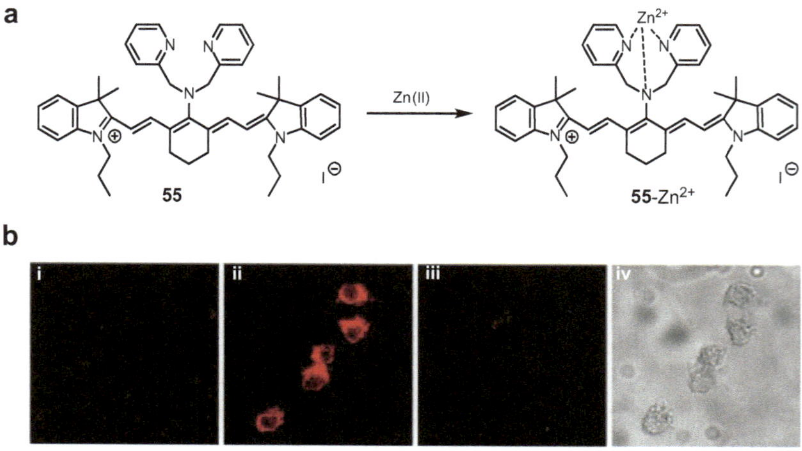

FIGURE 45 (a) Structure of **55** and illustration of the Zn^{2+}-sensing process. (b) Confocal fluorescence microscope images of live macrophage cells. (i) Cells incubated with 5 μM **55** for 30 minutes at 37°C. (ii) Cells supplemented with 100 μM Zn^{2+} (10:2 = Zn^{2+}/pyrithione ratio) to the **55**-treated macrophage cells and the cells were incubated for 30 minutes at 37°C. (iii) Treatment with 100 μM TPEN for 5 minutes at 37°C. (iv) Bright-field image of live macrophage cells shown in panel (ii). Adapted from Ref. [111] with permission. Copyright 2006, Chemical Communications.

nicotinamide adenine dinucleotides occurs mainly in the visible light region. In contrast, near-infrared light (NIR, 650–900 nm) can penetrate more deeply into organisms [109]. Moreover, NIR light causes minimal photogenic damage to biological samples, and NIR photons cause reduced photobleaching, which facilitates long-term tracking [110]. Therefore, near-infrared fluorescent probes used for *in situ* detection and imaging of analytes in deep in tissues have a higher resolution and signal-to-noise ratio.

Although the excitation wavelengths of typical two-photon fluorescent probes are in the NIR region, the emission wavelengths of a part of them are still in the visible region, hindering deeper tissue imaging. In addition, two-photon fluorescence imaging has certain limitations in its application to live imaging, such as the requirements for sensitive instrumentation and a femtosecond laser. Therefore, the development of NIR fluorescent probes for biological analysis can be complementary to two-photon fluorescent probes and enable the imaging of analytes in different environments. Zinc-ion NIR fluorescent probes mostly use cyanine dyes, phenoxazinium, and malononitrile derivatives as signal fluorophores.

One example is the NIR fluorescent probe **55** that contains a cyanine dye and has high selectivity and sensitivity for Zn^{2+} (Figure 45a) [111]. Tricarbocyanine possessing two propyl groups is the fluorophore in this probe and the DPA group is employed as the metal ion ligand. Before coordination to Zn^{2+}, fluorescence emission at 780 nm of **55** is weak (excitation at 731 nm) due to the PET quenching. After complexation with Zn^{2+}, fluorescence of the probe is significantly enhanced due to the inhibition of PET. The fluorescence quantum yield of **55**-Zn^{2+} is nearly 20-fold higher than that **55**. The NIR fluorescence probe can be applied to imaging in cells (Figure 45b).

Probe **56** is a novel NIR fluorescent probe that employs a dicyanoisophorone group as a recognition moiety and DPA group as a chelating ligand (Figure 46a) [112]. Upon Zn^{2+} chelation with **56**, the phenolic hydroxyl group undergoes deprotonation to form a phenolate ion, which participates in ICT to the dicyanoisophorone group. This ICT effect leads to greatly enhanced emissions. In the ethanol-HEPES buffer (pH 7.4, v/v = 3/7), the probe shows high selectivity for Zn^{2+} with a linear response in the 0–15 μM range, and Zn^{2+} binding constant (K_a) of 2.15×10^5 M^{-1}. Owing to its excellent photostability, good biocompatibility, probe **56** can be used for live cell staining to detect Zn^{2+} changes (Figure 46b).

Two kinds of NIR fluorescent probes, **57** and **58** (Figure 47a), containing Rhodol counterpart fluorophores were developed for highly selective detection of Zn^{2+} [113]. The dipicolylamine group in these probes not only acts as a ligand for Zn^{2+}, but also as an electron donor quencher of the excited state of the fluorophores. After coordination with Zn^{2+}, the PET effect is inhibited and the fluorescence at around 700 nm increase. In HEPES buffer (pH 7.0) and 1% EtOH solution, the probes display high selectivity for Zn^{2+} over other metal ions. Furthermore, probes **57** and **58** show linear fluorescence responses to Zn^{2+} in the 0.3–4.0 μM range with respective detection limits of 0.19 and 0.086 μM. In addition, the fluorescent probes enable visualization of exogenously supplemented Zn^{2+} in living cells and effective detection of intracellular Zn^{2+} released from intracellular metalloproteins in the cells treated with 2,2′-dithiodipyridine (Figure 47b).

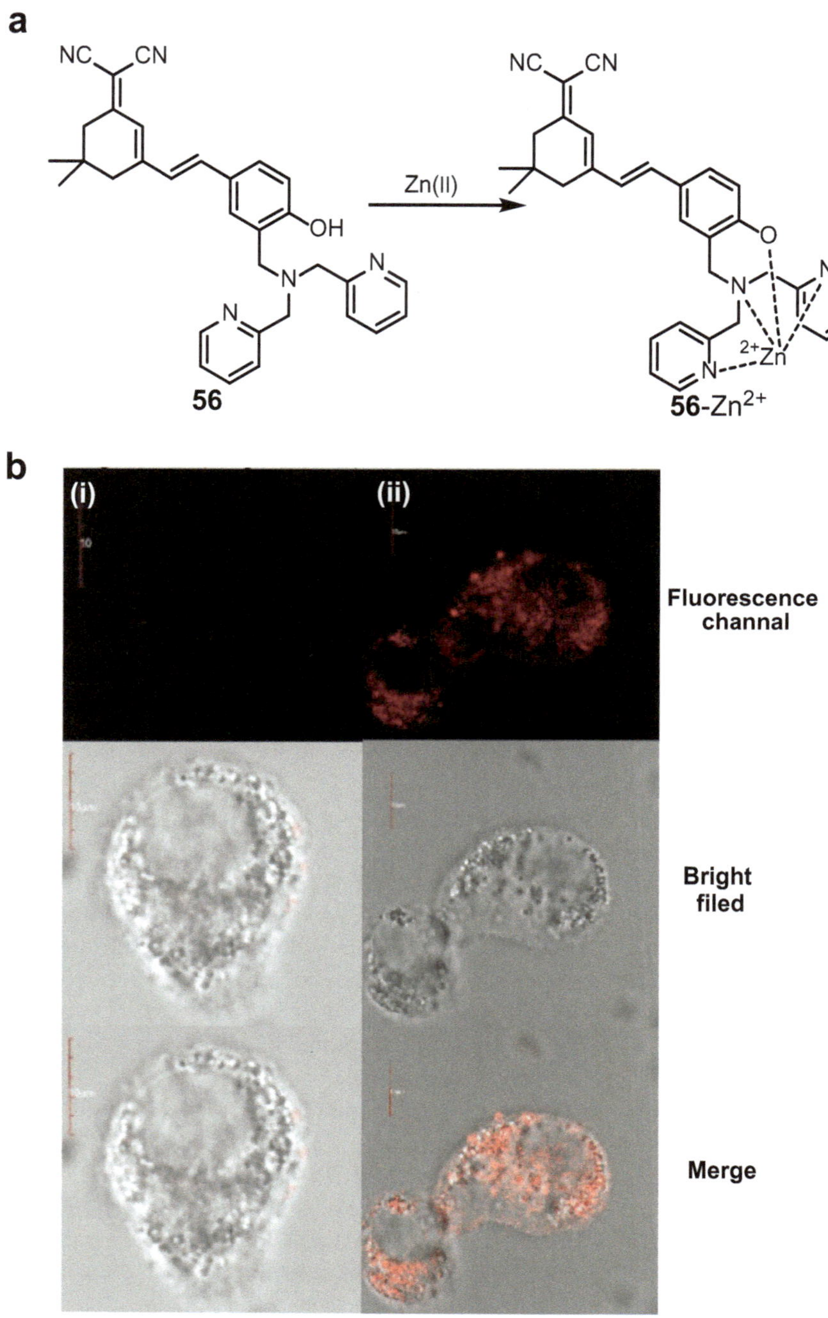

FIGURE 46 (a) Structure of **56** and illustration of the Zn^{2+}-sensing process. (b) Confocal fluorescence microscope images of HeLa cells. (i) Hela cells were incubated with **56** (10 μM) without DTDP (5 μM); (ii) Hela cells were incubated with DTDP (5 μM) for 1 hour and then detected by using **56** (10 μM). $\lambda_{ex} = 559$ nm. Adapted from Ref. [112] with permission. Copyright 2020, Inorganica Chimica Acta.

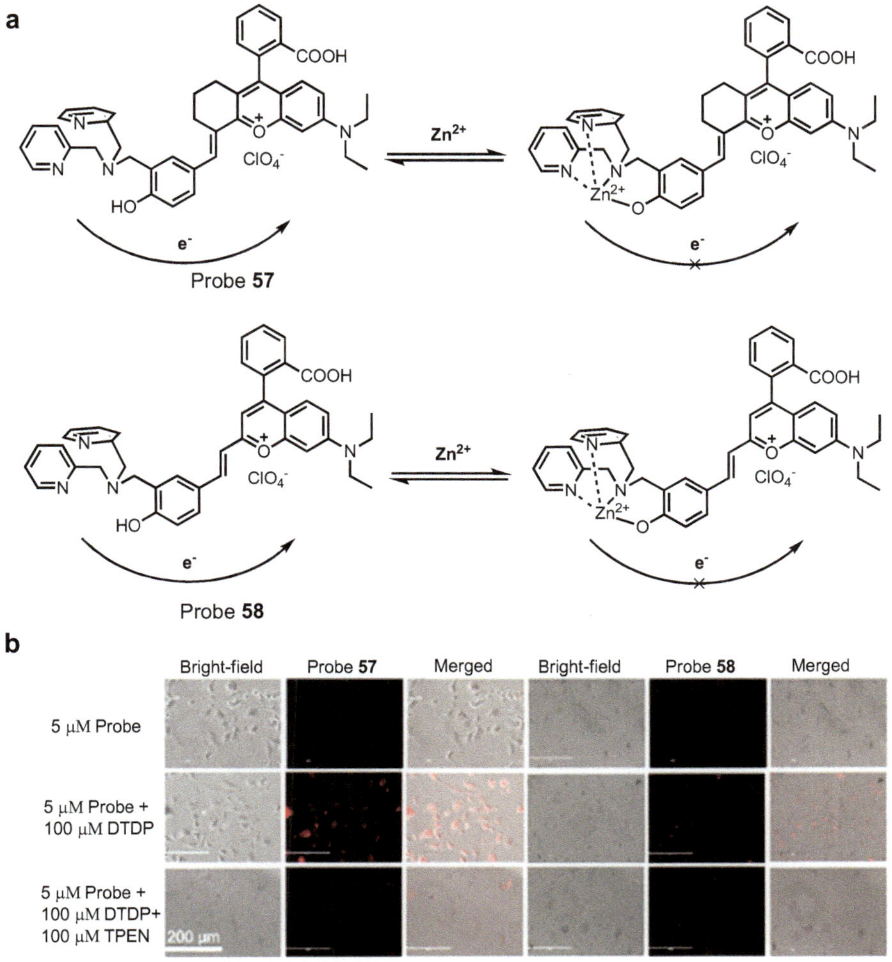

FIGURE 47 (a) Structures of **57** and **58** and illustration of the Zn^{2+}-sensing process. (b) Fluorescence microscope images of HUVEC-C cells in the presence of 5 μM each of probes **57** and **58**. Cells were supplemented with 100 μM 2,2′-dithiodipyridine (DTDP) for 30 minutes before acquiring images. Then 100 μM TPEN (zinc chelator) was added and cells were further incubated for 10 minutes before acquiring the images. Adapted from Ref. [113] with permission. Copyright 2016, ACS Sensors.

Probe **59** is a NIR fluorescent probe containing a hemicyanine fluorophore that is highly selective for Zn^{2+} (Figure 48a) [114]. The dipicolylamine moiety in **59** acts as both an electron donor in PET quenching and a ligand for Zn^{2+}. In HEPES buffer (pH 7.0) and 1% EtOH solution, **59** is selective for Zn^{2+} over other metal ions. Compared with other reported probes, **59** is far less interfered with by Cd^{2+}. Before coordination with Zn^{2+}, fluorescence of the probe is quenched by PET from the ligand to hemicyanine fluorophore. After binding with Zn^{2+}, electron transfer is inhibited and fluorescence is restored. In addition, probe **59** displays a linear fluorescence response to Zn^{2+} from

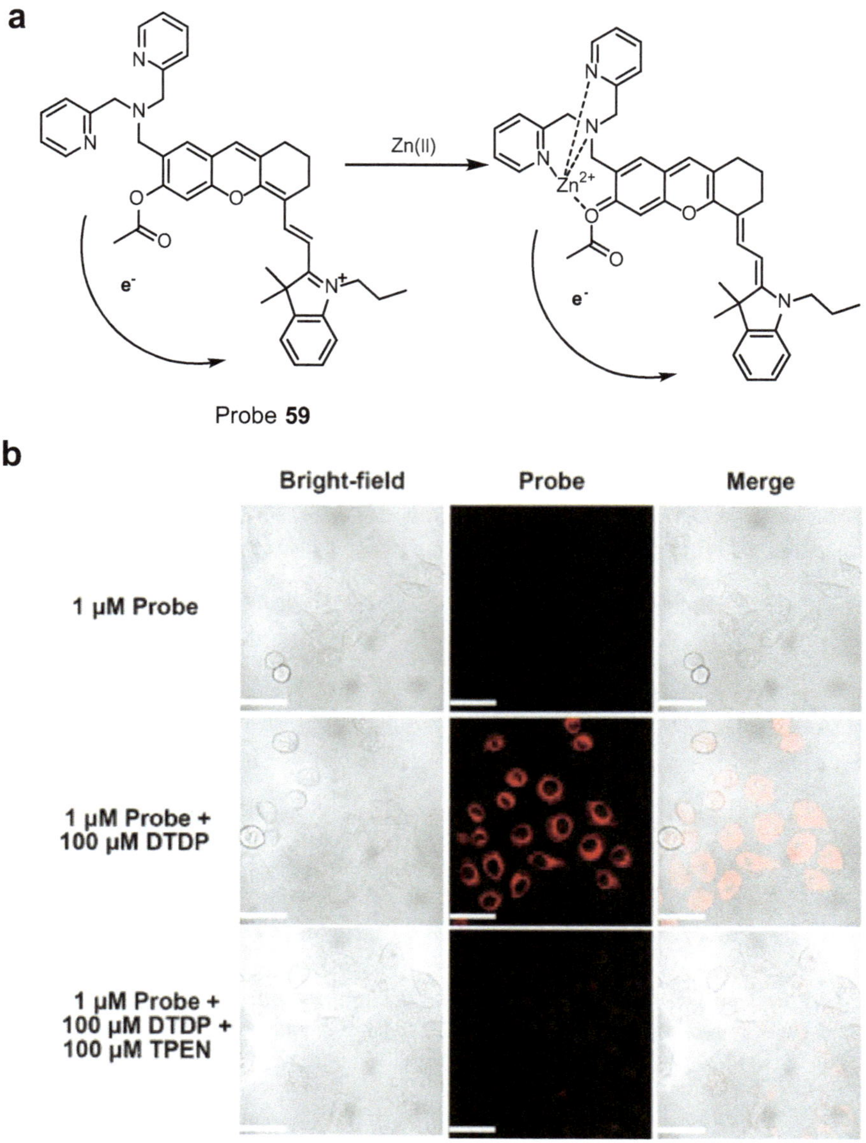

FIGURE 48 (a) Structure of **59** and illustration of the Zn^{2+}-sensing process. (b) Fluorescence microscope images of HeLa cells incubated with probe **59**. Cells were first incubated with 1.0 μM of probe **59** for 1 hour and then were supplemented with 100 μM of 2,2′-dithiodipyridine (DTDP) for 30 minutes before images were taken. Scale bar: 50 μm. λ_{ex}: 635 nm. Adapted from Ref. [114] with permission. Copyright 2019, Molecules.

0.1 to 1.5 μM with a detection limit of 0.45 nM. Owing to its high selectivity, good photostability, biocompatibility, and non-cytotoxicity, **59** has excellent applications to live cell imaging of both exogenously supplemented Zn^{2+} and free endogenous Zn^{2+} released from intracellular metalloproteins (Figure 48b).

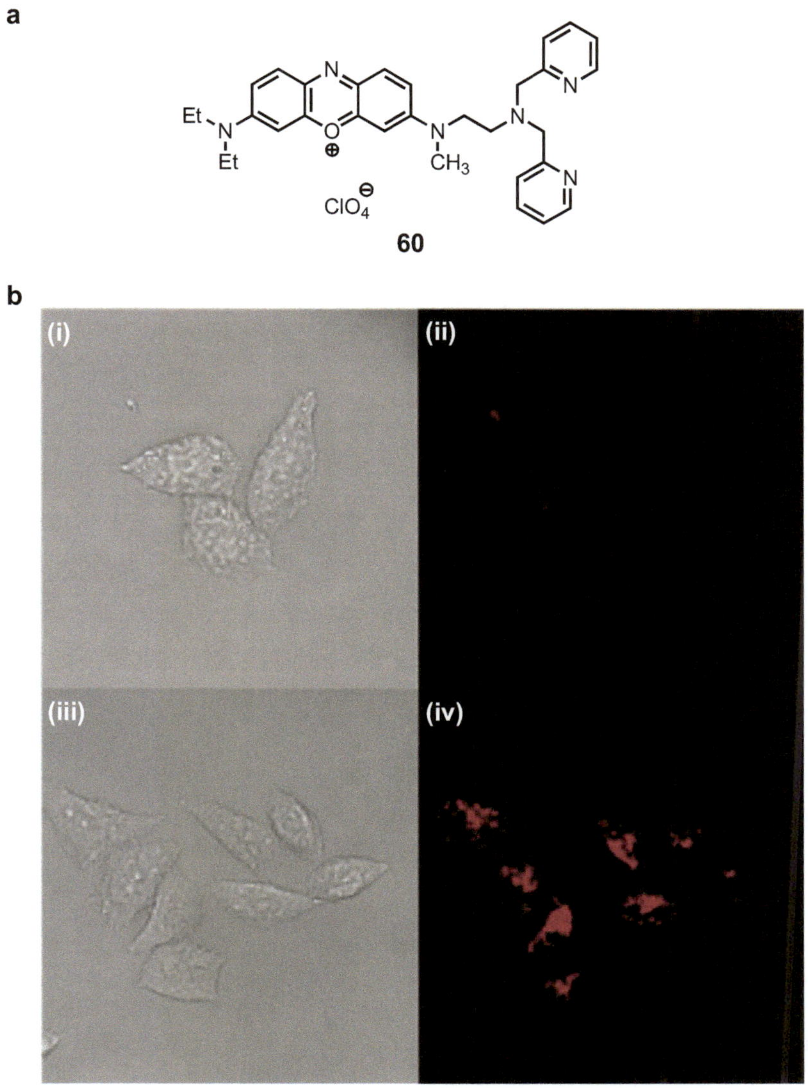

FIGURE 49 (a) Structure of probe **60**. (b) (i) Bright-field image in KB cells pre-incubated with **60** (10 μM) for 30 minutes; (ii) Fluorescence image of (i); (iii) Bright-field image of KB cells pre-incubated with $ZnCl_2$ (100 μM) for 1 hour, washed three times, and then treated with **60** (10 μM) for 30 minutes; and (iv) Fluorescence image of (iii). Emission was collected at 650–750 nm upon excitation at 633 nm. Adapted from Ref. [115] with permission. Copyright 2012, Sensors and Actuators B: Chemical.

Another NIR fluorescent probe for Zn^{2+} is the phenoxazinium-based compound **60** (Figure 49a) [115]. Phenoxazine derivatives have good water solubility, excellent membrane permeability, low toxicity, and long-wavelength emission. The presence of N-methyl-N′,N′-bis(pyridin-2-ylmethyl)ethane-1,2-diamine (DPEA) in the probe not only improves the specificity for chelating Zn^{2+}, but it also increases water solubility.

Upon addition of Zn^{2+} to a buffered aqueous solution of **60** at physiological pH, the maximum absorption shifts to 647 from 657 nm, and the emission intensity at 677 nm (650 nm excitation) increases in a zinc-ion concentration-dependent manner. The fluorescence intensity is enhanced about seven-fold. This phenomenon is attributed to Zn^{2+} blockage of PET from the lone electron pair of the tertiary nitrogen atom in the dipicolylamine group to the excited fluorophore (Figure 49a). The probe has good specificity for the detection of Zn^{2+} in a biological medium with a K_d value of 50 nM and a 1.5:1 stoichiometry. Although Co^{2+} can also enhance the fluorescence of this probe, the effect can be inhibited by adding cysteine. It is worth noting that Cu^{2+} and Ni^{2+} quench the fluorescence of complexes composed of probe and Zn^{2+}, and anion effects are negligibly small. The probe remains stable at pH 6.0–10.0, and binding of **60** with Zn^{2+} can be reversed by adding EDTA. The application of KB cell (human nasopharyngeal epidermal carcinoma cell) imaging shows that probe **60** penetrates the membrane and successfully images Zn^{2+} *in vivo* in real time (Figure 49b).

4.5 Bioprotein-Based Fluorescent Probes for Zinc Ions

Genetically encoded probes are hybrid proteins composed of fluorescence proteins and peptide linkers [116]. Most of these sensors respond to Zn^{2+} based on a change in the efficiency of FRET between two fluorophores [117]. Binding with Zn^{2+} affects folding of the peptides in these probes and the conformation changes result in substantial changes in the distances between fluorophores and a corresponding alteration of FRET efficiencies. These sensors achieve more precise targeting in subcellular organelles and more accurate quantification of Zn^{2+} due to their high specificity, selectivity, and affinity.

Qiao et al. described a novel probe to study Zn^{2+} trafficking *in vivo* by monitoring the binding of Zapl and Zn^{2+} in yeast [118]. As shown in Figure 50, sensor **61** is a hybrid protein consisting of a pair of zinc fingers (ZF, Cys-2His-2-type), and enhanced yellow (eYFP) and cyan fluorescent proteins (eCFP) fused to the respective N and C terminals of the fingers. When the two fluorescent proteins, eCFP and eYFP, are spatially close, emission of yellow fluorescent protein (FP) at 545 nm (excitation at 440 nm) occurs due to the FRET. Therefore, the intensity of FRET reflects the distance between the two FPs. Moreover, the apo form of the ZF is more likely to exist with folded ZF domains that are necessary for zinc-ion binding. Thus, zinc-ion binding induces a FRET effect in **61** because it alters the distance between two FPs. Sensor **61** could be used to examine or target zinc ions in variable subcellular compartments in different organisms. Two genetically encoded sensors **62** and **63** were devised on the basis of **61**. Different from **61**, eYFP in probes **62** and **63** contains the more pH-stable citrine (Figure 50) [119, 120]. Probes **62** and **63** target the endoplasmic reticulum (ER) and Golgi in HeLa cells and were used to estimate levels of trace Zn^{2+} are 0.9 and 0.6 pM, respectively. In a subsequent study, the difference between genetically encoded sensors and traditional small molecule indicators was assessed. The protein-based sensor **63** for Zn^{2+} has greater advantages in cell biology, because it provides a more accurate method of Zn^{2+} localization in different cells, reduces the interference of cellular ion pool, and facilitates Zn^{2+} quantification.

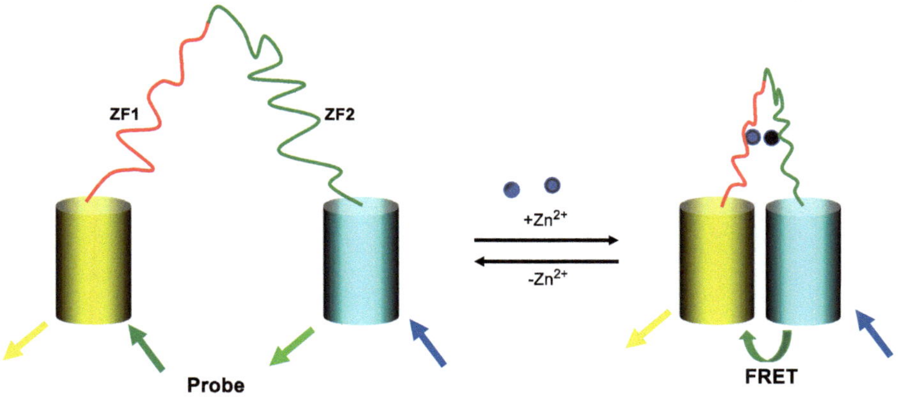

FIGURE 50 Schematic diagram of zinc-ion detection by probes **61–63**. ZF1 is shown in red, and ZF2 is shown in green.

To address the problem of inconsistency in the quantification of Zn^{2+} associated with the use of different genetically encoded sensors, the multifunctional FRET sensor **64** was developed (Figure 51) [121]. The probe contains a peptide linked to self-associating variants of cerulean and citrine, and a Cys_2His_2 binding pocket is incorporated into the dimerization interface, which improves Zn^{2+} affinity by over 1,000-fold and causes a large change in the emission ratio. This new binding mechanism ensures that the sensor realizes a more sensitive and precise determination of Zn^{2+} concentration in the ER or cytoplasm through the FRET process. Additionally, **64** can be used to determine the concentration of mitochondrial zinc ions and elucidate the influence of pH.

Because most FRET sensors contain two fluorophores, the existence of broad spectral bands limits performance in multi-analyte imaging. To enable signal acquisition at a single wavelength and to have a larger dynamic range, the single-FP (single-FP)-based zinc sensor **65** was designed (Figure 52) [122]. This sensor consists of two zinc fingers (ZF) fused in tandem with the C and N termini of a circularly

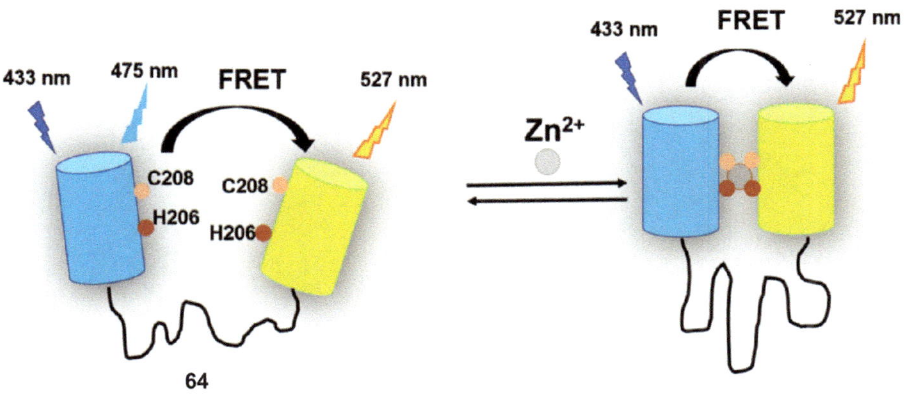

FIGURE 51 Schematic diagram of sensor **64** and illustration of the Zn^{2+}-sensing process. Adapted from Ref. [121] with permission. Copyright 2015, ACS Chemical Biology.

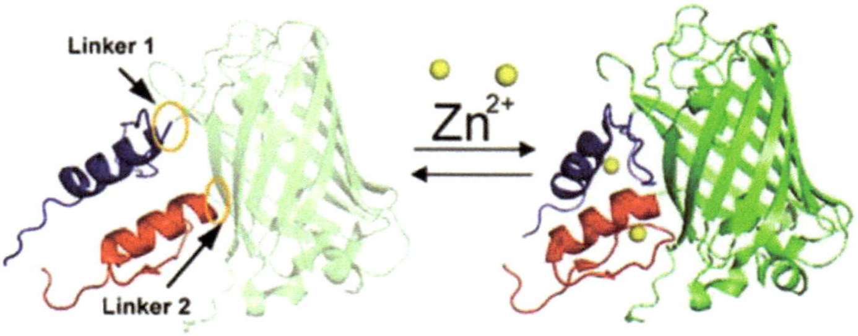

FIGURE 52 Schematic diagram of sensor **65** and illustration of Zn^{2+} sensing. Adapted from Ref. [122] with permission. Copyright 2016, ACS Chemical Biology.

permuted cGFP. Upon binding to Zn^{2+}, folding of ZF takes place to induce an interaction between the two ZFs, which causes cGFP to change location. In response to Zn^{2+}, the fluorescence intensity increases about 2.6-fold. However, the fluorescent intensity is pH dependent because the pK_a of **65** is 7.4. The stability and sensitivity of **65** enable its use for monitoring Zn^{2+} in the cytoplasm, mitochondria, and plasma membrane.

Two novel probes, **66** and **67**, based on green FP (GFP), were developed for studying Zn^{2+} secretion dynamics *in vivo* (Figure 53) [123]. Probe **67** precisely targets to the extracellular side of the plasma membrane, and it detects Zn^{2+} secreted from intracellular vesicles. The analog **66** was obtained by screening a library of substances containing two linkers at both ends of the Zn^{2+} binding domain. Sensor **67** was constructed by fusing the zinc hook peptide from *Pyrococcus furiosus* Rad 50 to the N- and C-termini of cpGFP derived from G-GECO1. As an excitation ratiometric indicator, **66** has a dynamic range of 2.5-fold and **67** has a dynamic range of 7-fold when excited at 480 nm. Meanwhile, **67** has a three-fold intensiometric turn-on fluorescence response to Zn^{2+}. These screening and optimization tools are expected to be key to bringing induced islet biologics from laboratories to clinics. Because it is completely genetically coded, probe **67** can be used *in vivo* models for long-term monitoring of the dynamics of Zn^{2+}.

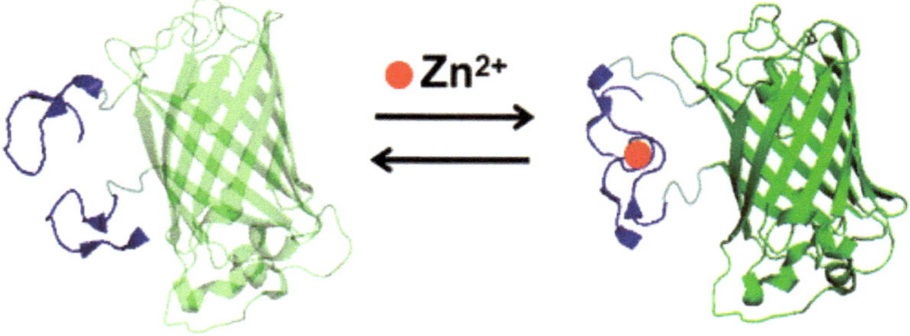

FIGURE 53 Schematic diagram of sensor **66**, **67** and illustration of Zn^{2+} sensing. Adapted from Ref. [123] with permission. Copyright 2019, Analytical Chemistry.

5 TARGETED Zn^{2+} IMAGING

Introduction of groups targeting subcellular organelles into zinc-ion probes can enable visualization of the distribution of Zn^{2+} in organelles in the cell. For example, fluorescence probes with high selectivity for Zn^{2+} can be modified by attaching specific targeting units, such as triphenylphosphonium for mitochondria recognition, according to the selective accumulation of lipophilic cations in mitochondria. Cholesterol as the targeting moiety can be employed to locate the probe on the outer surfaces of the plasma membranes of mammalian cells, morpholine can be used to target lysosomes, and glibenclamide can be used owing to its highly selective binding to the endoplasmic reticulum (ER). The above types of targeting units should not influence the binding efficiencies and responses of probes to Zn^{2+} and they have potential prospects for applications in studies designed to understand Zn^{2+} biology in cells.

5.1 Mitochondrial Zinc-Ion Imaging

Mitochondria are the energy factories of cells, and they are also involved in cell differentiation and death, and the control of cell cycle and growth [124]. Mitochondria have been implicated in a number of human diseases, including mitochondrial disease [125] and heart dysfunction [126]. Many Zn^{2+} enzymes are present in the mitochondria, and this metal ion is essential for maintaining mitochondrial protein function. Therefore, it is important to have the ability to detect, image, and trace zinc ions in the mitochondria. Because the transmembrane potential of the mitochondria is negative (ca. −180 mV, from outer to inner), fluorescent sensors with lipophilic cationic properties can more readily accumulate in mitochondria. For example, the phosphonium ion-containing substance **68** (Figure 54a) is a mitochondrial targeting probe for ratiometric detection of zinc ion [127]. In neutral or even weakly alkaline solutions, this probe is protonated to form a highly conjugated fluorophore that emits at 550 nm. Upon addition of zinc ion, a metal complex forms having a different conjugated fluorophore that emits blue-shifted fluorescence at 504 nm. The selectivity of the probe for Zn^{2+} was investigated in an aqueous solution. Fe^{2+}, Co^{2+}, Ni^{2+}, Cu^{2+}, especially Cd^{2+}, failed to cause changes in the fluorescence emission, indicating that the probe has a good selectivity for Zn^{2+}. Probe **68** can be used for imaging and quantifying zinc ions in the mitochondria of NIH3T3 cells (Figure 54b).

The novel dual excitation ratiometric fluorescent sensor **69** (Figure 55a), possessing a 4-amino-7-sulfamoyl-benzo[c]-[1,2,5]oxadiazole fluorophore and a mitochondria targeting phosphonium group, was designed for the detection of zinc ions [128]. In the presence of Zn^{2+}, the absorption maximum of **69** undergoes a blue shift from 453 nm to 391 nm because Zn^{2+} binding reduces the electron-donating ability of the polyamine moiety and associated ICT effect. Upon titration with Zn^{2+}, the ratio of emission intensity at 577 nm obtained by excitation at 397 and 488 nm changes. The binding constant of **69** with Zn^{2+} was determined to be 8.2 ± 0.2 nM. Moreover, a study of ratiometric Zn^{2+} imaging demonstrated that **69** can be used to monitor Zn^{2+} reversibly and quantitatively in mitochondria in cells stimulated by using H_2O_2 or SNOC (Figure 55b).

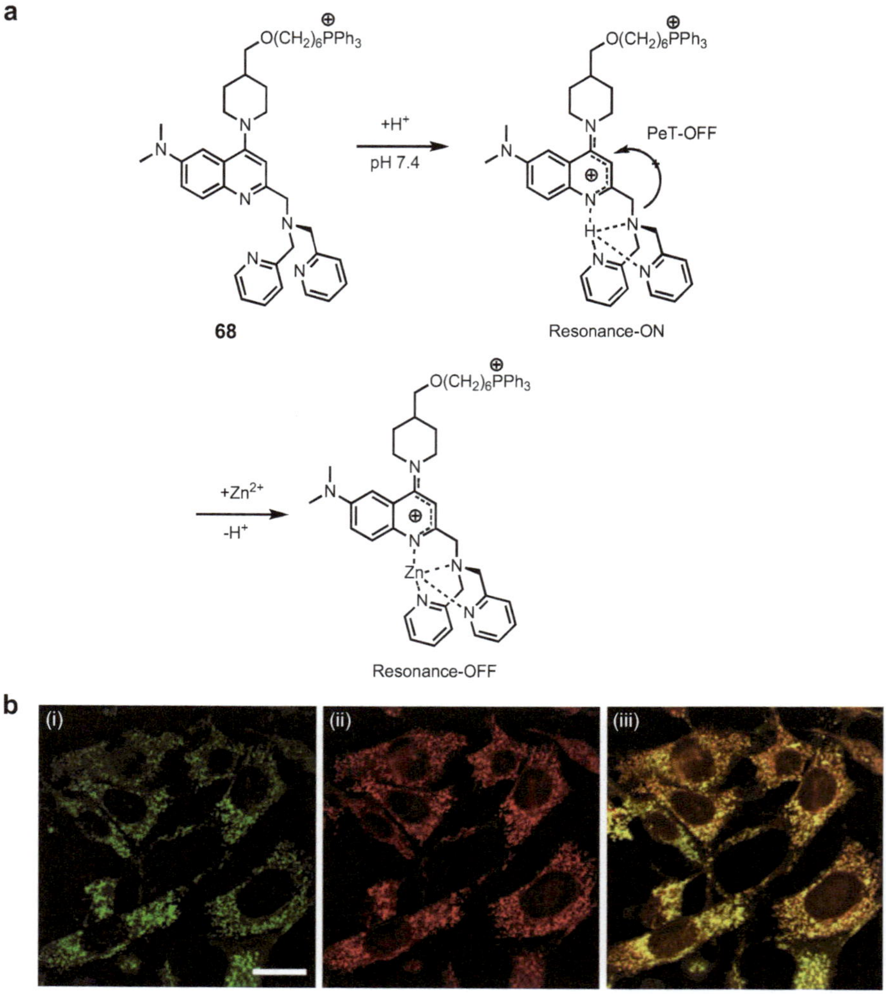

FIGURE 54 (a) Proposed ratiometric sensing strategy of **68**. (b) Probe **68** colocalizes to mitochondria in live NIH3T3 cells. Cells were stained with: (i) **68** (1.5 μM), and (ii) MitoTracker Deep Red (50 nM) for 30 minutes at 25°C in DMEM; (iii) Overlay of (i) and (ii); scale bar: 20 μm. Adapted from Ref. [127] with permission. Copyright 2012, Chemistry-A European Journal.

5.2 Endoplasmic Reticulum (ER) Zinc-Ion Imaging

The ER is an organelle responsible for the synthesis of proteins, lipids, and secretory and transmembrane proteins. It is well known that increased ER stress is associated with heart disease, stroke, neurodegenerative diseases, and cancer. Also, zinc-ion transport and consumption affect ER stress [129–131]. In view of these findings, probes, such as **70** and **71** (Figure 56a), were developed for ER zinc-ion imaging and detection [132]. The glibenclamide group is employed as the targeting unit in **70** and **71**, which have dissociation constants for zinc ions of 3.5 and 4.7 nM, respectively,

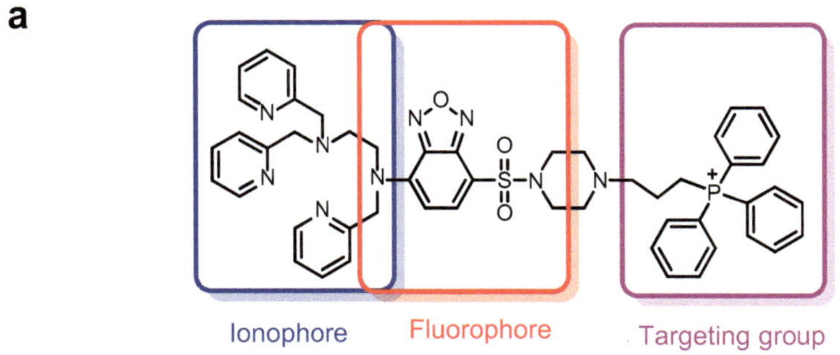

69

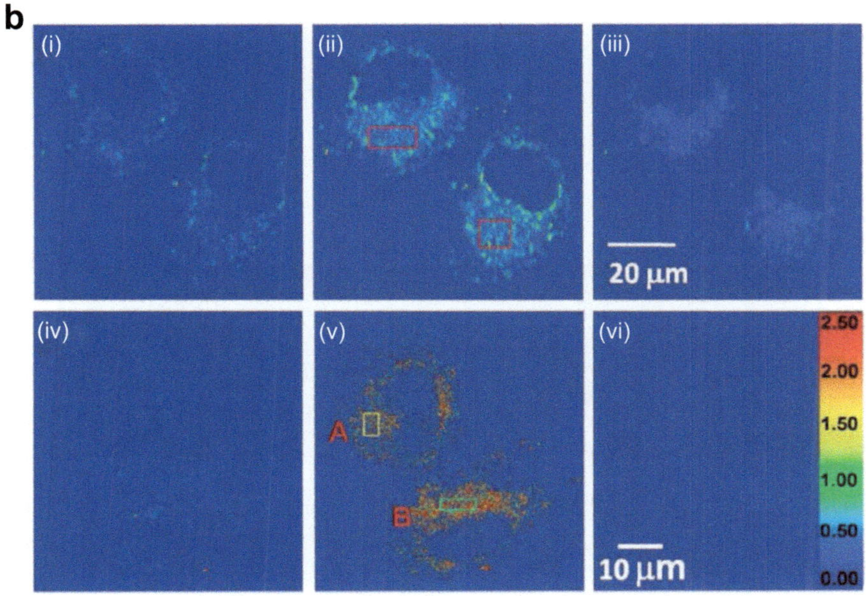

FIGURE 55 (a) Structure of **69**. (b) Fluorescence dual excitation ratiometric imaging of mitochondrial Zn^{2+} release in **69**-stained (20 μM, 30 minutes, 25°C) MCF-7 cells stimulated by 10 mM H_2O_2 (i–iii) or by 10 mM SNOC at 25°C (iv–vi). (i) Image of the stained cells before H_2O_2-stimulation; (ii) Image of cells in (i) exposed to 10 mM H_2O_2 solution (17 minutes); (iii) Image of cells in (ii) treated by TPEN solution (50 μM, 8 minutes); (iv) Image of the stained cells; (v) Image of cells in (iv) exposed to 10 mM SNOC (1 hour); (vi) Image of cells in (vi) treated by TPEN solution (50 μM, 10 minutes). The ratio bar is shown in (vi). Adapted from Ref. [128] with permission. Copyright 2012, Chemical Communications.

and respective detection limits of 47 pM and 0.71 nM. Probe **70** locates well in the ER (Figure 56b), and its fluorescence is enhanced by adding the exogenous membrane-permeable zinc source zinc-pyridinone. Subsequent addition of TPEN (*N,N,N′,N′*-tetrakis(2-pyridinylmethyl)-1,2-ethanediamine), a strong Zn^{2+} chelator, causes nearly complete quenching of fluorescence. Similarly, the same findings were obtained in studies with probe **71**.

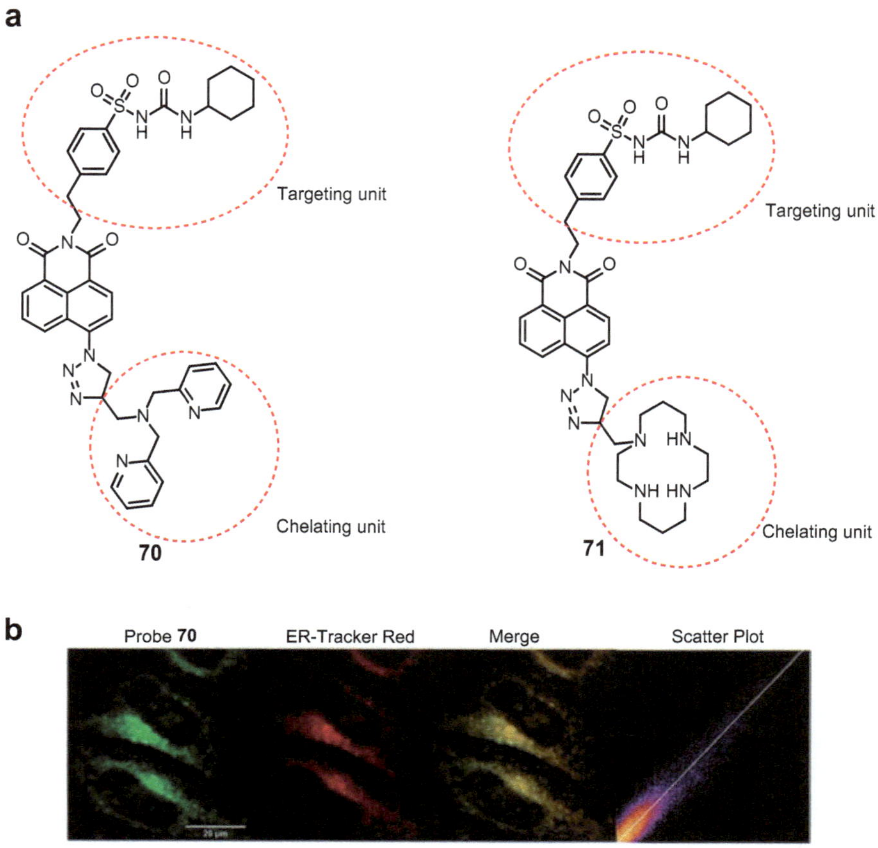

FIGURE 56 (a) Structures of **70** and **71**. (b) Co-localization images of HeLa cells incubated with **70** (20 μM, GFP filter: λ_{ex} = 470/30 nm, λ_{em} = 530/50 nm) and commercial red organelle tracker dyes (RFP filter: λ_{ex} = 530/40 nm, λ_{em} = 605/55 nm). (Scale bars = 20 μm.) Adapted from Ref. [132] with permission. Copyright 2019, Chemical Science.

5.3 Lysosome Zinc-Ion Imaging

The zinc-ion levels in cells significantly correlate with prostate cancer [133], where it is significantly lower than that in normal tissues [134, 135]. Lysosome is an acidic organelle in cells that participates in various metabolic processes, which are closely associated with the pathogenesis of cells. Detection and imaging of intra lysosomal zinc ions has significance in studies of relevant cellular metabolic processes. Since lysosomes are acidic (pH 4.5–6.0), weakly alkaline aliphatic amines tend to accumulate in this organelle. To minimize the interference caused by protonation, sensors for lysosomal Zn^{2+} imaging should have a low pK_a so that their fluorescence remains unchanged in the absence of the metal ion even in a micro acidic environment. Probe **72** is a targeted and specific probe for the detection of zinc ions in lysosomes (Figure 57a) [136]. The morpholine and DPEN groups in **72** act as lysosome targeting and Zn^{2+} chelating groups, respectively. The probe operates by two independent PET processes. When zinc ions are complexed, one of the PET effects involving the

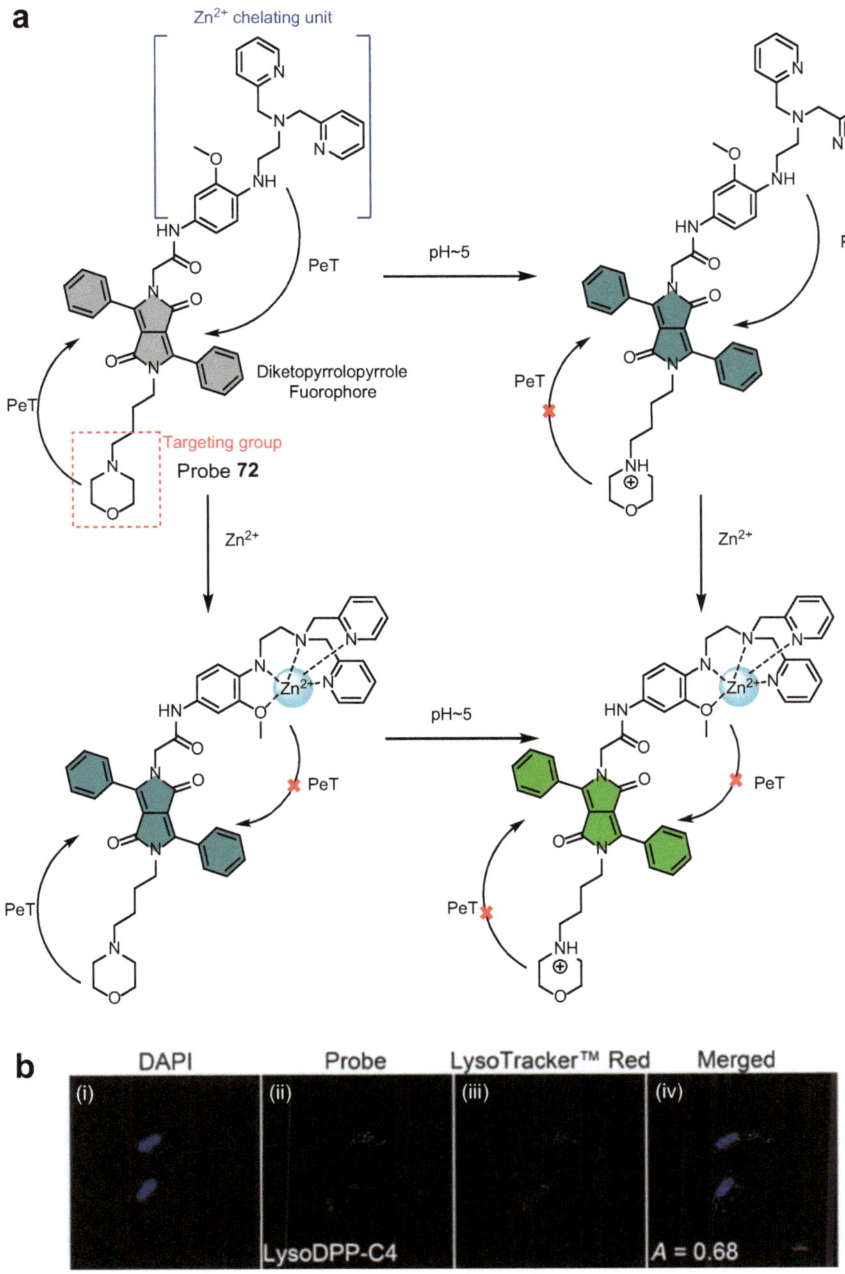

FIGURE 57 (a) Fluorescence 'turn on' mechanism allowing for concurrent detection of low pH and Zn^{2+}. Gray-weak fluorescence emission intensity. Green-strong fluorescence emission intensity. (b) Co-localization images of HeLa cells incubated with **72** (10 µM, 24 hours) and LysoTracker™ Red DND-99 (50 nM, 2.5 hours). (i) The nuclei of the cells (blue) were stained by the blue fluorescent dye, 4,6-diamidino-2-phenylindole (DAPI). (ii) Stained by **72**. (iii) Stained by LysoTracker™ Red DND-99. (iv) Merged image of (ii) and (iii). Adapted from Ref. [136] with permission. Copyright 2019, Chemical Science.

DPEN moiety as an electron donor is inhibited and the fluorescence intensity partially increases. When present in an acidic environment, the morpholine amine center becomes protonated, leading to inhibition of the second PET quenching pathway and full recovery of fluorescence. The probe has low cytotoxicity and can be used to monitor changes in intracellular Zn^{2+} concentrations in lysosomes and to distinguish cancer cells from normal prostate cells (Figure 57b).

5.4 Plasma Membrane Zinc-Ion Imaging

The plasma membrane plays an important role in maintaining cell metabolism and signal transduction. Zinc ions are closely related to cell proliferation, differentiation, and apoptosis. Therefore, the ability to detect zinc ions in cell membranes aids understanding of biochemical processes taking place in cells and could help diagnose diseases. Probe **73** is a cholesterol-conjugated fluorescence Zn^{2+} sensor that is based on the fluorescein platform (Figure 58a) [137]. The cholesterol moiety enables **73** to be specifically located in the plasma membrane. To verify its localization ability, **73** and its analogs **73a** and **73b** were subjected to confocal fluorescence microscope imaging in HeLa cells [137]. Compared with the non-cholesterol containing **73b**, analog **73a**, which contains a cholesterol moiety but no amine side chain, displays strong fluorescence emission in the plasma membrane region, where the cholesterol part strongly interacts with a subset of membrane lipids. The plasma membrane of cells stained with **73** emit only low fluorescence. However, the addition of zinc ion to cells containing **73** results in an increase in fluorescence in the same plasma membrane. Moreover, the fluorescence intensity decreases when the membrane-impermeable chelator for Zn^{2+} EDTA is added. The results show that binding to the zinc-ion center in **73** occurs on the extracellular side, and that the probe detects extracellular zinc ion reversibly in the vicinity of the plasma membrane (Figure 58b).

5.5 Presynaptic Terminal Zinc-Ion Imaging

Zinc ions have a neuroregulatory function. During synaptic vesicle transport, zinc ions are released into the synaptic cleft and are used by postsynaptic channel proteins, such as NMDA receptors.

Probe **74** is a novel two-photon fluorescence Zn^{2+} sensor (Figure 59a), composed of a DPA as a metal ion-binding moiety, a naphthalimide unit as a two-photon fluorophore, and the ifenprodil-like tail for targeting NMDA receptors in the presynaptic terminal [138]. Upon the addition of Zn^{2+}, the fluorescence intensity of probe **74** increases at 515 nm under excitation at 368 nm. The detection limit of **74** for Zn^{2+} is in the sub-nanomolar range with an apparent dissociation constant (K_d) for the complexation with Zn^{2+} of 1.2 ± 0.1 nM. Probe **74** has high sensitivity and selectivity for Zn^{2+} over a broad pH range, and it can be used to monitor the dynamics of the release of Zn^{2+} in biological neurons under physiological conditions (Figure 59b). Owing to its ability to bind Zn^{2+} near NMDA receptors in the presynaptic terminal, its two-photon and TPM imaging capability, **74** has a high potential for detecting neurogenic, neuron, and brain damage.

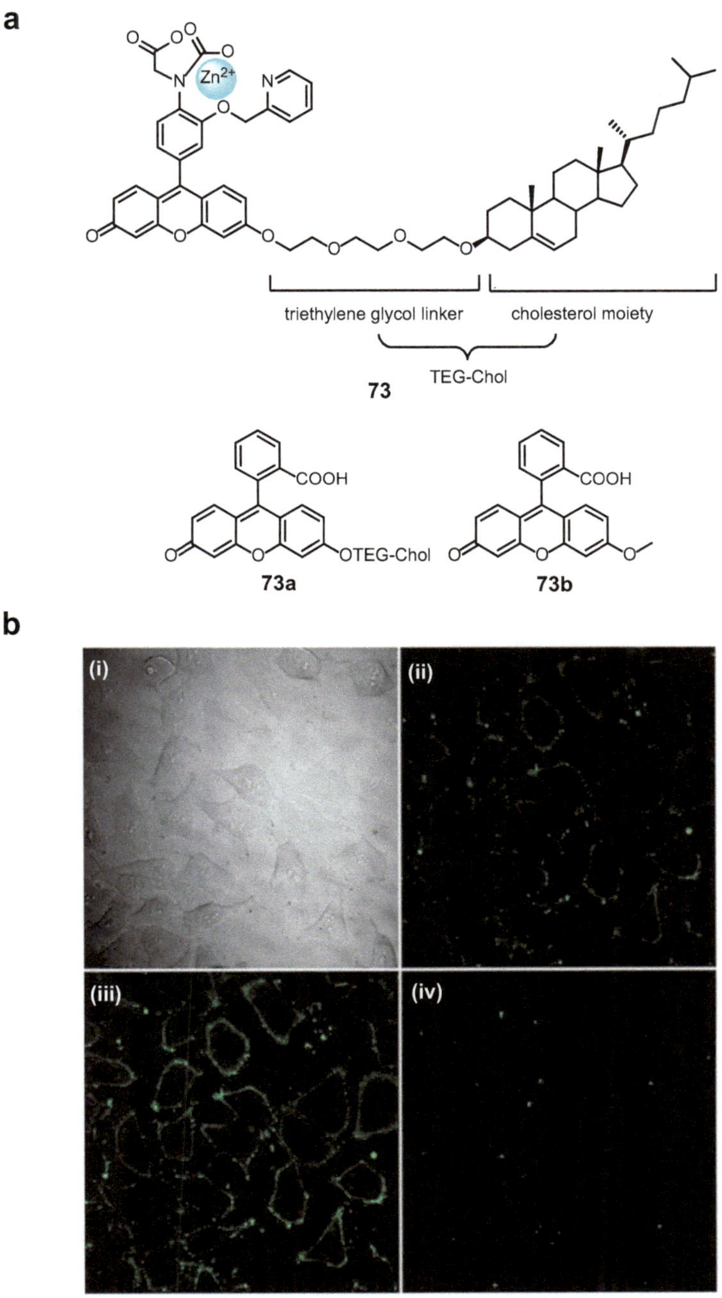

FIGURE 58 (a) Structures of **73** and reference compounds **73a** and **73b**. (b) Live cell fluorescence imaging of extracellular Zn^{2+} at the plasma membrane by using confocal microscopy. (i) Bright-field transmission image of HeLa cells. (ii) Cells stained with 2.5 μM **73**. (iii) Images obtained immediately after the addition of 20 μM Zn^{2+} to the same media shown in (ii). (iv) After addition of 100 μM EDTA to the same cells shown in (iii). Adapted from Ref. [137] with permission. Copyright 2011, Organic Letters.

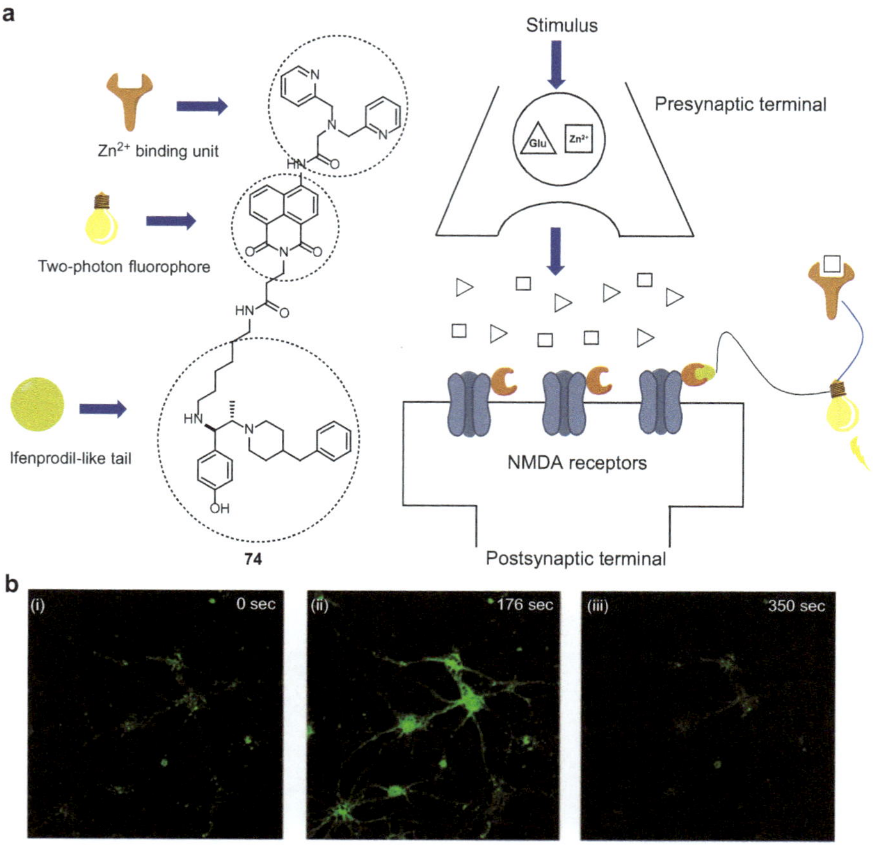

FIGURE 59 (a) Structure of **74** and the mechanism of binding zinc ions released by the presynaptic terminal upon neuronal excitation. (b) TPM images of 1 μM **74**-labeled primary cortical neuronal cells, before (i) and after (ii) the addition of 150 μM DTDP to the imaging solution, (iii) and after the addition of 10 μM TPEN to (ii). Scale bar, 60 μm. Cells shown are representative images from replicate experiments ($n = 5$). Adapted from Ref. [138] with permission. Copyright 2017, Biosensors & Bioelectronics.

6 APPLICATIONS OF Zn^{2+}-SENSITIVE FLUORESCENT PROBES

Information presented in the previous sections shows that zinc-ion fluorescent probes are widely used for qualitative and quantitative analyses, as well as cell, sub-organelle, tissue, and whole and organism imaging. The development of fluorescent probes that can be used to detect zinc ions in specific locations of organisms and to monitor specific zinc ion-related physiological processes has aided the understanding of relevant physiological processes and diseases [139]. Consequently, zinc-ion fluorescent probes with special applications have been developed.

An abundance of Zn^{2+} exists in some secretory granules of mammalian cells. However, at present, an efficient tool is lacking a highly sensitive and selective imaging Zn^{2+} in these granules. Probe **75** was developed to serve as a cell surface-targeted zinc sensor to monitor exocytotic release of the metal ion (Figure 60a) [140].

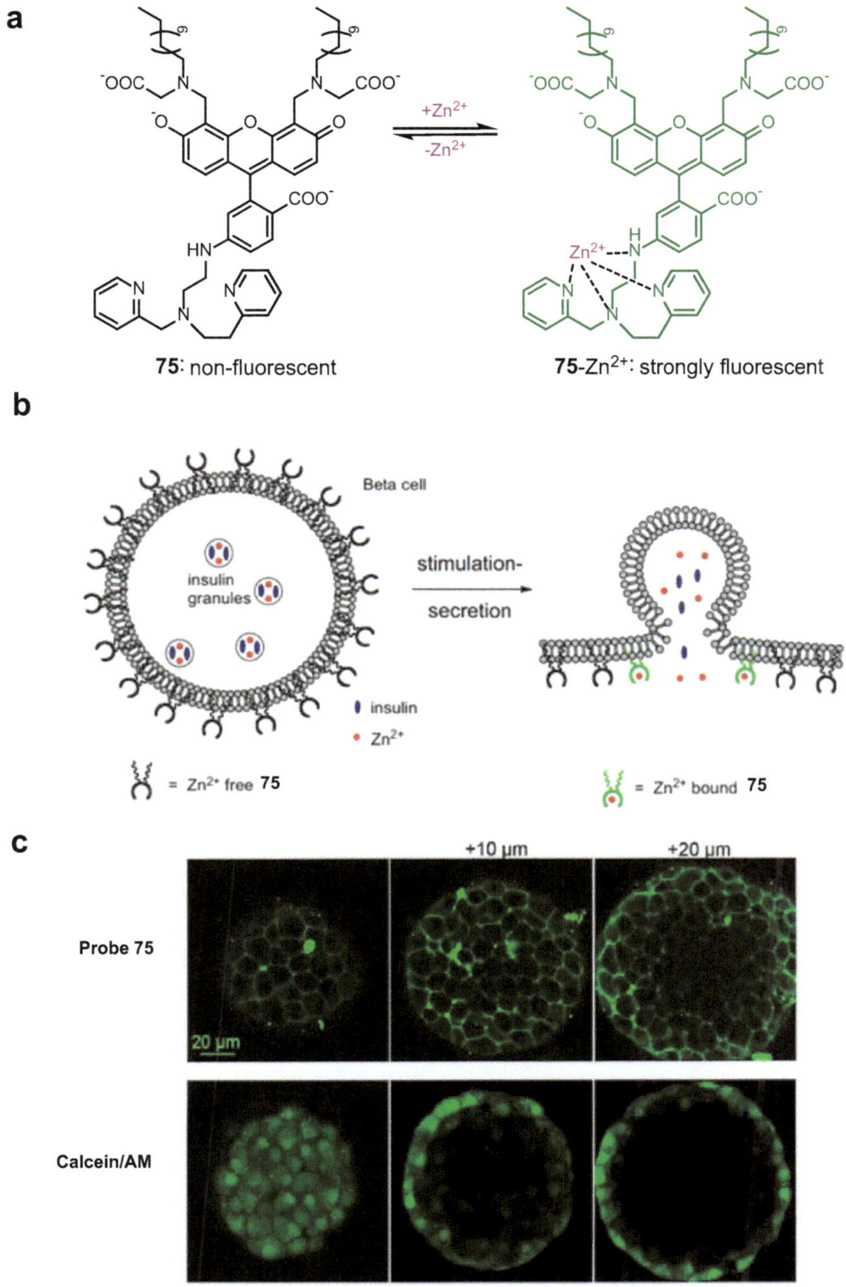

FIGURE 60 (a) Structure of **75** in the Zn^{2+}-free (nonfluorescent) and Zn^{2+}-bound (strongly fluorescent) states. (b) Mode of action of **75** for reporting local Zn^{2+} elevation at the membrane surface during exocytotic insulin granule fusion. The two lipophilic alkyl chains (wavy lines) anchor **75** to the outer leaflet of the membrane lipid bilayer. (c) Probe **75** labels cells throughout islets. Confocal microscope images of mouse islets labeled with **75** or calcein acetoxymethyl ester (calcein/AM). Scale bar, 20 μm. Adapted from Ref. [140] with permission. Copyright 2011, Proc. Natl. Acad. Sci. USA.

The probe bearing long alkyl chains targets the plasma membrane. When insulin in MIN6 cells is released from granules through stimulation by KCl, Zn^{2+} is co-released. Probe **75**, anchored to the membrane, can detect the co-released Zn^{2+} by emitting fluorescence (Figure 60b). The dynamics of insulin granule release can be monitored by **75** at a subcellular resolution by using simple epifluorescence microscopy. Furthermore, **75** was employed as an imaging tool in an investigation of the metabolic heterogeneity of β cells following the stimulation by glucose (Figure 60c). Based on the results of these studies, the related probe **76** was developed for fluorescence imaging of granular Zn^{2+} (Figure 61a) [10]. The new probe, which possesses good membrane permeability and high enrichment in acidic granules, exhibits a fluorescence intensity enhancement by more than 100-fold upon binding with Zn^{2+}. By using it to stain Zn^{2+}-rich insulin granule, **76** serves as a surrogate marker of the insulin content for sorting β cells (Figure 61b). Probe **76** can be employed to analyze the Zn^{2+} activities in primary mouse islet cells. By combining **76** and Ex4-Cy5 (a Cy5 dye conjugate of the Exendin-4 peptide), the authors developed one method to sort mouse islet α cells, β cells, and δ cells simultaneously.

Small molecule fluorescent probes serve as powerful tools in quantifying the levels and distributions of analytes in living cells. However, it is often difficult to determine accurately the concentration of analytes due to the cell-to-cell heterogeneity of fluorescence intensities and intracellular distributions of probes in subcellular organelles. Sensor **77** is a Zn^{2+} fluorescent probe, which emits the same intensity of fluorescence in different cells and localizes specifically in late endosome/lysosome vesicles (Figure 62a) [141]. These properties endow the probe with the capability of being used in uniformly quantifying labile Zn^{2+} levels in cells. Using this probe, the authors successfully monitored changes in the lysosomal Zn^{2+} levels in HC11 cells with lactation hormone treatment (Figure 62b).

Super-resolution fluorescence microscopic techniques can provide more accurate information about the distributions of biological species in living cells. A combination of the Zn^{2+}-specific fluorescent probe **78** and the structured illumination microscopy technique (Figure 63a) was used to image multiple organelles simultaneously (Figure 63b) [142]. Owing to its lipophilicity, the probe is enriched in multiple organelles, where its binding with Zn^{2+} enables imaging of typical morphologies, including mitochondria, lysosome and ER. Using probe **78**, the authors investigated how the carbonyl cyanide m-chlorophenylhydrazone-induced (mitochondrial damage inducer) autophagic behavior in HeLa cells is altered by changes in Zn^{2+} level in organelles. The probe **78** was also used to image organoids and identify specific organelles within liver organoids.

In summary, after several decades of intense investigations, a large variety of zinc-ion fluorescent probes have been developed. The performances of these probes have been significantly improved so that many of them are now being used for *in vitro* zinc-ion quantification, and subcellular, cellular, tissue and organism imaging. However, because zinc ions are involved in many physiological processes that are closely associated with a variety of diseases, including type 2 diabetes, Alzheimer's disease, breast cancer, and prostate cancer, future efforts in this area are needed to enhance the ability to detect zinc ion in specific cells, and identify its roles in physiological processes and diseases.

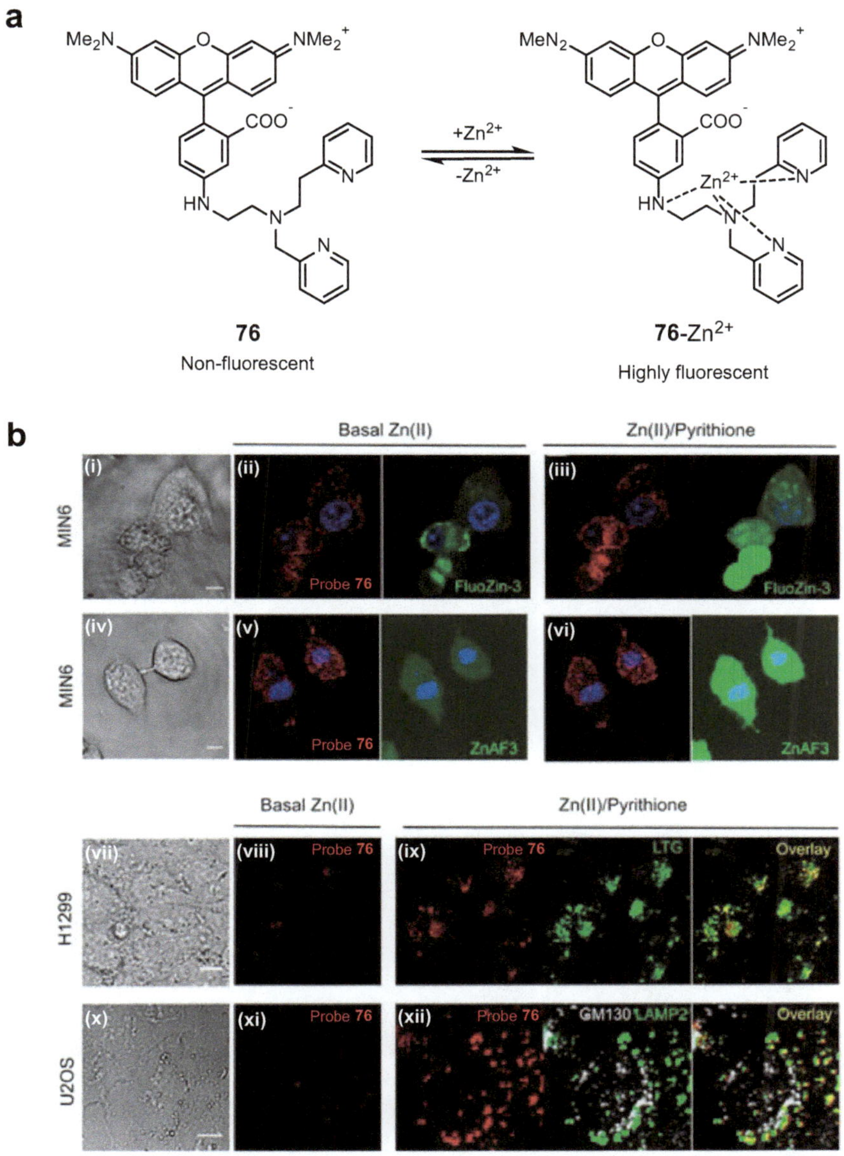

FIGURE 61 (a) Structure of **76** and its mode of action for sensing Zn^{2+}. (b) MIN6 cells were co-loaded with **76** (1 μM) and FluoZin-3/AM (5 μM, i–iii) or with **76** and ZnAF3-Ac (2 μM, iv–vi) and imaged by confocal microscopy first in the basal SAB solution and subsequently in a high Zn^{2+} solution (Zn/pyrithione, 20/10 μM). Cell nuclei are shown in blue with Hoechst 33342. Scale bar, 5 μm. (vii–xii) **76** is acidotropic and accumulates in acidic granules. Confocal images of H1299 cells (vii–ix) or U2OS cells (x–xii) labeled with **76** in the basal SAB (viii and xi) or after adding Zn/pyrithione (ix and xii). LTG (0.4 μM) was added to H1299 cells in the high Zn^{2+} buffer (ix, middle image). After **76** imaging, U2OS cells were fixed and stained with antibodies against LAMP2 and GM130 (xii, middle image). Scale bar, 10 μm. Adapted from Ref. [10] with permission. Copyright 2020, Cell Reports.

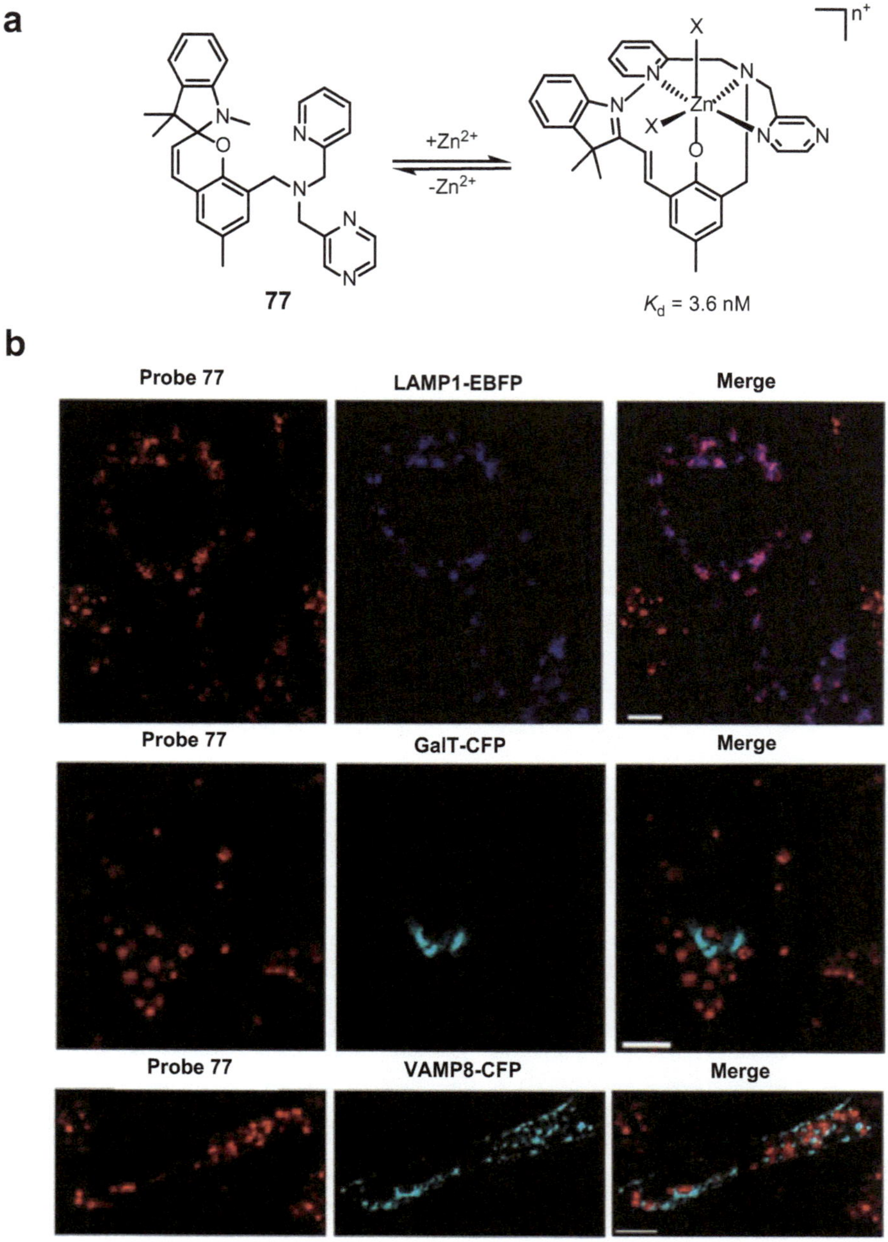

FIGURE 62 (a) Structure of **77** and zinc-bound **77**-Zn^{2+}. (b) Co-localization analysis of **77** and three vesicular markers demonstrates that **77** resides in lysosomal vesicles. Representative images of **77**-stained cells, vesicular markers-FP, and merged channel. Adapted from Ref. [141] with permission. Copyright 2018, Scientific Reports.

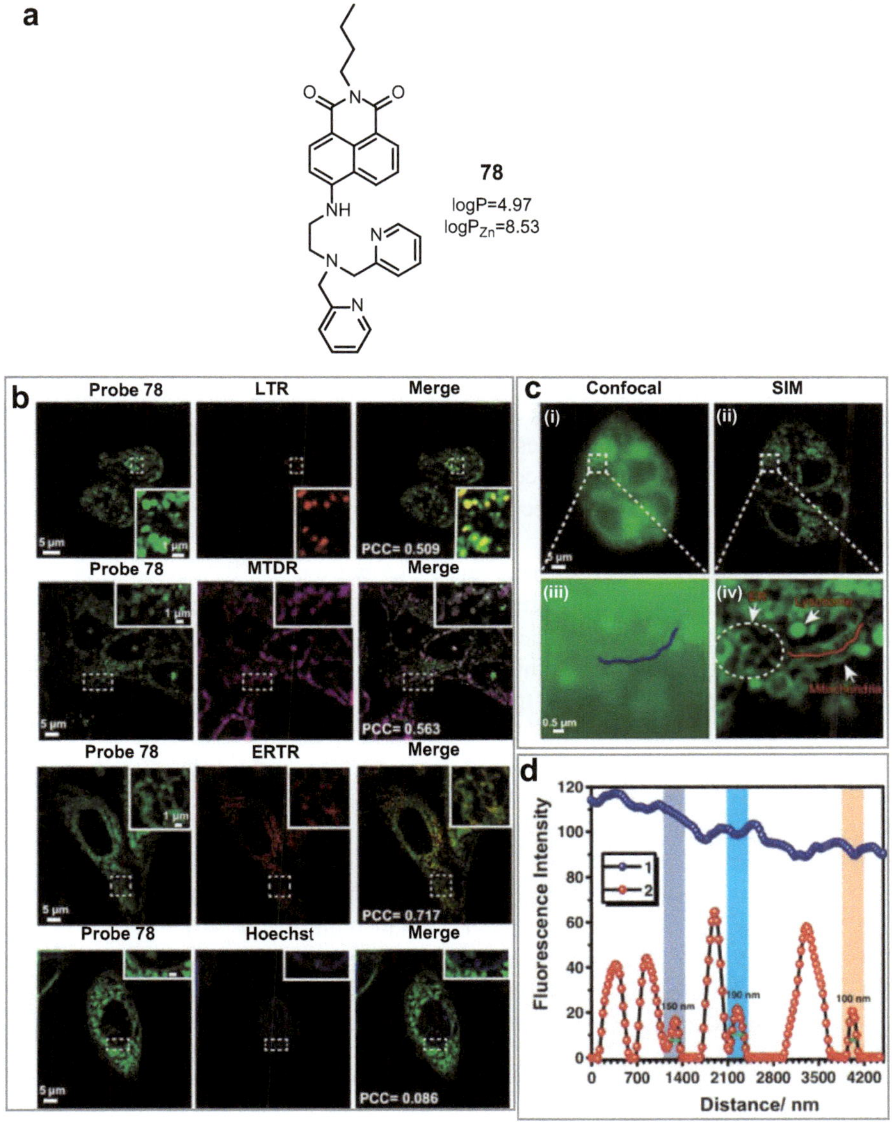

FIGURE 63 (a) Structure of fluorescent Zn^{2+} probe **78**. Log*P* values are for free probes, while logP_{Zn} values are for their zinc complexes formed with $ZnCl_2$. (b) SIM images for HeLa cells co-stained by **78** (10 μM) with LysoTracker Red (LTR, 0.1 μM), MitoTracker Deep Red (MTDR, 0.5 μM), ER-Tracker Red (ERTR, 1 μM), Nuclear dye Hoechst 33258 (Hoechst, 1 μg/mL). All insets give the enlarged images for the dashed pane. (c) Confocal (i and iii) and SIM (ii and iv) images for the **78**-stained HeLa cells, and (d) the fluorescence signal detected on the lines across a specific mitochondrion in (iii) and (iv). Scale bar: 5 μm, enlarged images scale bar: 0.5 μm. Adapted from Ref. [142] with permission. Copyright 2021, Nature Communications.

ABBREVIATIONS AND DEFINITIONS

AIE	aggregation induced emission
ASBD	4-amine-7-sulfamoylbenzo[c] [1,2,5]-oxadiazole
DPA	Di-2-picolylamine
BODIPY	Boron-dipyrromethene
CHEF	chelation-enhanced fluorescence
CDs	cyclodextrin
DFT	density functional theory
K_d	dissociation constant
D-A	donor-acceptor
SET	Single electron transfer
ER	endoplasmic reticulum
eCFP	enhanced cyan fluorescent proteins
eYFP	enhanced yellow fluorescent proteins
EDTA	ethylenediaminetetraacetic acid
ESIPT	excited state intramolecular proton transfer
ERTR	ER-Tracker Red
FRET	Förster (fluorescence) resonance energy transfer
GFP	green fluorescent protein
HTM	heavy and transition metal
HNBO	2-(2′-hydroxy-3′-naphthyl)benzoxazole
ICT	intramolecular charge transfer
LMCT	ligand to metal charge transfer
LTR	LysoTracker Red
MTDR	MitoTracker Deep Red
NPs	nanoparticles
NIR	near-infrared light
DPEA	N-methyl-*N*',*N*'-bis(pyridin-2-ylmethyl)ethane-1,2-diamine
DPEN	*N,N*-bis(2-pyridinylmethyl)-1,2-ethanediamine
TPEA	*N,N,N*′-tris(pyridin-2-ylmethyl)ethane-1,2-diamine
HQAS	*N*-(8-quinolyl)-p-aminobenzene-sulfonamide
NADH	nicotinamide adenine dinucleotides
PET	photoinduced electron transfer
PEI	polyethyleneimine
TPE	tetraphenylethylene
TICT	twisted intramolecular charge transfer
2PA	two-photon absorption
UCNPs	upconversion nanoparticles
SiNPs	silica nanoparticles
SIM	structure illumination microscopy
ZF	zinc fingers

ACKNOWLEDGEMENTS

This work is supported by the National Research Foundation of Korea (NRF) grant funded by the Korea government (MSIT) (No. 2022R1A2C3005420), the National Key R&D Program of China (2018YFA0902200) and the National Natural Science Foundation of China (21978131).

REFERENCES

1. J. Lamesch, *Rev. Metall.-Cah. Inf. Technol.* 2005, *102* (2), 119–+.
2. A. R. Mainar, L. C. Colmenares, J. A. Blázquez, I. Urdampilleta, *Int. J. Energy Res.* **2018**, *42* (3), 903–918.
3. J. J. D. Venezuela, S. Johnston, M. S. Dargusch, *Adv. Healthc. Mater.* **2019**, *8* (16), e1900408.
4. A. Moezzi, A. M. McDonagh, M. B. Cortie, *Chem. Eng. J.* **2012**, *185–186*, 1–22.
5. A. Gergely, I. Bertoti, T. Torok, E. Pfeifer, E. Kalman, *Prog. Org. Coat.* **2013**, *76* (1), 17–32.
6. E. Kabir, S. Ray, K. H. Kim, H. O. Yoon, E. C. Jeon, Y. S. Kim, Y. S. Cho, S. T. Yun, R. J. Brown, *Sci. World J.* **2012**, *2012*, 916705.
7. C. T. Chasapis, A. C. Loutsidou, C. A. Spiliopoulou, M. E. Stefanidou, *Arch. Toxicol.* **2012**, *86* (4), 521–534.
8. W. Maret, H. H. Sandstead, *J. Trace Elem. Med. Biol.* **2006**, *20* (1), 3–18.
9. A. P. Orlov, M. A. Orlova, T. P. Trofimova, S. N. Kalmykov, D. A. Kuznetsov, *J. Biol. Inorg. Chem.* **2018**, *23* (3), 347–362.
10. E. H. G. Zadeh, Z. J. Huang, J. Xia, D. L. Li, H. W. Davidson, W. H. Li, *Cell Rep.* **2020**, *32* (2), 107904.
11. F. A. Peralta, J. P. Huidobro-Toro, *Int. J. Mol. Sci.* **2016**, *17* (7), 1059.
12. J. Molina-López, *Int. J. Electrochem. Sci.* **2016**, 4470–4496.
13. L. Lai, C. Zhao, M. Su, X. Li, X. Liu, H. Jiang, C. Amatore, X. Wang, *Biomater. Sci.* **2016**, *4* (7), 1085–1091.
14. L. G. Sondergaard, M. Stoltenberg, A. Flyvbjerg, B. Brock, O. Schmitz, G. Danscher, J. Rungby, *Apmis* **2003**, *111* (12), 1147–1154.
15. A. Hassan, K. K. Sada, S. Ketheeswaran, A. K. Dubey, M. S. Bhat, *Cureus* **2020**, *12* (5), e8197.
16. P. Jiang, Z. Guo, *Coord. Chem. Rev.* **2004**, *248* (1–2), 205–229.
17. Y. C. Chen, Y. Bai, Z. Han, W. J. He, Z. J. Guo, *Chem. Soc. Rev.* **2015**, *44* (14), 4517–4546.
18. Y. Ma, F. Wang, S. Kambam, X. Chen, *Sensors Actuat. B Chem.* **2013**, *188*, 1116–1122.
19. Z. Guo, S. Park, J. Yoon, I. Shin, *Chem. Soc. Rev.* **2014**, *43* (1), 16–29.
20. N. S. Mohamad, N. H. Zakaria, N. Daud, L. L. Tan, G. C. Ta, L. Y. Heng, N. I. Hassan, *Sensors* **2021**, *21* (1), 311.
21. Z. Xu, J. Yoon, D. R. Spring, *Chem. Soc. Rev.* **2010**, *39* (6), 1996–2006.
22. J. Linling, D. LiPing, H. U. DaoDao, F. Yu, *Chin. J. Appl. Chem.* **2006**, *23* (10), 1069–1075.
23. W. Chyan, D. Y. Zhang, S. J. Lippard, R. J. Radford, *Proc. Natl. Acad. Sci. U. S. A.* **2014**, *111* (1), 143–148.

24. J. Wang, L. Lu, C. Wang, M. Wang, J. Ju, J. Zhu, T. Sun, *N. J. Chem.* **2020**, *44* (36), 15426–15431.
25. F. Sun, G. X. Zhang, D. Q. Zhang, L. Xue, H. Jiang, *Org. Lett.* **2011**, *13* (24), 6378–6381.
26. G. Yang, S. Li, S. Wang, R. Hu, J. Feng, Y. Li, Y. Qian, *Pure Appl. Chem.* **2013**, *85* (7), 1465–1478.
27. K. Komatsu, Y. Urano, H. Kojima, T. Nagano, *J. Am. Chem. Soc.* **2007**, *129* (44), 13447–13454.
28. L. Xue, C. Liu, H. Jiang, *Chem. Commun. 2009*, (9), 1061–1063.
29. Z. Liu, C. Zhang, Y. Chen, F. Qian, Y. Bai, W. He, Z. Guo, *Chem. Commun.* **2014**, *50* (10), 1253–1255.
30. T. Ha, *Methods* **2001**, *25* (1), 78–86.
31. H. Xu, C. Zhu, Y. Chen, Y. Bai, Z. Han, S. Yao, Y. Jiao, H. Yuan, W. He, Z. Guo, *Chem. Sci.* **2020**, *11* (40), 11037–11041.
32. J. Peng, W. Xu, C. L. Teoh, S. Han, B. Kim, A. Samanta, J. C. Er, L. Wang, L. Yuan, X. Liu, Y. T. Chang, *J. Am. Chem. Soc.* **2015**, *137* (6), 2336–2342.
33. G. Yang, S. Li, S. Wang, Y. Li, *C. R. Chim.* **2011**, *14* (9), 789–798.
34. A. C. Sedgwick, L. Wu, H. H. Han, S. D. Bull, X. P. He, T. D. James, J. L. Sessler, B. Z. Tang, H. Tian, J. Yoon, *Chem. Soc. Rev.* **2018**, *47* (23), 8842–8880.
35. L. Tang, X. Dai, K. Zhong, D. Wu, X. Wen, *Sensors Actuat. B Chem.* **2014**, *203*, 557–564.
36. D. Udhayakumari, V. Inbaraj, *J. Fluoresc.* **2020**, *30* (5), 1203–1223.
37. T. J. Saritha, P. Metilda, *J. Saudi Chem. Soc.* **2021**, *25* (6), 101245.
38. A. L. Berhanu, I. Mohiuddin, A. K. Malik, J. S. Aulakh, V. Kumar, K.-H. Kim, *TrAC Trends Anal. Chem.* **2019**, *116*, 74–91.
39. G. Consiglio, I. P. Oliveri, S. Failla, S. Di Bella, *Molecules* **2019**, *24* (13), 2514.
40. C.-G. Liu, Y.-Q. Qiu, S.-L. Sun, H. Chen, N. Li, Z.-M. Su, *Chem. Phys. Lett.* **2006**, *429* (4–6), 570–574.
41. B. J. Gangani, P. H. Parsania, *Spectrosc. Lett.* **2007**, *40* (1), 97–112.
42. S. K. Menon, S. K. Jogani, Y. K. Agrawal, *Rev. Anal. Chem.* **2000**, *19* (5), 361–412.
43. Z. H. Chohan, M. Praveen, *Appl. Organomet. Chem.* **2000**, *14* (7), 376–382.
44. J. Yang, R. Shi, P. Zhou, Q. Qiu, H. Li, *J. Mol. Struc.* **2016**, *1106*, 242–258.
45. X. Yang, F. Zhu, W. Shi, Y. Li, Y. Zeng, M. Yan, Y. Cui, G. Sun, *Spectrochim. Acta A Mol. Biomol. Spectrosc.* **2020**, *226*, 117608.
46. A. Roy, U. Shee, A. Mukherjee, S. K. Mandal, P. Roy, *ACS Omega* **2019**, *4* (4), 6864–6875.
47. L. Yan, R. Li, F. Ma, Z. Qi, *Anal. Methods* **2017**, *9* (7), 1119–1124.
48. M. Budri, P. Kadolkar, K. Gudasi, S. Inamdar, *J. Mol. Liq.* **2019**, *283*, 346–358.
49. H. Liu, T. Liu, J. Li, Y. Zhang, J. Li, J. Song, J. Qu, W. Y. Wong, *J. Mater. Chem. B* **2018**, *6* (34), 5435–5442.
50. W.-K. Dong, S. F. Akogun, Y. Zhang, Y.-X. Sun, X.-Y. Dong, *Sensors Actuat. B Chem.* **2017**, *238*, 723–734.
51. J. Li, Y. Chen, T. Chen, J. Qiang, Z. Zhang, T. Wei, W. Zhang, F. Wang, X. Chen, *Sensors Actuat. B Chem.* **2018**, *268*, 446–455.
52. J. Lv, G. Liu, C. Fan, S. Pu, *Spectrochim. Acta A Mol. Biomol. Spectrosc.* **2020**, *227*, 117581.
53. X. Tang, J. Han, Y. Wang, X. Bao, L. Ni, L. Wang, L. Li, *Spectrochim. Acta A Mol. Biomol. Spectrosc.* **2017**, *184*, 177–183.
54. S. Erdemir, M. Yuksekogul, S. Karakurt, O. Kocyigit, *Sensors Actuat. B Chem.* **2017**, *241*, 230–238.
55. B. Tabakci, H. M. A. Ahmed, S. Erdemir, *J. Fluoresc.* **2019**, *29* (5), 1079–1087.
56. C. Patra, A. K. Bhanja, C. Sen, D. Ojha, D. Chattopadhyay, A. Mahapatra, C. Sinha, *Sensors Actuat. B Chem.* **2016**, *228*, 287–294.

57. S. Ullmann, R. Schnorr, M. Handke, C. Laube, B. Abel, J. Matysik, M. Findeisen, R. Ruger, T. Heine, B. Kersting, *Chemistry* **2017**, *23* (16), 3824–3827.
58. L. Xue, H. H. Wang, X. J. Wang, H. Jiang, *Inorg. Chem.* **2008**, *47* (10), 4310–4318.
59. P. Alreja, D. Saini, S. S. Gautam, N. Kaur, *Inorg. Chem. Commun.* **2016**, *70*, 125–128.
60. B. A. Wong, S. Friedle, S. J. Lippard, *J. Am. Chem. Soc.* **2009**, *131* (20), 7142–7152.
61. K. Komatsu, K. Kikuchi, H. Kojima, Y. Urano, T. Nagano, *J. Am. Chem. Soc.* **2005**, *127* (29), 10197–10204.
62. H. H. Wang, Q. Gan, X. J. Wang, L. Xue, S. H. Liu, H. Jiang, *Org. Lett.* **2007**, *9* (24), 4995–4998.
63. X. Li, J. Li, X. Dong, X. Gao, D. Zhang, C. Liu, *Sensors Actuat. B Chem.* **2017**, *245*, 129–136.
64. J. E. Kwon, S. Lee, Y. You, K. H. Baek, K. Ohkubo, J. Cho, S. Fukuzumi, I. Shin, S. Y. Park, W. Nam, *Inorg. Chem.* **2012**, *51* (16), 8760–8774.
65. M. L. Zastrow, R. J. Radford, W. Chyan, C. T. Anderson, D. Y. Zhang, A. Loas, T. Tzounopoulos, S. J. Lippard, *ACS Sensors* **2016**, *1* (1), 32–39.
66. Z. C. Xu, K. H. Baek, H. N. Kim, J. N. Cui, X. H. Qian, D. R. Spring, I. Shin, J. Yoon, *J. Am. Chem. Soc.* **2010**, *132* (2), 601–610.
67. T. Hirano, K. Kikuchi, Y. Urano, T. Nagano, *J. Am. Chem. Soc.* **2002**, *124* (23), 6555–6562.
68. M. Y. Jia, Y. Wang, Y. Liu, L. Y. Niu, L. Feng, *Biosens. Bioelectron.* **2016**, *85*, 515–521.
69. Z. Guo, G. H. Kim, I. Shin, J. Yoon, *Biomaterials* **2012**, *33* (31), 7818–7827.
70. Y. Yin, G. Liu, *J. Photochem. Photobiol. A Chem.* **2021**, *421,* 113528.
71. Q. Miao, X. Huang, Y. Cheng, Y. Liu, L. Zong, Y. Cheng, *J. Appl. Polym. Sci.* **2009**, *111* (6), 3137–3143.
72. H. Zhang, Z. Shao, K. Zhao, *Chin. Phys. B* **2020**, *29* (8), 083304.
73. W. Y. Lin, J. B. Feng, L. Yuan, W. Tan, *Sens. Actuat. B Chem.* **2009**, *135* (2), 512–515.
74. D. S. Kopchuk, A. M. Prokhorov, P. A. Slepukhin, D. N. Kozhevnikov, *Tetrahedron Lett.* **2012**, *53* (46), 6265–6268.
75. H. Santhakumar, R. V. Nair, D. S. Philips, S. J. Shenoy, A. Thekkuveettil, A. Ajayaghosh, R. S. Jayasree, *Sci. Rep.* **2018**, *8* (1), 9069.
76. W. Li, B. Fang, M. Jin, Y. Tian, *Anal. Chem.* **2017**, *89* (4), 2553–2560.
77. F. Wang, K. Wang, Q. Kong, J. Wang, D. Xi, B. Gu, S. Lu, T. Wei, X. Chen, *Coord. Chem. Rev.* **2021**, *429,* 213636.
78. H. Kim, J. Kang, K. B. Kim, E. J. Song, C. Kim, *Spectrochim. Acta A Mol. Biomol. Spectrosc.* **2014**, *118*, 883–887.
79. B. Czaplinska, E. Spaczynska, R. Musiol, *Med. Chem.* **2018**, *14* (1), 19–33.
80. J. M. Jung, J. H. Kang, J. Han, H. Lee, M. H. Lim, K.-T. Kim, C. Kim, *Sensors Actuat. B Chem.* **2018**, *267*, 58–69.
81. X. Meng, S. Wang, Y. Li, M. Zhu, Q. Guo, *Chem. Commun.* **2012**, *48* (35), 4196–4198.
82. K. Vongnam, T. Aree, M. Sukwattanasinitt, P. Rashatasakhon, *ChemistrySelect* **2018**, *3* (12), 3495–3499.
83. Y. Liu, N. Zhang, Y. Chen, L. H. Wang, *Org. Lett.* **2007**, *9* (2), 315–318.
84. Y. Zhang, X. F. Guo, W. X. Si, L. H. Jia, X. H. Qian, *Org. Lett.* **2008**, *10* (3), 473–476.
85. X. Tian, X. Guo, F. Yu, L. Jia, *Sensors Actuat. B Chem.* **2016**, *232*, 181–187.
86. Y. Shi, Z. Chen, X. Cheng, Y. Pan, H. Zhang, Z. Zhang, C.-W. Li, C. Yi, *Biosens. Bioelectron.* **2014**, *61*, 397–403.
87. Z. Liu, W. Yang, Y. Li, F. Tian, W. Zhu, *RSC Adv.* **2015**, *5* (122), 100482–100487.
88. L. Fang, G. Trigiante, C. J. Kousseff, R. Crespo-Otero, M. P. Philpott, M. Watkinson, *Chem. Commun.* **2018**, *54* (69), 9619–9622.
89. H. C. Kolb, M. G. Finn, K. B. Sharpless, *Angew. Chem. Int. Ed.* **2001**, *40* (11), 2004–+.
90. J. E. Moses, A. D. Moorhouse, *Chem. Soc. Rev.* **2007**, *36* (8), 1249–1262.

91. K. Wechakorn, K. Suksen, P. Piyachaturawat, P. Kongsaeree, *Sensors Actuat. B Chem.* **2016**, *228*, 270–277.
92. G. Wu, Q. Gao, M. Li, X. Tang, K. W. C. Lai, Q. Tong, *J. Photochem. Photobiol. A Chem.* **2018**, *355*, 487–495.
93. H. J. Kim, J. H. Han, M. K. Kim, C. S. Lim, H. M. Kim, B. R. Cho, *Angew. Chem.* **2010**, *49* (38), 6786–6789.
94. Y. Hiruta, Y. Shindo, K. Oka, D. Citterio, *Chem. Lett.* **2021**, *50* (5), 870–887.
95. D. Liu, X. Deng, X. Yin, Y. Wang, J. Guo, J. Chen, G. Yang, H. He, *Inorg. Chem. Commun.* **2019**, *101*, 117–120.
96. E. P. S. Pratt, L. J. Damon, K. J. Anson, A. E. Palmer, *Biochim. Biophys. Acta Mol. Cell. Res.* **2021**, *1868* (1), 118865.
97. Y. Hai, J. J. Chen, P. Zhao, H. Lv, Y. Yu, P. Xu, J. L. Zhang, *Chem. Commun.* **2011**, *47* (8), 2435–2437.
98. P. D. Zalewski, I. J. Forbes, W. H. Betts, *Biochem. J.* **1993**, *296*, 403–408.
99. X. Chen, J. Xu, F. Suo, C. Yu, D. Zhang, J. Chen, Q. Wu, S. Jing, L. Li, W. Huang, *Spectrochim. Acta A Mol. Biomol. Spectrosc.* **2020**, *229*, 117949.
100. M. H. Lee, J. S. Kim, J. L. Sessler, *Chem. Soc. Rev.* **2015**, *44* (13), 4185–4191.
101. K. Du, S. Niu, L. Qiao, Y. Dou, Q. Zhu, X. Chen, P. Zhang, *RSC Adv.* **2017**, *7* (64), 40615–40620.
102. X. He, Q. Xie, J. Fan, C. Xu, W. Xu, Y. Li, F. Ding, H. Deng, H. Chen, J. Shen, *Dyes Pigm.* **2020**, *177,* 108255.
103. M. V. Karmegam, S. Karuppannan, D. B. Christopher Leslee, S. Subramanian, S. Gandhi, *Luminescence* **2020**, *35* (1), 90–97.
104. M. Pawlicki, H. A. Collins, R. G. Denning, H. L. Anderson, *Angew. Chem.* **2009**, *48* (18), 3244–3266.
105. L. Xu, J. Zhang, L. Yin, X. Long, W. Zhang, Q. Zhang, *J. Mater. Chem. C* **2020**, *8* (19), 6342–6349.
106. M. Khan, C. R. Goldsmith, Z. Huang, J. Georgiou, T. T. Luyben, J. C. Roder, S. J. Lippard, K. Okamoto, *Proc. Natl. Acad. Sci. U. S. A.* **2014**, *111* (18), 6786–6791.
107. K. P. Divya, S. Sreejith, P. Ashokkumar, K. Yuzhan, Q. Peng, S. K. Maji, Y. Tong, H. Yu, Y. Zhao, P. Ramamurthy, A. Ajayaghosh, *Chem. Sci.* **2014**, *5* (9), 3469–3474.
108. W. Li, Z. Liu, B. Fang, M. Jin, Y. Tian, *Biosens. Bioelectron.* **2020**, *148*, 111666.
109. G. Kamiya, N. Kitada, R. Saito-Moriya, R. Obata, S. Iwano, A. Miyawaki, T. Hirano, S. A. Maki, *Chem. Lett.* **2021**, *50* (8), 1523–1525.
110. K. D. Piatkevich, F. V. Subach, V. V. Verkhusha, *Chem. Soc. Rev.* **2013**, *42* (8), 3441–3452.
111. B. Tang, H. Huang, K. Xu, L. Tong, G. Yang, X. Liu, L. An, *Chem. Commun.* ***2006***, 34, 3609–3611.
112. Y. Zhang, B. Yuan, D. Ma, *Inorg. Chim. Acta* **2020**, 508, 119640.
113. S. Zhang, R. Adhikari, M. Fang, N. Dorh, C. Li, M. Jaishi, J. Zhang; A. Tiwari, R. Pati, F. T. Luo, H. Liu, *ACS Sensors* **2016**, *1* (12), 1408–1415.
114. M. Fang, S. Xia, J. Bi, T. P. Wigstrom, L. Valenzano, J. Wang, M. Tanasova, R. L. Luck, H. Liu, *Molecules* **2019**, *24* (8), 1592.
115. Q.-Q. Zhang, B.-X. Yang, R. Sun, J.-F. Ge, Y.-J. Xu, N.-J. Li, J.-M. Lu, *Sensors Actuat. B Chem.* **2012**, *171–172*, 1001–1006.
116. E. S. Potekhina, D. Y. Bass, I. V. Kelmanson, E. S. Fetisova, A. V. Ivanenko, V. V. Belousov, D. S. Bilan, *Int. J. Mol. Sci.* **2021**, *22* (1), 148.
117. B. T. Bajar, E. S. Wang, S. Zhang, M. Z. Lin, J. Chu, *Sensors* **2016**, *16* (9), 1488.
118. W. Qiao, M. Mooney, A. J. Bird, D. R. Winge, D. J. Eide, *Proc. Natl. Acad. Sci. U. S. A.* **2006**, *103* (23), 8674–8679.
119. Y. Qin, P. J. Dittmer, J. G. Park, K. B. Jansen, A. E. Palmer, *Proc. Natl. Acad. Sci. U. S. A.* **2011**, *108* (18), 7351–7356.

120. Y. Qin, J. G. Miranda, C. I. Stoddard, K. M. Dean, D. F. Galati, A. E. Palmer, *ACS Chem. Biol.* **2013**, *8* (11), 2366–2371.
121. A. M. Hessels, P. Chabosseau, M. H. Bakker, W. Engelen, G. A. Rutter, K. M. Taylor, M. Merkx, *ACS Chem. Biol.* **2015**, *10* (9), 2126–2134.
122. Y. Qin, D. W. Sammond, E. Braselmann, M. C. Carpenter, A. E. Palmer, *ACS Chem. Biol.* **2016**, *11* (10), 2744–2751.
123. M. Chen, S. Zhang, Y. Xing, X. Li, Y. He, Y. Wang, J. Oberholzer, H. W. Ai, *Anal. Chem.* **2019**, *91* (19), 12212–12219.
124. H. M. McBride, M. Neuspiel, S. Wasiak, *Curr. Biol.* **2006**, *16* (14), R551–R560.
125. E. J. Lesnefsky, S. Moghaddas, B. Tandler, J. Kerner, C. L. Hoppel, *J. Mol. Cell Cardiol.* **2001**, *33* (6), 1065–1089.
126. A. Gardner, R. G. Boles, *Curr. Psychiat. Rev.* **2005**, *1* (3), 255–271.
127. L. Xue, G. Li, C. Yu, H. Jiang, *Chem. Eur. J.* **2012**, *18* (4), 1050–1054.
128. Z. Liu, C. Zhang, Y. Chen, W. He, Z. Guo, *Chem. Commun.* **2012**, *48* (67), 8365–8367.
129. M. Wang, Q. Xu, M. Yuan, *Plant Signal Behav.* **2011**, *6* (1), 77–79.
130. T. S. Nguyen, K. Kohno, Y. Kimata, *Biosci. Biotechnol. Biochem.* **2013**, *77* (6), 1337–1339.
131. Q. Sun, W. Zhong, W. Zhang, Q. Li, X. Sun, X. Tan, X. Sun, D. Dong, Z. Zhou, *Am. J. Physiol. Gastrointest. Liver Physiol.* **2015**, *308* (9), G757–G766.
132. L. Fang, G. Trigiante, R. Crespo-Otero, C. S. Hawes, M. P. Philpott, C. R. Jones, M. Watkinson, *Chem. Sci.* **2019**, *10* (47), 10881–10887.
133. L. C. Costello, P. Feng, B. Milon, M. Tan, R. B. Franklin, *Prostate Cancer Prostat. Dis.* **2004**, *7* (2), 111–117.
134. V. Kolenko, E. Teper, A. Kutikov, R. Uzzo, *Nat. Rev. Urol.* **2013**, *10* (4), 219–226.
135. L. C. Costello, R. B. Franklin, P. Feng, *Mitochondrion* **2005**, *5* (3), 143–153.
136. C. Du, S. Fu, X. Wang, A. C. Sedgwick, W. Zhen, M. Li, X. Li, J. Zhou, Z. Wang, H. Wang, J. L. Sessler, *Chem. Sci.* **2019**, *10* (22), 5699–5704.
137. S. Iyoshi, M. Taki, Y. Yamamoto, *Org. Lett.* **2011**, *13* (17), 4558–4561.
138. X. Chen, C. S. Lim, D. Lee, S. Lee, S. J. Park, H. M. Kim, J. Yoon, *Biosens. Bioelectron.* **2017**, *91*, 770–779.
139. L. Fang, M. Watkinson, *Chem. Sci.* **2020**, *11* (42), 11366–11379.
140. D. Li, S. Chen, E. A. Bellomo, A. I. Tarasov, C. Kaut, G. A. Rutter, W. H. Li, *Proc. Natl. Acad. Sci. U. S. A.* **2011**, *108* (52), 21063–21068.
141. Y. Han, J. M. Goldberg, S. J. Lippard, A. E. Palmer, *Sci. Rep.* **2018**, *8* (1), 15034.
142. H. Fang, S. Geng, M. Hao, Q. Chen, M. Liu, C. Liu, Z. Tian, C. Wang, T. Takebe, J. L. Guan, Y. Chen, Z. Guo, W. He, J. Diao, *Nat. Commun.* **2021**, *12* (1), 109.

6 Chemo- and Bio-Sensors for Copper Ions

*Paramesh Jangili, Nem Singh, Ilwha Kim, Zehra Zunbul, and Jong Seung Kim**
Department of Chemistry, Korea University, Seoul 02841, South Korea

CONTENTS

* Corresponding author.

DOI: 10.1201/9781003229971-6

Abstract

Copper (Cu) ions play a pivotal role in chemical, environmental, and biological systems and are essential for plants, animals, and humans. Being a primary element of the cytochrome C oxidase and required for extensive cellular functions by distributing all over the body, Cu is crucial for all living organisms. Cu ions could be present in both forms (Cu^+ and Cu^{2+}) in living systems and mediate many critical enzyme processes. However, although it is crucial for optimal health, the aberrations in normal Cu levels (both excess and deficient levels) can lead to a severe wide range of diseases. Therefore, the detection of Cu ions in biological systems and all the environmental samples is vital for controlling the harsh consequences of Cu on human health and the environment. The conventional analytical techniques to trace the Cu ions provide reasonable detection limits; however, they require highly expensive analytical instruments and cause operational difficulties. Chemo- and bio-sensors are a powerful alternative to such conventional techniques for tracing and monitoring the Cu levels in different samples, including water, food, living samples, *etc.* Different kinds of chemo- and bio-sensors, such as fluorescent, colorimetric, electrometric, *etc.*, have been successfully applied to determine Cu ions in different samples. This chapter will underscore the essentiality and monitoring of Cu ions in living organisms and provide an overview of their determination using different chemo- and bio-sensors in a wide range of samples.

KEYWORDS

Bio-sensor; Chemosensor; Colorimetric Sensor; Detection of Copper; Electrochemical Sensor; Fluorescent Sensor

1 INTRODUCTION

Copper (Cu), a chemical element with an atomic number of 29, is naturally abundant in a metallic form [1]. Cu can be used as a conductor of heat and electricity due to its high thermal and electrical conductivity [2]. Cu is essential to all living organisms because it is a primary constituent of cytochrome C oxidase, a respiratory enzyme complex. The most commonly available Cu ions in all living organisms are Cu^+ and Cu^{2+}.

1.1 The Physiological Role of Copper Ions

As an essential trace metal to living organisms, Cu is required for vast cellular functions, such as adequate growth, lung elasticity, cardiovascular integrity, wound healing, neuroendocrine function, neovascularization, and iron metabolism [3, 4]. The average intake of Cu from the diet by human adults varies from 0.6 to 1.6 mg/day [5]. Although the higher concentrated levels of Cu ions can impair cell growth and function, the controlled quantity is essential for optimal health,

TABLE 1
The Health and Nutrition Organizations Recommended Dietary Reference Values (mg/day) According to the Ages of Humans

	Adequate Cu Intake (mg/day)	
Age	**Female**	**Male**
7–11 months	0.4	0.4
1–<3 years	0.7	0.7
3–<10 years	1.0	1.0
10–<18 years	1.1	1.3
≥18 years	1.3	1.6
Pregnant woman	1.5	
Lactating woman	1.5	

Source: Redrawn from Ref. [10].

along with other micronutrients, including calcium, iron, and zinc ions [6]. While Cu ions are found all over the body, they are mainly concentrated in the highly metabolic kidney, liver, heart, and brain organs. The ingested Cu by humans is absorbed into the bloodstream, binds to transport proteins, and is finally delivered into the liver. From then, it is distributed around the body or excreted through bile [7]. Cu ions mediate the many enzymes' critical functions and play a pivotal role in human metabolism [8]. They are significant catalytic cofactors in proteins' redox chemistry, a fundamental biological process in human growth and development [9].

As the human body cannot produce Cu, various health and nutrition organizations worldwide have recommended dietary reference values according to the ages of humans (Table 1) [10], which shows the significance of Cu for optimal health.

As one of the trio of minerals, Cu is indispensable for brain, nervous, and cardiovascular system functions. It involves the crucial brain and nervous system processes, such as fetal and post-natal growth, production and maintenance of myelin, which is important for transmitting nerve impulses, and the synthesis of neurotransmitters [11, 12]. Cu ions are needed for collagen (a protein found in the skin, muscles, bones, *etc.*), elastin (provides support and elasticity for the skin), and melanin (protects the skin from UV radiation) syntheses [13]. Cu^{2+} ions are catalytic cofactors for diverse metalloenzymes, such as tyrosinase, superoxide dismutase (SOD), dopamine-hydroxylase, lysyl oxidase, cytochrome C oxidase, and other enzymes involved in blood coagulation [14]. They can also act as cofactors for neutralizing free radicals that might be otherwise oxidized and destroy healthy tissues.

1.1.1 Deficiency of Cu Ions

As Cu ions are indispensable to human life, the consequence of deficiency of Cu could lead to many health complications. Cu plays a vital role in the iron transport around the body as well as in the Fe conversion to Fe^{3+}; hence, its deficiency results in anemia and iron overload of tissue [15]. Cu ions are necessary for the homeostasis of white blood cell count; Cu deficiency can lead to disturbance in the immune system, reduction in the white blood cells, and also promotes the incidence of pneumonia. [16]. Also, as discussed above, Cu ions are the central cofactors in many enzymes and proteins, which are necessary for growth and survival of the human body. Their deficiency could lead to degeneration of the nervous system, heart failure, low blood pressure, circulatory problems, skin degeneration and loss of pigmentation, skeletal abnormalities, and osteoporosis [17, 18].

1.1.2 Excess of Cu Ions

Although humans can tolerate Cu at a relatively large concentration, excess levels of Cu ions can lead to complications for optimal health. For example, overload and long-term exposure of Cu ions might cause irritation of the eyes, mouth, and nose, head and stomach aches, vomiting, dizziness, diarrhea, dyslexia, and liver damage in infants [19].

Hence, it is crucial to maintain the regulation of Cu ion levels; on the contrary, aberrations in normal levels lead to many health risks, *i.e.*, genetic disorders: Menkes [20] and Wilson's diseases [21], neurodegenerative diseases: Alzheimer's [22], Parkinson's [23], Huntington's [24], and Prion diseases [25], familial amyotrophic lateral sclerosis [26], and metabolic disorders: diabetes and obesity [27].

1.2 Toxicity of Copper Ions

Cu ions are non-biodegradable and toxic. They might have fast accumulation even at low concentrations in most living organisms, cause severe illnesses and be deadly at higher concentrations [28]. Because Cu is the most commonly used metal in many industrial applications, such as metal finishing, plastics, electroplating, and etching, it is usually found in high concentrations in industrial wastewater, which is very harmful to the environment. Therefore, the United States Environmental Protection Agency (USEPA) stated that the permissible Cu ions limit is 1.3 mg/L in industrial waste effluents [29]. As discussed above, Cu ions are vital for electron-transfer reactions in biological systems. They could interact with oxygen in their free, unbound state by catalyzing Haber–Weiss or Fenton reactions, which engender ROS to cause oxidative damage to biomolecules [30]. Therefore, in order to keep the Cu ions in a bound form, all living systems have evolved several transport and storage mechanisms [31]. Physiological plasma total Cu levels (free and bound concentrations) range between 0.9 and 1.2 mg/L [32].

1.3 Significance of the Detection of Copper Ions

Cu ions play a pivotal role in chemical, environmental, and biological systems and are essential for plants, animals, and humans. However, these are severely toxic to some living organisms like bacteria and viruses. Owing to their toxic behavior to bacteria, high concentrations of Cu ions hinder the self-purification capacity of the rivers or sea and destroy the water's biological reprocessing system [33]. In biological systems, the unbound Cu ions in cells are sensitive to specific environmental changes, like pH and oxidative and immunological stresses. As a consequence of free labile Cu ions, induced Fenton reaction followed by the release of harmful ROS in cells [34]. Hence, it is essential to detect and monitor the concentrations of Cu ions (both Cu^{+} and Cu^{2+}) in environmental samples and their subcellular distribution in physiological processes. Being a borderline Lewis acid, Cu^{2+} preferentially binds to hard ligands, such as oxygen and nitrogen, whereas Cu^{+} is a soft Lewis acid and shows preferential binding to soft ligands like thiols [35]. Several traditional methods have been proposed for the detection of Cu ions, including atomic absorption spectrometry, inductively coupled plasma atomic emission spectrometry, inductively coupled plasma mass spectroscopy, voltammetry, and anodic stripping voltammetry (ASV) [36]. While these methods have provided reasonable detection limits, they require highly expensive analytical instruments and cause operational difficulties. These methods are applicable to biological systems but may be lethal to cells. Alternatively, chemo- and bio-sensors have been developed for this purpose.

2 DETECTION OF COPPER IONS USING DIFFERENT CHEMOSENSORS

A chemosensor refers to a species used to sense/detect an analyte with a measurable and analytical signal [37]. The chemosensor changes its color, fluorescence, or redox potential upon binding with metal ions present in an analyte sample, which are conveniently measurable by easy instrumentation (Figure 1) [38–40]. Therefore, chemosensors are one of the most promising tools for Cu detection in various samples. Different kinds of chemosensors have been categorized based on their changing properties in response to Cu ions in multiple samples.

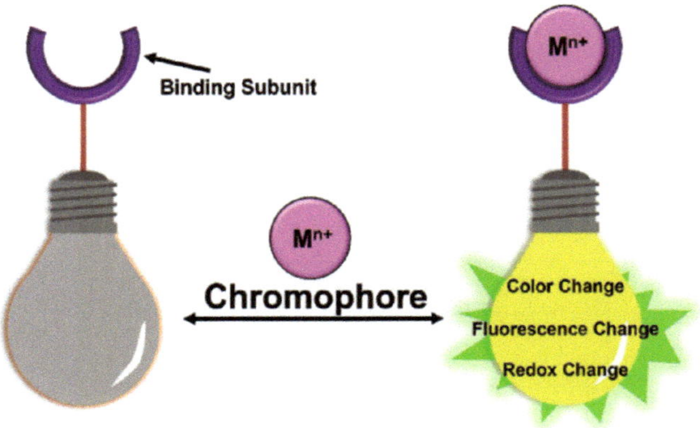

FIGURE 1 Illustration of the working principle of a chemosensor. The binding/cleaving of the metal ion at the binding subunit results in a change in chemosensor color or fluorescence or redox potential, *etc.*

2.1 Detection of Copper Ions Using Fluorescent Chemosensors

To safeguard human health and the environment, the detection of Cu in various samples is essential. Therefore, several analytical tools have been applied for the detection of Cu. However, the method of analysis can be chosen based on the analysis demand of the sample in terms of detection limit, accuracy, availability of sample preparation, instrumentation, *etc.* Fluorescent chemosensors are employed to detect various metal ions across the science discipline, including chemistry, pharmacology, environmental sciences, biology, and physiology [41–43]. These sensors estimate metal ions based on fluorescence changes upon binding with the metal ions present in the sample. Though fluorescence optical sensing techniques require spectroscopic instruments, this method delivers very precise and easily understandable results. The advantages, such as selectivity with high precision, make the future of fluorescent chemosensors very bright for detecting numerous metal ions in unsanitary water and food as well as in biological systems.

Numerous fluorescent chemosensors were developed to estimate Cu ions with high selectivity. For example, organic dye-based chemosensors commonly use a fluorescent dye core and another moiety with a chelating metal site. Cyanine, rhodamine, quinoline, coumarin, difluoro-boron-dipyrromethene (BODIPY), *etc.,* can be used as a fluorescent dye core [44]. When the Cu ion-containing sample interacts with the chelating moiety, it recognizes and selectively binds to Cu ions. As a result, the electronic structure of fluorophore alters to induce changes in the fluorescence intensity of the sensor. The mechanism of change in fluorescence can be classified in terms of

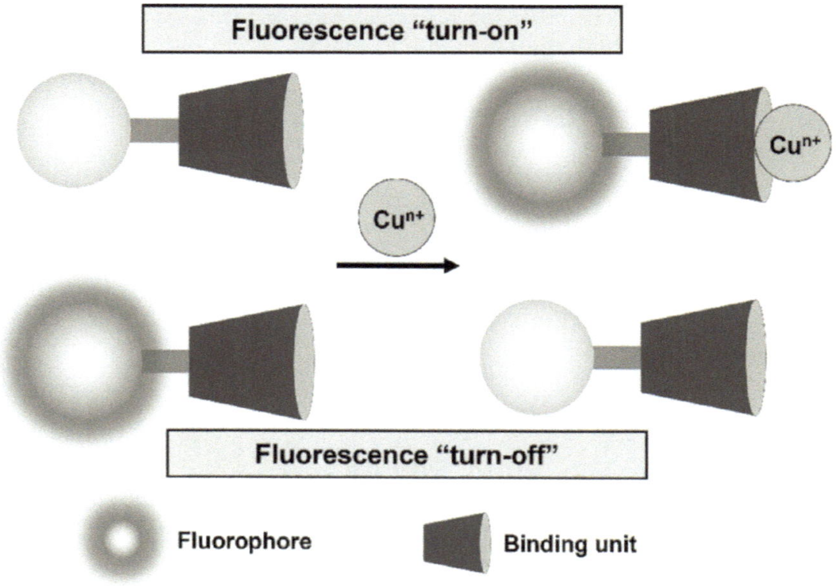

FIGURE 2 Illustrated representation of "turn-on" and "turn-off"-type fluorescent chemosensors.

intramolecular charge transfer (ICT), photoinduced electron transfer (PET), fluorescence resonance energy transfer (FRET), aggregation-induced emission (AIE), *etc.* [45–47]. Two kinds of fluorescence changes can be observed upon Cu ion binding to the fluorescent sensor, *i.e.*, "turn-on" and "turn-off" (Figure 2) [48]. The majority of fluorescent chemosensors are "turn-on" type, enhancing fluorescence, whereas "turn-off" chemosensors, indicating the subtraction of fluorescence upon metal binding, are relatively lower in numbers. Numerous fluorescent sensors have been developed that are suitable for detecting Cu in a particular sample type. Here, we discuss some commonly studied fluorescent chemosensors to detect Cu ions.

2.1.1 Organic Dye-Based Fluorescent Chemosensors

Traditional organic dyes, such as rhodamine, quinoline, coumarin, *etc.*, have been extensively studied as fluorescent probes because they possess high stability and fluorescence quantum yields [49–52]. Also, these fluorescent dye-based chemosensors have been widely used for Cu ion detection in all kinds of samples, including biomolecules, such as carbohydrates, proteins, hormones, and nucleic acids. For example, the Schiff base of organic dyes with metal receptor moieties is one of the common chemosensor design strategies. These dye-based sensing probes provide a wide range of emission spectrum from UV to near IR, commonly follow turn-off mechanism, and the fluorescence gets quenched upon chelation with Cu ions. Also, few chemosensors show a turn-on response to Cu ions. For example, a salicylaldehyde rhodamine B hydrazone chemosensor selectively recognizes Cu^{2+} ions by amplifying the absorbance and the fluorescence emission above 500 nm in neutral buffered media. Upon interaction with Cu^{2+}, the spirolactam ring of 1 opens to form a 1:1 metal–ligand complex (Figure 3a), resulting in fluorescence enhancement [53]. The sugar–rhodamine-based turn-on fluorescent chemosensor selectively determines Cu^{2+} ions in CH_3CN using Cu^{2+}-prompted spirolactam ring-opening reaction. Incorporating a sugar molecule into rhodamine improves the solubility and efficiency of the probe. The detection limit of this fluorescent chemosensor is about 10 μg/L, which is about 200 times lower than that of the WHO-recommended Cu levels (2.0 mg/L) in drinking water [54].

Few organic dye-based chemosensors can recognize Cu^+ ions with a turn-on response. For example, a cyanine dye (Cy7)-derived NIR fluorescent chemosensor, **CS790** (Figure 3b), is used for the selective detection of Cu^+ ($K_d = 3.0 \times 10^{-11}$ M). The probe **CS790** exhibits weak NIR fluorescence using the PET quenching mechanism; however, upon selective Cu^+ binding, it shows a 15-fold enhancement of fluorescence with an increase in quantum yield from 0.0042 to 0.072. The acetoxymethyl ester derivative of this probe, namely **CS790AM** (Figure 3b), could efficiently monitor the fluctuations in exchangeable Cu in living cells and mice under basal conditions and detect the deficiency or overload of Cu. In addition, it is also used to determine the levels of labile Cu in a murine model of Wilson disease [55].

Apart from the traditional organic dye-based chemosensors, there are many other organic chemosensors based on pyrazoline, BODIPY, calixarene, and commercial amine-based chemosensors that have selective responses toward Cu ions in various samples [56, 57].

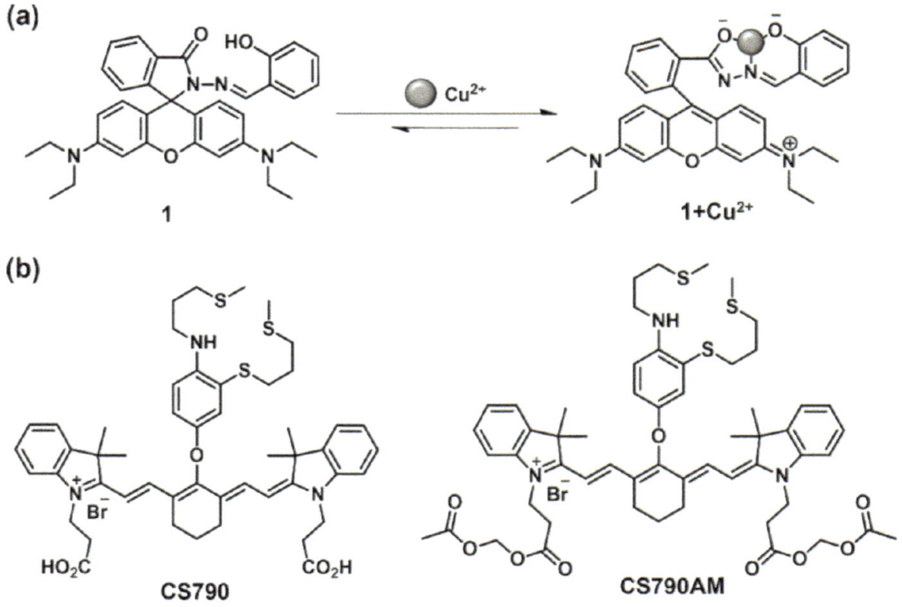

FIGURE 3 (a) The chemical structure and its binding complex ($1+Cu^{2+}$) of fluorescent chemosensor 1. (Redrawn from Ref. [53].); (b) The chemical structure of fluorescent chemosensor CS790 and its ester derivative CS790AM. (Redrawn from Ref. [55].)

2.1.2 Transition Metal Complex-Based Fluorescent Chemosensors

Transition metal-based chemosensors are generally luminescent metal complexes comprising a chelating receptor unit. Design and synthesis of luminescence complex chemosensors are relatively straightforward; however, many metal complexes need to be screened for fine-tuning to detect Cu ions selectively. So far, only a few examples of chemosensors for Cu ions based on transition metals Ir^{3+}, Ru^{2+}, Re^{+}, Au^{+}, and Zn^{2+} are known [6]. However, since these sensors can be prepared in a straightforward few-step synthesis, the future of developing fine-tuned chemosensors in this category for selective Cu ion detection is promising.

Four transition-metal-based chemosensors, *i.e.*, imine functionalized rhenium(I) polypyridine complexes containing phenol derivatized 2,2′-bipyridine ligands, **L1–L4**, show selective response toward Cu ions. A significant enhancement in fluorescence quantum yield, emission intensity, and luminescence lifetime is observed in these chemosensors after binding to Cu^{2+} ions with a constant binding range of 1.1×10^{4}–2.4×10^{4} M^{-1}. The fluorescence turn-on response for Cu^{2+} ions is attributed to the restriction of the C=N isomerization mechanism [58]. Pyrrole and thiophene derived two bimetallic Ru(II) polypyridyl complexes with chelating coordination sites to detect Cu^{2+} ions. Among these two complexes, the pyrrole-derived complex demonstrates a significant fluorescence quenching upon Cu^{2+} binding (association constant is about 5.50×10^4 M^{-1}), whereas the thiophene-based complex shows no such fluorescence quenching. The fluorescence changes of the pyrrole-based complex are highly selective for Cu^{2+} ions by forming a 1:1 complex with a low detection limit of 5.78×10^{-7} mol/L [59].

2.1.3 Covalent Organic Frameworks and Metal-Organic Frameworks-Based Fluorescent Chemosensors

To overcome the limitations of organic dye-based chemosensors, such as poor solubility, toxicity, reusability, and stability, various types of covalent organic frameworks (COFs)- and metal-organic frameworks (MOFs)-based chemosensors have been developed [60, 61]. A few MOF- and COF-based fluorescent chemosensors show high selectivity to detect Cu ions. For example, an AIE active zirconium-based luminescent MOF shows selective binding with Cu^{2+} ions over other metal ions with significantly quenched fluorescence at a fluorescence intensity of around 500 nm. This MOF-based chemosensor detects Cu^{2+} ions in the nanomolar concentration range (1–100 nM) with a detection limit of 550 pM [62]. The first example of a COF-based fluorescent chemosensor for Cu^{2+} ions could be azine-linked hydrogen bond-assisted COF (COF-JLU3). The heteroatoms present in the pore wall of COF can act as the binding site for the Cu^{2+} ions. It detects the Cu^{2+} ions by quenching the fluorescence intensity at 601 nm in the concentration range of 0–0.4 μM with a detection limit is about 0.31 μM [63]. Another COF-based chemosensor that contains hydroxyl and imine groups in its COF structure (COFs-DT) acts as a selective fluorescent sensor for Cu^{2+} ions. The blue fluorescence of the COFs-DT can be selectively quenched only upon binding with the Cu^{2+} ions (Figure 4). The fluorescence quenching occurs through the PET mechanism arising from the COF-DT donor and Cu^{2+} acceptor, which can be confirmed by a series of characterizing techniques, including cyclic voltammogram, ICP, and XPS [64].

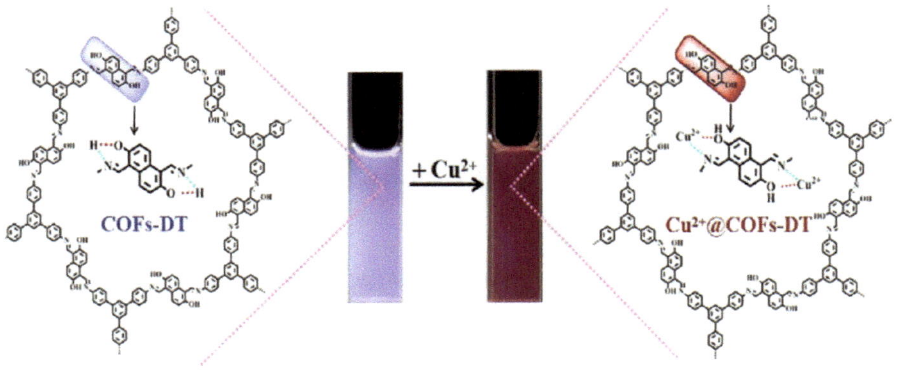

FIGURE 4 Covalent organic framework-based chemosensor COFs-DT-binding mechanism and corresponding color change (from blue to purple) with Cu^{2+} chelation. (Reproduced by permission from Ref. [64]; copyright/2020/Elsevier.)

2.1.4 Nanoparticle-Based Fluorescent Chemosensors

Noble metal-based ultra-small fluorescent nanoparticles have exhibited their excellence as chemosensors [65]. Their unique properties and high photostability have made them a potent alternative to traditional organic dye molecules [66]. For example, silver nanoclusters (AgNCs) and their derivatives are used as precise chemical fluorescent sensors to determine Cu^{2+} ions in various samples [67]. As an excellent fluorescent chemosensor, hyperbranched polyethyleneimine (hPEI) stabilized AgNCs demonstrated selective Cu^{2+} detection in the ultra-low concentration of 10 nM with the energy transfer between the cupric amine complexes and the hPEI-AgNCs. Good water solubility, ionic strength, and incredible stability in extreme atmospheric conditions facilitate the hPEI-AgNCs-doped hydrogel as a reusable optical sensor for Cu^{2+} detection. This fluorescent hydrogel almost entirely stops emitting upon plunging in 5 μM of Cu^{2+} solution. In such a way, this hydrogel sensor might be useful to screen Cu^{2+} levels in various running and stored water sources [68].

Another nanoparticle-based chemosensor, an electrospun rhodamine dye-doped poly(ether sulfones) (PES) nanofiber film, is used to detect Cu^{2+} ions selectively in an aqueous medium. It shows a low detection limit of Cu^{2+} at 1.1×10^{-9M} without interference from other biologically relevant metal ions. This nanofibrous sensor can be helpful for better practical applicability due to its compatibility with aqueous samples [69]. In addition, CdSe quantum dots-based chemosensor shows good selectivity with a detection limit is about 5 nM and a fast response time of 5 min [70]. Also, trimethylammonium bromide (CTAB) modified CdSe/ZnS QDs demonstrate the lowest Cu^{2+} ion detection down to 0.15 nM in the presence of thiosulfate in the water sample [71].

2.2 Detection of Copper Ions Using Colorimetric Chemosensors

The most significant advantage of using colorimetric chemosensors is that they can display naked eye-visible color changes in response to the analyte. It is not only a cost-effective technique but is also highly selective and sensitive in metal ion detection. On the other hand, the ease of analysis and the most precise instrumentation requirement for metal detection make colorimetric chemosensors highly demanding. Another main advantage of colorimetric detections is their potential to develop an easy-to-handle test kit for on-site metal ion detection in authorized safety inspections [72, 73].

The colorimetric sensor consists of a receptor and chromophore; the receptor holds a chelating site to bind with metal and send signals to chromophores by altering the electronic structure. Several colorimetric sensors are used to determine the Cu ions selectively. For example, the heptamethine cyanine dye-based sensor, Cy-NB (Figure 5a), detects the Cu^{2+} ions with the inhibiting effect of Cu^{2+} ions on the reaction of Cy-NB with L-cysteine. It shows a significant change in color in response to Cu^{2+} concentration, which the naked eye can observe. Cy-NB shows an exquisitely selective response toward Cu^{2+} ions over other relevant metal ions with a detection limit of 8.6 nM. It is used to determine the Cu^{2+} ions in different samples, including water (tap and sea) and biosamples, and high recoveries of Cu^{2+} concentration levels ranged from 102% to 114% [74]. A quinoline-based colorimetric chemosensor (CQ,

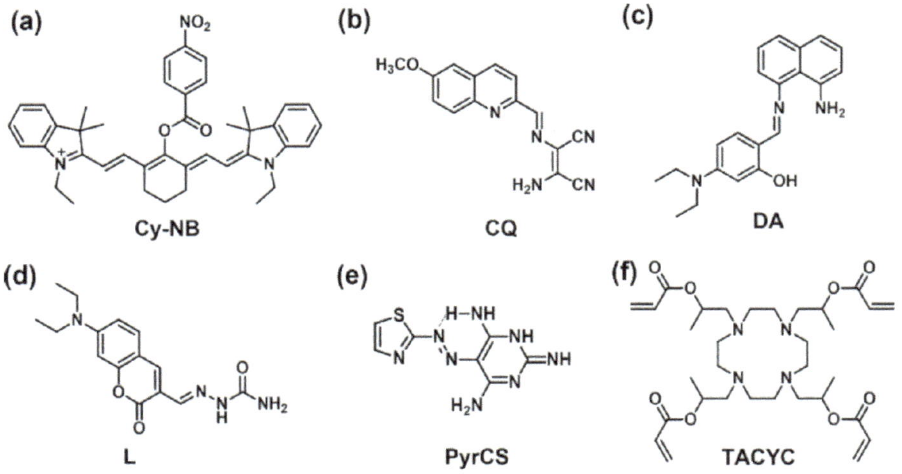

FIGURE 5 Chemical structures of representative examples of colorimetric chemosensors. (Redraw from (a) Ref. [68]; (b) Ref. [69]; (c) Ref. [70]; (d) Ref. [71]; (e) Ref. [72]; (f) Ref. [73].)

Figure 5b) detects Cu^{2+} ions with high precision and selectivity. It exhibits an apparent visible color change from yellow to red to establish the quantitative presence of Cu^{2+} in an aqueous solution with a detection limit of 10 mM. The color difference is assigned to the extended ICT in the Cu^{2+}complex of the sensor. It is also used to determine copper in the form of test strips [75].

A 1,8-diaminonaphthalene-based colorimetric and fluorometric sensor DA (Figure 5c) could determine the Cu^{2+} in food and water samples (drinking and environmental water). It can detect the Cu^{2+} ions using chelation-enhanced fluorescence quenching mechanism. It possesses many advantages: straightforward preparation, minimal preparation cost, high sensitivity and selectivity, quick response time, and reversible Cu^{2+} determination. Sensor DA can detect Cu ions selectively in environmental and biological samples with a nanomolar detection limit [76]. Another colorimetric chemosensor, a coumarin-derived probe L (Figure 5d), is used to detect Cu^{2+} ions at a low limit of 2×10^{-8} M. The retained response of the probe L to Cu^{2+} ions in broad pH ranges (4–10) can be allowed to detect Cu^{2+} in physiological conditions. Also, it exhibits negligible cytotoxicity in A375 cells and detects the Cu^{2+} ions in these cell lines with fluorescence imaging [77].

A pyrimidine-based colorimetric chemosensor, PyrCS (Figure 5e), could visually determine Cu^{2+} ions at the micromolar level in an aqueous solution. PyrCS forms a stable PyrCS-Cu^{2+} complex with Cu^{2+} ions; the consequence of this, a change in color of the solution, a bathochromic shift (UV), and quenching in fluorescence intensity at 507 nm of PyrCs, occurs. PyrCS shows a linear calibration range of 0.3–30 μM to Cu^{2+} with a detection limit is about 0.116 μM. In addition, the paper strips prepared by using the PyrCS sensor also detect Cu^{2+} ions in aqueous solutions of different concentrations from 0.1 to 50 μM. Moreover, PyrCS can also detect CN^- ions as the red color of PyrCS-Cu^{2+} readily changes to yellow when treated with CN^- ions (0.3 μM) [78]. Cyclen-based hydrogel colorimetric chemosensor, (p(HEMA-co-TACYC)) [p(2-hydroxyethyl methacrylate-co-tetraacrylic cyclen)] (Figure 5f), is used to detect Cu^{2+} ions in a wide range of water samples, including deionized, tap, river, and sea waters. This hydrogel sensor also works as a selective adsorbent material for Cu^{2+} ions with an excellent adsorption capacity of 15.85 mg/g. The formation of a stable complex with Cu^{2+} ions acts as an adsorbent material to Cu^{2+} ions, allowing (p(HEMA-co-TACYC)) hydrogel for the selective recognition of Cu^{2+} ions even in the mixture of metal ions [79].

2.3 Detection of Copper Ions Using Electrochemical Chemosensors

Among many other techniques, which are being used for tracing toxic heavy metals, electrochemical sensors have high potential owing to their quicker analysis, high sensitivity and selectivity, straightforward sampling, and inexpensive instrumentation. Electrochemical sensors' principle is based on their reaction with analytes to produce electrical signals directly proportional to their concentration [80]. Generally, an electrochemical sensor consists of sensing (working) and counter electrodes; a thin electrolytic layer separates them. A change in electrode potential can estimate the impact of the interaction between metal ions with the electrode [81]. For example, a modified electrode, which can be prepared by decorating thiolated calix[4] arene derivative on a screen-printed carbon electrode and gold nanoparticles (TC4/AuNPs/SPCE), is used for the detection of Cu^{2+} and lead (Pb^{2+}) ions. This electrode determines Cu^{2+} ions at a concentration as low as ppm levels using differential pulse voltammetry. The detection limit is found to be very decent at 1.3358×10^{-2} ppm. Moreover, this electrochemical sensor is used to recover the Cu^{2+} ions in river water with a reasonable recovery rate [82].

An electrode CdS/rGO/CC can be prepared based on carbon cloth (CC) modified with reduced graphene oxide (rGO), and cadmium sulfide (CdS) nanoparticles work as a visible-light-prompt photoelectrochemical sensor for Cu^{2+} detection. The decrease in photocurrent with the increase of Cu^{2+} ion concentration is based on the feasible binding of Cu^{2+} with CdS nanoparticles on the sensor's surface. The photocurrent response shows a good linear relationship with the concentration of Cu^{2+} ions with a detection limit is about 0.05 μM [83]. An electroanalytical chemosensor is prepared on the gold electrode surface by the self-assembly of 4-amino-6-hydroxy-2-mercaptopyrimidine monohydrate (AHMP) to determine Cu^{2+} ions. The sensor itself cannot block the ferrocene redox activity and $K_3[Fe(CN)_6]/K_4[Fe(CN)_6]$, but upon its further modification with 4-formylphenylboronic acid in 1,4-dioxane results in BA/AHMP/Au electrode, which can block ferrocene redox activity and $K_3[Fe(CN)_6]/K_4[Fe(CN)_6]$ completely. The selective voltammetric response of the BA/AHMP/Au electrode sensor for Cu^{2+} ions can be observed in the presence of interfering ions. This selectivity is based on the formation of a stable complex and electrostatic interaction between the BA/AHMP layer and Cu^{2+} ions [84].

In addition, single-walled carbon nanotubes (SWCNTs)-based nanosensor (PANI/SWCNTs) [85], silver (Ag) nanoparticles-based [86], gold (Au) decorated Ag (Au@

Ag) nanoparticles-based [87], 4-mercaptobenzoic acid-derived Ag nanoparticles-based sensors [88], and SnO_2 decorated reduced GO-based sensor [89] can be used as electrochemical chemosensors to detect Cu ions selectively in a wide range of samples.

2.4 Detection of Copper Ions Using Other Chemosensors

Apart from the chemosensors described above, numerous kinds of chemosensors are available. For example, luminescent, phosphorescent, membrane chemosensors, *etc.* PVC membranes-based sensors developed by using 1% of bis[acetylacetonato] Cu(II) (i), bis[ethylacetoacetate] Cu(II) (ii) or bis[salicyldehyde] Cu(II) (iii) can selectively detect Cu ions at a concentration limit as low as 0.1 ppm. The sensor prepared by a combination of A (1%), PVC (33%), tri-n-butylphosphate (TBP) (65%), and NaTPB (1%) shows the best linear potential response to Cu^{2+} over a wide range of concentration from 2 μM to 100 mM in 2.6–6.0 pH solutions with a fast response time (~9 s). This sensor can be employed for the potentiometric titration of Cu^{2+} with EDTA and estimate copper in multivitamin capsules and vegetable foliar [90].

Terbium (III)-based bimacrocyclic luminescent chemosensor **TCS1** selectively detects Cu^{2+} ions in aqueous solutions by its turn-off response to Cu^{2+}. **TCS1** irreversibly binds to Cu^{2+}; therefore, it is a potential chemically trap for Cu^{2+} ions. It shows several advantages, such as excellent solubility, biocompatibility, low energy excitation wavelength (350 nm), and the detection limit is as low as 1.7 nM [91]. A multichromophoric iridium(III) complex (ZIr_2)-based phosphorescent sensor is used for the ratiometric detection of Cu^{2+} ions in aqueous and biological media. The free complex displays a green and red dual phosphorescence; the red band selectively quenches upon binding with Cu^{2+} to increase the green/red emission intensity ratio to four-fold with an excellent K_d of 16 μM. It also affects the phosphorescence lifetime of the sensor (decrease in a lifetime). This phosphorescent probe, $ZIr_{2,}$ can determine the intracellular Cu^{2+} in live HeLa cells [92].

3 DETECTION OF COPPER IONS USING DIFFERENT BIO-SENSORS

Efficient detection of Cu ions in biological samples requires specific compatibility. The limitations of many chemosensors, such as poor water solubility, limited stability, and toxicity, make them incompatible for detecting Cu ions in biological samples. However, biomolecule-driven sensors can overcome these constraints for Cu detection in biological samples [42, 93]. A bio-sensor is designed by combining three crucial components, *i.e.*, recognition unit, transduction unit, and signal-processing unit. When a sample is applied to the sensor, it identifies the target analytes, and then the transducer transforms the information into measurable signals by the signal-processing unit [94]. The design principle of a bio-sensor is presented in Figure 6. Various kinds of modified biological molecules are used as bio-sensors, including antibodies, enzymes, proteins, DNA, and even whole cells, to detect Cu ions selectively [95, 96].

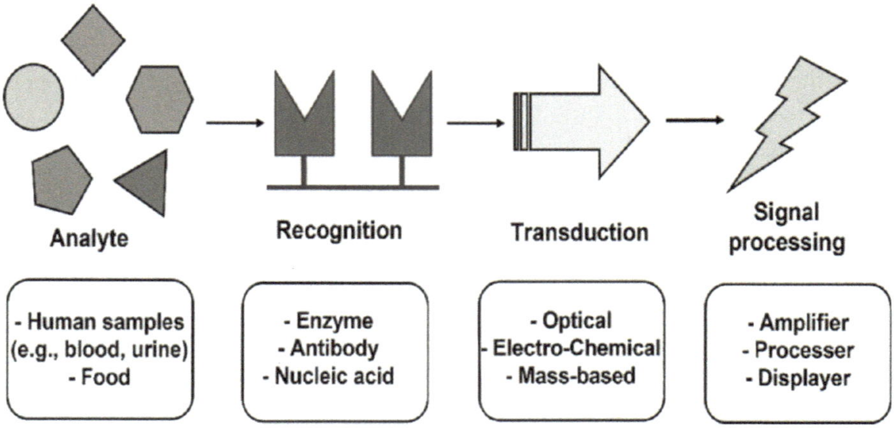

FIGURE 6 Illustrated representation for collective functioning of different components of a bio-sensor.

3.1 Detection of Copper Ions Using Fluorescent Bio-Sensors

Fluorescent bio-sensors are vital tools for the noninvasive detection of metal ions in biological samples with high precision and selectivity. These bio-sensors allow researchers to efficiently screen and quantify Cu ions in living and non-living organisms utilizing the changes in their fluorescence [97, 98].

Graphene can serve as a potential construct for fluorescent bio-sensors as it is an efficient quencher of luminescent nanomaterial using an electron or energy transfer process [99]. For example, the QDs complex-based fluorescent bio-sensor GO-dsDNA-CdTe can successfully detect Cu ions through fluorescence turn-on response. The quenched fluorescence of CdTe QDs in GO-dsDNA-CdTe through FRET from the two-dimensional GO can be recovered by the GO separating from CdTe QDs with Cu^{2+} ions mediated breaking of the DNA chain. With the aid of this property, the fluorescent bio-sensor detects Cu^{2+} ions in a linear concentration-based manner [100]. Apart from the Cu^{2+}-induced DNA cleavage bio-sensors, a few Cu^{2+}-promoted RNA-cleavable bio-sensors have also been developed for Cu ion recognition. For example, a phosphorothioate (PS) RNA-derived small DNAzyme, PSCu10, is identified to design a fluorescent bio-sensor for detecting Cu^{2+} ions in the ultra-low concentration of 1.6 nM. PSCu10 shows a highly selective response to Cu^{2+} with a turn-on fluorescence by the specific cleavage of RNA. But in the presence of ascorbate, no RNA-cleavage can occur because ascorbate reduces Cu^{2+} to Cu^{+}. From other divalent metals, only Hg^{2+} due to its high lipophilicity could cleave the PSCu10. Therefore, this DNAzyme bio-sensor can be helpful as a probe for detecting Cu^{2+} in water samples and studying Cu^{2+} binding to DNA [101].

ssDNA-templated AgNCs developed through anti-galvanic reduction works as a highly sensitive fluorescent bio-sensor for Cu^{2+} detection. This bio-sensor functions based on turn-on resonance light scattering (RLS) by reducing Cu^{2+} ions to Cu^{0}; the resulting Cu^{0} can deposit Cu nanoparticles on the sensor surface. This process

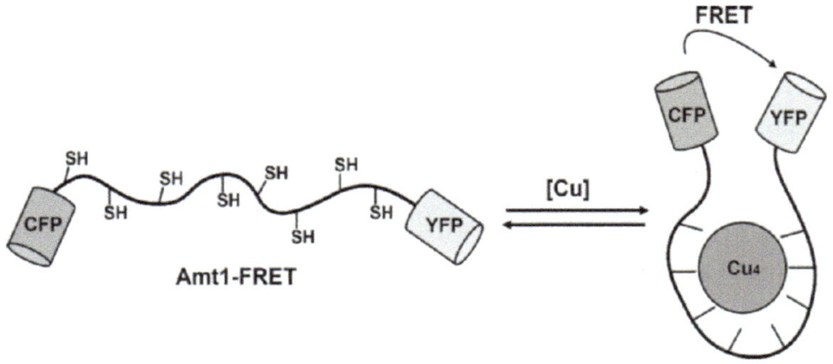

FIGURE 7 The representation of a FRET-based fluorescent bio-sensor, Amtl-FRET, for detecting Cu^+ ions. (Redrawn by permission from Ref. [105]; Copyright/2010/American Chemical Society.)

generates Ag/Cu alloy nanoparticles of larger diameters and turns on the RLS signal. The signal gradually enhances upon the increasing concentration of Cu^{2+}, which attains a linear range of 5×10^{-9}–7.5×10^{-7} M, and the detection limit is about 2 nM [102]. In most analytic samples, Cu occurs in its stable state of Cu^{2+}, but it is quickly reduced to Cu^+ states. Therefore, selective recognition of Cu^+ is also equally important. However, only a few fluorescent bio-sensors have been used to detect Cu^+ so far [103]. Chang et al. developed a fluorescent bio-sensor consisting of tris[(2-pyridyl)-methyl]amine (TPA) as the Cu^+ receptor site and a luminescent biomolecule d-luciferin for the sensitive response to Cu^+ ion based on Cu^+-dependent oxidative cleavage of d-luciferin. This bio-sensor can detect Cu^+ ions in both *in vitro* living cells and *in vivo* mice models of copper deficiency or overload situations [104]. Another fluorescent bio-sensor, Amtl-FRET, was developed by inserting a copper-binding domain between the two FRET partners, *i.e.*, a yellow fluorescent protein and a cyan fluorescent protein (Figure 7). The fluorescence variation occurs by the Cu^+ binding-induced conformational shift of Amtl, which estimates the Cu^+ concentration in the analyte sample. The sensor Amtl-FRET is highly selective and sensitive even at ultra-low concentrations of Cu^+ with $K_d=2.5\times10^{-18}$ M. This bio-sensor can estimate the meager availability of Cu^+ in yeast and activates copper detoxification genes at the upper limit of Cu^+ ions. It can also report the dynamic fluctuations of the Cu^+ ions in mammalian cells [105].

3.2 Detection of Copper Ions Using Colorimetric Bio-Sensors

Among the bio-sensors, colorimetric methods are more attractive and appreciable because of seamless handling. Nacked-eye tracing of visible color changes without sophisticated requirements makes the colorimetric bio-sensors a powerful tool for the applications of Cu ions sensing in a wide range of samples [106]. The click chemistry-based bio-sensors are popular for detecting Cu ions in colorimetric approaches. For example, the colorimetric bio-sensor based on click chemistry of terminal alkyne- and azide group-functionalized AgNPs is used to detect Cu^{2+} ions in aqueous solutions. The catalyst, Cu^+, which is generated by reducing Cu^{2+} in the presence of

Na-ascorbate, can catalyze the click chemistry reaction. The presence of Cu^{2+} is confirmed by visible color loss upon aggregation of AgNPs due to click adduct formation [107]. An AuNPs-based DNA-functionalized colorimetric bio-sensor, in which the Cu^{2+} catalyzed click reaction between the hexynyl or azide terminal of oligonucleotide strands on the surface of DNA. AuNPs increase the melting temperature (T_m) of the polymeric aggregate. The increased molecular motion at higher T_m develops the color change, which can be examined using ultraviolet-visible (UV-Vis) absorption spectroscopy or the naked eye. Therefore, it can be applied for detecting Cu^{2+} in biomolecules and water samples [108]. Another clickable DNA-modified AuNPs probe consists of alkyne and azide groups. Cu^{2+}-catalyzed click reaction induces dispersion or aggregation of AuNPs with a color change of red to blue. This optimized bio-sensor platform can detect Cu^{2+} in as low as 250 nM concentration with a linear range of 0.5–10 μM, and it is promising for low-cost Cu^{2+} detection, particularly without any advanced instrumentation [109].

3.3 Detection of Copper Ions Using Electrochemical Bio-Sensors

The electrochemical bio-sensors provide high selectivity with accuracy in quickly detecting the presence of metal ions in low concentrations, especially for real-time analysis [110–112]. The electrode is utilized as a support base in these bio-sensors to immobilize biomolecules and the electron movement [113]. The design of an electrochemical bio-sensor is based on connecting the redox-active biomolecule containing a receptor site that can bind with Cu ions resulting in alteration of the redox potential. Many types of biomolecule-modified electrodes can be used for the electrochemical sensing of Cu ions [114]. For example, L-cysteine-based electrodes demonstrate excellent efficiency as electrochemical bio-sensors for Cu ions. An electrochemical bio-sensor based on a screen-printed carbon electrode modified with L-cysteine self-assembled AuNPs can selectively detect Cu^{2+} ions at low concentrations. This bio-sensor demonstrates efficient detection of Cu^{2+} ions as low as the concentration of parts per billion (ppb) levels within a short response time of 10 min. The detection limit is found to be 8 ppb with a linear range of 10–0.005 ppm ($r = 0.9870$). It is a less costly method and suitable for using trace Cu ions in water samples [115].

L-cysteine and Cu^{2+} ions are closely associated with the pathological and physiological states of AD. Therefore, a typical sensor that can simultaneously detect both Cu^{2+} ions and L-cysteine can play a crucial role in finding an affordable solution for AD. A *N*, *N*-di-(2-picoly)ethylenediamine (DPEA)-based single electrochemical bio-sensor can be used to detect both Cu^{2+} and L-cysteine simultaneously *in vivo*. DPEA forms the DPEA–Cu^{2+} complex and estimates Cu^{2+} ions; this complex selectivity determines L-cysteine by binding with the released Cu^{2+} ions from the DPEA–Cu^{2+} complex (Figure 8). DPEA and DPEA–Cu^{2+} can determine the accurate concentration levels of Cu^{2+} and L-cysteine in the AD live rat brains [116]. Selective grafting of biological species, such as enzymes, antibiotics, peptides, and DNA, onto conducting polymers makes promising electrochemical bio-sensors [117]. A conducting polymer functionalized electrochemical bio-sensor consisting of a Cu ion-specific probe Gly–Gly–His tripeptide attached with a poly(3-thiopheneacetic acid) transducer film on a gold electrode is used to determine the trace amounts of Cu^{2+} ions. It shows a

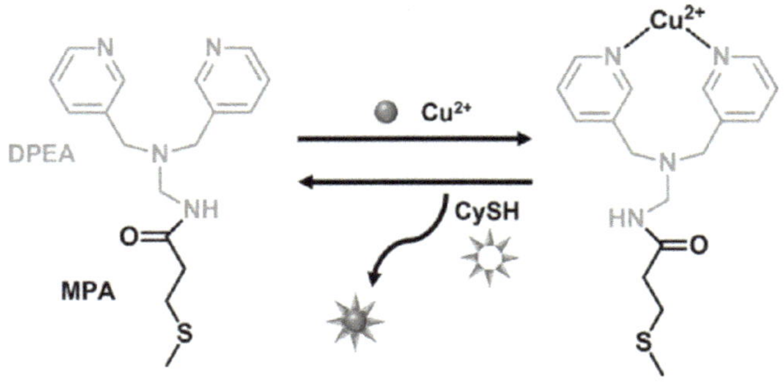

FIGURE 8 DPEA-based single electrochemical bio-sensor-detecting mechanism of Cu^{2+} and L-cysteine (CySH). DPEA, *N,N*-di-(2-picoly)ethylenediamine; MPA, mercaptopropionic acid. (Redrawn from Ref. [116].)

superior affinity for Cu^{2+} ions compared to the other relevant metal ions. This electrode has a high sensitivity for Cu^{2+} ions (0.02–20 μM), and it can be reusable just by washing with EDTA solution [118].

3.4 Detection of Copper Ions Using Other Bio-Sensors

As discussed above, fluorescent, colorimetric, and electrochemical bio-sensors have promising high sensitivity and selectivity toward Cu ions. In addition, many other bio-sensors demonstrate precise Cu ion detection over a wide range in various analyte samples. For example, enzymatic luminescent systems can be used as promising bio-sensors for Cu detection in water samples. The disposable microfluidic chips made of poly(methyl methacrylate) can be integrated into a bioluminescent system for enzyme-inhibition-based assay with a handheld luminometer. This bioluminescent system with microfluidic chips can detect copper sulfate up to 2.5 μmol/mL. It is a highly promising bio-sensor for toxic chemicals and heavy metal salts, which impact the bioluminescent signal of enzymatic reactions [119]. A highly bioluminescent bacterial strain, *Vibrio sp.* MM1, isolated from the Caspian Sea, detects Cu ions in water samples based on bioluminescence inhibition assay. However, this bacterial strain is also sensitive to the low concentrations of other heavy metals, including Zn^{2+}, Ni^{2+}, Co^{2+}, Cd^{2+}, and Pb^{2+}, which quenches the luminescence of the bacteria MM1. This bacterial strain has a high potential as a bioluminescent sensor for detecting heavy metals in water samples; however, more study is needed to find higher selectivity for Cu ions [120].

Growing heavy metal exposure to bacteria increases the risk of creating a highly resistant microbe to antibiotics and is a threat to human health. Therefore, monitoring metal intake by bacteria is considered an essential requirement [121]. In addition, the Cu metal can enter bacterial cells without a specific uptake mechanism. As a result, a whole-cell bacterial bio-sensor can be used to quantitatively detect the uptake of specific copper coming from organic or inorganic sources. This bio-sensor identified that organic Cu has higher uptake than inorganic Cu and found a significant difference between CuO and Cu_2O [122].

Electrochemical impedance spectroscopy (EIS) is also a suitable technique for metal ion detection in biological samples. Impedimetric bio-sensors using EIS determine the variations in charge conductance and capacitance at its surface upon selective binding of the Cu ions [123]. The presence of copper ions is known to control the activity of oxytocin (OT) neuropeptide. Inspired by this intriguing property, an electrochemical bio-sensor is prepared by immobilizing OT onto a glassy carbon electrode governed by copper ion binding. The bio-sensor selectively detects Cu^{2+} ions at physiological pH ranges. The binding of Cu^{2+} ions with OT induces conformational variations of OT that can translate to quantifiable impedimetric data. This OT-based sensor can effectively estimate the Zn/Cu ratio, distinguishing between multiple sclerosis patients and healthy control sera samples [124].

4 DETECTION OF COPPER IONS IN DIFFERENT SAMPLES

Because Cu has an essential role in environmental and biological systems, it is necessary to detect Cu ions (both Cu^{+} and Cu^{2+}) in different samples, including water, environment, biological, and other samples.

4.1 In Biological Samples

In 2013, Yang's research group prepared E_2Zn_2SOD, a Cu-free derivative of SOD that can be used as an electrochemical bio-sensor for Cu^{2+} ions. As Cu^{2+} ions have a specific interaction behavior with E_2Zn_2SOD to reconstitute SOD, the Cu^{2+} ions of a rat brain are determined *in vivo*. This electrode exhibits selective electrochemical detection of Cu^{2+} ions over other metal ions with a lower detection limit of 3 nM. Another electrode, 6-(ferrocenyl) hexanethiol (FcHT)-amended one, is employed by the combination of an E_2Zn_2SOD electrode to produce two-channel bio-sensor results in improved accuracy of the bio-sensor in determining the rat brain' Cu^{2+} ions [125]. Another bio-sensor, 2,2′,2″-(2,2′,2″-nitrilotris(ethane-2,1-diyl)-tris((pyridine-2-ylmethyl)azanediyl)triethanethiol (TPAASH) based molecule is used for the purpose of selective and accurate detection of Cu^{2+} in a rat brain *in vivo*. Combining an FcHT electrode with TPAASH improved the accuracy of the electrochemical bio-sensor [126]. An online optical detection platform (OODP) with two colorimetric sensors, such as an oxidized form of tetramethylbenzidine (oxTMB) and dopamine-derived AgNPs (DA-AgNPs), is used to determine the ascorbic acid and Cu^{2+} ions simultaneously in a rat brain. This dual-channel OODP, combined with *in vivo* microdialysis sampling, exhibits a continuous detection of Cu^{2+} ions in a range of 0.1–10 μM [127].

A silica-coated carbon quantum dots (APTES-CDs)-based fluorescent sensor is used to determine the Cu^{2+} ions selectively in the rat brain striatum (Figure 9). This sensor can detect the Cu ions by following the process of cerebral calm/sepsis. The probe, which can be produced through a reliable ratio of APTES and CDs (*i.e.*, 3.33% APTES-0.9 mg/mL CDs), exhibits an enhanced fluorescence signal of Cu^{2+} detection, ascribed to chelation and electrostatic interaction between Cu^{2+} ions and N and O atoms of APTES and CDs [128]. An ICT-based fluorescent sensor (*R*) recognizes the Cu^{2+} ions in living macrophage (RAW 264.7) cells by forming a 1:1 stoichiometric complex between *R* and Cu^{2+} ions [129]. The two quinoline-based fluorescent sensors,

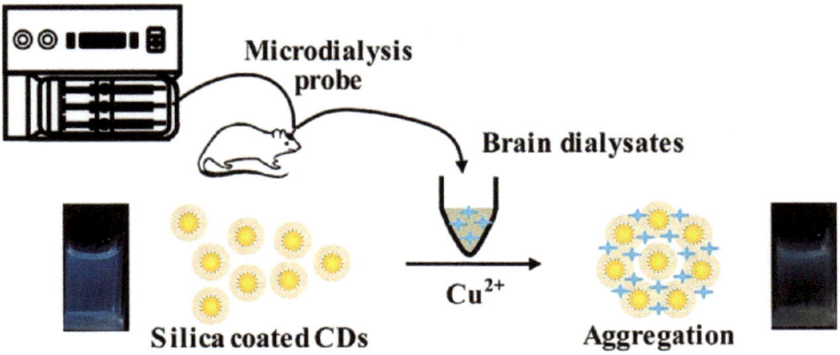

FIGURE 9 Schematic illustration for Cu^{2+}-induced aggregated fluorescent silica-coated CDs (APTES-CDs) and their fluorometric sensing of cerebral Cu^{2+} in rat brain microdialysates. (Reproduced by permission from Ref. [128]; Copyright/2015/American Chemical Society.)

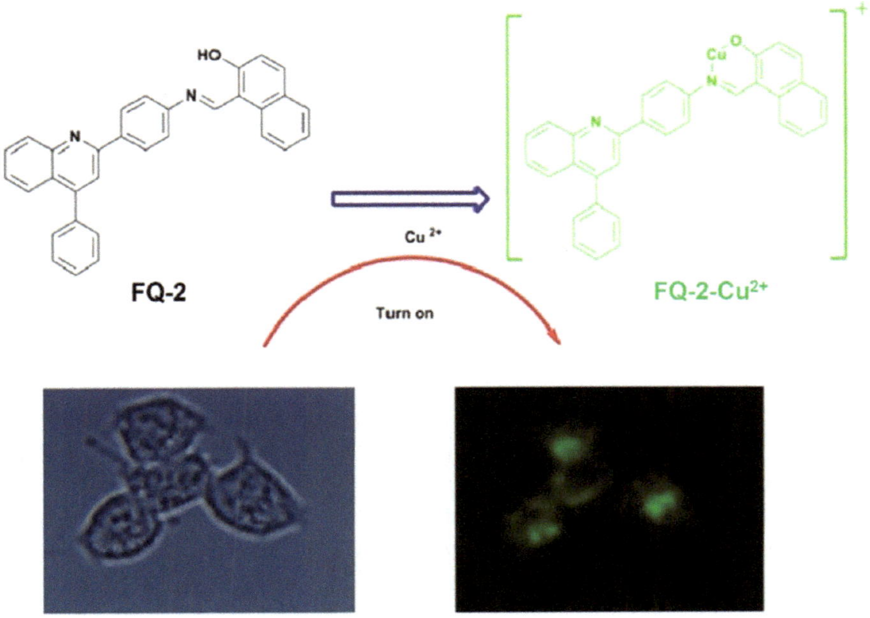

FIGURE 10 Structure, complexation with Cu^{2+} ions of FQ-2, corresponding bright-field images after incubation with FQ-2 (5 μM) for 30 minutes at 37°C, and fluorescence microscopic images of HeLa cells after incubation with FQ-2 (5 μM) and $CuCl_2$ (50 μM) for 10 minutes. (Reproduced by permission from Ref. [130]; copyright/2017/Elsevier.)

namely FQ-1 and FQ-2, can detect Cu^{2+} ions selectively in HeLa cells with its property of turn-on response to Cu^{2+} ions by the 1:1 complexation between FQ-1 or FQ-2 and Cu^{2+} ions. (Figure 10; FQ-2 binding and microscope imaging.) The detection limits are 1.8 and 44.2 nM for FQ-1 and FQ-2, respectively [130]. An anthraquinone-derivatized colorimetric and fluorimetric sensor (*L*) recognizes Cu^{2+} ions in SMMC-7721 cells, evidenced by the confocal images. The pink color of the 1:1 tetrahydrofuran:Tris–HCl buffer solution turns blue, which is ascribed to a 1:1 stoichiometric complexation between Cu^{2+} ions and *L*, resulting in fluorescence quenching at 604 nm [131]. A tetra-peptide, *i.e.*, Ser-Pro-Gly-His (SPGH), could specifically enhance the Cu^{2+} catalytic performance is due to the increased Cu^{2+}-induced·OH generation by SPGH. The formed $Cu(SPGH)_2$ complex can efficiently catalyze the luminol-H_2O_2 chemiluminescence. This chemiluminescence acts as a light source to enhance the fluorescence of QDs using CRET to determine the Cu^{2+} selectively within 15 min. This CRET-based sensor has a lower RSD, down to 6.38%, can determine the human serum's total Cu [132].

It is known that Cu^{2+} ions are directly involved in the aggregation of Aβ, a phenomenon that corresponds to the development of AD [133]. Therefore, combined detection of Cu^{2+} ions and Aβ is vital to understanding the nature of AD pathology. The composite of an 2,2′-azinobis-(3-ethylbenzothiazoline-6-sulphonate (ABTS) with poly(diallyldimethylammonium chloride (PDDA)-derived carbon nanotubes (ABTS-PDDA/CNTs) can be used as an electrochemical bio-sensor for the dual detection of Cu^{2+} ions and $A\beta_{1-42}$. The coatings of neurokinin B (NKB) on this probe result in forming a $[Cu^{II}(NKB)_2]$ complex with Cu^{2+}. It responds to concentration-dependent Cu^{2+} and $A\beta_{1-42,}$ and the detection limits are 0.04 μM and 0.5 ng/mL for Cu^{2+} and $A\beta_{1-42}$, respectively [134].

A thioether-rich tetrathia-azacrown derivatized naphthalimide-based fluorescent sensor, Naphthyl-CS1, is used to determine the Cu^+ ions in neuronal living SH-SY5Y cells. Naphthyl-CS1 exhibits a selective binding of Cu^+ ions with a binding constant of 12.1 ± 0.6. The Cu^+ ion corresponding confocal microscopy images of SH-SY5Y cells highlight the probe's detection capability for Cu^+ ions [135].

4.2 In Water Samples

Due to the Cu ions' toxicity to the environment and animals, including humans, USPEA stated the guidelines for maximum Cu concentration in drinking water is 1.3 ppm. At the same time, as per WHO, it should not exceed 2 ppm [136]. Hence, it is crucial to determine the Cu ions in drinking and environmental water samples. The competitive adsorption of proteins combined with Surface Plasmon Resonance bio-sensor for Cu^{2+} is called the Vroman effect, in which Cu^{2+} ions interact with native protein bio-receptors (albumin). Consequently, albumin denaturation occurs, which has less absorption affinity than the native albumin. Due to the competitive adsorption among the denatured and the native albumins, this sensor detects Cu^{2+} ion concentration down to 0.1 mg/L in different water samples, including PBS, deionized, tap, and bottled water [137]. Histidine has a strong interaction with Cu^{2+} ions. An electrochemical sensor of histidine and gold-labeled carbon nanotubes is used to estimate the Cu ions in river water. An enhancement in the signal of this sensor occurs after binding Cu^{2+} ions to histidine and exhibits a linear curve between peak current and Cu^{2+} concentration (10^{-11}–10^{-7} mol/L). The detection limit is found to be 10^{-12} mol/L [138]. A sandwich-type structured L-cysteine-gold nanoparticle-based optical–electrochemical bio-sensor, ITO/GNPs/L-cys··Cu^{2+}·L-cys/GNPs is prepared from the L-cys/GNPs and ITO/GNPs/L-cys with an open-circuit technique, which is used for the quantification of Cu^{2+} in water samples. This bio-sensor demonstrates the detection limit: <5 nM (0.31 ppb) with a total calibration linear range: 10–100,000 nM (0.64–6354.6 ppb) [139].

A cellulose-based colorimetric sensor, DAC-PDH, is formed from the amino-functional groups of 2,6-pyridine dihydrazide (PDH) and aldehyde functional groups of dialdehyde cellulose (DAC) by following the Schiff base-type reaction. Its visual color changes from white to green with Cu^{2+} ions by the chelation of carbonyl (–C=O) and the amino-functional groups (–NH–, $-NH_2$) of DAC-PDH with Cu^{2+} ions. It shows a low response time of 30 s with 10^{-7} mol/L detection limit [140]. A smartphone android application is used with the colorimetric indicator of a natural food pigment, *i.e.*, red beet pigment, as a sensor to determine Cu^{2+} in drinking water. This indicator color changes from purple to orange-red with Cu^{2+} ions by the chelation and a redox reaction. It shows a linear calibration curve with the Cu^{2+} concentration (4–20 μM), and a detection limit is about 0.84 μM [141]. Another colorimetric sensor, *i.e.*, amide and morpholine groups derivatized calix[4]arene, determines Cu^{2+} ions in water samples by changing color from colorless to yellow. The detection limit is about 0.1 ppb [142]. Considering the N and O atoms of dopamine have affinities toward Cu^{2+} ions, a dopamine-functionalized AgNP nanosensor is used to determine the Cu^{2+} ions in the tap water. The binding of Cu^{2+} ions with dopamine caused the aggregation of AgNPs, followed by a response of color change. It shows a rapid response to Cu^{2+} ions with a concentration range of about 3.2–512 ppb [143].

An "ON–OFF" colori- and fluoro-metric chemosensor, 1*N*-allyl-2-(2,5-dimethoxyphenyl)-4,5-diphenyl-1*H*-imidazole (ADPPI), detects Cu^{2+} ions in water samples (environmental) based on the interaction between ADPPI and Cu^{2+} ions. The complexation between Cu^{2+} and ADPPI results in a blue-colored solution with a 610 nm absorbance and 417 nm emission peaks. The detection limit is about 1.01 nM [144]. A fluorescent assay is used to quantitatively detect Cu^{2+} ions based on Cu^{2+} acting as a catalyst in *p*-cresol oxidation by H_2O_2 under basic conditions resulting in an intense fluorescent *p*-cresol oxidation product. It responds below to 1.0×10^{-8} mol/L of Cu^{2+} with the linear calibration range between 3.0×10^{-7} and 5.0×10^{-5} mol/L. It is used to determine the trace of Cu^{2+} in water samples from the river, lake, and swimming pool [145]. A ligand-capped CdTe QDs fluorescent nanosensor determines the Cu^{2+} in environmental water samples with the fluorescence quenching property of CdTe-L QDs in the presence of Cu^{2+} ions. The detection limit is about $1.55\pm0.05\times10^{-8}$ mol/L, and the linear calibration Cu^{2+} concentration range is between $5.16\pm0.07\times10^{-8}$ and $1.50\pm0.03\times10^{-5}$ mol/L. It is used to determine the Cu^{2+} in various water samples, including tap, sea, well, flume, and wellhead water [146]. In addition, a pigment-based whole-cell bio-sensor in a fresh pond and tap water [147] and solid-phase extraction coupled with a microfluidic paper-based technique in drinking water is used to determine Cu^{2+} ions [148].

4.3 In Biological and Water Samples

Few sensors are also used to determine Cu ions in water and biological samples. For example, under UV excitation, *O*-phenylenediamine (OPD)-functionalized CDs (OPD-CDs) exhibit a yellow photoluminescent emission. The Cu^{2+} ions further enhance this photoluminescent emission due to the coordination affinities between Cu^{2+} ions and amino groups of OPD-CDs, which causes the induced aggregation of OPD-CDs. This specific AIE-based enhancement feature of OPD-CDs allows detecting Cu content in river closure reservoirs and living Hep-2 cells. The dynamic detection range is from 0.5 to 40 μmol/L and has a detection limit of 0.28 μmol/L [149]. A PDH-conjugated macrocyclic Schiff base fluorescent probe (BP-MSB) determines Cu^{2+} ions based on the quenching property of BP-MSB toward Cu^{2+}. After interacting Cu^{2+} ions with BP-MSB, the strong fluorescence of BP-MSB shows a quenched response. This behavior of BP-MSB is used to detect the Cu^{2+} ions in the tap water and Xiang Jiang water samples and in living GM12878 cells [150].

4.4 In the Environment

Some sensors are used to determine Cu^{2+} ions in different environmental samples other than the environmental water samples discussed in Section 4.2. For example, the nanoporous anodic alumina (NAA) interferometers functionalized with poly-ethylenimine double-layered glutaraldehyde (PEI-GA-PEI) are used with reflectometric interference spectroscopy (RIfS) to detect Cu^{2+} ions in environmental samples. The schematic setup, working principle, and sensing process of this RIfS with (PEI-GA-PEI) sensor are shown in Figure 11. The detection limit is 0.007 ± 0.001 mg/L

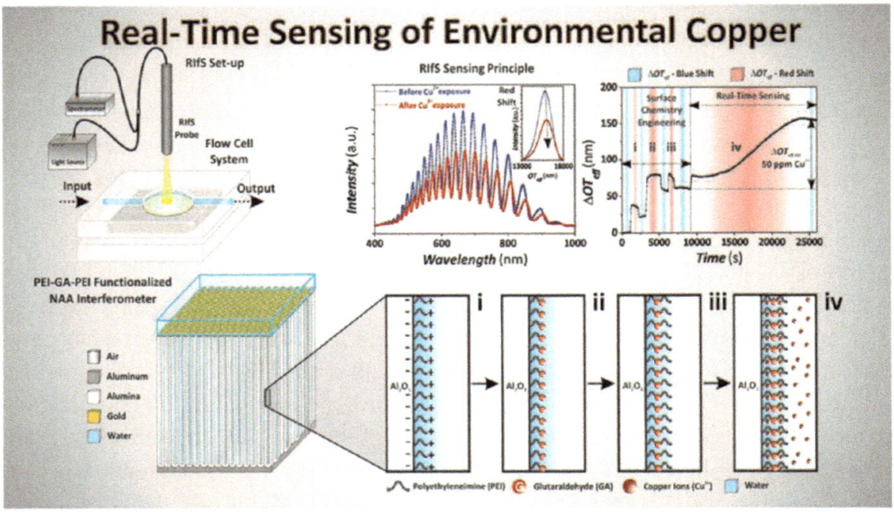

FIGURE 11 Schematic representation of RIfS setup used to monitor-binding interactions between PEI-GA-PEI-modified NAA interferometers and Cu^{2+} ions in real-time under dynamic flow conditions; Representative RIfS spectra before and after exposure to Cu^{2+} ions; Real-time effective optical thickness changes (ΔOTeff) associated with the surface chemistry engineering and sensing stages: surface chemistry engineering (i–iii) and real-time sensing (iv); Schematic showing the structure of PEI-GA-PEI-functionalized NAA interferometers; Illustration showing details of the inner surface chemistry of gold-coated PEI-GA-PEI-functionalized NAA interferometers during the different stages of the sensing process (i–iv). (Reproduced by permission from Ref. [151]; Copyright/2019/American Chemical Society.)

(~0.007 ppm), and the dynamic linear detection range is about 1–100 mg/L. This PEI-GA-PEI-NAA sensor shows an exquisite specificity toward Cu^{2+} ions over other metal ions. It demonstrates the quantitative detection of Cu^{2+} ions in various environmental samples, including tap water and acid mine drainage liquid [151]. A cloud point extraction method combined with flame atomic absorption spectroscopy (FAAS) can determine the Cu^{2+} ions in various environmental and biological samples. Cu^{2+} ions are recovered quantitatively from the chrysoidine-Cu^{2+} complex after centrifugation in Triton X-114. The detection limit is 0.6 ng/mL, and RSD is 1.0% toward Cu^{2+}. This sensor effectively detects Cu^{2+} ions quantitatively in environmental samples, including vegetables, lotus (tree), and soil, as well as in biological samples, including blood, liver, and meat [152]. In an electrochemical system, the surface of carbon tips of the disposable carbon tip electrodes is electrodeposited with mercury to give thin-film mercury electrodes (TFME), which are used to automatically detect Cu^{2+} ions along with Cd^{2+} and Pb^{2+} ions simultaneously. The bare electrode (unmodified carbon tip) shows the electrochemical response at a peak maximum of −0.29 V and a detection limit of 0.1 µg/mL. At the same time, TFME shows the electrochemical response at a peak maximum of −0.29 V and a detection limit of 0.02 µg/mL. The accuracy results are consistent with the AAS and XRF results. This electrochemical sensor detects the Cu^{2+} ions in different samples, including mineralized rock, blood plasma, and chicken embryo organ samples [153].

4.5 In Other Samples

Apart from the above, a few sensors are also used to determine the Cu species in different samples. For instance, a gold microelectrode is used in a linear sweep ASV method to determine the Cu^{2+} ions in ethanol fuel samples (commercial). This sensor shows a linear range between 0.05 and 1.0 µM with a 22 nM detection limit. The results obtained by this method are consistent with those produced by FAAS [154]. Luminol is used as a chelating agent in the adsorptive stripping voltammetric method to detect Cu^{2+} in water and food samples. Cu^{2+} forms a stable complex with luminol, which has adsorptive characteristics on hanging mercury drop electrodes. The linear calibration range is between 0.5 and 105.0 ng/mL with a 0.04 ng/mL detection limit. This method is used to determine the Cu^{2+} content in food samples, including tomato, rice, tea, spinach, and water samples, including tap, mineral, river, sea, and lab water [155].

A colorimetric sensor 2-chloroquinoline-3-carbaldehyde-conjugated rhodamine B hydrazide (RB-CQC) detects Cu^{2+} ions by changing visual color. The favorable interactions between RB-CQC and Cu^{2+} give rise to a stable complex, changing the color from yellow to pink. Consequently, a new absorbance peak appears at 566 nm. This sensor selectively detects Cu^{2+} ions without interfering with other physiological metal ions, such as Fe^{2+}, Ca^{2+}, Mg^{2+}, Zn^{2+}, Pb^{2+}, Cd^{2+}, Mn^{2+}, Co^{2+}, Hg^{2+}, and Fe^{3+}. It shows a linear range from 0.05 to 5.00 µg/mL with a 0.016 µg/mL detection limit. It determines the Cu^{2+} content in different food samples, including rice, tomato, cucumber, and lettuce, and different water samples, including river, drinking, and tap water [156]. The surface of AgNPs modified by 2,2′-thiodiacetic acid (TDA) with the aggregation feature in the presence of Cu^{2+} ions based on the interaction between Cu^{2+} and donating groups (–COO^-) of TDA is used as a colorimetric sensor for the determination of Cu^{2+} in water and urine samples. UV-Vis spectrum shows the decrease in absorbance peak at 393 nm, and a new peak appears at 570 nm with the visual color of TDA-AgNPs solution changing from yellow to brown with the addition of Cu^{2+} ions. The detection limit is about 17 nM with a linear calibration range from 0.3 to 6.0 μM. It can be used to recover Cu^{2+} ions in water and urine samples ranging between 97.5% and 101.1%, with RSD values of 0.9%–3.4% [157]. Another nanosensor with star-like tetragonal morphology of polyaniline, in which star-like nanostructure could be used as a sorbent to extract and determine Cu^{2+} along with Pb^{2+} in water samples, seafood, and agricultural products. The optimized pH for the tendency of this sorbent to effective adsorption of Cu^{2+} is found to be 7. This sensor shows a linear calibration range of 1–120 μg/L, a detection limit of 0.4 μg/L, an adsorption capacity of 84 mg/g, extraction efficiency of >96%, and an RSD of <4%. It determines the Cu^{2+} ions in different water samples, including tap, lake, river, sea, distilled water, seafood, including fish, shrimp, and crab, and agricultural foods, including potatoes, tomatoes, apples, and mushrooms [158].

5 GENERAL CONCLUSION

As described above, tightly maintaining the environmental and living systems' Cu levels is crucial for controlling the severe consequences of Cu on human health and the environment. Because the abnormalities in Cu ion levels (both Cu^+ and Cu^{2+}) could lead to several diseases, numerous research has been devoted to developing sensitive and selective Cu ions sensors. Chemo- and bio-sensors are used to overcome the certain drawbacks of conventional analytical techniques. Both chemo- and bio-sensors have a vast range of applications, including environmental monitoring, food safety, and the detection of various diseases. As described above, different sensors categorized based on their changes in color, fluorescence, luminescence, phosphorescence, and electrical signal in response to the analyte sample determined the Cu ions from a wide range of samples. Among these, fluorescent, colorimetric, and electrochemical sensors (in the case of both chemo- and bio-sensors) have demonstrated a wide range of applications for sensitive and selective detection of Cu ions from water, environment, biological, and other samples. The selective detection of a particular metal ion over other relevant metal cations from the environmental and biological systems must be complex; however, designing a molecular sensor with the appropriate material will undoubtedly help for the selective determination of Cu ions and any other metal ions.

ACKNOWLEDGMENT

This work was supported by the National Research Foundation of Korea (CRI Project No. 2018R1A3B1052702).

ABBREVIATIONS AND DEFINITIONS

AAS	atomic absorption spectrometer
Aβ	β-amyloid peptide
AD	Alzheimer's disease
AgNCs	silver nanoclusters
AgNPs	silver nanoparticles
AIE	aggregation-induced emission
ASV	anodic stripping voltammetry
AuNPS	gold nanoparticles
BODIPY	difluoro-boron-dipyrromethene
CC	carbon cloth
CDs	carbon quantum dots
COFs	covalent organic frameworks
CRET	chemiluminescence resonance energy transfer
EDTA	ethylenediamine tetraacetic acid
EIS	electrochemical impedance spectroscopy
FAAS	flame atomic absorption spectroscopy
FRET	fluorescence resonance energy transfer
GO	graphene oxide
ICT	intramolecular charge transfer
IR	infrared
K_d	dissociation constant
MOFs	metal-organic frameworks
mol/L	moles per liter
μM	micromolar
NIR	near-infrared
nM	nanomolar
OT	oxytocin
PBS	phosphate saline buffer
PET	photoinduced electron transfer
PD	Parkinson's disease
ppb	parts per billion
ppm	parts per million
QDs	quantum dots
RIfS	reflectometric interference spectroscopy
RLS	resonance light scattering
ROS	reactive oxygen species
RSD	relative standard deviation
s	reconds
SOD	ruperoxide dismutase
TFME	thin-film mercury electrodes
THF	tetrahydrofuran
T_m	melting temperature
USPEA	The United States Environmental Protection Agency
UV	ultraviolet

UV-Vis	ultraviolet-visible
WHO	World Health Organization
XPS	X-ray photoelectron spectroscopy
XRF	X-ray fluorescence

REFERENCES

1. D. G. Barceloux, *J. Toxicol. Clin. Toxicol.* **1999**, *37*, 217–230.
2. A. Luyt, J. Molefi, H. Krump, *Polym. Degrad. Stab.* **2006**, *91*, 1629–1636.
3. R. A. Festa, D. J. Thiele, *Curr. Biol.* **2011**, *21*, R877–R883.
4. A. P. Kornblatt, V. G. Nicoletti, A. Travaglia, *J. Inorg. Biochem.* **2016**, *161*, 1–8.
5. P. Trumbo, A. A. Yates, S. Schlicker, M. Poos, *J. Am. Diet. Assoc.* **2001**, *101*, 294–301.
6. A. Ramdass, V. Sathish, E. Babu, M. Velayudham, P. Thanasekaran, S. Rajagopal, *Coord. Chem. Rev.* **2017**, *343*, 278–307.
7. P. Verwilst, K. Sunwoo, J. S. Kim, *Chem. Commun.* **2015**, *51*, 5556–5571.
8. C. F. Mills, *Food Chem.* **1992**, *43*, 239–240.
9. H. Tapiero, K. D. Tew, *Biomed. Pharmacother.* **2003**, *57*, 399–411.
10. N. EFSA Panel on Dietetic Products, Allergies, *EFSA J.* **2015**, *13*, 4253.
11. C. M. Ackerman, C. J. Chang, *Int. J. Biol. Chem.* **2018**, *293*, 4628–4635.
12. C. M. Opazo, M. A. Greenough, A. I. Bush, *Front. Aging Neurosci.* **2014**, *6*, 143.
13. G. Borkow, *Curr. Chem. Biol.* **2014**, *8*, 89–102.
14. R. M. Llanos, J. F. Mercer, *DNA Cell Biol.* **2002**, *21*, 259–270.
15. J. F. Collins, J. R. Prohaska, M. D. Knutson, *Nutr. Rev.* **2010**, *68*, 133–147.
16. G. Nussey, J. Van Vuren, H. Du Preez, *Comp. Biochem. Physiol. C-Pharmacol. Toxicol. Endocrinol.* **1995**, *111*, 381–388.
17. D. M. Medeiros, *Exp. Biol. Med.* **2016**, *241*, 1316–1322.
18. J. R. Prohaska, *J. Nutr. Biochem.* **1990**, *1*, 452–461.
19. M. Saleem, K. H. Lee, *RSC Adv.* **2015**, *5*, 72150–72287.
20. B. E. Ziegler, R. A. Marta, M. B. Burt, T. B. McMahon, *Inorg. Chem.* **2014**, *53*, 2349–2351.
21. S. Lutsenko, *Biochem. Soc. Trans.* **2008**, *36*, 1233–1238.
22. M. G. Savelieff, S. Lee, Y. Liu, M. H. Lim, *ACS Chem. Biol.* **2013**, *8*, 856–865.
23. M. Bisaglia, L. Bubacco, *Biomolecules* **2020**, *10*, 195.
24. G. Xiao, Q. Fan, X. Wang, B. Zhou, *Proc. Natl. Acad. Sci. U.S.A.* **2013**, *110*, 14995–15000.
25. A. J. McDonald, J. P. Dibble, E. G. Evans, G. L. Millhauser, *J. Biol. Chem.* **2014**, *289*, 803–813.
26. L. J. Hayward, J. A. Rodriguez, J. W. Kim, A. Tiwari, J. J. Goto, D. E. Cabelli, J. S. Valentine, R. H. Brown, Jr., *J. Biol. Chem.* **2002**, *277*, 15923–15931.
27. T. S. Nielsen, N. Jessen, J. O. Jorgensen, N. Moller, S. Lund, *J. Mol. Endocrinol.* **2014**, *52*, R199–R222.
28. H. Tapiero, D. á. Townsend, K. Tew, *Biomed. Pharmacother.* **2003**, *57*, 386–398.
29. S. A. Al-Saydeh, M. H. El-Naas, S. J. Zaidi, *J. Ind. Eng. Chem.* **2017**, *56*, 35–44.
30. M. E. Letelier, S. Sanchez-Jofre, L. Peredo-Silva, J. Cortes-Troncoso, P. Aracena-Parks, *Chem-Biol Interact* **2010**, *188*, 220–227.
31. J. H. Kaplan, S. Lutsenko, *J. Biol. Chem.* **2009**, *284*, 25461–25465.
32. H. Kodama, C. Fujisawa, *Metallomics* **2009**, *1*, 42–52.
33. C. Barranguet, F. P. van den Ende, M. Rutgers, A. M. Breure, M. Greijdanus, J. J. Sinke, W. Admiraal, *Environ. Toxicol. Chem.* **2003**, *22*, 1340–1349.
34. M. Saleem, M. Rafiq, M. Hanif, M. A. Shaheen, S. Y. Seo, *J. Fluoresc.* **2018**, *28*, 97–165.

35. G. Sivaraman, M. Iniya, T. Anand, N. G. Kotla, O. Sunnapu, S. Singaravadivel, A. Gulyani, D. Chellappa, *Coord. Chem. Rev.* **2018**, *357*, 50–104.
36. D. Vlascici, I. Popa, V. A. Chiriac, G. Fagadar-Cosma, H. Popovici, E. Fagadar-Cosma, *Chem. Cent. J.* **2013**, *7*, 111.
37. A. Roy, M. Nandi, P. Roy, *Trends Anal. Chem.* **2021**, *138,* 116204.
38. S. Sharma, K. S. Ghosh, *Spectrochim. Acta A Mol. Biomol. Spectrosc.* **2021**, *254*, 119610.
39. C. Guo, A. C. Sedgwick, T. Hirao, J. L. Sessler, *Coord. Chem. Rev.* **2021**, *427*, 213560.
40. B. Kaur, N. Kaur, S. Kumar, *Coord. Chem. Rev.* **2018**, *358*, 13–69.
41. J. F. Zhang, Y. Zhou, J. Yoon, J. S. Kim, *Chem. Soc. Rev.* **2011**, *40*, 3416–3429.
42. J. A. Cotruvo Jr, A. T. Aron, K. M. Ramos-Torres, C. J. Chang, *Chem. Soc. Rev.* **2015**, *44*, 4400–4414.
43. C. Shen, E. J. New, *Metallomics* **2015**, *7*, 56–65.
44. D. Wu, A. C. Sedgwick, T. Gunnlaugsson, E. U. Akkaya, J. Yoon, T. D. James, *Chem. Soc. Rev.* **2017**, *46*, 7105–7123.
45. J. Wu, W. Liu, J. Ge, H. Zhang, P. Wang, *Chem. Soc. Rev.* **2011**, *40*, 3483–3495.
46. K. P. Carter, A. M. Young, A. E. Palmer, *Chem. Rev.* **2014**, *114*, 4564–4601.
47. L. Wu, C. Huang, B. P. Emery, A. C. Sedgwick, S. D. Bull, X.-P. He, H. Tian, J. Yoon, J. L. Sessler, T. D. James, *Chem. Soc. Rev.* **2020**, *49*, 5110–5139.
48. S. K. Sahoo, G. Crisponi, *Molecules* **2019**, *24*, 3267.
49. D. Udhayakumari, S. Naha, S. Velmathi, *Anal. Methods* **2017**, *9*, 552–578.
50. X. Chen, T. Pradhan, F. Wang, J. S. Kim, J. Yoon, *Chem. Rev.* **2012**, *112*, 1910–1956.
51. H. N. Kim, M. H. Lee, H. J. Kim, J. S. Kim, J. Yoon, *Chem. Soc. Rev.* **2008**, *37*, 1465–1472.
52. H. Xu, X. Wang, C. Zhang, Y. Wu, Z. Liu, *Inorg. Chem. Commun.* **2013**, *34*, 8–11.
53. Y. Xiang, A. Tong, P. Jin, Y. Ju, *Org. Lett.* **2006**, *8*, 2863–2866.
54. J. Yin, X. Ma, G. Wei, D. Wei, Y. Du, *Sens. Actuators B Chem.* **2013**, *177*, 213–217.
55. T. Hirayama, G. C. Van de Bittner, L. W. Gray, S. Lutsenko, C. J. Chang, *Proc. Natl. Acad. Sci. U.S.A.* **2012**, *109*, 2228–2233.
56. Y. Yang, Q. Zhao, W. Feng, F. Li, *Chem. Rev.* **2013**, *113*, 192–270.
57. M. Formica, V. Fusi, L. Giorgi, M. Micheloni, *Coord. Chem. Rev.* **2012**, *256*, 170–192.
58. A. Ramdass, V. Sathish, M. Velayudham, P. Thanasekaran, S. Umapathy, S. Rajagopal, *Sens. Actuator B Chem.* **2017**, *240*, 1216–1225.
59. F. Cheng, C. He, M. Ren, F. Wang, Y. Yang, *Spectrochim. Acta A Mol. Biomol. Spectrosc.* **2015**, *136*, 845–851.
60. L. Li, S. Shen, R. Lin, Y. Bai, H. Liu, *Chem. Commun.* **2017**, *53*, 9986–9989.
61. Y. Li, M. Chen, Y. Han, Y. Feng, Z. Zhang, B. Zhang, *Chem. Mater.* **2020**, *32*, 2532–2540.
62. S. Xie, Q. Liu, F. Zhu, M. Chen, L. Wang, Y. Xiong, Y. Zhu, Y. Zheng, X. Chen, *J. Mater. Chem. C.* **2020**, *8*, 10408–10415.
63. Z. Li, Y. Zhang, H. Xia, Y. Mu, X. Liu, *Chem. Commun.* **2016**, *52*, 6613–6616.
64. C. Cui, Q. Wang, C. Xin, Q. Liu, X. Deng, T. Liu, X. Xu, X. Zhang, *Microporous Mesoporous Mat.* 2020, *299*, 110122.
65. L. Shang, S. Dong, G. U. Nienhaus, *Nano Today* **2011**, *6*, 401–418.
66. Y. Lu, W. Chen, *Chem. Soc. Rev.* **2012**, *41*, 3594–3623.
67. I. Díez, R. H. Ras, *Nanoscale* **2011**, *3*, 1963–1970.
68. Z. Yuan, N. Cai, Y. Du, Y. He, E. S. Yeung, *Anal. Chem.* **2014**, *86*, 419–426.
69. M. Min, X. Wang, Y. Chen, L. Wang, H. Huang, J. Shi, *Sens. Actuator B Chem.* **2013**, *188*, 360–366.
70. Y.-H. Chan, J. Chen, Q. Liu, S. E. Wark, D. H. Son, J. D. Batteas, *Anal. Chem.* **2010**, *82*, 3671–3678.
71. L.-H. Jin, C.-S. Han, *Anal. Chem.* **2014**, *86*, 7209–7213.
72. I. Abdulazeez, C. Basheer, A. A. Al-Saadi, *RSC Adv.* **2018**, *8*, 39983–39991.

73. S. Chaiyo, W. Siangproh, A. Apilux, O. Chailapakul, *Anal. Chim. Acta* **2015**, *866*, 75–83.
74. Q. Gao, L. Ji, Q. Wang, K. Yin, J. Li, L. Chen, *Anal. Methods* **2017**, *9*, 5094–5100.
75. C. Wu, J. Wang, J. Shen, C. Zhang, Z. Wu, H. Zhou, *Tetrahedron* **2017**, *73*, 5715–5719.
76. Z. Guo, Q. Niu, T. Li, E. Wang, *Tetrahedron* **2019**, *75*, 3982–3992.
77. Y. Feng, Y. Yang, Y. Wang, F. Qiu, X. Song, X. Tang, G. Zhang, W. Liu, *Sens. Actuator B Chem.* **2019**, *288*, 27–37.
78. A. Mohammadi, Z. Ghasemi, *Spectrochim. Acta A Mol. Biomol. Spectrosc.* **2020**, *228*, 117730.
79. H. Ozay, Z. Gungor, B. Yilmaz, P. Ilgin, O. Ozay, *J. Hazard. Mater.* **2020**, *389*, 121848.
80. R. Gui, H. Guo, H. Jin, *Nanoscale Adv.* **2019**, *1*, 3325–3363.
81. L. Yang, N. Huang, L. Huang, M. Liu, H. Li, Y. Zhang, S. Yao, *Anal. Methods* **2017**, *9*, 618–624.
82. C. J. Mei, N. A. Yusof, S. A. Alang Ahmad, *Chemosensors* **2021**, *9*, 157.
83. C. Foo, H. Lim, A. Pandikumar, N. Huang, Y. Ng, *J. Hazard. Mater.* **2016**, *304*, 400–408.
84. Y. Oztekin, A. Ramanaviciene, A. Ramanavicius, *Sens. Actuator B Chem.* **2011**, *155*, 612–617.
85. M. A. Deshmukh, R. Celiesiute, A. Ramanaviciene, M. D. Shirsat, A. Ramanavicius, *Electrochim Acta* **2018**, *259*, 930–938.
86. L. Cui, J. Wu, J. Li, Y. Ge, H. Ju, *Biosens. Bioelectron.* **2014**, *55*, 272–277.
87. T. Wu, T. Xu, Z. Ma, *Analyst.* **2015**, *140*, 8041–8047.
88. M. Zhou, L. Han, D. Deng, Z. Zhang, H. He, L. Zhang, L. Luo, *Sens. Actuator B-Chem.* **2019**, *291*, 164–169.
89. Y. Wei, C. Gao, F.-L. Meng, H.-H. Li, L. Wang, J.-H. Liu, X.-J. Huang, *J. Phys. Chem. C.* **2012**, *116*, 1034–1041.
90. A. Jain, V. Gupta, L. Singh, J. Raisoni, *Talanta* **2005**, *66*, 1355–1361.
91. M. Clerc, F. Heinemann, B. Spingler, G. Gasser, *Inorg. Chem.* **2019**, *59*, 669–677.
92. Y. You, Y. Han, Y.-M. Lee, S. Y. Park, W. Nam, S. J. Lippard, *J. Am. Chem. Soc.* **2011**, *133*, 11488–11491.
93. A. P. Turner, *Chem. Soc. Rev.* **2013**, *42*, 3184–3196.
94. S. M. Borisov, O. S. Wolfbeis, *Chem. Rev.* **2008**, *108*, 423–461.
95. L. Su, W. Jia, C. Hou, Y. Lei, *Biosens. Bioelectron.* **2011**, *26*, 1788–1799.
96. W. Zhou, R. Saran, J. Liu, *Chem. Rev.* **2017**, *117*, 8272–8325.
97. F. Ma, Y. Li, B. Tang, C.-y. Zhang, *Acc. Chem. Res.* **2016**, *49*, 1722–1730.
98. S. B. VanEngelenburg, A. E. Palmer, *Curr. Opin. Chem. Biol.* **2008**, *12*, 60–65.
99. H. Dong, W. Gao, F. Yan, H. Ji, H. Ju, *Anal. Chem.* **2010**, *82*, 5511–5517.
100. L. Ding, B. Xu, T. Li, J. Huang, W. Bai, *Sensors* **2018**, *18*, 2605.
101. P.-J. J. Huang, J. Liu, *Anal. Chem.* **2016**, *88*, 3341–3347.
102. D.-Q. Feng, G. Liu, W. Wang, *J. Mat. Chem. B* **2015**, *3*, 2083–2088.
103. B.-E. Kim, T. Nevitt, D. J. Thiele, *Nat. Chem. Biol.* **2008**, *4*, 176–185.
104. M. C. Heffern, H. M. Park, H. Y. Au-Yeung, G. C. Van de Bittner, C. M. Ackerman, A. Stahl, C. J. Chang, *Proc. Natl. Acad. Sci. U.S.A.* **2016**, *113*, 14219–14224.
105. S. V. Wegner, H. Arslan, M. Sunbul, J. Yin, C. He, *J. Am. Chem. Soc.* **2010**, *132*, 2567–2569.
106. N. L. Rosi, C. A. Mirkin, *Chem. Rev.* **2005**, *105*, 1547–1562.
107. Y. Zhou, S. Wang, K. Zhang, X. Jiang, *Angew. Chem. Int. Ed.* **2008**, *47*, 7454–7456.
108. X. Xu, W. L. Daniel, W. Wei, C. A. Mirkin, *Small* **2010**, *6*, 623–626.
109. Q. Shen, W. Li, S. Tang, Y. Hu, Z. Nie, Y. Huang, S. Yao, *Biosens. Bioelectron.* **2013**, *41*, 663–668.
110. D. Kato, K. Goto, S.-i. Fujii, A. Takatsu, S. Hirono, O. Niwa, *Anal. Chem.* **2011**, *83*, 7595–7599.

111. Y. Yu, L. Zhang, C. Li, X. Sun, D. Tang, G. Shi, *Angew. Chem. Int. Ed.* **2014**, *126*, 13046–13049.
112. H. Pei, N. Lu, Y. Wen, S. Song, Y. Liu, H. Yan, C. Fan, *Adv. Mater.* **2010**, *22*, 4754–4758.
113. I.-H. Cho, D. H. Kim, S. Park, *Biomater. Res.* **2020**, *24*, 1–12.
114. N. J. Ronkainen, H. B. Halsall, W. R. Heineman, *Chem. Soc. Rev.* **2010**, *39*, 1747–1763.
115. W. Pooi See, S. Nathan, L. Yook Heng, *J. Sens.* **2011**, *2011*, 230535.
116. Y. Luo, L. Zhang, W. Liu, Y. Yu, Y. Tian, *Angew. Chem. Int. Ed.* **2015**, *54*, 14053–14056.
117. T.-Y. Lee, Y.-B. Shim, S. C. Shin, *Synth. Met.* **2002**, *126*, 105–110.
118. M. Lin, M. Cho, W. Choe, Y. Lee, *Biosens. Bioelectron.* **2009**, *25*, 28–33.
119. K. A. Lukyanenko, I. A. Denisov, V. V. Sorokin, A. S. Yakimov, E. N. Esimbekova, P. I. Belobrov, *Chemosensors* **2019**, *7*, 16.
120. M. Mohseni, J. Abbaszadeh, S. S. Maghool, M. J. Chaichi, *Ecotox. Environ. Safe.* **2018**, *148*, 555–560.
121. C. Pal, K. Asiani, S. Arya, C. Rensing, D. J. Stekel, D. J. Larsson, J. L. Hobman, *Adv. Microb. Physiol.* **2017**, *70*, 261–313.
122. A. Costello, J. Parker, M. Clynes, R. Murphy, *Metallomics* **2020**, *12*, 1729–1734.
123. E. B. Bahadır, M. K. Sezgintürk, *Artif. Cell. Nanomed. Biotechnol.* **2016**, *44*, 248–262.
124. K. K. Tadi, I. Alshanski, E. Mervinetsky, G. Marx, P. Petrou, K. M. Dimitrios, C. Gilon, M. Hurevich, S. Yitzchaik, *ACS Omega* **2017**, *2*, 8770–8778.
125. X. Chai, X. Zhou, A. Zhu, L. Zhang, Y. Qin, G. Shi, Y. Tian, *Angew. Chem. Int. Ed.* **2013**, *52*, 8129–8133.
126. L. Zhang, Y. Han, F. Zhao, G. Shi, Y. Tian, *Anal. Chem.* **2015**, *87*, 2931–2936.
127. C. Wang, X. Bi, M. Wang, X. Zhao, Y. Lin, *Anal. Chem.* **2019**, *91*, 16010–16016.
128. Y. Lin, C. Wang, L. Li, H. Wang, K. Liu, K. Wang, B. Li, *ACS Appl. Mater. Interfaces* **2015**, *7*, 27262–27270.
129. T. Ebaston, G. Balamurugan, S. Velmathi, *Anal. Methods* **2016**, *8*, 6909–6915.
130. S. J. Ranee, G. Sivaraman, A. M. Pushpalatha, S. Muthusubramanian, *Sens. Actuator B Chem.* **2018**, *255*, 630–637.
131. L. Hou, X. Kong, Y. Wang, J. Chao, C. Li, C. Dong, Y. Wang, S. Shuang, *J. Mat. Chem. B* **2017**, *5*, 8957–8966.
132. Y. Xiong, L. Zhou, X. Peng, H. Li, H. Wang, L. He, P. Huang, *Sens. Actuator B Chem.* **2020**, *320*, 128411.
133. F. Hane, G. Tran, S. J. Attwood, Z. Leonenko, *PLoS One* **2013**, *8*, e59005.
134. Y. Yu, P. Wang, X. Zhu, Q. Peng, Y. Zhou, T. Yin, Y. Liang, X. Yin, *Analyst* **2018**, *143*, 323–331.
135. C. Satriano, G. T. Sfrazzetto, M. E. Amato, F. P. Ballistreri, A. Copani, M. L. Giuffrida, G. Grasso, A. Pappalardo, E. Rizzarelli, G. A. Tomaselli, *Chem. Commun.* **2013**, *49*, 5565–5567.
136. D. J. Fitzgerald, *Am. J. Clin. Nutr.* **1998**, *67*, 1098S–1102S.
137. R. Wang, W. Wang, H. Ren, J. Chae, *Biosens. Bioelectron.* **2014**, *57*, 179–185.
138. R. Zhu, G. Zhou, F. Tang, C. Tong, Y. Wang, J. Wang, *Int. J. Anal. Chem.* **2017**, *2017*, 1727126.
139. M. Atapour, G. Amoabediny, M. Ahmadzadeh-Raji, *RSC Adv.* **2019**, *9*, 8882–8893
140. S. Zhou, H. He, W. Guo, H. Zhu, F. Xue, M. Cheng, J. Lin, L. Wang, S. Wang, *Carbohydr. Polym.* **2019**, *219*, 95–104.
141. Y. Cao, Y. Liu, F. Li, S. Guo, Y. Shui, H. Xue, L. Wang, *Microchem. J.* **2019**, *150*, 104176.
142. G. Vyas, S. Bhatt, M. K. Si, S. Jindani, E. Suresh, B. Ganguly, P. Paul, *Spectrochim. Acta A Mol. Biomol. Spectrosc.* **2020**, *230*, 118052.
143. Y.-r. Ma, H.-y. Niu, Y.-q. Cai, *Chem. Commun.* **2011**, *47*, 12643–12645.
144. M. H. Mahnashi, A. M. Mahmoud, S. A. Alkahtani, R. Ali, M. M. El-Wekil, *Spectrochim. Acta A Mol. Biomol. Spectrosc.* **2020**, *228*, 117846.
145. H. Cao, W. Shi, J. Xie, Y. Huang, *Anal. Methods* **2011**, *3*, 2102–2107.

146. H. Elmizadeh, M. Soleimani, F. Faridbod, G. R. Bardajee, *J. Fluoresc.* **2017**, *27*, 2323–2333.
147. P.-H. Chen, C. Lin, K.-H. Guo, Y.-C. Yeh, *RSC Adv.* **2017**, *7*, 29302–29305.
148. C. W. Quinn, D. M. Cate, D. D. Miller-Lionberg, T. Reilly III, J. Volckens, C. S. Henry, *Environ. Sci. Technol.* **2018**, *52*, 3567–3573.
149. W. Lv, M. Lin, R. Li, Q. Zhang, H. Liu, J. Wang, C. Huang, *Chin. Chem. Lett.* **2019**, *30*, 1410–1414.
150. D. Zhang, Z. Wang, J. Yang, L. Yi, L. Liao, X. Xiao, *Biosens. Bioelectron.* **2021**, *182*, 113174.
151. S. Kaur, C. S. Law, N. H. Williamson, I. Kempson, A. Popat, T. Kumeria, A. Santos, *Anal. Chem.* **2019**, *91*, 5011–5020.
152. A. Shokrollahi, M. Ghaedi, O. Hossaini, N. Khanjari, M. Soylak, *J. Hazard. Mater.* **2008**, *160*, 435–440.
153. J. Kudr, H. V. Nguyen, J. Gumulec, L. Nejdl, I. Blazkova, B. Ruttkay-Nedecky, D. Hynek, J. Kynicky, V. Adam, R. Kizek, *Sensors* **2015**, *15*, 592–610.
154. R. M. Takeuchi, A. L. Santos, M. J. Medeiros, N. R. Stradiotto, *Microchim. Acta* **2009**, *164*, 101–106.
155. S. Abbasi, A. Bahiraei, F. Abbasai, *Food Chem.* **2011**, *129*, 1274–1280.
156. M. Mohammadnejad, M. Shiri, M. Heydari, Z. Faghihi, L. Afshinpoor, *ChemistrySelect* **2020**, *5*, 13690–13693.
157. M. Rasul Fallahi, G. Khayatian, *Curr. Anal. Chem.* **2017**, *13*, 167–173.
158. M. Behbahani, Y. Bide, M. Salarian, M. Niknezhad, S. Bagheri, A. Bagheri, M. R. Nabid, *Food Chem.* 2014, *158*, 14–19.

7 Molecular Design for Cadmium-Specific Fluorescent Sensors

Yuji Mikata
Laboratory for Molecular & Functional Design, Department of Engineering, Nara Women's University, Nara 630-8506, Japan

CONTENTS

ABSTRACT

Cadmium is a toxic metal for living systems. Detection, quantification, and imaging of cadmium are of central importance in life sciences or toxicology. This chapter updates current knowledge with regard to plausible molecular mechanisms of cadmium toxicity, followed by an overview of recent progress of cadmium-specific fluorescent sensors with respect to signal-switching machinery and metal ion selectivity. Sensors containing fluorophore(s), including coumarin, fluorescein/rhodamine, BODIPY, quinoline, and others, are also described. Small changes in probe structure reverse fluorescence

DOI: 10.1201/9781003229971-7

selectivity between zinc and cadmium, both of which are group 12 elements and difficult to distinguish, and metal ion discrimination mechanisms in Zn/Cd dual fluorescent probes are emphasized. Based on the molecular structure of reported sensors, a molecular design strategy for cadmium-specific fluorescent probes is discussed.

KEYWORDS

Cadmium; Sensor; Probe; Toxicity; Zinc; Fluorescence; Quinoline; Molecular Design

1 INTRODUCTION

Cadmium is a useful element that has been utilized in the industry in pigments, batteries, metal plating, and many additional applications. Extensive use of cadmium in human activities leads to the ubiquity of cadmium in our environment. Usually, cadmium enters the body through food, house dust, and cigarettes. Vegetable and agricultural products account for 70%–90% of cadmium exposure in humans, which can accumulate at a rate of 10–30 μg/day [1]. Cadmium exhibits high toxicity and ranks seventh in the 2019 Substance Priority List in the Agency for Toxic Substances and Disease Registry (ATSDR) [2]. The 'Itai-itai' diseases is one of the most severe examples of cadmium-induced disruption in human health. Cadmium accumulates in the kidneys, which induces tubular dysfunction and decreases calcium absorption, which results in serious damage to the bone accompanied by extreme pain in the whole body ('itai' means pain in Japanese; patients repeatedly screamed this word) [3]. Other cadmium-accumulating organs include the lungs, liver, and pancreas. Most cadmium absorbed in the human body binds to thiol groups in metallothionein (MT) proteins and is removed from bloodstream [4, 5]. MT expression is upregulated by cadmium exposure, and this reduces the toxicity of cadmium. Because our body has no extraction pathway for cadmium, the protein-bound cadmium stays in the kidneys, liver, pancreas, and lungs [6]. Cadmium exhibits an extremely long clearance half-life (10–30 years) from the human body [7], which is caused in part by MT overexpression.

1.1 Toxicity of Cadmium

Although cadmium is stable in its +2 oxidation state and redox inert, it induces oxidative stress indirectly. Binding of cadmium to reductive thiols, such as glutathione (GSH), results in disruption of the antioxidant system, and therefore, accumulation of reactive oxygen species (ROS) [8]. Cadmium induces severe hepatotoxicity because GSH is abundant in the liver. In addition, cadmium-induced inhibition of superoxide dismutase, catalase, glutathione peroxidase, glutathione reductase, and glutathione-*S*-transferase leads to hydroxy radical formation [9]. Another mechanism of cadmium-induced ROS formation is a displacement of protein-bound iron and copper by cadmium [6]. Liberated free iron and copper ions generate ROS from hydrogen peroxide using Fenton reactions. Antioxidants, including vitamin C and/or E, reduce the ROS-related damage in testes. Melatonin is also effective in the protection of testes from lipid peroxidation induced by cadmium.

Cadmium is also carcinogenic. Deoxyribonucleic acid (DNA) lesions induced by ultraviolet light, chemicals, and cellular ROS are treated using DNA repair machinery, including base excision repair, nucleotide excision repair, and mismatch repair. All such DNA repair systems are disrupted by cadmium [10]. Moreover, the damaged and un-repaired DNA can be amplified because cadmium activates zinc-dependent transcription factors [11]. Proliferation of cells containing DNA damage promotes tumor growth. Cadmium also induces protein phosphorylation, which disrupts E-cadherin-mediated cell-cell adhesion, blocking the signal transduction pathways [12]. These processes further generate tumor tissues. Cadmium also induces apoptosis using mitochondria-dependent pathways.

1.2 Detection of Cadmium

Because of its high toxicity, cadmium in living systems and the environment needs to be detected, quantified, and possibly remediated [13]. Due to recent human activities, cadmium now exists in multiple environments, including air, water, and soil. Several spectroscopic techniques for detection of cadmium, including atomic absorption spectroscopy, atomic fluorescence spectroscopy, inductively coupled plasma mass spectroscopy, are currently available. However, these methods require large and expensive facilities, well-trained operators, and difficult sample preparation, including time-consuming pre-treatment, etc. The on-site and real-time analysis are not applicable to these methods.

Ion chromatography is developed for the detection of heavy metal ions [14]. Ion exchange, ion interaction, and chelation ion exchange using oxalate or dipicolinate are used for the separation of metal ions. The post-column reaction with organic dyes, such as 4-(2-pyridylazo)resorcinol (PAR) or 2-(5-bromo-2-pyridylazo)-5-diethylaminophenol (5-Br-PADAP), can be used to visualize metal ions using 500–570 nm optical bands.

One of the most convenient detection methods for heavy metals, including cadmium, is the electrochemical method utilizing double-stranded DNA, carbon nanotubes, graphene, and carbon nanoparticles [9, 15]. Electrochemical bio-sensors based on enzyme inhibition or whole cell microbial bio-sensors offer portable and on-site analysis. Cadmium response regulator *cadR*, for example, upregulates the expression of reporter genes, such as *gfp* and *lacZ*, which exhibit green fluorescence and pH changes in output signals, respectively [16]. Logic gate strategies utilizing regulatory genes allow dual- or higher order sensing of analytes. Limitations for these analyses that deal with transgenic cells include their sensitivity to cadmium toxicity and accidental gene transfer/release to the environment.

Finally, fluorometric and colorimetric detections of cadmium using appropriate small molecule probes are both convenient and reliable. This methodology offers sensitive, non-invasive, and rapid and real-time analysis of concentration change images of cadmium in living cells and even whole animals. Chemical modification of the probe molecules affords fine tuning of sensing conditions, such as working concentration range, pH profile, cadmium selectivity, and other necessary properties. This chapter hereafter focuses on the molecular design of small molecule fluorescent probes for cadmium. Several review articles dealing with fluorescent and colorimetric sensing of cadmium and other toxic heavy metal ions have been extensively published in this decade [17–23].

2 MOLECULAR DESIGN FOR FLUORESCENT CADMIUM SENSORS

Development of fluorescence switching mechanisms with specificity for cadmium is a prerequisite for the molecular design of cadmium-specific fluorescent sensors. Fluorescence enhancement (turn-on), fluorescence quenching (turn-off), and fluorescence color change (ratiometric) are typical fluorescence readouts. From the viewpoint of sensitivity, turn-on and ratiometric responses are preferred over turn-off mechanisms, although many quenching sensors have been developed. For application in biological samples, water solubility and cell-membrane permeability of probes also need to be considered.

2.1 Mechanism of Fluorescence Enhancement and Spectral Changes

There are several established methods to trigger fluorescence changes by analyte binding. The classical and most versatile mechanism is the inhibition of photoinduced electron transfer (PET) [24–26]. In PET probes, the fluorescence of the fluorophore is quenched in its free form because an electron in a photoexcited state migrates from the HOMO of a donor moiety (usually an aliphatic nitrogen atom) into a partially filled orbital in the excited-state fluorophore, which prevents radiative decay of the fluorophore to the ground state. As a result, the energy of photoexcited electron is lost by non-radiative decay. Upon binding with target molecules or a metal ion, the HOMO level of the receptor is lowered below the HOMO level of the fluorophore, which prevents PET quenching.

Chelation-enhanced fluorescence (CHEF) is another basic switching mechanism [27]. Quinoline fluorescence is best described by this process. The transition from the most stable excited state of type $n\pi^*$ to the ground state of quinoline fluorophore accounts for the nonfluorescent nature of the free probe. Metal or proton binding to the lone pair of the ring nitrogen atom of the quinoline raises the energy level of $n\pi^*$-type excited state and leads to population of a $\pi\pi^*$-type excited state that exhibits enhanced radiative decay upon excitation of the chromophore. The combined influence of CHEF and PET mechanisms is frequently used in actual probe design.

Internal charge transfer (ICT) and photoinduced charge transfer are effective mechanisms to construct ratiometric probes [26]. Ratiometric responses are preferred because they facilitate quantitative analysis. The dipole moment of the probe molecule based on the ICT at the ground/excited state precisely reflects the microenvironment of the fluorophore. Large perturbations of electron distributions can be achieved upon metal binding to the donor-acceptor coupled probes. Metal binding at the donor site induces expansion of the HOMO-LUMO energy gap for photoexcitation, leading to blue-shifted absorbance and fluorescence, therefore preventing ICT. Alternatively, metal binding to the acceptor site can reduce excitation energy to afford a red shift of absorption/emission maxima. Fluorescence enhancement of metal-bound probe may be accompanied by the inhibition of partial PET quenching in the free probe.

Conformational changes upon metal ion binding can induce or inhibit intramolecular excimer formation. Bis(pyrene) probe **1** (Figure 1) exhibits only monomer emission in its free form, but cadmium complex forms an intramolecular excimer that emits at the long-wavelength region [28]. In contrast, the excimer emission of free probe **2** is disrupted by addition of cadmium [29]. The ratiometric analysis based on monomer/excimer emission maxima is often achieved.

Another interaction of two fluorophores is the Förster (or fluorescence) resonance energy transfer (FRET) mechanism [30, 31]. Two different fluorophores designated as donor and acceptor are connected in proximity (<35 Å) in FRET sensors. Upon donor excitation, the energy transfer from donor to acceptor induces emission from the acceptor fluorophore. The metal ion-induced bond cleavage of tether chain between donor and acceptor shuts down the FRET and donor emission is observed at short wavelengths. In contrast, a well-separated donor-acceptor pair can be forced to interact with FRET contact by binding of metal ion, where donor fluorescence shifts to longer wavelengths. In either case, an efficient ratiometric response is constructed.

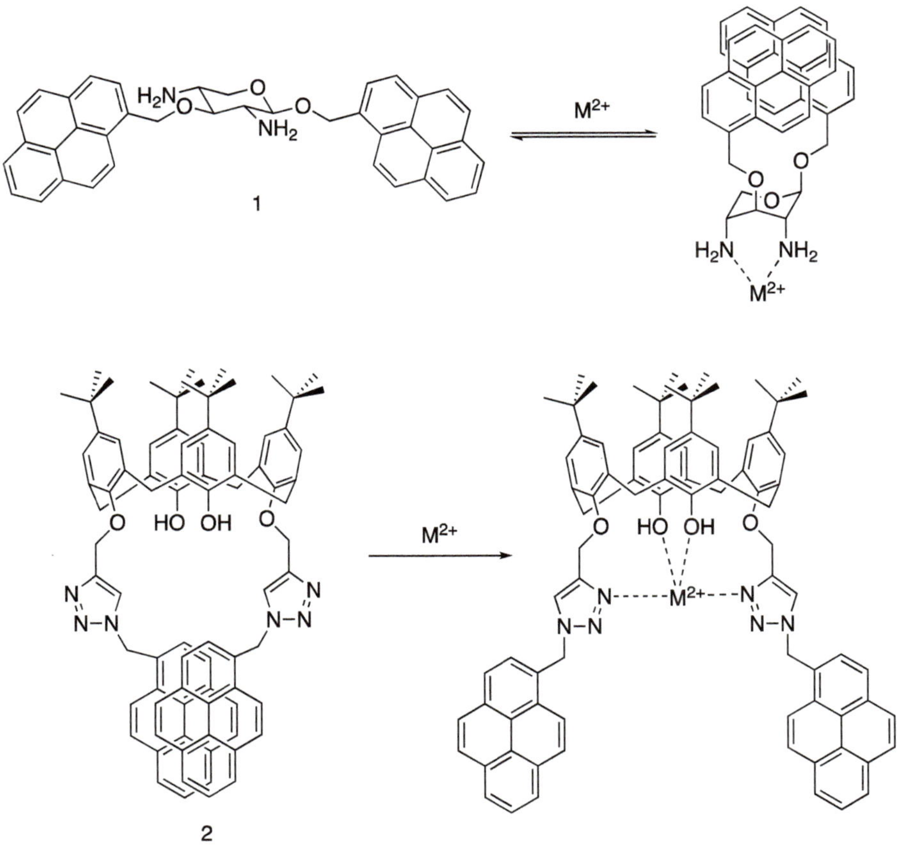

2

FIGURE 1 EXCIMER-BASED Cd^{2+} sensors.

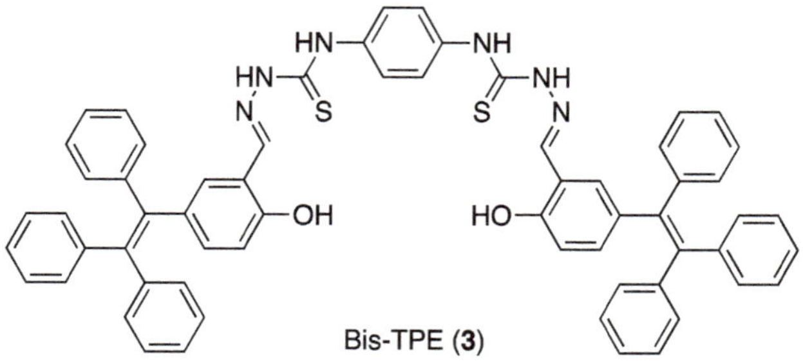

FIGURE 2 AIE-BASED Cd^{2+} sensors.

Aggregation-induced emission (AIE) has been utilized to detect Cd^{2+} by employing tetraphenylethylene (TPE) and glutathione-stabilized gold nanoclusters (GSH-AuNCs) as functional devices [29]. Fluorescence quenching was observed by the addition of Cu^{2+} to GSH-AuNCs, but Cd^{2+} induced aggregation of Cu^{2+}-GSH-AuNCs, exhibiting orange fluorescence. Combined with blue-emitting ethylenediamine-functionalized graphene oxide, the fluorescent colorimetric/ratiometric system was applied to the paper strips and rice samples. The bis-TPE sensor **3** (Figure 2) detects Zn^{2+}, Cd^{2+}, and Hg^{2+} simultaneously using AIE in H_2O-THF (90:10) solution, with a slightly sharp and intense emission band at the long-wavelength region for Hg^{2+} [32]. Individual metal ion detection can be achieved by masking of Hg^{2+} with Cl^- and Zn^{2+} with adenosine triphosphate (ATP).

Reaction-based fluorescence changes are an effective, sensitive, and selective recognition method [33]. Spirolactam ring opening reaction generating rhodamine is often utilized [34]. A hydrazide compound was reacted with an activated carboxylate moiety of rhodamine to construct a sensor molecule. The non-fluorescent spirolactam precursor undergoes a ring-opening reaction upon binding with a specific metal ion, generating fluorescent rhodamine. FRET sensors can be designed by introduction of appropriate fluorophore in the hydrazide moiety [35–37].

2.2 Cadmium Specificity

Another prerequisite for fluorescent cadmium sensors is a strict specificity of output signals to the target metal ion. Discrimination between cadmium and zinc is highly desired because these two metals are in the same group of the periodic table. The primary difference between cadmium and zinc is their ionic radii, in which only 21 pm is different.

Despite the small difference in ionic radii between Cd^{2+} and Zn^{2+}, using quinoline as a metal-binding ligand is an effective selectivity strategy. In most cases, quinoline functions as a chromophore through both CHEF and PET mechanisms. The reduced nitrogen basicity and steric hindrance due to the peri hydrogen atom at the 8-position forces quinoline complexes to exhibit elongated metal-nitrogen distance in comparison to pyridine. The long coordination distance of quinoline is preferable for cadmium over zinc binding. Several examples of quinoline-based fluorescent cadmium sensors are discussed in Section 3.4.

The cadmium softness can also be advantageous for ligand design. According to the hard and soft acids and bases theory [38], the introduction of sulfur atoms as binding sites for metal ions will impact preferable for selective detection of soft metal ions, such as Cd^{2+} and Hg^{2+}. Anthraquinone macrocyclic compounds **4a–c** (Figure 3) exhibit different metal ion selectivity for their fluorescence enhancement in acetonitrile, depending on the number of sulfur atoms in the receptor site [39]. The Pb^{2+}-selective fluorescent response for compound **4a**, which has all oxygen in its macrocyclic binding site, was changed to Cd^{2+}-selective response for **4b** and Hg^{2+}-selective response for **4c** by replacing two or three oxygen atoms with sulfur atoms.

A new approach for designing Cd^{2+}-specific fluorescent sensors by focusing on the coordination number of the metal center was recently demonstrated. Larger cadmium ions require a higher number of coordinating atoms. Most of the cadmium complexes in the Cambridge crystallographic database centre exhibit 6-, 7-, or 8-coordinated structures, whereas zinc complexes are found as 4-, 5-, and 6-coordinate complexes. In this context, polydentate ligands that possess 7- or 8-coordinating atoms are effective for accommodating cadmium ions. Thus, the heptadentate ligand TQOPEN (*N,N,N′,N′*-tetrakis(2-quinolylmethyl)-3-oxa-1,5-pentanediamine, **5a**, Figure 3) exhibits high Cd^{2+}-specificity [40]. The thia- and aza-derivatives TQSPEN (**5b**) and TQNPEN (**5c**) also exhibit consistent Cd^{2+}-specificity in their fluorescence enhancement with slightly enhanced Cd^{2+} selectivity and metal-binding affinity compared to TQOPEN. Carbohydrate-appended TQNPEN derivatives are utilized for intracellular Cd^{2+} detection in HeLa cells [41].

The above approaches can be understood whether they manipulate the metal-binding affinity of the ligand (K_d) or the quantum yield of the metal complex (ϕ). The difference in the structure of the metal complexes affects both parameters. Construction of a sensor molecule that exhibits unique structures for cadmium complex should afford a unique fluorescence response for cadmium ion. Generalization of previous knowledge to create a new strategy is the goal of this chapter.

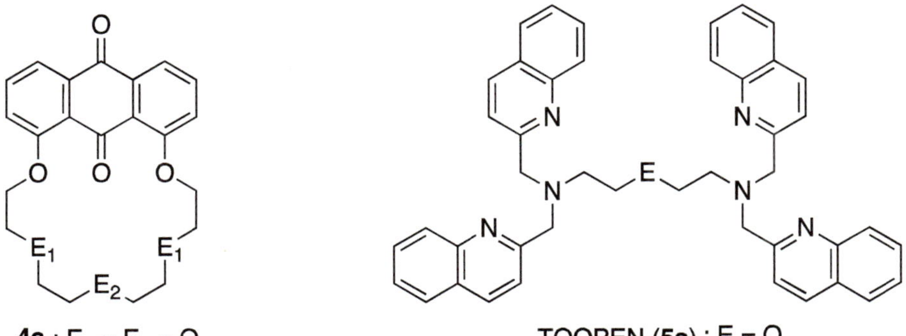

4a : $E_1 = E_2 = O$
4b : $E_1 = S; E_2 = O$
4c : $E_1 = E_2 = S$

TQOPEN (**5a**) : E = O
TQSPEN (**5b**) : E = S
TQNPEN (**5c**) : E = NH

FIGURE 3 MACROCYCLIC and heptadentate Cd^{2+} sensors containing sulfur derivatives.

3 CADMIUM-SPECIFIC FLUORESCENT SENSORS

The previous section overviewed cadmium-specific fluorescent sensors in terms of the mechanism of fluorescence switching and cadmium specificity. This section gives more examples of previous achievements that are categorized based on the structure of the fluorophore. In addition to the concept of molecular design, some mechanistic insights are also discussed.

3.1 Coumarin Derivatives

The coumarin derivative CadMQ (**6**) ratiometrically visualizes the intracellular Cd^{2+} (Figure 4) [42]. The ICT structure of **6** in HEPES buffer is destabilized upon metal ion coordination owing to the nitrogen atom at the 7-position, leading to a blue shift of the absorbance spectra and thereby changing the excitation spectra of the probe molecule. The ^{1}H NMR analysis reveals the interaction of the 7-amino group with Cd^{2+}, but no binding with Zn^{2+} is detected. The fluorescence intensity ratio (340 nm/387 nm) in Cd^{2+}-incorporated HeLa cell changes immediately after the addition of TPEN (*N,N,N′,N′*-tetrakis(2-pyridylmethyl)ethylenediamine), confirmed by the intracellular Cd^{2+} imaging. High quantum yield ($\phi_{Cd}=0.70$) and high binding affinity ($K_d=0.16$ nM) were reported for this sensor.

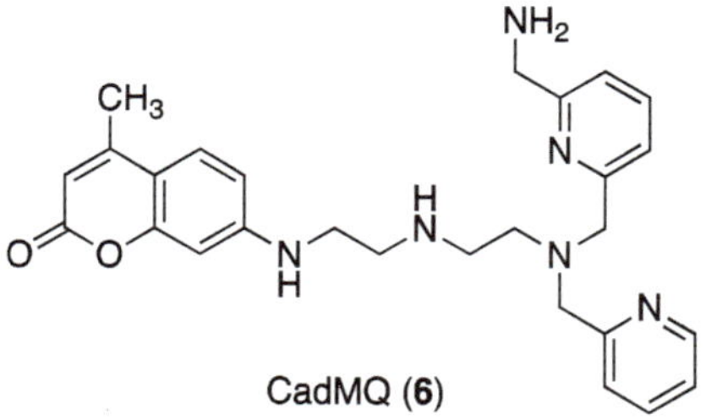

CadMQ (**6**)

FIGURE 4 COUMARIN-BASED Cd^{2+} sensor.

3.2 Fluorescein/Rhodamine Derivatives

Fluorescein is a highly fluorescent and low-cost fluorophore that has been utilized in many applications. Water-solubility of fluorescein and fluorescence switching by the formation of its poorly fluorescent spirolactone form make it useful as a fluorescent sensor platform. Restriction of C=N bond rotation in fluorescein-thiosemicarbazide conjugate **7** (Figure 5) induced by cation binding was applied to Cd^{2+} sensing in HEPES buffer [43]. The acyclic C=N bond in free **7** significantly reduces the fluorescence intensity, in which a PET mechanism may also contribute to the response. Upon chelation with Cd^{2+}, the rotation of acyclic C=N bond in the complex is inhibited, which enhances the fluorescence. Other open-shell transition metal ions do not induce fluorescence through the electron or energy transfer between chromophore and metal ions. An equimolar amount of such metal ions, including Fe^{2+}, Co^{2+}, Ni^{2+}, Cu^{2+}, Hg^{2+}, and Zn^{2+}, inhibit the formation of Cd^{2+} complex. The intracellular Cd^{2+} in HK-2 cells was successfully imaged with the use of confocal microscopy.

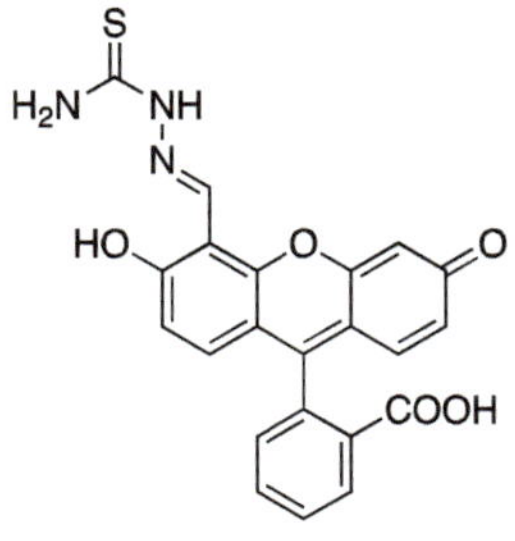

7

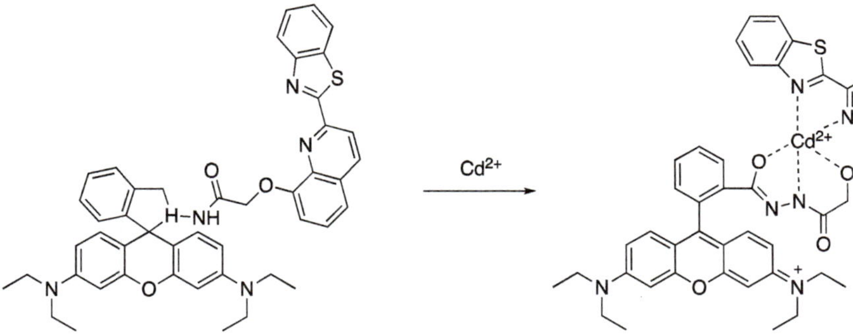

RQBT (**8**)

FIGURE 5 FLUORESCEIN/RHODAMINE-BASED Cd^{2+} sensors.

The rhodamine-based FRET sensor RQBT (**8**, Figure 5) possesses quinoline-conjugated benzothiazole as a donor fluorophore [44]. In its free form, **8** emits fluorescence around 470 nm mainly from donor components in 80% aqueous methanol. Upon addition of Cd^{2+}, a spirolactam ring-opening reaction induces the formation of a new absorption band around 565 nm, which allows pink colorimetric sensing, and a strong emission attributed to rhodamine using the FRET process around 585 nm ($\phi_{Cd}=0.37$). The structures of free **8** and its Cd^{2+} complex were characterized by X-ray crystallography. Since the metal binding cavity comprised of benzothiazole, 8-alkoxyquinoline, and hydrazide moieties after ring-opening of **8** is suitable for the ionic radius of Cd^{2+} (association constant $K_a = \sim 10^5$ M), Zn^{2+} and Hg^{2+} do not significantly inhibit Cd^{2+} sensing.

3.3 BODIPY Derivatives

The ICT-based fluorescent Cd^{2+} sensor **9** (Figure 6) was developed using a boron-dipyrromethane (BODIPY) scaffold [45]. BODIPY is a stable small molecule that has high fluorescence quantum yield upon excitation with visible light. A dipicolylamine (DPA) unit was synthetically added to a BODIPY-type backbone and was shown to induce shorter wavelength absorption upon acting as a metal binding and ICT donor moiety. In acetone–Tris-HCl buffer solution (9:1), the emission band at 656 nm for free **9** ($\phi=0.12$) shifts to 597 nm ($\phi=0.59$) upon Cd^{2+} binding by

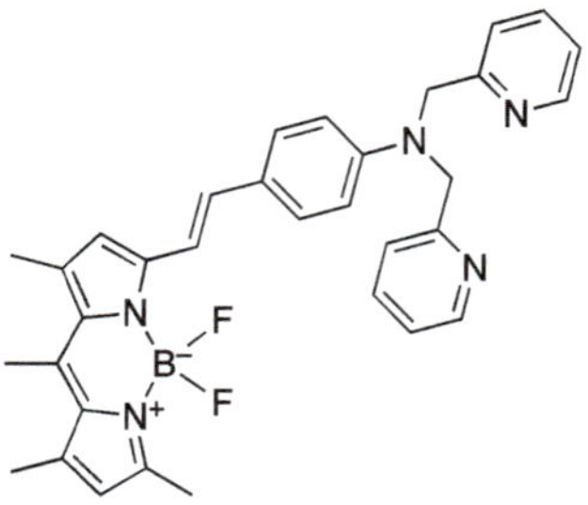

FIGURE 6 BODIPY-BASED Cd^{2+} sensor.

inhibition of ICT process. This blue-shifting and emission enhancement is suitable for normal intensity-based fluorescence responses and ratiometric sensing of Cd^{2+}. The dissociation constant with Cd^{2+} is around $5–7\times10^{-5}$ M; however, no detectable interaction with Zn^{2+} was observed. This probe was applied to PC12 and DC cells for intracellular Cd^{2+} imaging in general fluorescence and ratiometric fluorescence microscopic manners.

3.4 Quinoline Derivatives

Quinoline derivatives having 8-hydroxy, 8-alkoxy, and appropriate Cd^{2+}-binding motifs are extensively utilized as fluorescent Cd^{2+} sensors. For example, 8-methoxyquinoline derivatives bearing a DPA moiety in its 2-position create a N4O1 binding pocket suitable for Cd^{2+} coordination over Zn^{2+}. Introduction of electron-donating substituent in 6- or 7-position also affords ICT characteristics. Compound **10** (Figure 7) having an acetamide group in the 8-position exhibits fluorescence enhancement at 422 nm upon binding with Cd^{2+} by restriction of PET quenching ($\phi_{Cd}=0.28$), whereas Zn^{2+} induces ICT emission ($\lambda_{em}=476$ nm) by deprotonation of amide group in Tris-HCl buffer solution containing 20% dimethyl sulfoxide [46]. The binding affinity with Cd^{2+} is quite high ($K_d=0.25$ pM). The weak emission intensity of Zn^{2+} complex of **10** ($\phi_{Zn}=0.019$) is improved in **11** ($\phi_{Zn}=0.23$; $\lambda_{em}=458$ nm), while maintaining the Cd^{2+} response ($\phi_{Cd}=0.22$; $\lambda_{em}=425$ nm) in Tris-HCl buffer containing 50% acetonitrile [47]. Based on the higher binding affinity of **11** for Cd^{2+} over Zn^{2+}, a ratiometric displacement system is proposed for **11**-Zn^{2+} complex for Cd^{2+} detection. In contrast to the weak fluorescence of **10** and **11** in their free forms, strong fluorescence of free **12** ($\phi_{free}=0.15$; $\lambda_{em}=558$ nm) is attributed to the protonation of the quinoline nitrogen and surrounding donor atoms at neutral pH in HEPES buffer, which induces resonance by electron donation from the 4-alkoxy group [48]. This resonance form is disrupted by Cd^{2+} binding, which results in blue-shifted emission ($\phi_{Cd}=0.11$; $\lambda_{em}=495$ nm), affording a ratiometric response. Zinc ion exhibits moderate fluorescence intensity at a slightly different wavelength from Cd^{2+} ($\phi_{Zn}=0.055$; $\lambda_{em}=510$ nm). The high binding affinity with Cd^{2+} ($K_d=41$ pM) and low LOD (limit of detection, 9.6 pM) of **12** allow cellular detection in NIH 3T3 cells and confocal imaging in HEK 293 cells.

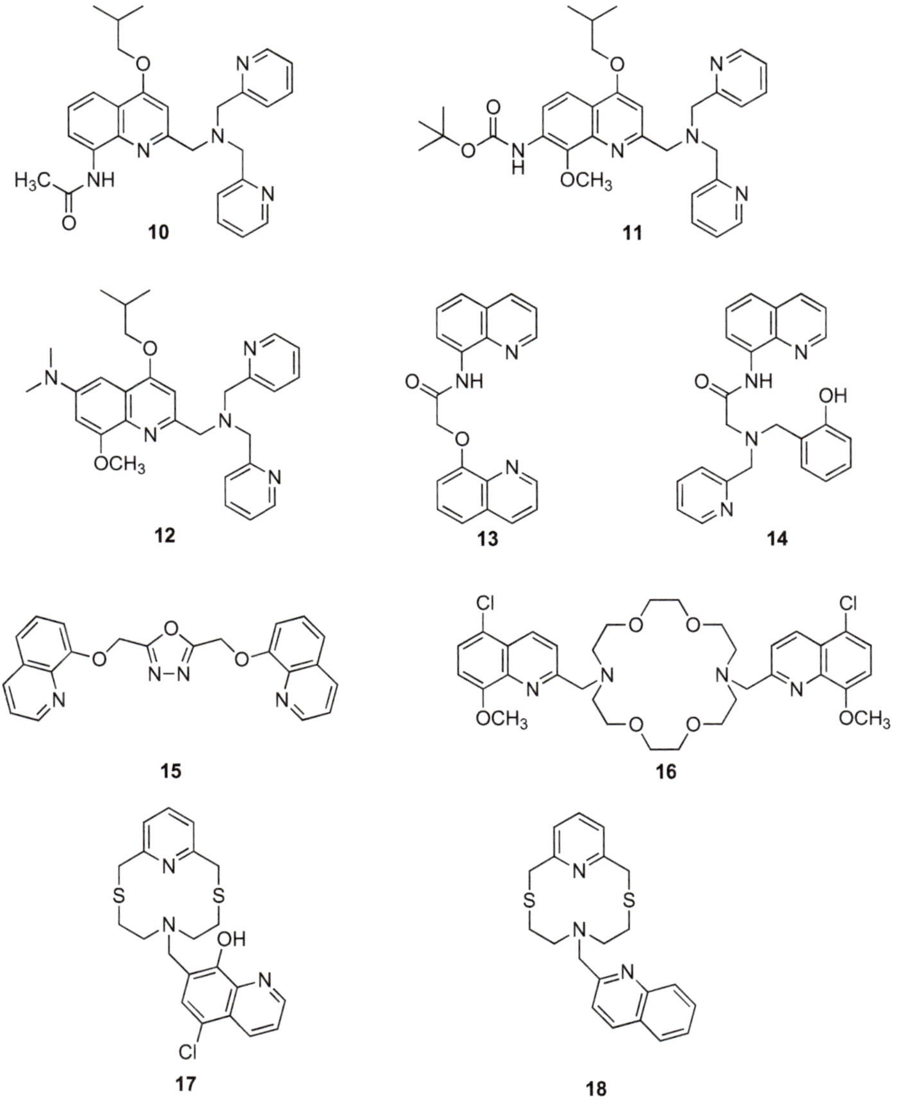

FIGURE 7 QUINOLINE-BASED Cd^{2+} sensors.

As described for compound **10**, a conjugate of 8-aminoquinoline and 8-hydroxyquinoline (**13**) exhibits a Cd^{2+}-specific fluorescence enhancement by CHEF ($\phi_{Cd}=0.16$; $\lambda_{em}=410\,nm$) and Zn^{2+}-specific red shift of emission band by an ICT mechanism ($\phi_{Zn}=0.019$; $\lambda_{em}=486\,nm$) in ethanol [49]. X-ray crystallography of Cd^{2+} complex of **13** reveals that the carbonyl oxygen atom of the amide moiety coordinates to the cadmium center and the aminoquinoline nitrogen atom remains unbound, where PET inhibition of aminoquinoline, as well as CHEF for alkoxyquinoline, is invoked. In addition, the amide nitrogen atom coordinates to Zn^{2+} by deprotonation of amide NH and both quinoline nitrogen and alkoxy oxygen atoms are bound to the metal center, leading to an ICT mechanism as discussed for **10**-Zn^{2+} complex. For Cd^{2+} sensing with **13**, acetate anion quenches the fluorescence.

The sensor **14** also has an 8-amidoquinoline moiety and exhibits Zn^{2+}/Cd^{2+}-selective fluorescence enhancement depending on the solvent. In acetonitrile solution, **14** acts as a Zn^{2+} sensor at 510 nm (I_{Cd}/I_{Zn} = ~1/3; ϕ_{Zn} = 0.24) through amide nitrogen coordination, which is supported by ^{1}H NMR and infrared analyses [50]. In contrast, fluorescence of the Zn^{2+} complex diminishes in 1:1 acetonitrile–HEPES buffer solution, which changes the ratio of the metal ion selectivity (I_{Zn}/I_{Cd} = ~1/3; ϕ_{Cd} = 0.039). Zn^{2+} binding in aqueous buffer-acetonitrile solution remains because one equiv. of Zn^{2+} interferes with the Cd^{2+}-induced fluorescence of **14**. The aminoquinoline nitrogen atom in **14** does not bind to Cd^{2+} in the crystal structure, similar to that found in **13**-Cd^{2+}. Another bis(alkoxyquinoline) sensor, **15**, possesses a semi-rigid cavity suitable for two cadmium ions [51]. The CHEF is responsible for Cd^{2+}-specific fluorescence enhancement at 410 nm in 5% H_2O-MeOH solution. The fluorescence response of the tridentate chelator **15** is highly dependent on the counter-anion and solvent. In contrast to **13**, acetate enhances the fluorescence intensity of **15**-Cd^{2+} system and Zn^{2+} complex with **15** is emissive in acetonitrile solution.

Macrocyclic ligands were utilized as ionophores for quinoline-based fluorescent Cd^{2+} sensors. Sensor **16** has two 5-chloro-8-methoxyquinolines with a diaza-18-crown-6 scaffold (log K_{assoc} = 6.1; ϕ_{Cd} = 0.66 in methanol) [52]. Related compounds with 8-hydroxy substituents in the quinoline moiety connected through the 7-position to the macrocycle are fluorescent Mg^{2+} and Hg^{2+} probes [53, 54]. This 5-chloro-8-hydroxyquinoline chromophore is attached to N2S2 pyridine-containing macrocycle through the 7-position to exhibit Cd^{2+}-specific fluorescence response of **17** in acetonitrile-HEPES buffer (1:1) [55]. The weak fluorescence intensity of **17** is attributed to both photoinduced proton transfer (PPT) and PET mechanisms. The fluorescence enhancement upon the addition of Cd^{2+} results from a CHEF mechanism. The small fluorescence intensity of **17**-Zn^{2+} system (I_{Zn}/I_{Cd} = 1/4) is effectively reduced in 3-*N*-morpholinopropanesulfonic acid-buffered aqueous solution containing sodium dodecyl sulfate. Intracellular Cd^{2+} in HL-60, Cos-7, and Saos-2 cells was successfully visualized. The crystal structure of **17**-Cd^{2+} complex [Cd(**17**)-(H_2O)]$^{2+}$ reveals the deprotonation of the 8-hydroxy group and protonation of quinoline nitrogen atom. Interestingly, compound **18**, which possesses an unsubstituted quinoline attached through the 2-position to the same macrocycle, exhibits Zn^{2+} specificity in fluorescence enhancement (I_{Cd}/I_{Zn} = 9%) in acetonitrile-HEPES buffer (1:1) [56].

Replacement of carboxylate groups with quinoline moieties in widely used metal ion chelators, such as EDTA (ethylenediamine-*N,N,N′,N′*-tetraacetic acid), EGTA (ethylene glycol bis(2-aminoethyl ether)-*N,N,N′,N′*-tetraacetic acid), and BAPTA (1,2-bis(2-aminophenoxy)ethane-*N,N,N′,N′*-tetraacetic acid), affords fluorescent metal ion binders, TQEN (*N,N,N′,N′*-tetrakis(2-quinolylmethyl)ethylenediamine, **19**), EGTQ (**20a**), and BAPTQ (**21a**), respectively (Figure 8). The TQEN (**19**) exhibits moderate Zn^{2+} selectivity over Cd^{2+} (I_{Cd}/I_{Zn} = 64%; ϕ_{Zn} = 0.007) in *N,N*-dimethylformamide (DMF)–water (1:1) mixed solvent [57], but its isoquinoline derivative 1-isoTQEN (*N,N,N′,N′*-tetrakis(1-isoquinolylmethyl)ethylenediamine)

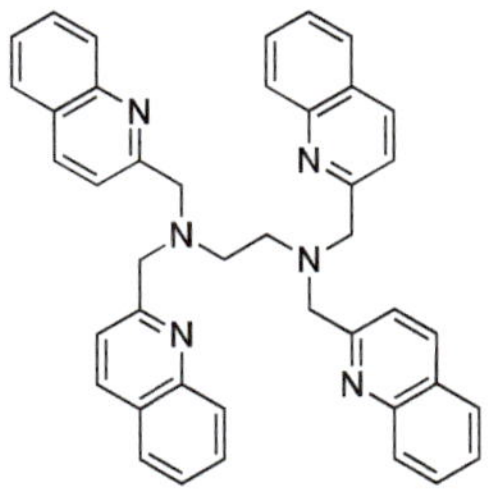

TQEN (**19**)

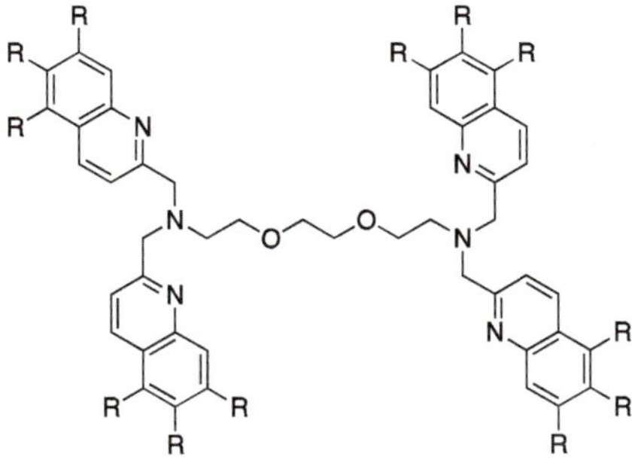

EGTQ (**20a**) : R =H
TriMeOEGTQ (**20b**) : R = OCH_3

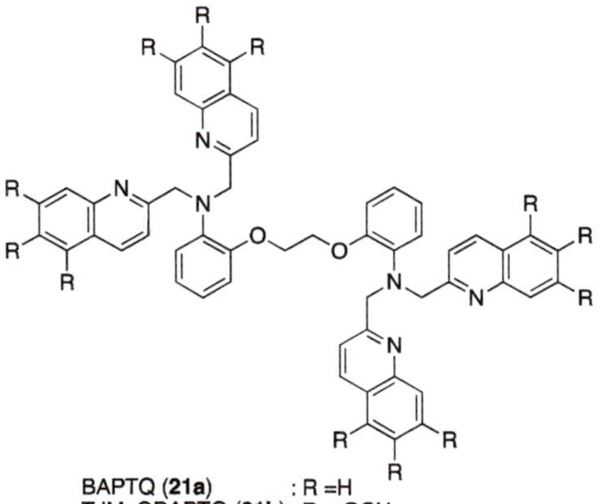

BAPTQ (**21a**) : R =H
TriMeOBAPTQ (**21b**) : R = OCH_3

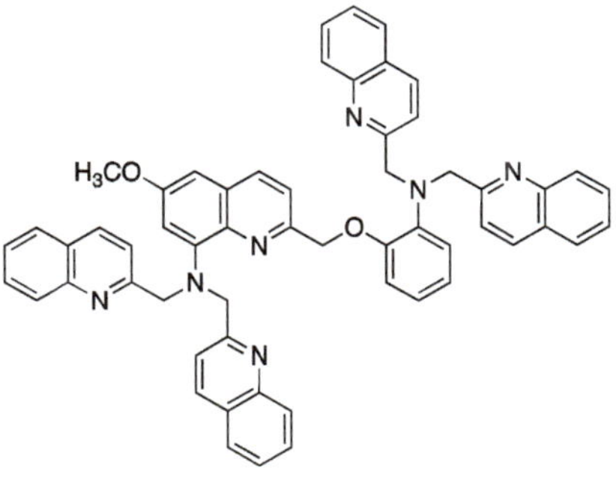

TQquin2 (**22**)

FIGURE 8 TERAKISQUINOLINE derivatives of EDTA, EGTA, BAPTA, and quin2.

shows strict Zn^{2+} specificity in 450 nm emission due to the Zn^{2+}-specific intramolecular excimer formation ($I_{Cd}/I_{Zn} = 14\%$; $\phi_{Zn} = 0.034$) [58]. Conversely, as described above, heptadentate ligand TQOPEN (**5a**, Figure 3) exhibits Cd^{2+} specificity for its chelation-enhanced fluorescence enhancement ($I_{Zn}/I_{Cd} = 10\%$; $\phi_{Cd} = 0.017$) in the same solvent system [40]. The crystal structure of Zn^{2+} complex with **5a** reveals the formation of a poorly fluorescent, hydroxide-bridged dizinc complex. The seven donor atoms in the mononuclear cadmium complex of **5a** afford flexible and dynamic coordination environment suitable for the intramolecular excimer emission. Although the octadentate derivatives of tetrakisquinoline ligands **20a** and **21a** do not exhibit enough fluorescent response in DMF–water (1:1), a trimethoxyquinoline analog **21b** shows strict Cd^{2+} specificity and enhanced fluorescence intensity ($I_{Zn}/I_{Cd} = 5\%$; $\phi_{Cd} = 0.19$) in methanol-HEPES buffer (9:1) [59]. Corresponding EGTA analog **20b** responded less selectively to Cd^{2+} than Zn^{2+} ($I_{Zn}/I_{Cd} = 56\%$) under the same experimental conditions. A tetrakisquinoline derivative (TQquin2, **22**) of fluorescent Ca^{2+}-specific probe quin2 (8-amino-2-((2-amino-5-methylphenoxy)-methyl)-6-methoxyquinoline-*N,N,N′,N′*-tetraacetic acid) [60] also exhibits excellent Cd^{2+} specificity ($I_{Zn}/I_{Cd} = 2\%$; $I_{Cd}/I_0 = 170$; $\phi_{Cd} = 0.058$) in methanol; however, several drawbacks, such as formation of non-fluorescent ML_2 species, in the initial stage of titration and sensitivity to water were admitted for **22** [61].

3.5 Others

Anthracene derivative **23** was reported as a Cd^{2+}-specific chemosensor in HEPES buffer solution containing 0.135 M NaCl [62]. The PET mechanism on the aniline nitrogen atom in the iminodiacetate binding site quenches the fluorescence found in anthracene. However, addition of Zn^{2+} restores the monomer emission as found in the highly acidic solution. In the presence of Cd^{2+}, broad and structureless emission was observed in the 400–600 nm region. This band was assigned to the formation of charge-transfer complexes (exciplexes) between the anthracene moieties and the ion-receptor complexes, or π-complexes. The possibility of intermolecular excimer formation was ruled out because of low concentration (1 μM) of the probe (Figure 9).

Near-Infrared absorption/emission has been utilized by tricarbocyanine sensors CYP-1 (**24a**) and CYP-2 (**24b**) [63]. The *o*-phenylenediamine nitrogen atoms are responsible for PET quenching of the fluorophore. The two pentadentate metal binding sites of **24b** afford a 1:2 (LM_2) complex with ~two-fold fluorescence enhancement at 793 nm ($\phi_{Cd}=0.0145$) in Tris-HCl buffer solution containing 0.05 mM sodium phosphate. The association constants are $K_{11}=8.8\times10^3$ and $K_{21}=1.9\times10^5\ M^{-1}$. The presence of equimolar amount of other metal ions, including Zn^{2+}, Cu^{2+}, Co^{2+}, Pb^{2+}, and Hg^{2+}, does not affect the detection of Cd^{2+} by **24b**. The LOD was estimated to be 2.3 μM for **24b** and 3.1 μM for **24a** (in acetonitrile/water = 9/1 with 0.05 mM sodium phosphate). Intracellular Cd^{2+} was detected by using **24a** because sulfonate probe **24b** failed to penetrate the cell membrane.

Strict discrimination of Cd^{2+} from Zn^{2+} is achieved in a π-conjugated ligand HPDQ (**25**) with polypyridyl arms in CH_2Cl_2/CH_3CN (1:9) [64]. The addition of 0–2 equiv. of Cd^{2+} induces a very small change in fluorescence spectrum of **25**, but more than 2 equiv. of Cd^{2+} causes a distinct fluorescence around 456 nm, which saturates at ~20 equiv. of Cd^{2+} ($\phi_{Cd}=0.048$). This fluorescence is specific to Cd^{2+} without any interference in the presence of equimolar amount of other metal ions, such as Cu^{2+}, Zn^{2+}, and Hg^{2+}. The structures of tricadmium and tetrazinc complexes with **25** were revealed by X-ray crystallography. The mechanism of fluorescence enhancement is hypothesized to result from ICT and/or increased conjugation.

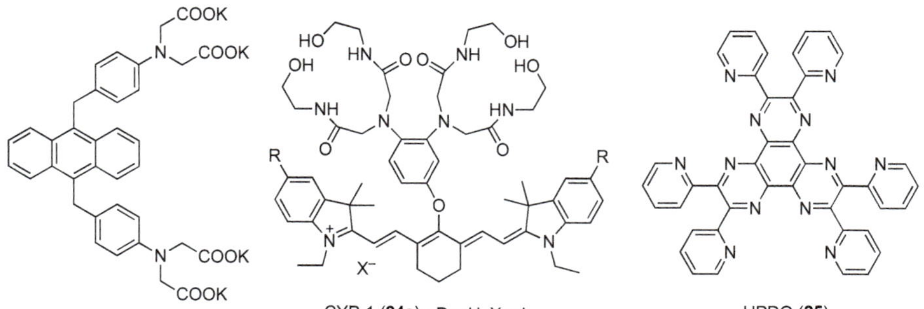

23

CYP-1 (**24a**) : R = H, X = I
CYP-2 (**24b**) : R = SO_3H, SO_3^-

HPDQ (**25**)

FIGURE 9 VARIOUS Cd^{2+} sensors.

4 MANIPULATION OF FLUORESCENT ZINC/CADMIUM SELECTIVITY

As described in Section 2.2, strict discrimination between cadmium and zinc ions is important but challenging because of the similarity of these group 12 metal ions. The previous section provides several examples of selective fluorescent detection of Cd^{2+} with minimized response to Zn^{2+}. Other probe designs highlight Zn^{2+} selectivity over Cd^{2+} (*cf.* Chapter 5 by Yoon et al.). Interestingly, some works have succeeded in developing a Zn^{2+}/Cd^{2+} dual sensor that responds to cadmium and zinc ions at different wavelengths. Moreover, there are some examples of reversing Zn^{2+}/Cd^{2+} fluorescence selectivity by introducing small changes into the structure of the probe molecule. This section picks up some examples of such molecular design strategies in manipulating fluorescent Zn^{2+}/Cd^{2+} selectivity.

4.1 Zinc/Cadmium Dual Sensors

Amide-bearing fluorophores often exhibit differential fluorescent responses toward Zn^{2+} and Cd^{2+}. The higher Lewis acidity of Zn^{2+} compared to Cd^{2+} promotes deprotonation of the amide hydrogen atom, which alters the electron distribution of the metal-bound probe. Therefore, the ICT process upon metal binding results in opposite behavior for Zn^{2+} and Cd^{2+}. Zn^{2+} enhances ICT process through increased electron density at the donor moiety, whereas Cd^{2+} reduces ICT character of the fluorophore but inhibits PET to enhance the fluorescence at a shorter-wavelength region than the ICT emission band. When the intensity of Cd^{2+}-induced fluorescence by PET inhibition is higher than the Zn^{2+}-promoted ICT emission, the probe becomes essentially Cd^{2+} sensor as discussed earlier for **10** and **13** (Figure 7) [46, 49].

The 1,8-naphthalimide fluorophore ZTRS (**26**, Figure 10) exhibits a fluorescent Zn^{2+}/Cd^{2+} dual-sensing property using differential tautomeric forms of the amide moiety in acetonitrile-HEPES buffer (1:1) [65]. In this case, Zn^{2+} binds to the nitrogen atom of the imidic acid tautomer without deprotonation, exhibiting green emission ($\lambda_{em}=514$ nm; $I_{Zn}/I_0=22$; $\phi_{Zn}=0.36$). The Cd^{2+} is coordinated by the oxygen atom of the amide tautomer, showing blue emission ($\lambda_{em}=446$ nm; $I_{Cd}/I_0=21$; $\phi_{Cd}=0.34$). The red-shifted emission wavelength is achieved in the Zn^{2+}-**26** complex due to the expanded conjugation of the fluorophore. Such differential sensing is possible in 100% HEPES buffer solution, but the signal intensity is diminished. The cell-permeability of **26**, as well as its strong and specific binding with Zn^{2+}, is suitable for imaging of Zn^{2+} in living cells and zebrafish.

Other than the amide functionalities, the 4,5-diamino-1,8-naphthalimide skeleton in **27** (Figure 10) also discriminates between Zn^{2+} and Cd^{2+} by deprotonation by Zn^{2+} [66]. The fluorescent properties of **27** upon coordination to Zn^{2+} ($\lambda_{em}=558$ nm; $\phi_{Zn}=0.23$) and Cd^{2+} ($\lambda_{em}=487$ nm; $\phi_{Cd}=0.60$) is analogous to the amide derivatives described above. Interestingly, related compound **28** remains ICT-based ratiometric fluorescence change upon Zn^{2+} binding accompanied deprotonation ($\lambda_{em}=593$ nm; $\phi_{Zn}=0.14$) but does not respond to Cd^{2+} probably due to the low affinity [67].

The Zn^{2+}-specific intramolecular excimer formation is utilized for Zn^{2+}/Cd^{2+} dual sensing with 3-isoTQLN (*N,N,N′,N′*-tetrakis(3-isoquinolylmethyl)-2,6-

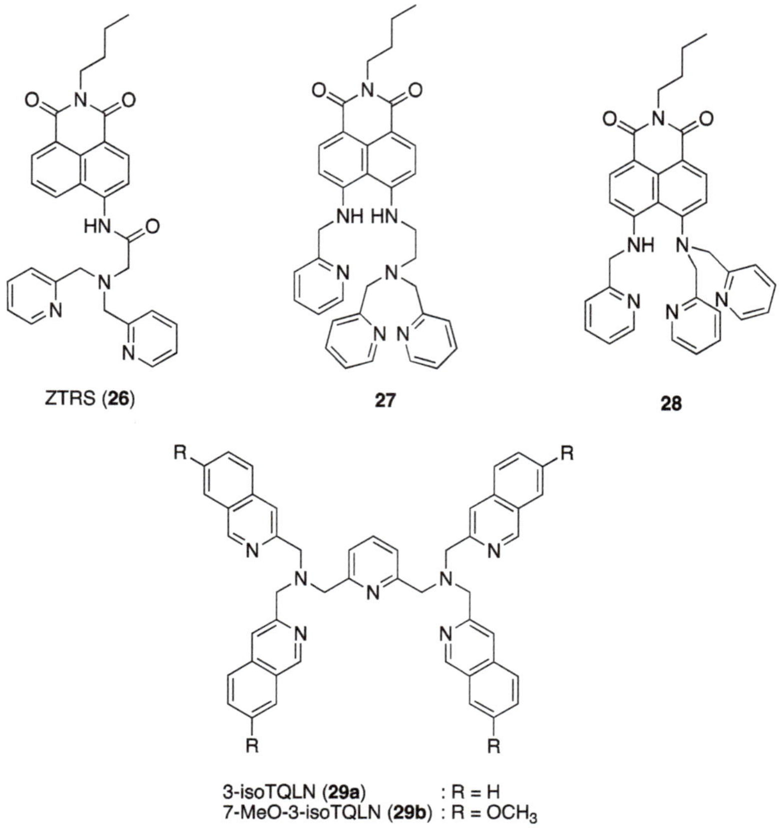

FIGURE 10 **Zn^{2+}/Cd^{2+}** dual sensors **26**, **27**, and **29a** and related compounds.

lutidylenediamine, **29a**) in DMF–water (1:1) [68]. Considerable fluorescence intensity of **29a** in the presence of Cd^{2+} due to the CHEF mechanism ($\lambda_{em}=356$ nm; $I_{Zn}/I_{Cd}=21\%$; $\phi_{Cd}=0.012$) is comparable to that of Zn^{2+}-induced excimer emission ($\lambda_{em}=428$ nm; $I_{Cd}/I_{Zn}=25\%$; $\phi_{Zn}=0.014$). The 20-fold enhancement of fluorescence intensity of Zn^{2+}/Cd^{2+} complexes was achieved in **29b** by introducing methoxy substituents; however, significant overlap of fluorescent signals of Zn^{2+} ($\lambda_{em}=422$ nm; $I_{Cd}/I_{Zn}=140\%$; $\phi_{Zn}=0.032$) and Cd^{2+} ($\lambda_{em}=402$ nm; $I_{Zn}/I_{Cd}=43\%$; $\phi_{Cd}=0.033$) complexes eliminates the selectivity.

4.2 Reversal of Zinc/Cadmium Selectivity

The alteration of fluorescent target metal selectivity by a small change of the probe structure is of significant interest. One such example has been described already for compound **17**, in which the primary target metal ion for fluorescent response is changed from Cd^{2+} to Zn^{2+} by replacement of the 5-chloro-8-hydroxy-7-quinolyl moiety of **17** with an 8-hydroxy-2-quinolyl group of **18** (Figure 7) [55, 56].

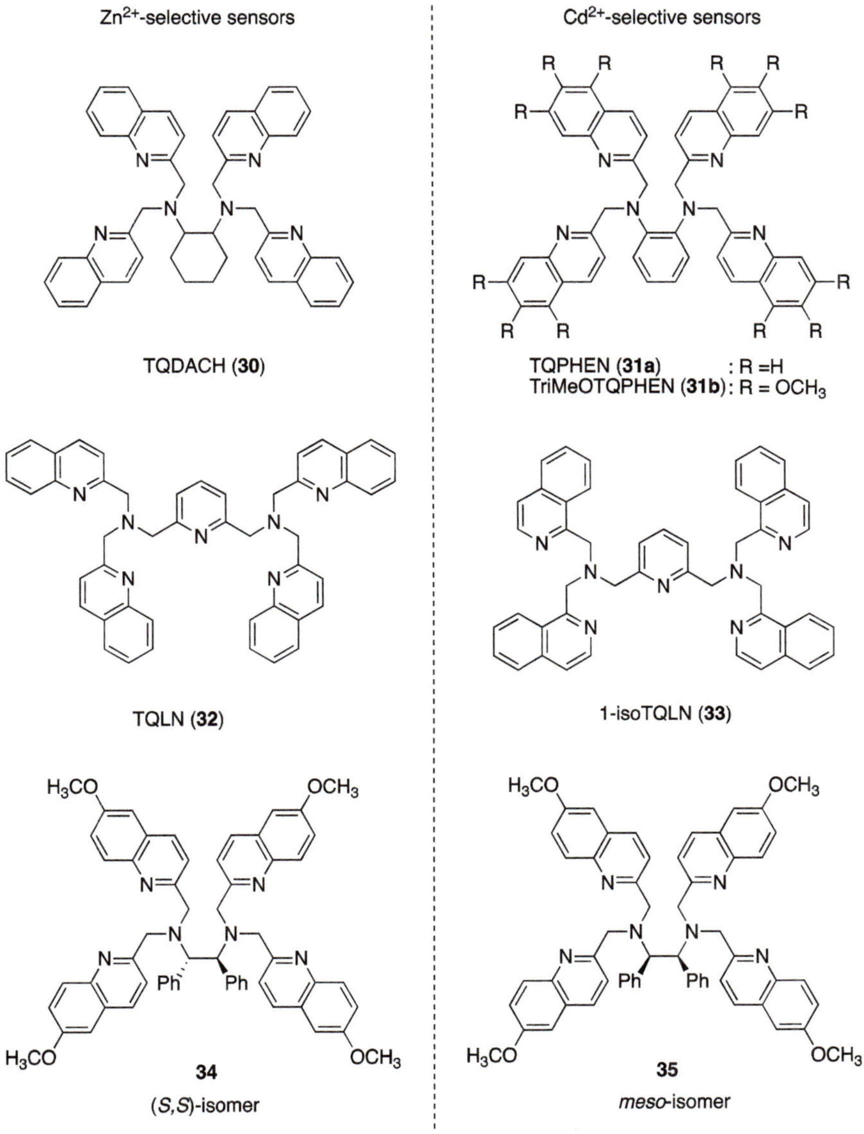

FIGURE 11 REVERSAL of Zn^{2+}/Cd^{2+} selectivity by small changes in compound structure. (Reproduced from Ref. [69] with permission from the Royal Society of Chemistry.)

Introduction of cyclohexane or benzene ring to TQEN (**19**, Figure 8) induces opposite change in fluorescent metal ion specificity in DMF–water (1:1) (Figure 11) [69]. Moderate Zn^{2+} selectivity of TQEN ($I_{Cd}/I_{Zn} = 64\%$) was improved significantly in a *trans*-1,2-cyclohexanediamine derivative **30** ($\lambda_{em} = 455$ nm; $I_{Cd}/I_{Zn} = 19\%$; $\phi_{Zn} = 0.010$) [70]. The long-wavelength emission indicates the Zn^{2+}-specific intramolecular excimer formation served by conformational restriction of **30**. On the other hand, corresponding 1,2-phenylenediamine derivative **31a** exhibits distinct Cd^{2+}-selectivity in fluorescence enhancement ($\lambda_{em}=392$ nm; $I_{Zn}/I_{Cd}=27\%$; $\phi_{Cd}=0.006$) [71]. The weak donating ability of aniline nitrogen atom of **31a** significantly inhibits binding with Zn^{2+}. The trimethoxyquinoline derivative **31b** effectively improved poor fluorescence intensity and Cd^{2+} selectivity ($\lambda_{em}=499$ nm; $I_{Zn}/I_{Cd}=5\%$; $\phi_{Cd}=0.026$). The X-ray crystallography of Zn^{2+} and Cd^{2+} complexes suggests the significant distortion of the bond angles for Cd^{2+}-**30** and Zn^{2+}-**31a** complexes, consistent with their poor fluorescent properties.

The 2-quinolyl (TQLN, **32**) and 1-isoquinolyl (1-isoTQLN, **33**) derivatives of 3-isoquinolyl probe **29a** exhibit Zn^{2+}- and Cd^{2+}-selective fluorescence enhancement in DMF–water (1:1) (Figure 11) [72]. The quinoline derivative **32** exhibits Zn^{2+}-specific intramolecular excimer emission ($\lambda_{em}=428$ nm; $I_{Cd}/I_{Zn}=24\%$; $\phi_{Zn}=0.069$) and 1-isoquinolyl analog **33** shows Cd^{2+}-specific fluorescence enhancement by a monomeric CHEF mechanism ($\lambda_{em}=365$ nm; $I_{Zn}/I_{Cd}=19\%$; $\phi_{Cd}=0.015$). The weakly fluorescent Zn^{2+}-induced fluorescence of **33** owes to the hydroxide-bridged dinuclear structure of Zn^{2+}-**33** complex with poor binding affinity. Careful inspection of the molecular structures of **29a**, **32**, and **33** reveals that the 3-isoquinolyl probe **29a** possesses an intermediate structure between **32** and **33**, exhibiting the hybrid nature of both compounds in its dual fluorescent metal sensing ability [68].

The reversal of Zn^{2+} and Cd^{2+} selectivity is achieved by a stereogenic inversion of one asymmetric carbon atom in a pair of diastereomers **34** and **35** (Figure 11) [73]. Due to limited solubility, the 6-methoxyquinoline derivatives were investigated in DMF–water (2:1). The (*S*,*S*)-isomer **34** exhibits Zn^{2+}-specific long wavelength emission by intramolecular excimer emission ($\lambda_{em}=498$ nm; $I_{Cd}/I_{Zn}=20\%$; $\phi_{Zn}=0.047$). Several trivalent metal ions, such as Al^{3+}, Fe^{3+}, and Cr^{3+}, also respond to **34** in the short-wavelength region, but this emission disappears in DMF–water (1:1) solution. Whereas meso isomer **35** highlights Cd^{2+} over Zn^{2+} in the short-wavelength region ($\lambda_{em}=405$ nm; $I_{Zn}/I_{Cd}=29\%$; $\phi_{Cd}=0.022$). The highly distorted structure of **35** upon metal binding suppresses interaction with Zn^{2+}, preserving the affinity with Cd^{2+} ($K_d=9.4\times10^{-6}$ M) in similar order with Zn^{2+}-**34** ($K_d=5.5\times10^{-6}$ M).

5 GENERAL CONCLUSIONS

This chapter briefly overviews recent progress of fluorescent cadmium sensor developments. Molecular design strategies for fluorescence signal switching and specific detection of cadmium, especially Zn/Cd selectivity, are extracted from literature during these two decades, in which many general ideas are utilized in common platforms. The use of quinolines as a device for fluorescent cadmium sensors is rationally reasonable because (i) the hybrid nature of fluorophore and binding motifs reduces molecular size of the probes, which is suitable for biological applications; (ii) weak basicity and long coordination distance of quinoline, compared with aliphatic nitrogen and pyridine derivatives, increase cadmium specificity over zinc; and (iii) rigid structure of bicyclic quinoline ring increases the crystallization ability of metal complexes to analyze binding/fluorescence mechanism. However, some drawbacks of quinoline-based probes, including poor fluorescence quantum yield, weak metal-binding affinity, and poor water solubility, need to be improved. Other general strategies and platforms are found in Zn/Cd discrimination mechanisms.

The most critical difference in the development of fluorescent cadmium sensors from zinc-targeting ones is that cadmium sensors require strict Zn/Cd selectivity because zinc is abundant in living systems and the environment, whereas less cadmium exists in normal conditions. In this context, extensive tuning of molecular structures to improve the target metal specificity is necessary for development of fluorescent cadmium sensors. It is obvious that high toxicity and many unexplored mechanisms of cadmium in the disruption of living functions need to be continuously investigated. Superior fluorescent cadmium sensors would reveal unknown functions and possible beneficial effects of cadmium in living systems in the future.

ACKNOWLEDGMENT

The author deeply thanks Prof. Shawn C. Burdette of Worchester Polytechnic Institute for his help in critical reading and valuable suggestions on the manuscript.

ABBREVIATIONS AND DEFINITIONS

AIE	aggregation-induced emission
ATP	adenosine triphosphate
ATSDR	Agency for Toxic Substances and Disease Registry
BAPTA	1,2-bis(2-aminophenoxy)ethane-*N,N,N',N'*-tetraacetic acid
BODIPY	boron-dipyrromethane
CHEF	chelation-enhanced fluorescence
DMF	*N,N*-dimethylformamide
DNA	deoxyribonucleic acid
DPA	2,2′-dipicolylamine
EDTA	ethylenediamine-*N,N,N′,N′*-tetraacetic acid
EGTA	ethylene glycol bis(2-aminoethyl ether)-*N,N,N',N'*-tetraacetic acid
FRET	Förster (or fluorescence) resonance energy transfer
GSH	glutathione
GSH-AuNC	glutathione-stabilized gold nanocluster
HEPES	2-(4-(2-hydroxyethyl)-1-piperazinyl)ethanesulfonic acid
HOMO	highest occupied molecular orbital
ICT	internal charge transfer
LOD	limit of detection
LUMO	lowest unoccupied molecular orbital
MT	metallothionein
NMR	nuclear magnetic resonance
PET	photoinduced electron transfer
PPT	photoinduced proton transfer

quin2 8-amino-2-((2-amino-5-methylphenoxy)methyl)-6-methoxyquinoline-*N,N,N',N'*-tetraacetic acid
ROS reactive oxygen species
THF tetrahydrofuran
TPE tetraphenylethylene
TPEN *N,N,N',N'*-tetrakis(2-pyridylmethyl)ethylenediamine
TQEN *N,N,N',N'*-tetrakis(2-quinolylmethyl)ethylenediamine
Tris tris(hydroxymethyl)aminomethane

REFERENCES

1. Y. Huang, C. He, C. Shen, J. Guo, S. Mubeen, J. Yuan, Z. Yang, *Food Funct.* **2017**, *8*, 1373–1401.
2. Agency for Toxic Substances and Disease Registry. *2019 Substance Priority List*, 2019, https://www.atsdr.cdc.gov/SPL/#2019spl.
3. M. Berglund, A. Åkesson, P. Bjellerup, M. Vahter, *Toxicol. Lett.* **2000**, *112–113*, 219–225.
4. C. D. Klaassen, J. Liu, B. A. Diwan, *Toxicol. Appl. Pharmacol.* **2009**, *238*, 215–220.
5. D. R. Winge, K.-A. Miklossy, *J. Biol. Chem.* **1982**, *257*, 3471–3476.
6. K. Jomova, M. Valko, *Toxicology* **2011**, *283*, 65–87.
7. L. Järup, *Nephrol. Dial. Transplant.* **2002**, *17* (Suppl 2), 35–39.
8. J. Liu, W. Qu, M. B. Kadiiska, *Toxicol. Appl. Pharmacol.* **2009**, *238*, 209–214.
9. M. B. Gumpu, S. Sethuraman, U. M. Krishnan, J. B. B. Rayappan, *Sens. Actuators B* **2015**, *213*, 515–533.
10. C. Giaginis, E. Gatzidou, S. Theocharis, *Toxicol. Appl. Pharmacol.* **2006**, *213*, 282–290.
11. M. P. Waalkes, *Mutat. Res.* **2003**, *533*, 107–120.
12. M. Waisberg, P. Joseph, B. Hale, D. Beyersmann, *Toxicology* **2003**, *192*, 95–117.
13. M. Khairy, S. A. El-Safty, M. A. Shenashen, *Trends Anal. Chem.* **2014**, *62*, 56–68.
14. M. J. Shaw, P. R. Haddad, *Envion. Int.* **2004**, *30*, 403–431.
15. K. Duarte, C. I. L. Justino, A. C. Freitas, A. M. P. Gomes, A. C. Duarte, T. A. P. Rocha-Santos, *Trends Anal. Chem.* **2015**, *64*, 183–190.
16. L. T. Bereza-Malcolm, G. Mann, A. E. Franks, *ACS Synth. Biol.* **2015**, *4*, 535–546.
17. H. N. Kim, W. X. Ren, J. S. Kim, J. Yoon, *Chem. Soc. Rev.* **2012**, *41*, 3210–3244.
18. M. Taki, *Imaging and Sensing of Cadmium in Cells*, in "*Cadmium: From Toxicity to Essentiality*", Metal Ions in Life Sciences 11, Eds A. Sigel, H. Sigel, R. K. O. Sigel, Springer Science+Business Media Dordrecht 2013.
19. B. Kaur, N. Kaur, S. Kumar, *Coord. Chem. Rev.* **2018**, *358*, 13–69.
20. T. Rasheed, M. Bilal, F. Nabeel, H. M. N. Iqbal, C. Li, Y. Zhou, *Sci. Total Environ.* **2018**, *615*, 476–485.
21. X. Wang, C. Shen, C. Zhou, Y. Bu, X. Yan, *Chem. Eng. J.* **2021**, *417*, 129125.
22. S.-Y. Chen, Z. Li, K. Li, X.-Q. Yu, *Coord. Chem. Rev.* **2021**, *429*, 213691.
23. C.-T. Shi, Z.-Y. Huang, A.-B. Wu, Y.-X. Hu, N.-C. Wang, Y. Zhang, W.-M. Shu, W.-C. Yu, *RSC Adv.* **2021**, *11*, 29632–29660.
24. B. Daly, J. Ling, A. P. de Silva, *Chem. Soc. Rev.* **2015**, *44*, 4203–4211.
25. D. Escudero, *Acc. Chem. Res.* **2016**, *49*, 1816–1824.
26. B. Valeur, I. Leray, *Coord. Chem. Rev.* **2000**, *205*, 3–40.
27. A. P. de Silva, H. Q. N. Gunaratne, T. Gunnlaugsson, A. J. M. Huxley, C. P. McCoy, J. T. Rademacher, T. E. Rice, *Chem. Rev.* **1997**, *97*, 1515–1566.
28. H. Yuasa, N. Miyagawa, T. Izumi, M. Nakatani, M. Izumi, H. Hashimoto, *Org. Lett.* **2004**, *6*, 1489–1492.

29. S. Y. Park, J. H. Yoon, C. S. Hong, R. Souane, J. S. Kim, S. E. Matthews, J. Vicens, *J. Org. Chem.* **2008**, *73*, 8212–8218.
30. L. Yuan, W. Lin, K. Zheng, S. Zhu, *Acc. Chem. Res.* **2013**, *46*, 1462–1473.
31. K. Kikuchi, *Chem. Soc. Rev.* **2010**, *39*, 2048–2053.
32. S. Jiang, S. Chen, Z. Wang, H. Guo, F. Yang, *Sens. Actuators B* **2020**, *308*, 127734.
33. J. Chan, S. C. Dodani, C. J. Chang, *Nat. Chem.* **2012**, *4*, 973–984.
34. X. Chen, T. Pradhan, F. Wang, J. S. Kim, J. Yoon, *Chem. Rev.* **2012**, *112*, 1910–1956.
35. X. Zhang, Y. Xiao, X. Qian, *Angew. Chem. Int. Ed.* **2008**, *47*, 8025–8029.
36. M. H. Lee, H. J. Kim, S. Yoon, N. Park, J. S. Kim, *Org. Lett.* **2008**, *10*, 213–216.
37. Z.-X. Han, X.-B. Zhang, Z. Li, Y.-J. Gong, X.-Y. Wu, Z. Jin, C.-M. He, L.-X. Jian, J. Zhang, G.-L. Shen, R.-Q. Yu, *Anal. Chem.* **2010**, *82*, 3108–3113.
38. R. G. Pearson, *J. Am. Chem. Soc.* **1963**, *85*, 3533–3539.
39. M. Kadarkaraisamy, A. G. Sykes, *Polyhedron* **2007**, *26*, 1323–1330.
40. Y. Mikata, A. Kizu, K. Nozaki, H. Konno, H. Ono, S. Mizutani, S. Sato, *Inorg. Chem.* **2017**, *56*, 7404–7415.
41. Y. Mikata, K. Nozaki, M. Kaneda, K. Yasuda, M. Aoyama, S. Tamotsu, A. Matsumoto, *Eur. J. Inorg. Chem.* **2018**, 2755–2761.
42. M. Taki, M. Desaki, A. Ojida, S. Iyoshi, T. Hirayama, I. Hamachi, Y. Yamamoto, *J. Am. Chem. Soc.* **2008**, *130*, 12564–12565.
43. W. Liu, L. Xu, R. Sheng, P. Wang, H. Li, S. Wu, *Org. Lett.* **2007**, *9*, 3829–3832.
44. K. Aich, S. Goswami, S. Das, C. D. Mukhopadhyay, C. K. Quah, H.-K. Fun, *Inorg. Chem.* **2015**, *54*, 7309–7315.
45. X. Peng, J. Du, J. Fan, J. Wang, Y. Wu, J. Zhao, S. Sun, T. Xu, *J. Am. Chem. Soc.* **2007**, *129*, 1500–1501.
46. L. Xue, C. Liu, H. Jiang, *Org. Lett.* **2009**, *11*, 1655–1658.
47. L. Xue, Q. Liu, H. Jiang, *Org. Lett.* **2009**, *11*, 3454–3457.
48. L. Xue, G. Li, Q. Liu, H. Wang, C. Liu, X. Ding, S. He, H. Jiang, *Inorg. Chem.* **2011**, *50*, 3680–3690.
49. X. Zhou, P. Li, Z. Shi, X. Tang, C. Chen, W. Liu, *Inorg. Chem.* **2012**, *51*, 9226–9231.
50. E. J. Song, J. Kang, G. R. You, G. J. Park, Y. Kim, S.-J. Kim, C. Kim, R. G. Harrison, *Dalton Trans.* **2013**, *42*, 15514–15520.
51. X.-L. Tang, X.-H. Peng, W. Dou, J. Mao, J.-R. Zheng, W.-W. Qin, W.-S. Liu, J. Chang, X.-J. Yao, *Org. Lett.* **2008**, *10*, 3653–3656.
52. L. Prodi, M. Montalti, N. Zaccheroni, J. S. Bradshaw, R. M. Izatt, P. B. Savage, *Tetrahedron Lett.* **2001**, *42*, 2941–2944.
53. L. Prodi, F. Bolletta, M. Montalti, N. Zaccheroni, P. B. Savage, J. S. Bradshaw, R. M. Izatt, *Tetrahedron Lett.* **1998**, *39*, 5451–5454.
54. L. Prodi, C. Bargossi, M. Montalti, N. Zaccheroni, N. Su, J. S. Bradshaw, R. M. Izatt, P. B. Savage, *J. Am. Chem. Soc.* **2000**, *122*, 6769–6770.
55. M. Mameli, M. C. Aragoni, M. Arca, C. Caltagirone, F. Demartin, G. Farruggia, G. De Filippo, F. A. Devillanova, A. Garau, F. Isaia, V. Lippolis, S. Murgia, L. Prodi, A. Pintus, N. Zaccheroni, *Chem. Eur. J.* **2010**, *16*, 919–930.
56. M. C. Aragoni, M. Arca, A. Bencini, A. J. Blake, C. Caltagirone, G. De Filippo, F. A. Devillanova, A. Garau, T. Gelbrich, M. B. Hursthouse, F. Isaia, V. Lippolis, M. Mameli, P. Mariani, B. Valtancoli, C. Wilson, *Inorg. Chem.* **2007**, *46*, 4548–4559.
57. Y. Mikata, M. Wakamatsu, S. Yano, *Dalton Trans.* **2005**, 545–550.
58. Y. Mikata, A. Yamanaka, A. Yamashita, S. Yano, *Inorg. Chem.* **2008**, *47*, 7295–7301.
59. Y. Mikata, M. Kaneda, H. Konno, A. Matsumoto, S. Sato, M. Kawamura, S. Iwatsuki, *Dalton Trans.* **2019**, *48*, 3840–3852.
60. R. Y. Tsien, T. Pozzan, T. J. Rink, *J. Cell. Biol.* **1982**, *94*, 325–334.
61. Y. Mikata, M. Kaneda, M. Tanaka, S. Iwatsuki, H. Konno, A. Matsumoto, *Eur. J. Inorg. Chem.* **2020**, 757–763.

62. T. Gunnlaugsson, T. C. Lee, R. Parkesh, *Tetrahedron* **2004**, *60*, 11239–11249.
63. Y. Yang, T. Cheng, W. Zhu, Y. Xu, X. Qian, *Org. Lett.* **2011**, *13*, 264–267.
64. Q. Zhao, R.-F. Li, S.-K. Xing, X.-M. Liu, T.-L. Hu, X.-H. Bu, *Inorg. Chem.* **2011**, *50*, 10041–10046.
65. Z. Xu, K.-H. Baek, H. N. Kim, J. Cui, X. Qian, D. R. Spring, I. Shin, J. Yoon, *J. Am. Chem. Soc.* **2010**, *132*, 601–610.
66. C. Lu, Z. Xu, J. Cui, R. Zhang, X. Qian, *J. Org. Chem.* **2007**, *72*, 3554–3557.
67. Z. Xu, X. Qian, J. Cui, R. Zhang, *Tetrahedron* **2006**, *62*, 10117–10122.
68. Y. Mikata, A. Takekoshi, M. Kaneda, S. Yonemura, Y. Aono, A. Matsumoto, H. Konno, S. C. Burdette, *Eur. J. Inorg. Chem.* **2021**, 1287–1296.
69. Y. Mikata, *Dalton Trans.* **2020**, *49*, 17494–17504.
70. Y. Mikata, Y. Sato, S. Takeuchi, Y. Kuroda, H. Konno, S. Iwatsuki, *Dalton Trans.* **2013**, *42*, 9688–9698.
71. Y. Mikata, A. Kizu, H. Konno, *Dalton Trans.* **2015**, *44*, 104–109.
72. Y. Mikata, A. Takekoshi, M. Kaneda, H. Konno, K. Yasuda, M. Aoyama, S. Tamotsu, *Dalton Trans.* **2017**, *46*, 632–637.
73. Y. Mikata, K. Nozaki, M. Tanaka, H. Konno, A. Matsumoto, M. Kawamura, S. Sato, *Inorg. Chem.* **2020**, *59*, 5313–5324.

8 Molecular Bio-Sensors and the Biological and Biomedical Activities of Vanadium

Nuttaporn Samart
Department of Chemistry, Colorado State University, Fort Collins, CO 80523, USA
Department of Chemistry, Rajabhat Rajanagarindra University, Chachoengsao, Thailand

Deborah A. Roess
Department of Biomedical Sciences, Colorado State University, Fort Collins, CO 80523, USA

Debbie C. Crans
Department of Chemistry, Colorado State University, Fort Collins, CO 80523, USA
Cell and Molecular Biology Program, Colorado State University, Fort Collins, CO 80523, USA

CONTENTS

DOI: 10.1201/9781003229971-8

Abstract

Vanadium is one of the ten first-row transition metals that is generally not recognized as an essential element in humans but is essential in other organisms, such as some tunicates, algae, and worms. Because vanadium has both beneficial biological effects to humans and the biosphere in general as well as toxic effects, more knowledge about vanadium's mechanism of actions in either positive or negative ways remains important. Uncovering fundamental mechanisms for observed activities of vanadium is complicated by vanadium's ability to interact and form complexes with both anions and cations as well as complex speciation chemistry. Since speciation is important to the effects of vanadium, studies about vanadium have depended, and will continue to depend, on the ability to measure vanadium content in vanadium-containing systems and to evaluate the speciation of this vanadium content. Vanadium has also proved useful in the development of sensors that include effective devices based on a combination of enzymatic and electrochemical detection. The integration of vanadium oxides into sophisticated bio-sensors, particularly for environmental, biological, and biomedical samples, also involves a variety of systems that take advantage of, mainly, vanadium oxide's desirable electrochemical properties. The widespread use of such vanadium-containing systems documents the impact of vanadium chemistry on a range of important medical and societal problems.

KEYWORDS

Bioactive Element; Bio-sensors Containing Vanadium; Bio-sensors Measuring Vanadium; Both Beneficial And Toxic; Glucose Bio-sensors; Speciation Chemistry; Vanadium

1 INTRODUCTION

This chapter focuses on the beneficial and toxic effects of vanadium in biological systems, the detection of vanadium, and its role in the construction of bio-sensors. Vanadium, the element with atomic number 23, was discovered over two centuries ago. This predates Dmitri Mendeleev's creation of the framework for the periodic table and involved scientists associated with Spain, Mexico, France, Sweden, and Germany [1,2]. Vanadium was first isolated in 1801 from ore containing vanadinite, ($Pb_5[VO_4]_3Cl$) by Andrés Manuel del Rio, who, at the time, worked at an academy located in Spanish territory in what is today Mexico [3]. Del Rio named the new element Erythronium. Unfortunately, his letter to the Institute de France in Paris, France, supporting his claim and requesting the analysis of his samples, was lost in a shipwreck. When Del Rio's samples eventually reached Paris, they were incorrectly analyzed and described as an impure form of chromium leading del Rio to withdraw his discovery of a new element.

Twenty-nine years later, in 1830, vanadium was rediscovered by Nils Gabriel Sefström from Sweden, who was analyzing iron samples from the Taberg Mine in Småland, Sweden [4]. Sefström named the newly discovered element vanadium after Vanadis, the Nordic goddess of beauty and fertility, because of vanadium's many colorful compounds. Sefström, in collaboration with his mentor J. J. Berzelius in Sweden and the German chemist Friedrich Wöhler, prepared vanadium chlorides and demonstrated that their vanadium compounds were derived from the same element, erythronium, previously discovered by del Rio. Credit for vanadium's discovery is often attributed to research in Mexico, despite Spanish colonization of Mexico at the time, and Sweden, where vanadium was officially named. Ultimately, vanadium's discovery involves contributions from scientists associated with five different countries over a period of 30 years.

Vanadium is one of the ten first-row transition metals that is generally not recognized as an essential element in humans but is essential in other organisms. Vanadium, nonetheless, has some beneficial biological effects on humans and the biosphere in general [5], including but not limited to the normalization of thyroid function [6], blood glucose levels, and beneficial effects on lipid metabolisms [5,7–10]. However, there are also forms of vanadium that are potentially toxic [11,12]. Studies probing vanadium's mechanism of action, in either positive or negative ways, remain essential, particularly given the increasing interest in the use of vanadium compounds to treat various forms of cancer [13–15] and reports of toxic effects due to environmental exposures, including vanadium effects on preterm births [11,16]. Uncovering fundamental mechanisms for observed activities of vanadium is complicated by vanadium's ability to interact and form complexes with both anions and cations [17], which can both reduce the toxicity of parent compounds [17] and increase the beneficial effects of this element [13–15]. Given the interest in vanadium compounds and their complex chemistry and biology, vanadium studies have depended on the ability to measure V-content in vanadium-containing systems. Furthermore, the integration of vanadium oxides into sophisticated bio-sensors, particularly for environmental, biological, and biomedical samples, means that vanadium chemistry can impact a range of biologically relevant societal problems.

2 ESSENTIAL AND BENEFICIAL EFFECTS OF VANADIUM

Whether vanadium is essential for humans remains controversial. Although it is generally agreed that vanadium is not one of the five essential first-row transition metals, it is one of the three additional elements with biological activity [18,19]. Interestingly, the requirement for vanadium has been demonstrated in organisms, such as tunicates (sea squirts) [20–23], algae [24–27], and mussels [28,29]. An exploration of why vanadium is needed in these species has led to the discovery of chloroperoxidase and bromoperoxidase in the algae and the vanabin proteins in tunicate enzymes, where vanadium functions as a co-factor. The oxidation state of vanadium depends on the distribution of proteins in these organisms; for chloroperoxidase and bromoperoxidases, vanadium is in the form of vanadium(V) [24–26], and for vanabin proteins in tunicates, vanadium is generally in oxidation state IV or III [20,22–23]. However, this analysis is complicated by vanadium's ability to cycle between oxidation states and the identification of vanadium in an oxidation state that is generally compatible with its local cellular environment. For example, vanadium in human blood, made up largely of water-containing proteins and ions, is primarily found as vanadium(IV). However, in the cytoplasm of cells or near the cell membrane, vanadium can exist in both the IV and V oxidation states. Numerous salts and complexes have been identified as potential pharmacologic agents in the treatment of diseases, such as diabetes and various forms of cancer [5, 7–10, 30–37]. Salts and a single vanadium complex (bismaltolatooxovanadium(IV), abbreviated as BMOV) were used in early clinical trials for the treatment of diabetes but were subsequently discontinued [9]. This was due, in part, to the small therapeutic index for vanadium salts and the loss of patent protection for bismaltolato oxovanadium(IV) on September 30, 2011, which, unfortunately, reduced the financial incentive for its further development as a drug [9, 10]. Nevertheless, a number of vanadium coordination complexes have been investigated with promising insulin enhancing properties, [35], and studies continue to be published using new systems [7, 38, 39]. For example, oxometalates (specifically decavanadate derivatives) have been reported to possess functional diabetic properties in animal model systems [38, 40].

Current interest focuses on new highly specific phosphatase inhibitors developed using a new design paradigm to create a vanadium derivative with specificity for the protein phosphatase 1B (PTP1B), Figure 1 [41]. Investigators coordinated vanadate to a peptide that was complementary to the ligand-binding site adjacent to the phosphorylation site in PTP1B. The vanadium compound was stably integrated onto a membrane-permeable graphene quantum dot platform where it showed selectivity *in vivo* in the db/db mouse model for liver and muscle tissue. In addition, specificity for PTP1B was changed by exchanging the peptide coordinated to vanadate with one specific for the T-lymphocyte tyrosine phosphatase (TCPTP) [41]. These studies demonstrated that specific vanadium-based inhibitors could be developed for phosphatases.

Low levels of vanadium have been linked to a lower incidence of diabetes in a human clinical study in China involving about 1,500 subjects and the measurement of blood levels of vanadium. Plasma vanadate levels were significantly lower in subjects with newly diagnosed non-insulin-dependent diabetes mellitus (NIDDM) than in control subjects ($p=0.001$) [42]. Furthermore, lower levels of vanadium found in the blood were correlated with NIDDM diagnosis compared with vanadium levels in healthy individuals.

In addition to the insulin-like activity of vanadium compounds, the anticancer properties of several vanadium complexes have been reported. For example, simple vanadium compounds have been shown to enhance the activity of oncolytic viruses, an emerging class of anticancer therapeutics that induce antitumor immunity through their selective replication in tumor cells [15, 43]. One strategy that boosts the therapeutic efficacy of oncolytic viruses is to combine their activity with immuno-modulating, small molecule protein tyrosine phosphatase inhibitors. Both vanadium salts and simple coordination complexes can enhance oncolytic infection *in vitro* and *ex vivo* in otherwise resistant tumor cell lines [15]. Mechanistically, this involves subverting the antiviral type I interferon (IFN) response toward a death-inducing, pro-inflammatory type II IFN response to produce improved oncolytic spread with increased bystander killing of cancer cells and enhanced antitumor immune

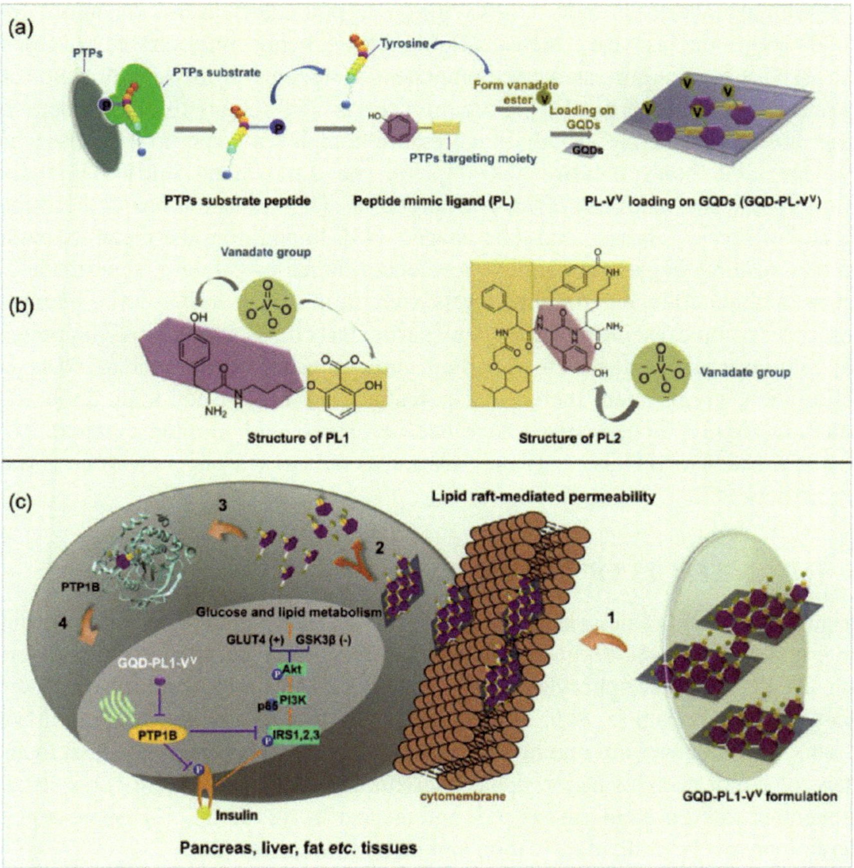

FIGURE 1 Design and action of PTPs selective GQD-(peptide mimic ligand)-vanadate complex. (a) Schematic diagram of the design of GQD-(peptide mimic ligand)-vanadate complex; (b) The structure of PTP1B selective mimic ligand PL1 and TCPTP selective mimic ligand PL2; (c) Schematic of cellular uptake and PTP1B inhibition of GQD-PL1-VV that were absorbed through lipid raft-mediated permeation (step 1). Then PL1-V(V) was released into the cytoplasm (step 2) and selectively bound to PTP1B (step 3) causing inhibition of PTP1B, which trigger insulin signal transduction and downstream effects (step 4). (Adapted with permission from Ref. [41].)

stimulation [15]. In another study using brain cancer cells (glioblastoma multiforme T98g cells), metastatic breast adenocarcinoma cells, epithelioid carcinoma of the pancreas cells, lung epithelial carcinoma cells, and normal human foreskin fibroblasts cells, non-innocent Schiff base oxovanadium catecholate complexes were shown to be at least one order of magnitude superior to cisplatin [14]. Interestingly, the lead compound was reactive but sufficiently stable to enter cells and exert anticarcinogenic actions. Similar results were obtained with human chondrosarcoma cells (SW1353) [44]. Subsequent studies show that the lead compound did not show any toxicity at concentrations as high as 300 mg/kg in mice and was less toxic than vanadate and the catechol making up the complex [39].

There have also been reports of exciting and potentially useful vanadium compounds within the last year. One vanadium complex being considered for anti-oncolytic activity is curcumin, an active component of rhizome turmeric with pleiotropic pharmacological properties [45]. Curcumin reacts with metals through a beta-diketone moiety to generate vanadium–curcumin complexes taken up by cells, which have favorable bioavailability and improve the antioxidant, anti-inflammatory, antimicrobial, and antiviral effects of curcumin. This molecule has demonstrated efficacy in cervical, breast, and liver cancers [45]. In addition, the metal–curcumin complex restores hepatic glutathione levels and increases catalase, superoxide dismutase, and glutathione peroxidase levels. Curcumin studies have revived interest in some older complexes, including Metvan, particularly their interactions with proteins [46]. Similarly, albumin–EDTA–vanadium complexes have been found to reduce cell proliferation, giving them therapeutic potential [13]. Lastly, non-toxic antioxidant vanadium-vitamin E derivatives have been prepared with limited cytotoxicity in NIH/3T3, Cal33, and HeLa cells and function as radical scavengers with no antioxidant activity [47].

3 TOXIC EFFECTS OF VANADIUM

Previous reviews have described the toxicity of vanadium salts, V_2O_5, and coordination complexes [11, 48, 49] and the inhibition of essential cellular enzyme systems, such as ATPases, phosphorylases, ribonucleases, and other phosphatases, which affect cell metabolism [5, 8, 9, 30, 31, 33]. In addition to direct effects on enzyme systems, several genes are regulated by vanadium salts or complexes, which include genes for tumor necrosis factor-alpha, Interleukin-8, activator protein-1, ras, c-raf-1, mitogen-activated protein kinase, p53, and nuclear factors-B [11, 50]. Some of these interactions involve radical formation and Fenton chemistry [51].

Preterm birth has also been linked to vanadium toxicity. A recent Chinese study involving 7,359 pregnant women enrolled at Wuhan Medical and Health Center investigated the occurrence of preterm birth (defined as 37 weeks of completed gestation or less according to the World Health Organization) [52] with accompanying metal exposure. The study measured 18 elements in the urine to determine metal mixture exposure and which metals contributed to preterm birth. The three elements most associated with preterm births are vanadium, chromium, and zinc [16], with the highest risk associated with vanadium. Skeletal malformations with vanadium exposure where vanadium was in the form of Na_3VO_4 (50 nM–1 mM) have been examined in *Paracentrotus lividus* sea urchin embryos to produce a vanadium toxicity profile [53]. At the morphological level, altered phenotypes and skeletal malformations were induced in a dose-dependent manner. At the molecular level, vanadium-exposed embryos showed the activation of a cellular stress response, in particular autophagy, and a high degree of cell-selective apoptosis. The stress response mediated by heat shock proteins counteracted the damage induced by low (50–100 nM) and intermediate (500–1,000 nM) concentrations of vanadium, while the high cytotoxic concentrations induced more frequent cell death mechanisms [53].

Arbuscular mycorrhizal fungi (AMF) has been shown to alleviate the adverse effects of V (350 mg V/kg soil) on shoots and roots of rye and sorghum, Figure 2 [17];

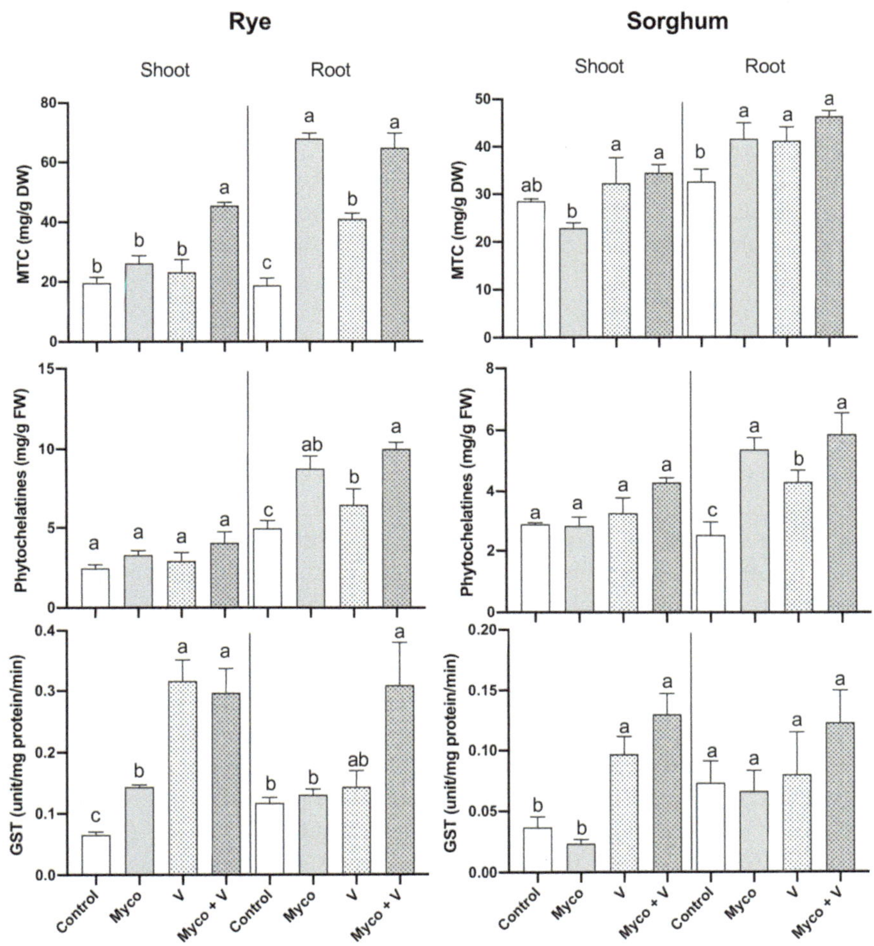

FIGURE 2 Changes in metallothionins (MTC) and phytochelatins content and Gluthathione-S-transferase (GST) activity in shoots and roots of 6-week-old rye and sorghum grown in soils with 0 and 350 mg/kg soil sodium vanadate and without soil enrichment with inoculum of the mycorrhizal *Phisophagus irregularis*. Data are mean values + sE ($n = 3$). Data were statistically analyzed by one-way ANOVA followed by Tukey's posthoc test for comparing the means. Different letters indicate statistically significant differences between means of the same plant species at significance level of at least ($p < 0.005$). (Adapted from Ref. [17]. This article is an open access article distributed under the terms and conditions of the Creative Comments. Attribution (CC BY) license (https://creativecommons.org/licences/by/4.0/).)

AMF-mitigated V-induced oxidative stress in rye and sorghum, which negatively impacted plant growth, photosynthesis, induced photorespiration, and oxidative damage. These protective effects were accompanied by increased acid and alkaline phosphatase activity in plant roots, increased organic acid and polyphenol extrusion into the soil, and increased absorption of mineral nutrients including Ca^{2+}, Mg^{2+}, and $H_2PO_4^-$ [17]. The plant detoxification strategies used to handle excessive vanadium resulted in sequestration in the root system, compartmentalization in vacuoles and cell walls, changing vanadium oxidation states, and plant uptake of vanadium. The plant detoxification strategies depend on vanadium concentration, pH, soil microorganisms and plant species, increased activity of the antioxidant enzymes and defense systems, decreased oxidative damage by reactive oxygen species, lipid peroxidation, and Fenton chemistry [50].

V_2O_5 is the most toxic vanadium compound and the one for which there are the strictest regulations. Extensive literature exists for this compound, and we refer the readers to reviews [11, 48, 50, 54–57] and the original literature for more information [56, 58–60].

4 SPECIATION AND SPECTROSCOPIC CHARACTERIZATION OF VANADIUM ACROSS OXIDATION STATES II–V

A detailed analysis of vanadium speciation, which is strongly dependent on the nature of the sample and the oxidation state of the vanadium compound, is essential before attempting to attribute observed biological effects to vanadium compounds. Numerous methods are available to measure vanadium levels, with some linked to vanadium's oxidation state and others to the biomedical sample and the processing needed before analysis. The two most abundant oxidation states in the biosphere are vanadium(V) and vanadium(IV), as shown in the E_h-pH diagram in Figure 3. However, recent reports [61, 62] suggest that vanadium(III) compounds may be more common than previously recognized. This recent development is likely due, in part, to high-field electron paramagnetic resonance (EPR) spectroscopy for observation of some forms of vanadium(III) and a recognition of the relatively mild conditions needed [61, 62].

Vanadium in oxidation state V is diamagnetic and forms oxidovanadate ions, characterized by solid-state and solution ^{51}V nuclear magnetic resonance (NMR) spectroscopy. In solution, the speciation of oxidovanadate ions is complex and depends on

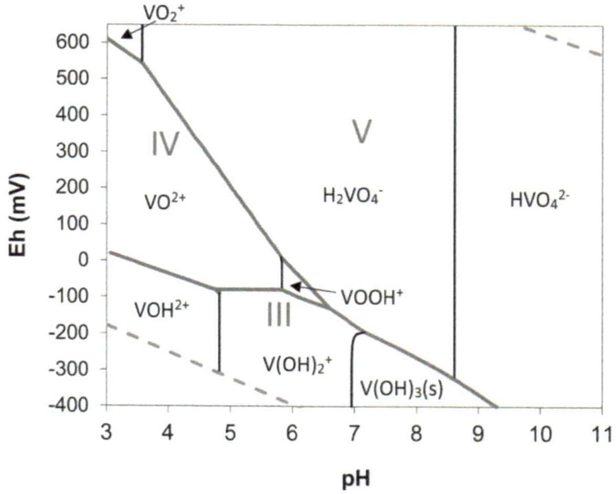

FIGURE 3 E_h-pH diagram of 0.01 mM V chemical speciation in water (0.01 M NaCl and 25°C). Transitions between oxidation states are indicated with full lines. The dotted lines indicate the stability limit for water. (Adapted with permission from Ref. [63].)

concentration, the presence of other sample components, pH, ionic strength, and temperature [63]. At low pH, the oxovanadyl ion exists as a cis-dioxovanadium(V) ions and, at neutral pH, analogs of phosphate anions. The situation is complicated by the fact that some vanadium(V) species, such as the yellow decavanadates, are stable [64] while others, such as the simple colorless oxovanadates, exchange rapidly [65]. This complicates these species' separation and isolation because rapid interconversion occurs [66].

Nevertheless, much has been done to characterize many coordination complexes as described in the original literature and reviews [47, 67–71]. Solid-state ^{51}V NMR has been used effectively to characterize the electronic properties of a series of coordination complexes, documenting this approach's usefulness. Unfortunately, the sensitivity of this method limits its use for the investigation of biological and biomedical samples [27, 69, 72–74].

Although vanadium in oxidation IV can be analyzed at ambient temperatures in solution using EPR spectroscopy, more information is obtained when the spectra are obtained from frozen films [62]. Free vanadium(IV)-oxido cationic species contain an unpaired electron at pH below 5 (as VO^{2+}) and pH > 10 (as $V(OH)_3^-$) [8, 75–78]. Vanadium(IV) ions dimerizes at neutral pH and can only be observed if complexed to a ligand. Extensive work has been done characterizing vanadium(IV) complexes in solution and the solid-state as detailed in the literature [26, 79, 80]. UV-vis spectroscopy has supplemented studies of these species.

Vanadium in oxidation state III is generally analyzed using UV-vis spectroscopy of pure samples because vanadium(III) species do not form oxido ions [81, 82]. However, aqueous V(III) solutions form several hydrolysis products, none containing the oxido group, that are simple cations hydrated by different water molecules and protonation states. The bare hexaaqua V(III) species is very sensitive to oxygen, and, in the absence of complexing ligands, these species oxidize to form V(IV) [81]. However, in the presence of ligands, V(III) can be more stable [83–86].

For the V(III) ion, there is more than a single unpaired electron, namely a 3d2 electron configuration, for which the spin ground state is usually a triplet, $S=1$. This non-Kramers system can occasionally be observed at X-band [87], but often zfs are too small to be observed. In such cases, high field electron paramagnetic resonance (HFEPR) is needed for its detailed characterization [62]. Aqueous vanadium(II) is even more reactive than V(III); this oxidation state exists in very reducing environments and is often stabilized by *N*-based ligands. V(II) is characterized by a 3d3 electron configuration, which yields an $S=3/2$ spin number and is, therefore, a Kramers (half-integer spin) species. Conventional EPR can often yield information on these complexes [62].

Additional methods used to characterize the oxidation state of vanadium in vanadium complexes include many electrochemical [88, 89] and magnetic [83] methods. Electrochemical methods describe the redox states of compounds in contrast to the methods in many biomedical studies where the emphasis is on measuring total metal [79].

5 ANALYSIS OF BIOLOGICAL AND BIOMEDICAL SAMPLES

In addition to the spectroscopic methods described above, ways to examine biological and biomedical samples that characterize samples containing vanadium in different oxidation states include mass spectrometry, inductively coupled plasma, and Synchrotron-radiation X-ray absorption spectroscopic (XAS) techniques [79]. These studies involve processing the samples and considering the properties of vanadium compounds, such as differences in lability as described for the kinetically stable decavanadate and the rapidly exchanging colorless oxovanadates [66].

Mass spectrometry (MS) can be used to study the behavior of metallodrugs in biological samples and characterize at the molecular level their interactions with, for example, proteins [90, 91]. ESI-MS (Electrospray Ionization) and MALDI (Matrix-Assisted Laser Desorption/Ionization) are powerful methods to document metallodrug-protein binding. Although MALDI generates mainly singly charged pseudo-molecular ions, ESI gives singly or multiply charged metal-protein adducts upon deprotonation or association with protons and alkali ions [90, 91]. An essential advantage of this method is that metal concentrations for ESI-MS experiments are in the range 1–100 μM and, as such, are close to those found under biological conditions. Recently, the ESI technique has been applied to metal species that form labile coordinative bonds with biomolecules, such as vanadium, in which proteins bind with a coordination bond [91–93]. However, there has been only limited application of this method to potential vanadium-containing drugs, one example being the inductively coupled plasma-mass spectroscopy (ICP-MS) investigation of biospeciation of some antidiabetic $V(IV)O^{2+}$ compounds in the serum from blood samples [94–96]. Recently, the interactions of V(IV) compounds with molecules of potential pharmacological or biological relevance [33, 97] were investigated using ESI-MS spectrometry [92]. In particular, the binding to lysozyme by four vanadium compounds with antidiabetic or anticancer activities has been demonstrated [92, 97].

X-ray absorption spectroscopy (XAS) is a method for speciation analysis with atomic resolution and information regarding the local chemical environment of vanadium [79, 99, 100]. XAS spectra are compared to extended X-ray absorption fine structure (EXAFS) (Figure 4) [98]. Hyphenated methods include *in situ* correlative imaging and speciation, *in vivo* X-ray Absorption Near Edge Spectroscopy (XANES), full-field XANES, and X-ray Free Electron Laser (XFEL) XAS have been developed. The speciation of vanadium-containing compounds has been studied in detail by the Levina and Lay team, including the generation of a comprehensive vanadium compound XAS database containing 23 compounds for use in the analysis of potential pharmaceuticals [101, 102]. The pre-edge peak is due to electronic transitions (mainly dipole allowed) to empty bound states near the Fermi level and is thus sensitive to vanadium atom oxidation state and coordination number. Therefore, the pre-edge features in the vanadium XAS spectrum simplify the speciation analysis of the biotransformation of potential vanadium-based antidiabetic compounds [101, 102].

As a consequence of this work, a XANES study of vanadium antidiabetic compounds under conditions that mimic oral administration mapped out the chemical changes in the gastrointestinal medium, including complex hydrolysis of V(IV) oxidation to V(V) in the absence of food and V(V) reduction to V(IV) and V(III) in the presence of food [102]. Furthermore, vanadium drug biotransformation has also been assessed in red blood cells; uptake of vanadium compounds, either by active transport [H_2VO_4], by passive diffusion V(IV), or due to V(V) complexes with organic

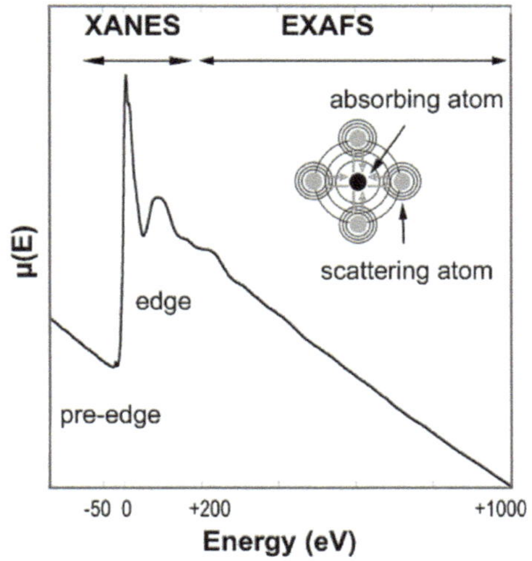

FIGURE 4 X-ray absorption spectroscopy. Schematic representation of a XAS spectrum showing the characteristic variation of the absorption coefficient $\mu(E)$ as a function of the X-ray incident energy. The sharp rise in $\mu(E)$ at the absorption edge energy, the oscillation of $\mu(E)$ after the edge, as well as the XANES are schematically depicted and compared to EXAFS spectra. (Adapted with permission from Ref. [98].)

ligands and their conversion into well-defined V(IV) and V(V) metabolites, occurs in a more reducing environment for vanadium than in blood plasma [103]. Importantly, XANES spectroscopic studies confirmed that mammalian cells could generate and maintain V(V) levels even when treated with V(IV) complexes [103].

Atomic absorption spectroscopy and inductively coupled plasma have been used for many biological studies of vanadium. When carefully calibrated for the media from which the sample is obtained or measured, these methods provide valuable information. In addition, they have been used for animal and human studies in which vanadium levels are measured [102, 104].

6 VANADIUM-CONTAINING BIO-SENSORS

Some types of bio-sensors are devices that use living organisms or biological molecules, such as enzymes or antibodies, to measure biological or chemical reactions by generating signals proportional to the concentration of an analyte with high specificity. Here we describe two classes of bio-sensors utilizing vanadium chemistry. One type of bio-sensor measures and quantifies vanadium, while the other bio-sensor contains some form of vanadium important to the bio-sensor structure and function.

6.1 Bio-Sensors Measuring Vanadium

Electrochemical methods in sensors and bio-sensors can be a useful and inexpensive approach for vanadium and many different biomolecules [105, 106]. For example, a bio-sensor has been developed based on batch-adsorptive stripping voltammetry and is used to measure trace levels of vanadium. In this electrochemical method, the analyte forms a complex with an added ligand, which can then adsorb onto an electrode surface, usually a mercury drop electrode [105, 106]. This step provides sample preconcentration, which gives the method its high sensitivity. The adsorbed complex can then be stripped off using various voltammetric methods. An example of the online electrochemical stripping procedure for trace determination of vanadium has been reported to combine the advantages of adsorptive stripping voltammetry with flow injection analysis [105, 106]. Under optimized conditions, the detection limit was 8 ng/I, and a mid-range standard deviation of 2%–3% was obtained. In addition, the response was linear up to 3.5 μg/L with a sensitivity of 0.1 mA L/μg.

An electrochemical method has also been developed to detect and separate V(V) in the presence of other ions, such as Cr(VI) in water samples [107]. The detection technique was based on the direct linear sweep voltammetric reduction of Cr(VI) and V(V) using multi-walled carbon nanotube-neutral red-gold nanoparticles (MWCNTs-NR-AuNPs) on a modified, commercially-available screen-printed carbon electrode. Neutral red was covalently bound to the carboxylates on the MWCNTs on the surfaces of NR-AuNPs. Total chromium and vanadium were determined after oxidation of Cr(III) to Cr(VI) and V(IV) to V(V) by potassium permanganate. The linear range of Cr(VI) and V(V) was 0.4–80 and 3–200 μM with the detection limits of 0.025 and 0.42 μM ($S/N = 3$), respectively [107].

Another strategy uses polysulfone hydrogels prepared as water-swellable powders or drop cast as thin films, which are electroactive and characterized by two well-resolved redox peaks, with a formal potential of 0.0867 V and a diffusion coefficient in an aqueous medium of $9.06e^{-9}$ cm^2/s [108]. Initial speciation studies were done for selenium and vanadium. As evaluated by using cyclic voltammetry, the analytical performance at the hydrogel electrodes was in a range of −0.7 to +0.0 V versus Ag/AgCl. Finally, the morphology, adsorption, and thin-film integrity were evaluated using high-resolution scanning electron microscopy, UV-Vis spectroscopy, and Raman spectroscopy [108].

Numerous electrochemical bio-sensors based on the principle of enzyme inhibition have been reported [109, 110]. Phosphatases are a class of enzymes that bind vanadium in the form of vanadate and vanadyl specifically [111]. Much work has been carried out on the details of the transition state structures [111, 112]. Although the hydrolysis of phosphate esters is thermodynamically favorable, the uncatalyzed hydrolysis of phosphate monoesters is extraordinarily slow, making phosphatases among the most catalytically efficient enzymes known, with rate enhancements of up to 10^{11}–10^{13} [113, 114]. The central role of protein phosphatases in signal transduction and their roles in disease has led to detailed mechanistic studies of different classes of protein phosphatases [111]. Recently, the origin of the order of magnitude difference in catalytic rates observed in NMR studies of the human protein tyrosine phosphatase 1B (PTP1B) and YopH protein tyrosine phosphatase from *Yersinia pestis* was investigated using computational studies linking the catalytic rates with the differing conformational dynamics of the closure of a protein loop, the WPD-Loop. These analyses identified key amino acid residues and structural features responsible for the rate differences, as well as the residues and pathways that facilitate allosteric communication within these enzymes [115, 116].

Alkaline phosphatase (ALP) is more abundant, less specific, and a significantly less expensive phosphatase than protein phosphatases [117]. The ability to use alkaline phosphatase to detect vanadate and vanadyl cation by observing the inhibition of p-nitrophenyl phosphate hydrolysis and the speciation of vanadium(V) and (IV) in buffer solutions was reported [118, 119]. These enzyme assays demonstrate that even a phosphatase, such as alkaline phosphatase with lower specificity than protein tyrosine phosphatases, has sufficient sensitivity to detect the subtle interactions of buffer with vanadate and vanadyl cations [118, 119]. In addition, an aqueous assay was reported where the phosphatase activity was observed by screen-printed carbon electrodes modified with gold nanoparticles [120]. The inhibition by vanadate resulted in a decrease in the chronoamperometric signal, which was proportional to the concentration of vanadate.

Considering the expense of phosphatases, bio-sensors are generally based on the enzyme activity of immobilized ALP, Figure 5 [108, 121–124]. Electrochemical ALP-based assays have been reported with substrates including phenyl phosphate

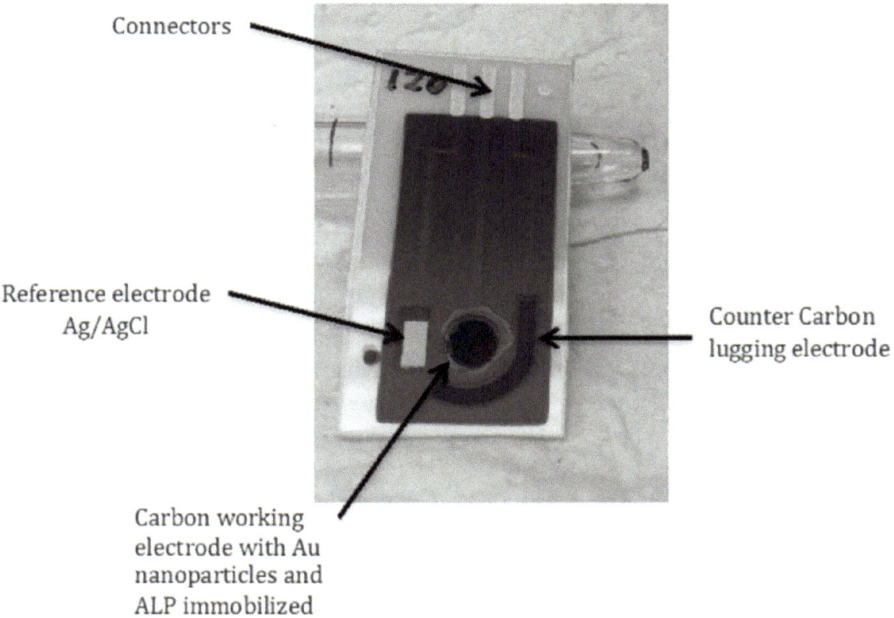

FIGURE 5 ALP-AuNPs-SPCE bio-sensor used for vanadium determination. Carbon working electrode area, 12.6 mm^2. The working electrode of these devices was electrochemically modified by AuNPs, using a 0.1 mM solution of $HAuCl_4$ in 0.5 M H_2SO_4. The deposition was performed by applying a potential of +0.18 V (*vs.* Ag/AgCl SPE) for 15 s under stirring conditions. The enzyme was then immobilized by cross-linking a mixture of ALP, BSA, and glutaraldehyde on the surface of AuNPs at the optimum conditions for best current response. The sensor was developed at 4°C and stored at this temperature. Samples were developed by reacting the test solution containing ALP-AuNPs-SPCEs and 4-p-nitrophenyl phosphate at 4°C for 1 h. The activity of the Bio-sensor was highest at pH following the enzyme reactivity profile (from pH 8 to 9). (Adapted with permission from Ref. [122].)

[125–127], naphthyl phosphate [127, 128], rivoflavin-5-monophosphate [120], ascorbic acid 2-phosphate [127, 129], and *p*-nitrophenyl phosphate [127]. The most widely used substrate in electrochemical assays with ALP is *p*-nitrophenyl phosphate, which gives the greatest electrochemical response [123]. In one study, a gold electrode was covalently modified with a hydrogel (Au-HGL) as a platform for the immobilization of ALP enzyme to produce an Au-HGL/ALP bio-sensor [121]. The analytical performance of this bio-sensor was assessed using square wave voltammetry, electrochemical impedance spectroscopy, and amperometry. The calibration curve displayed a dynamic linear range of 0.5–30 μM with a detection limit and quantification limit of 0.23 and 0.76 μM, respectively. This Au-HGL/ALP bio-sensor exhibited high sensitivity to vanadium in an aqueous medium with good reproducibility ($n=4$) and a relative standard deviation of 8.0% [121].

A chronoamperometric bio-sensor has also been reported to determine vanadate using alkaline phosphatase [122]. Screen-printed carbon electrodes modified with gold nanoparticles were used as transducers for the immobilized alkaline phosphatase. The alkaline phosphatase catalyzed hydrolysis of 4-nitrophenyl phosphate was inhibited by vanadate and resulted in a decrease in the chronoamperometric current registered with a detection limit of 0.39 ± 0.06 μM. Repeatability was 7.7% ($n=4$) and reproducibility was 8% ($n=3$). Although interferences were observed with Mo(VI), Cr(III), Ca(II), and W(VI) ions that may affect vanadate determinations at a concentration higher than 1.0 mM, this method has, nevertheless, been successfully applied to the determination of vanadate in spiked tap water [122].

6.2 Bio-Sensors Containing Vanadium and Vanadium Oxides

Several bio-sensors have been reported that contain some form of vanadium, usually vanadium oxide. Although these bio-sensors do not measure vanadium, it is part of the material from which the sensor is made and is essential to the function of the bio-sensor [130, 131]. Many of these sensors detect biologically important molecules and, for this reason, this topic is summarized here. The vanadium-containing materials sensor is a vanadium oxide that can be orthorhombic V_2O_5, rhombohedra V_2O_3, monoclinic V_3O_7, orthorhombic V_4O_9, monoclinic V_6O_{13}, or one of the allotropic VO_2, Figure 6 [131]. The VO_2 phase includes the tetragonal/rutile phase, P 42/mnm VO_2(R), monoclinic phase, P21/c VO_2 (M), monoclinic phase, C_2/m VO_2(B), or tetragonal phase, P 42/ncm VO_2(A). The most used vanadium oxides in sensors are V_2O_5 and VO_2(R) due to a metal-insulator transition at relatively low temperatures that may alter the resistivity by three orders of magnitude [132]. Combustible gas sensors are frequently based on nanostructured vanadium pentoxide, V_2O_5 [133]. Wu et al. [133] synthesized V_2O_5 hollow spheres and suggested that the high ratio of the volume-to-surface area significantly improved gas-sensing; V_2O_5 nanostructures increased the conductivity of the sensor because, when exposed to reducing gases, the V^{5+} species were partially reduced to V^{4+}, allowing for the formation of oxygen vacancies and thus increasing the conductivity of the sensor [133]. This phenomenon was observed visually by a color change of the material from yellow to dark blue. In semiconductor nanomaterials, water vapor is ionized to OH^- and H^+. The oxygen vacancies absorb oxygen ions at surface active sites, allowing for another H_2O

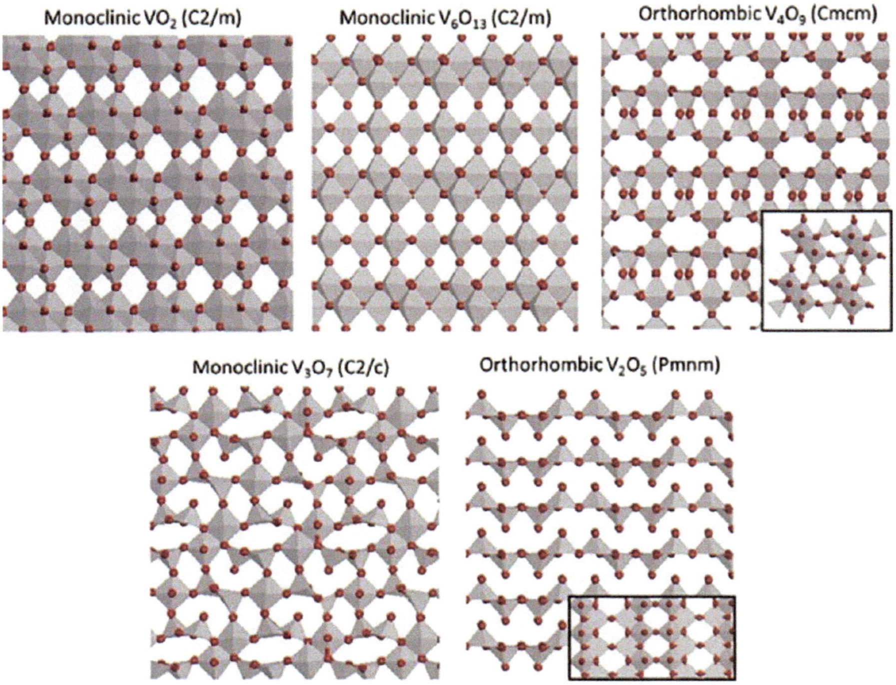

FIGURE 6 The crystal lattices of five different vanadium oxides: monoclinic VO_2, monoclinic V_6O_{13}, orthorhombic V_4O_9, monoclinic V_3O_7, and orthorhombic V_2O_5. (Adapted with permission from Ref. [131].)

molecule to be adsorbed to the H^+ bond between two adjacent OH^- groups [134, 135]. The metallic VO_2(B) phase has a lower resistance because the dominant charge carriers are electrons found at high humidity, where H_2O layers and H_3O^+ are formed at the VO_2(M) surface. This causes the depletion layer to decrease and increases surface conductivity. At high humidity and as a result of the polarity of the H_2O molecules, some of these e^- form H-bonds with the positively charged H^+ ions [136]; the density of free electrons at the VO_2(B) surface will decrease, and the resistance of the material in a sensor designed for humid conditions will increase [134]. Some of the reported vanadium-containing bio-sensors detecting glucose, dopamine, DNA, and several other metabolites are summarized in Table 1 and discussed below.

6.2.1 Glucose Sensors

Bio-sensors to detect glucose are the largest group of vanadium oxide-containing bio-sensors because of the need for sensitive detection of blood glucose levels by people with diabetes. For this application, non-invasive optical monitors are of particular interest. A few representative approaches to developing these devices include non-enzymatic vanadium-containing bio-sensors and two enzymatic bio-sensors of glucose and a non-enzymatic V_2O_5 NP sensor using a new marker, methylglyoxal (MG), that is detected using electrochemical methods. An electrochemical non-enzymatic bio-sensor based on a NiVP/Pi material was developed for the selective

TABLE 1
Bio-sensors Containing Vanadium or Vanadium Oxides

Bio-sensor	Detection	Material	Film/Immobilization Film	Substrate	Detection Limit	References
Glucose NiVP/Pi material	Electrochemical	NiVP/Pi material on Whatman filter paper	NiVP/Pi material	Glucose	3.7 nM	[137]
Glucose YVO4: Yb^{3+}, Er^{3+} @Nd^{3+} core/shell UCNPs La Ion	Fluorescent UCNPs with YVO_4 upconversion efficiency	Core: YVO_4: Yb^{3+}, Er^{3+} / UCNP upconversion efficiency	YVO_4/shell UCNPs particles doped with lanthanides	Glucose		[139]
Glucose GdVO4: Yb^{3+}, Er^{3+} @Nd^{3+} core/shell UCNPs La Ion	Fluorescent UCNPs with $GdVO_4$ upconversion efficiency	Core: $GaVO_4$: Yb^{3+}, Er^{3+} / UCNP containing $GdVO_4$ upconversion efficiency	$GdVO_4$/shell particles doped with lanthanides	Glucose		[139]
Glucose YVO4: Yb^{3+}, Er^{3+} @Nd^{3+} core/shell UCNPs La Ion	Fluorescent UCNPs with lanthanides doped upconversion efficiency	Core: YVO4: Yb^{3+}, Er^{3+} @ Nd^{3+}	YVO4: /shell UCNPs with Nd and Yb shell	Glucose		[140]
Glucose Yb^{3+}, Er^{3+} @Nd^{3+} core/shell UCNPs La ion	Fluorescent UCNPs with Lanthanide doped upconversion efficiency	Core: $GaVO_4$: Yb^{3+}, Er^{3+} @ Nd^{3+} core	GaVO4: /shell UCNPs with Nd and Yb shell	Glucose		[140]
Glucose V_2O_5/GO_x	Enzymatic processing/ amperometric detection	GO_x on V_2O_5 film PEG/ITO terephthalate	GO_x on V_2O_5 film	Glucose	1.621 A A/ (mmol/L cm^2)	[141]
Glucose VO_2/GO_x	Enzymatic processing/ amperometric detection	GO_x – citric acid-glutaraldehyde film PET/ITO substrate	GOx on VO_2 – citric acid- glutaraldehyde film	Glucose	33.9 mV	[142]

(Continued)

TABLE 1 (*Continued*)
Bio-sensors Containing Vanadium or Vanadium Oxides

Bio-sensor	Detection	Material	Film/Immobilization Film	Substrate	Detection Limit	References
MethylglyoxalV_2O_5 NPs detecting MG/Au	V_2O_5 NPs modified Au electrode	Vanadium containing NPs sensitive to MG	V_2O_5 NPs modified Au electrode	Methyl-glyoxal	0.24 μM	[143]
Dopamine PMo11V@ graphene	PMo11V@graphene enhanced electrochemistry	PMo11V@ graphene	PMo11V@ graphene	Dopamine	0.88 μmol/ dm^3	[144]
Dopamine V_2O_5 embedded on graphene oxide	V_2O_5/graphite oxide enhanced/amperometric sensed by MVGO50 electrodes	V_2O_5/graphite oxide on MVGO50 electrodes	V_2O_5/graphite oxide	Dopamine	0.07 μM	[145]
Urease immobilized on V_2O_5	Urea/detected by urease in V_2O_5 sol gel/ Amperometric detection	Urea/urease/V_2O_5 sol gel/ PET/ITO	Urease immobilized on V_2O_5 gel films	Urea		[146]
DNA chitosan (CTS)/nano-V_2O_5/ multi-walled carbon nanotubes (MWCNTs)/CILE	DPV with methylene blue as an indicator	DNA chitosan/nano-V_2O_5/ MWCNTs	ssDNA adsorbed increased by nano-V_2O_5 and MWCNTs	ss-DNA LAMP product of *Yersinia enterocolitica* gene sequence	1.76×10^{-12} mol/L	[147]
DNA-V-doped	Various methods characterizing system	V^{3+}-doped DX DNA lattices and DNA layers/ ITO/Glass	DNA doped with V^{3+}	DNA		[148]
Immuno bio-sensor for CEA	Electrochemistry	Stannic oxide/reduced graphene oxide/Au NPs/Pd NPs/V_2O_5/MWCNTs	Stannic oxide/reduced graphene oxide/Au NPs/ Pd NPs/V_2O_5/MWCNTs	CEA	0.17 pg/mL	[149]

PET/ITO, indium tin oxide coated PET plastic; UCNPs, fluorescent upconversion nanoparticles; MG, Methylglyoxal; MWCNTs, multiwalled carbon nanotubes; CILE, carbon ionic liquid electrode; DPV, Differential pulse voltammetry; Au NPs, gold nanoparticles; Pd NPs, palladium nanoparticles; CEA, embryonic antigen.

and sensitive determination of glucose [137]. The sensor showed a high sensitivity of 6.04 mA mM cm^2 with a detection limit of 3.7 nM in a wide detection range of 100 nM–10 mM. The approach resulted in a system that exhibited selectivity and showed no interference from O_2 evolution during glucose sensing.

Furthermore, negligible interference was observed in the presence of biological metabolites that were potentially interfering, including ascorbic acid, uric acid, dopamine, or sodium chloride [138]. The system was modified to prepare a flexible sensor by coating the NiVP/Pi onto Whatman filter paper, which exhibited two linear ranges in 0.1 M NaOH of 100 nM^{-1}–1.0 mM and 100 μM–10 mM with an ultra-sensitivity of 1.130 and 0.746 mA mM cm^2, respectively. The sensor was used to test glucose levels in human blood serum samples to confirm that an approach using filter paper was reasonable [138].

A sensitive, reversible, and selective bio-sensor has also been developed based on optical glucose detection by two different upconverting vanadium-containing nanoparticles (UCNP). The UCNPs are prepared in the presence of citric acid and do not require any surface functionalization or other modifications to detect glucose [139, 140]. The detection of glucose was accomplished by using YVO_4: Yb^{3+}, Er^{3+}@Nd^{3+} core/shell fluorescent UCNPs with Nd and Yb shell or alternatively $GdVO_4$: Yb^{3+}, Er^{3+}@Nd^{3+} core/shell fluorescent UCNPs with Nd and Yb shell; Table 1 [140].

Bio-sensors using a combination of V_2O_5, immobilized glucose oxidase (GO_x) [141, 142], and electrochemical detection have been reported. One bio-sensor used crosslinked GO_x on a V_2O_5/GO_x thin film deposited onto a layer of indium tin-oxide (ITO)-coated polyethylene terephthalate film. Immobilization of GO_x resulted in an increase in total charge and better performance in glucose detection. The rugosity factor indicated a 0.125 and subsequently 1.7×10^{-4} of an electrochemically active surface of V_2O_5/GO_x with low porosity. The sensitivity of the V_2O_5/GO_x sensor was initially 1.62 L A $(mmol/Lcm^2)^{-1}$ [141] and subsequently improved [142], Table 1.

MG is a predominant precursor for advanced glycation end products (AGEs). Due to protein glycation reactions, a significant cause of diabetic complications, MG has been explored both for its potential as a biomarker and as a bio-sensor target that predicts the onset of diabetic complications. This is accomplished by detecting MG in human blood plasma samples using V_2O_5-containing NPs [143]. The non-enzymatic electrochemical bio-sensor consists of orthorhombic V_2O_5 nanoplates prepared in a microwave as interface material to fabricate a modified gold Au working electrode for electrochemical sensing of MG. Cyclic voltammetry and amperometry studies have confirmed the electrocatalytic nature of a V_2O_5 nanoplate-modified Au electrode in detecting MG with a sensitivity of 4.519 μA/μM, with a linear range of 3–30 μM. The limit of detection was 0.24 μM and the limit of quantification was 0.80 μM, with a MG response time less than 8 s. The lifetime of the sensor was found to be 25 days (90%) and the percentage recovery was 102.5%–108.7%, respectively.

6.2.2 Dopamine Bio-Sensor

Detection of dopamine, an excitatory neurotransmitter, is also of considerable interest. When the tetrabutylammonium salt of phosphovanadylmolybdate was immobilized on graphene flakes (GFs), its electrochemical properties were enhanced, and the immobilized $PMo_{11}V$ was found to selectively detect dopamine [144]. Graphene is a two-dimensional material consisting of sheets of sp2-bonded carbon atoms arranged in a perfect honeycomb lattice. The lattice provides a large electrochemical potential window, good electronic properties, a large specific surface area, and an excellent platform on which different molecules can be anchored. Calculations showed that the electrochemical surface coverage was 20 times higher than the free POM. The $PMo_{11}V$@GF modified electrode showed excellent sensing performance for detecting dopamine in the presence of ascorbate [145]. A different approach was also reported to develop a bio-sensor for dopamine using melt-quenched vanadium oxide (V_2O_5) embedded in graphene oxide sheets as composite electrodes for amperometric dopamine sensing [145].

6.2.3 Urea Bio-Sensor

Management of kidney diseases requires the ability to measure high urea levels. The method used most commonly in the clinic is to assess urea and creatine clearance in 24-hour urine samples [145]. An amperometric bio-sensor has been developed for the detection of urea using the enzyme urease immobilized in vanadium pentoxide films. This V_2O_5/urease bio-sensor has high reproducibility, low cost, high effectiveness, and is easy to handle [146].

6.2.4 DNA Bio-Sensors

A deoxyribonucleic acid (DNA) electrochemical sensor based on the CTS–V_2O_5–MWCNTs/CILE has been reported and applied to detect the LAMP product from the *Yersinia enterocolitica* gene. The modified electrode surface was characterized using SEM, cyclic voltammetry, and electrochemical impedance spectroscopy. The addition of V_2O_5 nanobelts and MWCNTs to the electrode surface had synergistic effects on the immobilization of the ssDNA probe, increasing ssDNA loading quantity and improving the detection sensitivity for DNA hybridization. These results suggest that the CTS–V_2O_5–MWCNTs/CILE has the potential for detecting specific gene sequences [147]. DNA doped with transition metal ions shows excellent versatility and potential use as molecular-based bio-sensors. Generally, $[V^{3+}]_C$ provides crucial information on the structural stability and extremum physical characteristics of V^{3+}-doped DNA nanostructures by the optimum incorporation of V^{3+} into DNA and the formation of DNA lattices (formed by synthetic double-crossover tiles) and DNA layers with the V^{3+} ions between layers. DNA nanostructures doped with transition metal ions might be considered part of a new class of materials that are easy to fabricate, cost-effective, and feasible to construct via mass production [148].

6.2.5 Carcino-Embryonic Antigen Bio-Sensor

A sandwich-type non-enzymatic electrochemical immune-bio-sensor has been reported that measures levels of carcino-embryonic antigen (CEA) [149]. Stannic oxide on reduced graphene oxide was used as a substrate to enhance the conductivity of the glassy carbon electrode. Gold nanoparticles (AuNPs) were used to link substrate and primary antibodies (Ab1) by accelerating the electron transfer. Palladium nanoparticles (PdNPs)–V_2O_5/MWCNTs were used to label secondary antibodies (Ab2). This experimental design catalyzed the reduction of H_2O_2 and improved signal amplification to improve bio-sensor sensitivity and lower the limit of detection to 0.17 pg/mL [149].

7 CONCLUSIONS

The role of vanadium chemistry in biological systems continues to be an exciting and evolving field. Here we have described recent studies demonstrating vanadium's ability to induce beneficial effects on plants, fungi, mammals, and humans and toxic effects on these organisms. Generally, beneficial effects are observed at low levels, and exposures and harmful effects appear at higher levels of exposure. Although there is no strong evidence for vanadium being an essential element in humans, some tunicates, algae, and mussels require vanadium, which plays an interesting role in their metabolism.

Recently, specific vanadium-containing inhibitors for protein tyrosine phosphatases using the exchangeable appendage coordinated to the vanadium have, surprisingly, demonstrated that vanadate-inhibition of phosphatases is not specific. The change in protein phosphatase specificity from PTP1B to TCPTP was accomplished by binding to an adjacent and allosteric binding site in the phosphatase, which increased the inhibitor's affinity and incorporated protein phosphatase specificity. In comparison to the known insulin-enhancing effects of the vanadium compound BMOV, it was shown that BMOV had less impact than the PTP1B inhibitor in lowering elevated glucose levels and selectivity *in vivo* in the db/db mouse model. The inhibitor was highly effective when introduced into cells, which occurred readily using vanadium-peptide derivatives associated with graphene quantum dots. These studies demonstrate that specific vanadium-based inhibitors can be developed for particular phosphatases.

New clinical studies have shown that vanadium levels in pregnant mothers are associated with premature birth compared to other metals, such as chromium, zinc, copper, and other elements, such as selenium. This motivates a continued interest in identifying contaminated soils and exploring methods for soil remediation. The effects of vanadium on the roots of rye and sorghum have also been investigated. Low concentrations of vanadium appeared to be beneficial. Further, high concentrations of the AMF can counteract the inhibiting effects of high concentrations of vanadium, attributed to vanadium effects on heat shock proteins.

Complex aqueous speciation was briefly mentioned because of the different oxidation states and many protonation states of oxidovanadium ions for vanadium in oxidation states IV and V. The methods for systematic studies of the three common oxidation states of vanadium in biological and environmental samples include vanadium(V)–^{51}V NMR spectroscopy, vanadium(IV)–EPR spectroscopy, and vanadium(III, IV, V)–UV-VIS spectroscopy, which were summarized. Although vanadium(III) samples are less commonly observed, this may be because high-field EPR spectrometers are needed to observe the spin states of vanadium(III). Mass spectroscopic studies have been conducted to characterize protein-vanadium complexes. Some vanadium compounds, such as oxidometalates, have only recently been observed using mass spectroscopic methods, and this was also only observed using high-field mass spectroscopy. These methods may be most important when vanadium is extracted and isolated to determine the vanadium content in biological and environmental samples. In such studies, metal content is generally determined using atomic absorption spectroscopy and inductively induced plasma. These methods measure total vanadium content and are very sensitive to the matrices in which the compounds are found. The more experimentally demanding synchrotron-based methods require access to a synchrotron but will show speciation and distinguish several oxidation states of vanadium.

Two types of bio-sensors are described: bio-sensors measuring vanadium materials and bio-sensors containing vanadium. Bio-sensors that measure vanadium are divided into those using non-enzymatic and those using enzymatic methods. Both approaches generally use detection by electrochemical methods. Several different approaches have been reported, with several enzymatic bio-sensors using phosphatases to detect vanadium. Perhaps the greatest surprise was the number of bio-sensors reported using vanadium in their makeup. Most use vanadium oxides, but a few are based on other coordination complexes. The largest group of these bio-sensors are the glucose sensors that are of considerable biomedical interest. Only a few representative examples are described here. However, many other bio-sensors are being developed to detect other biomedically-relevant molecules, including dopamine, a neurotransmitter, urea, a biomarker for kidney function, DNA, gene products, and antigens. These bio-sensors underscore the desirable properties that vanadium oxides have as electron transfer agents and motivate the further development of novel sensors.

ACKNOWLEDGMENTS

DCC thanks CSU and a private donor for funding.

ABBREVIATIONS

ALP	alkaline phosphatase
AMF	arbuscular mycorrhizal fungi
AP-1	activator protein-1
BMOV	bismaltolatooxovanadium(IV)
CEA	carcino-embryonic antigen
CILE	carbon ionic liquid electrode
DNA	deoxyribonucleic acid
DPV	differential pulse voltammetry
EDTA	ethylenediaminetetraacetic acid
EPR	electron paramagnetic resonance
ESI-MS	electrospray ionization
EXAFS	extended X-ray absorption fine structure
Cal33	human tongue squamous cell carcinoma
GF	graphine flakes
GOx	glucose oxidase
GQD	graphite quantum dots
ICP-MS	inductively coupled plasma-mass spectroscopy
HeLa	human cell line used for testing cancer
HFEPR	high-field electron paramagnetic resonance
IFN	interferon
MALDI	matrix-assisted laser desorption/ionization

MAPK	mitogen-activated protein kinase
Metvan	4,7-dimethyl-1,10-phenanthroline;oxovanadium(2+);sulfate
MG	methylglyoxal
MWCNT	multi-walled carbon nanotube
MS	mass spectrometry
MWCNTs-NR	multi-walled carbon nanotube-neutral red
NIDDM	non-insulin dependent diabetes mellitus
NIH/3T3	cell line used in material compatibility testing
NMR	nuclear magnetic resonance
NP	nanoparticles
PET/ITO	indium tin oxide coated PET plastic
PEG	polyethylene
PTPs	protein tyrosine phosphatases
PTP1B	protein tyrosine phosphatase 1B
SW1353	human chondrosarcoma cells
TCPTP	T-lymphocyte tyrosine phosphatase
UCNP	upconverting containing nanoparticles
^{51}V NMR	vanadium-51 nuclear magnetic resonance
WHO	World Health Organization
XANES	X-ray absorption near edge spectroscopy
XAS	X-ray absorption spectroscopy
XFEL	X-ray free electron laser

REFERENCES

1. N. Samart, D. Althumairy, D. Zhang, D. A. Roess, D. C. Crans, *Coord. Chem. Rev.* **2020**, *416*, 213286.
2. C. Van Cleave, D. C. Crans, *Inorganics* **2019**, *7* (9), 111.
3. F. Collazo-Reyes, M. E. Luna-Morales, J. M. Russell, M. Á. Pérez-Angón, *Scientometrics* **2017**, *110* (3), 1505–1521.
4. N. N. Greenwood, *Catal. Today* **2003**, *78* (1–4), 5–11.
5. D. C. Crans, L. Henry, G. Cardiff, B. I. Posner, *Met. Ions Life Sci.* **2019**, *19*, 203–230.
6. K. Gruzewska, A. Michno, T. Pawelczyk, H. Bielarczyk, *J. Physiol. Pharmacol.* **2014**, *65* (5), 603–611.
7. S. Treviño, A. Díaz, E. Sánchez-Lara, B. L. Sanchez-Gaytan, J. M. Perez-Aguilar, E. González-Vergara, *Biolog. Trace Ele. Res.* **2019**, *188* (1), 68–98.
8. D. C. Crans, *J. Org. Chem.* **2015**, *80* (24), 11899–11915.
9. K. H. Thompson, J. Lichter, C. LeBel, M. C. Scaife, J. H. McNeill, C. Orvig, *J. Inorg. Biochem.* **2009**, *103* (4), 554–558.
10. K. H. Thompson, C. Orvig, *J. Inorg. Biochem.* **2006**, *100* (12), 1925–1935.
11. B. Mukherjee, B. Patra, S. Mahapatra, P. Banerjee, A. Tiwari, M. Chatterjee, *Toxicol. Lett.* **2004**, *150* (2), 135–143.
12. M. Imtiaz, M. S. Rizwan, S. Xiong, H. Li, M. Ashraf, S. M. Shahzad, M. Shahzad, M. Rizwan, S. Tu, *Environ. Int.* **2015**, *80*, 79–88.
13. I. Cooper, O. Ravid, D. Rand, D. Atrakchi, C. Shemesh, Y. Bresler, G. Ben-Nissan, M. Sharon, M. Fridkin, Y. Shechter, *Pharmaceutics* **2021**, *13* (10), 1557.

14. A. Levina, A. Pires Vieira, A. Wijetunga, R. Kaur, J. T. Koehn, D. C. Crans, P. A. Lay, *Ang. Chem.* **2020**, *132* (37), 15968–15972.
15. M. Selman, C. Rousso, A. Bergeron, H. Son, R. Krishnan, N. El-Sayes, O. Varette, A. Chen, F. Tzelepis, J. Bell, D. Crans, J. Diallo, *Mol. Ther.* **2018**, *26* (1): 56–69.
16. J. Liu, F. Ruan, S. Cao, Y. Li, S. Xu, W. Xia, *Chemosphere* **2022**, *289*, 133015.
17. S. Selim, W. Abuelsoud, S. S. Alsharari, B. F. Alowaiesh, M. M. Al-Sanea, S. Al Jaouni, M. M. Madany, H. AbdElgawad, *J. Fungi.* **2021**, *7* (11), 915.
18. H. Sigel, A. Sigel, *Z. für Naturforschung B* **2019**, *74* (6), 461–471.
19. D. C. Crans, K. Kostenkova, *Comm. Chem.* **2020**, *3* (1), 1–4.
20. Q. F. Kuang, A. Abebe, J. Evans, M. Sugumaran, *Bioorg. Chem.* **2017**, *73*, 53–62.
21. P. Frank, R. M. Carlson, E. J. Carlson, B. Hedman, K. O. Hodgson, *J. Inorg. Biochem.* **2020**, *205*, 110991.
22. T. Ueki, N. Yamaguchi, Y. Isago, H. Tanahashi, *Coor. Chem. Rev.* **2015**, *301*, 300–308.
23. S. Yamamoto, K. Matsuo, H. Michibata, T. Ueki, *Inorg. Chim. Acta* **2014**, *420*, 47–52.
24. V. Agarwal, Z. D. Miles, J. M. Winter, A. S. Eustáquio, A. A. El Gamal, B. S. Moore, *Chem. Rev.* **2017**, *117* (8), 5619–5674.
25. C. Leblanc, H. Vilter, J.-B. Fournier, L. Delage, P. Potin, E. Rebuffet, G. Michel, P. Solari, M. Feiters, M. Czjzek, *Coord. Chem. Rev.* **2015**, *301*, 134–146.
26. A. Butler, J. N. Carter-Franklin, *Nat. Prod. Rep.* **2004**, *21* (1), 180–188.
27. R. Gupta, G. Hou, R. Renirie, R. Wever, T. Polenova, *J. Am. Chem. Soc.* **2015**, *137* (16), 5618–5628.
28. M. Mesko, L. Xiang, S. Bohle, D. S. Hwang, H. Zeng, M. J. Harrington, *Chem. Mater.* **2021**, *33* (16), 6530–6540.
29. C. N. Schmitt, A. Winter, L. Bertinetti, A. Masic, P. Strauch, M. J. Harrington, *J. Roy. Soc. Interfac.* **2015**, *12* (110), 20150466.
30. M. Aureliano, N. I. Gumerova, G. Sciortino, E. Garribba, A. Rompel, D. C. Crans, *Coord. Chem. Rev.* **2021**, *447*, 214143.
31. J. C. Pessoa, M. F. Santos, I. Correia, D. Sanna, G. Sciortino, E. Garribba, *Coord. Chem. Rev.* **2021**, *449*, 214192.
32. D. Crans, L. Yang, A. Haase, X. Yang, *Met. Ions Life Sci* **2018**, *18*, 251–279.
33. J. C. Pessoa, S. Etcheverry, D. Gambino, *Coord. Chem. Rev.* **2015**, *301*, 24–48.
34. E. Kioseoglou, S. Petanidis, C. Gabriel, A. Salifoglou, *Coord. Chem. Rev.* **2015**, *301*, 87–105.
35. D. Rehder, J. Costa Pessoa, C. F. Geraldes, M. M. Castro, T. Kabanos, T. Kiss, B. Meier, G. Micera, L. Pettersson, M. Rangel, *J. Biol. Inorg. Chem.* **2002**, *7* (4), 384–396.
36. H. Sakurai, Y. Kojima, Y. Yoshikawa, K. Kawabe, H. Yasui, *Coord. Chem. Rev.* **2002**, *226* (1–2), 187–198.
37. U. Jungwirth, C. R. Kowol, B. K. Keppler, C. G. Hartinger, W. Berger, P. Heffeter, *Antioxid. Red. Sign.* **2011**, *15*, 1085–1127.
38. N. D. Corona-Motolinia, B. Martínez-Valencia, L. Noriega, B. L. Sánchez-Gaytán, F. J. Melendez, A. García-García, D. Choquesillo-Lazarte, A. Rodríguez-Diéguez, M. E. Castro, E. González-Vergara, *Front. Chem.* **2022**, *10*, 830511.
39. L. Lima, H. Murakami, D. J. Gaebler, W. E. Silva, M. F. Belian, E. C. Lira, D. C. Crans, *Inorganics* **2021**, *9* (6), 42.
40. S. Trevino, E. Brambila-Colombres, D. Velazquez-Vazquez, E. Sanchez-Lara, E. Gonzalez-Vergara, A. Diaz-Fonseca, J. A. Flores-Hernandez, A. Perez-Benitez, *Oxid. Med. Cell. Longev.* **2016**, *2016*, 6058705.
41. B. Feng, Y. Dong, B. Shang, B. Zhang, D. C. Crans, X. Yang, *Adv. Funct. Mat.* **2022**, *32* (5), 2108645.
42. X. Wang, T. Sun, J. Liu, Z. Shan, Y. Jin, S. Chen, W. Bao, F. B. Hu, L. Liu, *Am. J. Epidem.* **2014**, *180* (4), 378–384.

43. A. Bergeron, K. Kostenkova, M. Selman, H. A. Murakami, E. Owens, N. Haribabu, R. Arulanandam, J.-S. Diallo, D. C. Crans, *BioMetals* **2019**, *32* (3), 545–561.
44. D. C. Crans, J. T. Koehn, S. M. Petry, C. M. Glover, A. Wijetunga, R. Kaur, A. Levina, P. A. Lay, *Dalton Trans.* **2019**, *48* (19), 6383–6395.
45. S. Prasad, D. DuBourdieu, A. Srivastava, P. Kumar, R. Lall, *Inter. J. Mol. Sci.* **2021**, *22* (13), 7094.
46. D. Sanna, V. Ugone, G. Micera, P. Buglyo, L. Biro, E. Garribba, *Dalton Trans.* **2017**, *46*, 8950–8967.
47. M. Loizou, I. Hadjiadamou, C. Drouza, A. D. Keramidas, Y. V. Simos, D. Peschos, *Inorganics* **2021**, *9* (10), 73.
48. M. M. Altaf, X.-p. Diao, A. Shakoor, M. Imtiaz, A.-u. Rehman, M. A. Altaf, L. U. Khan, *J. Soil. Sci. Plant Nutri* **2021**, *22* (1), 121–39.
49. L. Chen, J.-R. Liu, W.-F. Hu, J. Gao, J.-Y. Yang, *J. Haz. Mat.* **2021**, *405*, 124200.
50. J.-B. Li, W.-S. Xi, S.-Y. Tan, Y.-Y. Liu, H. Wu, Y. Liu, A. Cao, H. Wang, *NanoImpact* **2021**, *24*, 100351.
51. M. Valko, H. Morris, M. Cronin, *Cur. Med. Chem.* **2005**, *12* (10), 1161–1208.
52. WHO, *Acta Obstet. Gynecol. Scand.* **1977**, *56*, 247–253.
53. R. Chiarelli, C. Martino, M. C. Roccheri, P. Cancemi, *Chemosphere* **2021**, *274*, 129843.
54. E. Montiel-Flores, O.A. Mejia-Garcia, J.L. Ordonez-Librado, A.L. Gutierrez-Valdez, J. Espinosa-Villanueva, C. Dorado-Martinez, L. Reynoso-Erazo, R. Tron-Alvarez, V. Rodriquez-Lara, M.R. Avila-Costa, *Heliyon*, **2022**, *7*, e07856.
55. H. Peng, J. Guo, B. Li, H. Huang, *Environ. Chem. Lett.* **2022**, *20* (2), 1249–1263.
56. D. C. Crans, K. Postal, J. A. MacGregor, in *Metal Toxicology Handbook*, CRC Press, Boca Raton, 2020, pp. 183–196.
57. S. M. Shaheen, D. S. Alessi, F. M. G. Tack, Y. S. Ok, K.-H. Kim, J. P. Gustafsson, D. L. Sparks, J. Rinklebe, *Adv. Coll. Interfac. Sci.* **2019**, *265*, 1–13.
58. M. A. Altaf, H. Shu, Y. Hao, Y. Zhou, M. A. Mumtaz, Z. Wang, *Horticulturae* **2022**, *8* (1), 28.
59. S. Das, A. Roy, A. K. Barui, M. M. A. Alabbasi, M. Kuncha, R. Sistla, B. Sreedhar, C.≈R. Patra, *Nanoscale* **2020**, *12* (14), 7604–7621.
60. A. Al-Qatati, F. L. Fontes, B. G. Barisas, D. Zhang, D. A. Roess, D. C. Crans, *Dalton Trans.* **2013**, *42* (33), 11912–11920.
61. D. C. Horton, D. VanDerveer, J. Krzystek, J. Telser, T. Pittman, D. C. Crans, A. A. Holder, *Inorg. Chim. Acta* **2014**, *420*, 112–119.
62. J. Krzystek, A. Ozarowski, J. Telser, D. C. Crans, *Coord. Chem. Rev.* **2015**, *301*, 123–133.
63. J. P. Gustafsson, *Appl. Geochem.* **2019**, *102*, 1–25.
64. M. Aureliano, D. C. Crans, *J. Inorg. Biochem.* **2009**, *103* (4), 536–546.
65. D. C. Crans, C. D. Rithner, L. A. Theisen, *J. Am. Chem. Soc.* **1990**, *112*, 2901–2908.
66. I. Boukhobza, D. C. Crans, *Arab. J. Chem.* **2020**, *13* (1), 1198–1228.
67. M. R. Maurya, B. Sarkar, F. Avecilla, I. Correia, *Eur. J. Inorg. Chem.* **2016**, *2016* (25), 4028–4044.
68. E. Halevas, T. Papadopoulos, C. Swanson, G. Smith, A. Hatzidimitriou, G. Katsipis, A. Pantazaki, I. Sanakis, G. Mitrikas, K. Ypsilantis, *J. Inorg. Biochem.* **2019**, *191*, 94–111.
69. C. Ahmani Ferdi, M. Belaiche, E. Iffer, *J. Solid. State. Electrochem.* **2021**, *25* (1), 301–313.
70. G. Sahu, A. Banerjee, R. Samanta, M. Mohanty, S. Lima, E. R. Tiekink, R. Dinda, *Inorg. Chem.* **2021**, *60* (20), 15291–15309.
71. S. Selvaraj, U. M. Krishnan, *J. Med. Chem.* **2021**, *64* (17), 12435–12452.
72. R. Gupta, J. Stringer, J. Struppe, D. Rehder, T. Polenova, *Solid State Nucl. Mag. Res.* **2018**, *91*, 15–20.

73. J. J. Smee, J. A. Epps, K. Ooms, S. E. Bolte, T. Polenova, B. Baruah, L. Yang, W. Ding, M. Li, G. R. Willsky, *J. Inorg. Biochem.* **2009**, *103* (4), 575–584.
74. L. H. B. Nguyen, J. Olchowka, S. Belin, P. Sanz Camacho, M. Duttine, A. Iadecola, F. Fauth, D. Carlier, C. Masquelier, L. Croguennec, *ACS Appl. Mat. Interfac.* **2019**, *11* (42), 38808–38818.
75. D. C. Crans, J. J. Smee, E. Gaidamauskas, L. Q. Yang, *Chem. Rev.* **2004**, *104* (2), 849–902.
76. I. Correia, S. Roy, C. P. Matos, S. Borovic, N. Butenko, I. Cavaco, F. Marques, J. Lorenzo, A. Rodriguez, V. Moreno, J. Costa Pessoa, *J. Inorg. Biochem.* **2015**, *147*, 134–146.
77. J. C. Pessoa, *J. Inorg. Biochem.* **2015**, *147*, 4–24.
78. T. Jakusch, J. C. Pessoa, T. Kiss, *Coord. Chem. Rev.* **2011**, *255* (19–20), 2218–2226.
79. F. Porcaro, S. Roudeau, A. Carmona, R. Ortega, *Trends Anal. Chem.* **2018**, *104*, 22–41.
80. Y. Yoshikawa, H. Sakurai, D. C. Crans, G. Micera, E. Garribba, *Dalton Trans.* **2014**, *43* (19), 6965–6972.
81. R. Meier, M. Boddin, S. Mitzenheim, K. Kanamori, *Metal Ions Biol. Systems*, *31*, 45–88.
82. R. Meier, M. Boddin, S. Mitzenheim, V. Schmid, T. Schoherr, *J. Inorg. Biochem.* **1998**, 249–252.
83. M. R. Bond, R. S. Czernuszewicz, B. C. Dave, Q. Yan, M. Mohan, R. Verastegue, C. J. Carrano, *Inorg. Chem.* **1995**, *34* (23), 5857–5869.
84. A. D. Keramidas, A. B. Papaioannou, A. Vlahos, T. A. Kabanos, G. Bonas, A. Makriyannis, C. P. Rapropoulou, A. Terzis, *Inorg. Chem.* **1996**, *35* (2), 357–367.
85. K. Kanamori, E. Kameda, K.-I. Okamoto, *Bull. Chem. Soc. Jpn.* **1996**, *69* (10), 2901–2909.
86. P. Buglyo, D. C. Crans, E. M. Nagy, R. L. Lindo, L. Q. Yang, J. J. Smee, W. Z. Jin, L. H. Chi, M. E. Godzala, G. R. Willsky, *Inorg. Chem.* **2005**, *44* (15), 5416–5427.
87. P. J. Alonso, J. Forniés, M. A. García-Monforte, A. Martín, B. Menjón, *Chem. Comm.* **2001**, *20*, 2138–2139.
88. W.-S. Xi, J.-B. Li, Y.-Y. Liu, H. Wu, A. Cao, H. Wang, *Toxicology* **2021**, *459*, 152859.
89. M. Zhang, Y. Niu, Y. Xu, *J. Coll. Interfac. Sci.* **2020**, *579*, 269–281.
90. C. G. Hartinger, M. Groessl, S. M. Meier, A. Casini, P. J. Dyson, *Chem. Soc. Rev.* **2013**, *42* (14), 6186–6199.
91. M. Wenzel, A. Casini, *Coor. Chem. Rev.* **2017**, *352*, 432–460.
92. V. Ugone, D. Sanna, G. Sciortino, D. C. Crans, E. Garribba, *Inorg. Chem.* **2020**, *59* (14), 9739–9755.
93. N. Farrell, *Chem. Soc. Rev.* **2015**, *44* (24), 8773–8785.
94. T. Iglesias-González, C. Sánchez-González, M. Montes-Bayón, J. Llopis-González, A. Sanz-Medel, *Anal. Bioanal. Chem.* **2012**, *402* (1), 277–285.
95. K. G. Fernandes, M. Montes-Bayón, E. B. González, E. Del Castillo-Busto, J. A. Nóbrega, A. Sanz-Medel, *J. Anal. Atom Spectr.* **2005**, *20* (3), 210–215.
96. M. H. Nagaoka, H. Akiyama, T. Maitani, *Analyst* **2004**, *129* (1), 51–54.
97. T. Jakusch, T. Kiss, *Coord. Chem. Rev.* **2017**, *351*, 118–126.
98. R. Ortega, A. Carmona, I. Llorens, P. L. Solari, *J. Anal. Atom Spectr.* **2012**, *27* (12), 2054–2065.
99. L. H. Nguyen, A. Iadecola, S. Belin, J. Olchowka, C. Masquelier, D. Carlier, L. Croguennec, *J. Phys. Chem. C* **2020**, *124* (43), 23511–23522.
100. K. Murota, E. Pachoud, J. P. Attfield, R. Glaum, R. Sutarto, K. Takubo, D. I. Khomskii, T. Mizokawa, *Phys. Rev. B* **2020**, *101* (24), 245106.
101. A. Levina, A. I. McLeod, P. A. Lay, *Chem. Eur. J.* **2014**, *20* (38), 12056–12060.
102. A. Levina, A. I. McLeod, L. E. Kremer, J. B. Aitken, C. J. Glover, B. Johannessen, P. A. Lay, *Metallomics* **2014**, *6* (10), 1880–1888.

103. A. Levina, A. I. McLeod, S. J. Gasparini, A. Nguyen, W. M. De Silva, J. B. Aitken, H. H. Harris, C. Glover, B. Johannessen, P. A. Lay, *Inorg. Chem.* **2015**, *54* (16), 7753–7766.
104. G. R. Willsky, K. Halvorsen, M. E. Godzala III, L.-H. Chi, M. J. Most, P. Kaszynski, D. C. Crans, A. B. Goldfine, P. J. Kostyniak, *Metallomics* **2013**, *5* (11), 1491–1502.
105. J. M. George, A. Antony, B. Mathew, *Microchim. Acta* **2018**, *185* (7), 1–26.
106. G. M. Greenway, G. Wolfbauer, *Anal. Chim. Acta* **1995**, *312* (1), 15–25.
107. H. Filik, A. Aslıhan Avan, *Microchem. J.* **2020**, *158*, 105242.
108. F. N. Muya, X. T. Ngema, P. G. L. Baker, E. I. Iwuoha, *J. Nano. Res.* **2016** *44*, 142–157.
109. A. Amine, H. Mohammadi, I. Bourais, G. Palleschi, *Biosens. Bioelectron.* **2006**, *21* (8), 1405–1423.
110. P. Butmee, G. Tumcharern, C. Songsiriritthigul, M. J. Durand, G. Thouand, M. Kerr, K. Kalcher, A. Samphao, *Anal. Bioanal. Chem.* **2021**, *413* (23), 5859–5869.
111. C. C. McLauchlan, B. J. Peters, G. R. Willsky, D. C. Crans, *Coord. Chem. Rev.* **2015**, 301–*302* (1), 163–199.
112. S. R. Akabayov, B. Akabayov, *Inorg. Chim. Acta* **2014**, *420*, 16–23.
113. A. C. Hengge, *Biochim. Biophys. Acta Prot. Proteom.* **2015**, *1854* (11), 1768–1775.
114. Y. Chu, N. H. Williams, A. C. Hengge, *Biochemistry* **2017**, *56* (30), 3923–3933.
115. J. Pinkston, J. Jo, K. J. Olsen, D. Comer, C. A. Glaittli, J. P. Loria, S. J. Johnson, A. C. Hengge, *Biochemistry* **2021**, *60* (38), 2888–2901.
116. R. M. Crean, M. Biler, M. W. van der Kamp, A. C. Hengge, S. C. Kamerlin, *J. Am. Chem. Soc.* **2021**, *143* (10), 3830–3845.
117. B. Le-Vinh, Z. B. Akkuş-Dağdeviren, N. M. N. Le, I. Nazir, A. Bernkop-Schnürch, *Adv. Therap.* **2022**, 2100219.
118. D. Crans, R. Bunch, L. Theisen, M. Gottlieb, *J. Inorg. Biochem.* **1989**, *36* (3–4), 352.
119. D. Crans, M. S. Gottlieb, J. Tawara, R. Bunch, L. Theisen, *Anal. Biochem.* **1990**, *188* (1), 53–64.
120. A. L. Alvarado-Gámez, M. Alonso-Lomillo, O. Domínguez-Renedo, M. Arcos-Martínez, *J. Electroanal. Chem.* **2013**, *693*, 51–55.
121. F. N. Muya, P. G. Baker, E. I. Iwuoha, *Electrocatalysis* **2020**, *11* (4), 374–382.
122. A. L. Alvarado-Gámez, M. A. Alonso-Lomillo, O. Domínguez-Renedo, M. J. Arcos-Martínez, *Sensors* **2014**, *14* (2), 3756–3767.
123. P. Fanjul-Bolado, M. B. González-García, A. Costa-García, *Anal. Bioanal. Chem.* **2006**, *385* (7), 1202–1208.
124. A. Berezhetskyy, O. Sosovska, C. Durrieu, J.-M. Chovelon, S. Dzyadevych, C. Tran-Minh, *IRBM* **2008**, *29* (2–3), 136–140.
125. S. Ito, S.-I. Yamazaki, K. Kano, T. Ikeda, *Anal. Chim. Acta* **2000**, *424* (1), 57–63.
126. B. Serra, M. Morales, A. Reviejo, E. Hall, J. Pingarron, *Anal. Biochem.* **2005**, *336* (2), 289–294.
127. A. Preechaworapun, Z. Dai, Y. Xiang, O. Chailapakul, J. Wang, *Talanta* **2008**, *76* (2), 424–431.
128. E. M. Abad-Villar, M. T. Fernández-Abedul, A. Costa-García, *Biosens. Bioelectron.* **2002**, *17* (9), 797–802.
129. A. Kokado, H. Arakawa, M. Maeda, *Anal. Chim. Acta* **2000**, *407* (1–2), 119–125.
130. J. Livage, *Chem. Mater.* **1991**, *3* (4), 578–593.
131. D. Nunes, A. Pimentel, A. Gonçalves, S. Pereira, R. Branquinho, P. Barquinha, E. Fortunato, R. Martins, *Semicond. Sci. Technol.* **2019**, *34* (4), 043001.
132. S. Lee, I. N. Ivanov, J. K. Keum, H. N. Lee, *Sci. Rep.* **2016**, *6* (1), 1–7.
133. M. Wu, X. Zhang, S. Gao, X. Cheng, Z. Rong, Y. Xu, H. Zhao, L. Huo, *Cryst. Eng. Comm.* **2013**, *15* (46), 10123–10131.
134. H. Yin, K. Yu, Z. Zhang, M. Zeng, L. Lou, Z. Zhu, *Electroanalysis* **2011**, *23* (7), 1752–1758.

135. W. Qu, J.-U. Meyer, *Sensor. Actuat.* **1997**, *40* (2–3), 175–182.
136. Y. Fukai, Y. Kondo, S. Mori, E. Suzuki, *Electrochem. Comm.* **2007**, *9* (7), 1439–1443.
137. M. W. Billard, H. A. Basantani, M. W. Horn, B. J. Gluckman, *IEEE Sensors J.* **2016**, *16* (8), 2211–2212.
138. N. Thakur, D. Mandal, T. C. Nagaiah, *J. Mat. Chem.* B, **2021**, *9* (40), 8399–8405.
139. M. H. Alkahtani, C. L. Gomes, P. R. Hemmer, *Opt. Lett.* **2017**, *42* (13), 2451–2454.
140. A. J. Talib, M. Alkahtani, L. Jiang, F. Alghannam, R. Brick, C. L. Gomes, M. O. Scully, A. V. Sokolov, P. R. Hemmer, *Opt. Mat. Exp.* **2018**, *8* (11), 3277–3287.
141. F. De Souza, R. Da Rocha, N. Vieira, D. Cestarolli, E. Guerra, *Bul. Mat. Sci.* **2020**, *43* (1), 1–6.
142. N. Vieira, F. de Souza, R. da Rocha, D. Cestarolli, E. Guerra, *Mat. Sci. Semicond. Proc.* **2021**, *121*, 105337.
143. L. R. Bhat, S. Vedantham, U. M. Krishnan, J. B. B. Rayappan, *Biosens. Bioelectron.* **2018**, *103*, 143–150.
144. D. M. Fernandes, C. Freire *Chem. Electro. Chem.* **2015**, *2*, 269–279.
145. M. Sreejesh, S. Shenoy, K. Sridharan, D. Kufian, A. Arof, H. Nagaraja, *App. Sur. Sci.* **2017**, *410*, 336–343.
146. R. C. F. da Rocha, F. A. de Souza, N. S. Vieira, D. T. Cestarolli, E. M. Guerra, *Biotech. Appl. Biochem.* **2020**.
147. W. Sun, P. Qin, H. Gao, G. Li, K. Jiao, *Biosens. Bioelectron.* **2010**, *25* (6), 1264–1270.
148. M. R. Kesama, S. R. Dugasani, S.-G. Jung, B. Gnapareddy, T. Park, S. H. Park, *Nanotechnology* **2019**, *31* (8), 085705.
149. J. Han, L. Jiang, F. Li, P. Wang, Q. Liu, Y. Dong, Y. Li, Q. Wei, *Biosens. Bioelectron.* **2016**, *77*, 1104–1111.

9 Non-Invasive Detection of Stem Cell Therapies Facilitated by Metal Ion-Based Contrast Agents

Meghan W. Dukes
Departments of Chemistry, Molecular Biosciences, Neurobiology, and Radiology, Northwestern University, 2145 Sheridan Rd, Evanston, IL 60208, USA

Michel Modo
University of Pittsburgh, Departments of Radiology and Bioengineering, McGowan Institute for Regenerative Medicine, Centre for Neural Basis of Cognition, 3025 East Carson Street, Pittsburgh, PA 15203, USA

Thomas J. Meade
Departments of Chemistry, Molecular Biosciences, Neurobiology, and Radiology, Northwestern University, 2145 Sheridan Rd, Evanston, IL 60208, USA

CONTENTS

DOI: 10.1201/9781003229971-9

ABSTRACT

Stem cell therapy is a unique approach to regenerative medicine as the fundamental pathways of cell lineage can be hijacked to renew damaged tissue in the body. Such therapies have clinical interest in numerous fields and disorders, such as oncology, cardiovascular disease, neurological disorders, spinal cord/bone injury, and more. However, translation of these therapies toward clinical approval necessitates the ability to accurately investigate in preclinical and clinical trials *where* cells are located following transplantation and *how* they behave. Clinically available non-invasive imaging techniques, such as magnetic resonance imaging, nuclear medicine imaging, and optical imaging, are the most promising methods to monitor these advanced therapeutic products. This chapter will evaluate each imaging modality for its utility in stem cell tracking, and describe metal-ion-containing contrast agents (CAs) that promote specific detection of stem cell populations *in vivo*. We will focus on how each modality and the associated CAs are designed to not interfere with the activity and function of native stem cells. We will report on the current strategies for longitudinal imaging studies that provide promising avenues to develop bio-markers of therapeutic efficacy.

KEYWORDS

Metal Atoms; Cell Therapy; Cell Tracking; Magnetic Resonance Imaging; Contrast Agent; Molecular Imaging; Nanoparticle

1 INTRODUCTION

Stem cell therapies have garnered attention in the past few decades as safe intervention strategies for several disease conditions characterized by a loss of cells [1, 2]. Essentially, pluripotent (or multipotent cells), also known as stem cells, can differentiate and restore damaged tissues when directed to the site of an injury, disease, or malfunction [3]. Embryonic stem cells (ESCs) are desirable for stem cell therapy for their ability to differentiate into all cell lineages, but a number of ethical concerns and the logistics of sourcing embryonic tissues have hindered clinical translation [4]. However, the advent of induced pluripotent stem cells (iPS) has provided a readily available source of human cells for clinical translation without the need to use embryonic tissue.

Herein, we focus primarily on regenerative therapies in the central nervous system for which multipotent neuronal stem cells (NSCs) provide a tissue-specific cell. NSCs can be harvested from fetal brain tissue with the same challenges as ESCs. Today, these cells can be derived from iPS. Several iPS-derived NSCs are commercially available. Human mesenchymal stem cells (MSCs), harvested most commonly from bone marrow, can also be used if the therapeutic target involves the secretion of growth factors rather than the replacement of brain cells [5]. Although few MSCs have received clinical approval, preclinical investigations of ESCs, MSCS, and NSCs have shown great promise for stem cell therapies [6, 7].

Cancer, cardiovascular diseases, inflammatory diseases, neurological disorders, and bone injuries are the most investigated conditions for stem cell therapy. In oncology, MSCs are used to target tumors to deliver anticancer drugs [8, 9], and treat leukemia and other blood cancers and disorders [10–12]. Cell therapies in cardiovascular disease can promote improved cardiac function and decreased infarct size following myocardial infarction [13–15]. Stem cells have also successfully aided in reducing inflammation in simple wound healing processes as well as more involved nephritis associated with autoimmune disorders [16, 17].

The primary focus of this chapter is on stem cell therapies for disorders and diseases of the central nervous system and bone regeneration. Preclinical investigations include stem cell therapy for multiple sclerosis, Alzheimer's disease, Motor Neuron Disease, Parkinson's disease, stroke, and spinal cord injuries [18–20]. In clinical trials, MSCs have promoted functional recovery in patients following a stroke with no adverse side effects out to 1 year [21]. A similar clinical trial saw no adverse side effects when evaluating MSCs to treat a spinal cord injury; however, the study could not draw any conclusions about treatment efficacy [1]. Preclinical investigations of bone regeneration have shown promise, and clinical trials are still ongoing [22]. In general, the safety of stem cell therapies has been established, but their potential for therapeutic efficacy remains unproven. A major challenge in establishing efficacy is our ability to ascertain in a clinical setting that cells were delivered to the target site, survived, and elicited their therapeutic effects.

Given such wide applicability for stem cell therapies, clinical translation will require the ability to accurately and reliably monitor stem cells non-invasively in deeply seated organs [23, 24]. Issues with therapeutic efficacy often arise from either incomplete cell differentiation or stem cells not honing to the injured or diseased part of the body. Inappropriate migration to undesired areas or even uncontrolled growth in the desired area can also lead to unwanted tumorigenesis [25]. As such, the ultimate clinical approval of stem cell therapy will require preclinical and clinical investigations of cellular tracking methodologies that can (i) accurately determine where stem cells are localized and (ii) report on their function. It is imperative for cell tracking methods to be nontoxic to stem cell viability and inert to its differentiation potential, to have reproducible and sufficient cellular internalization, to produce a strong signal *in vivo*, and to promote longitudinal studies that track cell location/function over time.

We will discuss current progress toward developing metal-ion-based non-invasive imaging strategies for tracking cellular based therapies. We will evaluate the utility of contrast agents (CAs) developed for Magnetic Resonance Imaging (MRI) (Section 2), Nuclear Medicine and Computed Tomography (CT, Section 3), and Optical Imaging

Modalities (Section 4), as they are most widely used in clinical and preclinical settings. Lastly, we will examine multi-modal combinations of available CAs. We will scrutinize each modality for stem cell tracking, as well as a few major advances in metal-ion-based CAs in the context of the above-mentioned design principles. We will explore both molecular and nanoparticle CAs. Nanoparticle agents are defined as having a core made of metal ions, but molecular agents are chelated complexes even if the study utilizes nano-encapsulation delivery methods.

2 MAGNETIC RESONANCE IMAGING

Since the early 1980s, MRI has become the top-choice diagnostic imaging technique for neurological, cardiovascular, and musculoskeletal diseases and disorders, and its use has expanded rapidly in utility in the past 20 years [26]. One reason for this is the safety profile of MRI, which uses non-ionizing radiation for imaging [27]. During an MRI scan, a magnetic field B_0 is applied to the body and the nuclear spins of biological protons (mostly water) preferentially align with the magnetic field. This produces a net magnetization vector along the z-axis, M_0. A radio frequency pulse is then applied to tip the net magnetization vector into the xy-plane. A signal is produced as detectors measure the radiofrequency relaxation of the nuclear spins back to M_0. Longitudinal relaxation time, T_1, is measured by the time it takes for the net magnetization to relax back to M_0 along the z-axis. The signal in T_1-weighted MR imaging is proportional to the inverse of the time T_1, also defined as the relaxation rate R_1. This makes shorter T_1 times desirable for T_1-weighted MR imaging. Transverse relaxation time, T_2, describes the time it takes for the loss of net magnetization in the xy-plane by dephasing [28]. The signal in T_2-weighted MR imaging is similarly proportional to the inverse of the transverse relaxation rate, R_2. Contrast agents can be used in MRI to modulate T_1 and T_2 times of analytes of interest and obtain more information from the experiment. The values of these relaxation times are then converted into a black-and-white signal scale to produce an image.

For most populations, MRI is a reliably safe non-invasive imaging modality. The use of non-ionizing radiofrequency pulses prevents the risks associated with multiple scans that would be necessary for tracking stem cells in a longitudinal study. Additionally, an MRI scan can be taken immediately after the administration of a CA, and has unlimited depth penetration and excellent spatial resolution [29, 30]. Although inherent differences in tissue environment create excellent soft-tissue contrast in MRI, paramagnetic CAs are used to produce even more significant differences in signals in an area of interest by influencing either T_1 or T_2 relaxation processes [29]. However, a major disadvantage of MRI is the lack of sensitivity of CAs in imaging experiments [31].

A positive CA will predominantly shorten the T_1 relaxation time, producing a brighter signal compared to surrounding tissue. A negative CA will predominantly shorten T_2, creating a negative contrast observed by a loss in signal compared to surrounding tissue. For a T_1 CA to be detectable, the tissue concentration needs to be in the high μM–mM range [32]. T_2 strategies tend to require a lower concentration of agents to create a distinguishable negative contrast [30]. T_1 CAs are still preferred, as a bright signal is visually more compelling than a loss of signal.

2.1 Molecular Contrast Agents

Since the approval of the first CA in 1988, ten molecular CAs have received full approval for clinical use in the United States. Among these, nine rely on paramagnetic Gd(III) to produce contrast and one is Mn(II)-based. In comparison, only two nanoparticle CAs have received full approval for clinical use by the US Food and Drug Administration (FDA) [33]. As such, molecular CAs for MRI are among the most heavily explored as stem cell tracking agents.

2.1.1 Gd(III) Agents

Gadolinium-based CAs are used in approximately 40% of all MRI scans, with almost 50 tons of gadolinium administered annually [29, 34]. Gadolinium is a T_1 CA that shortens the longitudinal relaxation time of hydrogen in water molecules directly bound to gadolinium itself, hydrogen bound to a chelating ligand, and hydrogen simply in the proximity of the strongly paramagnetic ion [35]. Signal enhancement comes predominately from inner-sphere water molecules that are exchanged with bulk water in proximity to the complex [35]. Gd(III) complexes discussed in this section are summarized in Table 1.

Gadolinium chelated by the linear molecule diethylenetriaminepantaacidic acid (forming Gd-DTPA, Magnevist®) was FDA approved for clinical use in 1988. However, Gd(III) complexes have poor cellular permeability, thus requiring creative strategies for delivery into stem cells before implantation back into the body to be detected by MRI. One study delivered Gd-DTPA into MSCs using non-liposomal transfection agents prior to direct stereotaxic injection into a murine model of stroke. The contrast was detectable out to 7 days, but the efficacy of the treatment or continued imaging capability over time was not evaluated [36]. In contrast, delivery of Gd-DTPA into MSCs by liposomal transfection agents promoted imaging out to 2 weeks [37]. Another study evaluated the efficacy of commercially available jetPolyehtyleneimine as a transfection agent for Gd-DTPA into MSCs, and also saw contrast out to 2 weeks in a model of spinal cord injury [38]. Although each of these studies reported a final imaging timepoint, they do not discuss at what point the signal could no longer be detected. This makes comparisons between methodologies difficult.

Although initial studies with Gd-DTPA were promising, linearly chelated Gd(III) complexes are less stable than macrocyclic alternatives and can result in unwanted gadolinium retention in a patient following administration. Dissociation of the metal ion from the ligand has been observed in the acidic environment of endosomes following internalization [39]. Because of this, the FDA has warned clinicians to exercise caution when administering linear Gd(III) chelates. This will be a common point of discussion throughout this chapter. Gadolinium retention is often benign, but free Gd(III)-ions can cause damage to tissues by blocking calcium channels and interfering with protein functions [40–42]. An additional concern is for patients with kidney disease, as unstable gadolinium chelates may be causative of nephrogenic systemic fibrosis [43]. One study's solution to this was to bind Gd(III) ions directly to melanin nanoparticles to deliver into MSCs. As a proof of concept, Gd(III)-melanin labeled MSCs were injected directly into the thigh muscle of rats and were imaged out to 4 weeks post injection. Cai et al. found these particles to be more stable than Gd-DTPA with greater signal contrast [44].

TABLE 1
Gd(III) Complexes for T_1-Weighted MR Imaging

Structure	Notes	Refs.
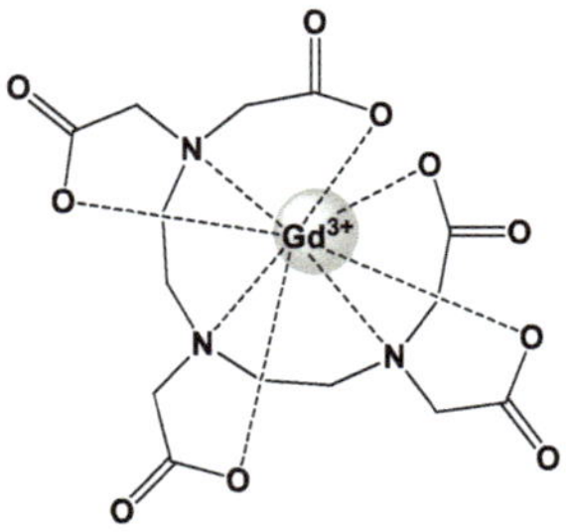	**Gd-DTPA** Magnevist® FDA approved in 1988 Linear chelate that is sensitive to Gd(III) displacement in acidic conditions	[36–38]
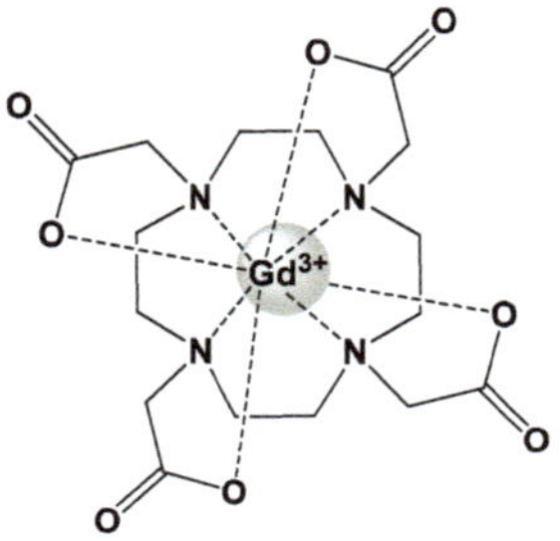	**Gd-DOTA** Dotarem® FDA Approved in 2013 Macrocyclic chelate with better thermodynamic stability	[46, 47]
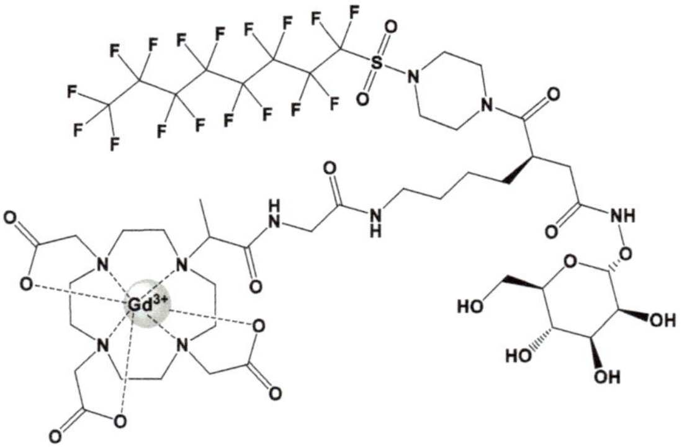	**Gadofluorine** Gd-DOTA derivative Further potential for use in ^{19}F MRI	[48]

Other studies have focused on developing macrocyclic chelating strategies for Gd(III) agents, as three macrocyclic agents are approved for clinical use and are known to bind Gd(III) more stably [45]. One example of this is FDA-approved 1,4,7,10-tetraazacyclododecane macrocyclic chelate (Gd-DOTA, Dotarem®) to label MSCs. Although Gd-DOTA has been explored in some stem cell tracking, cellular permeability remains a major issue [46, 47]. Giesel et al. evaluated Gadophrin-2, a functionalized derivative of Gd-DOTA, both with and without transfection agents and found the transfection agents to still be necessary for producing sufficient MRI signals. Labeled MSCs were stereotaxically injected into the brain, but a longitudinal study was not performed to see how far out signal contrast could still be detected [48].

An untapped benefit of this complex is the ability to simultaneously perform ^{19}F MRI. As fluorine has a negligible background signal, areas of appreciable ^{1}H signal contrast seen in T_1 images could be correlated with colocalization of fluorine MRI signals.

2.1.2 Mn(II) Agents

The rapid rate of clearance of molecular gadolinium CAs is the primary attraction for utility in most MRI applications since metal retention can be associated with negative side effects. As such, we question their long-term utility as tracking agents for cellular based therapies. One method to avoid this possible complication is the use of paramagnetic metal species that are found endogenously in healthy tissue, such as manganese. Mn(II) CAs present a unique substitute to Gd(III), as Mn(II) is an essential metal in biological function that increases its overall biocompatibility [49]. To date, the only clinically approved injectable Mn(II)-based CA is Teslascan, which is a Mn(II) ion chelated by dipyridoxaldiphosphate (DPDP). Mn(II)-DPDP is used clinically for liver imaging and the overall safety profile is significantly higher than that of Gd(III)-DTPA [49]. Even so, an over-exposure to manganese can present serious health concerns; as with gadolinium, chelate stability is an important consideration for these complexes [49–51]. Herein, we discuss the progression of a family of Mn(II)-porphyrin complexes (Table 2) and their application for T_1-weighted MR imaging for STEM cell tracking.

The first generation of Mn(II) porphyrin complexes designed by the Chen group was Mn(II)-[(acetoxymethoxycarbonyl)porphyrin] (MnAMP), an enzyme-activatable CA first explored *in vitro* in ESCs [52, 53]. As a major advantage compared to Gd(III) complexes, Mn(II)-porphyrin complexes do not require aid from cellular internalization agents. Initial investigations showed promise for these complexes as they did not impede ESC viability, proliferation, or differentiation potential [52]. However, synthetic challenges encouraged the group to make a second generation of Mn(II)-[5-(4-aminophenyl)-10,15,20-tris(4-sulfonatophenyl) porphyrin] ($MnPNH_2$) complexes. $MnPNH_2$ was synthetically more favorable than MnAMP and maintained the inert profile to cell function, but exhibited worse uptake in ESCs. Labeled cells were injected subcutaneously into rats and imaged, but a longitudinal study was not performed to assay signal strength over time [54]. A third generation of complexes, Mn(II)-[5,10,15,20-tetrakis(ethoxycarbonyl) porphyrin] (MnEtP), was found to satisfy both synthetic ease and cellular uptake profiles. The Cheng group indeed observed superior internalization with retained inertness to ESC cellular function. Subcutaneous injections of labeled ESCs revealed superior signal enhancement over unlabeled ESCs, but again a longitudinal study was not performed [55]. In order to assess the utility of this family of CAs more thoroughly, a longitudinal study is imperative. However, initial findings for these complexes show promise for further optimization and preclinical investigation as Gd(III) substitutes.

2.1.3 ParaCEST Agents

Chemical Exchange Saturation Transfer (CEST) is an alternative method to T_1 and T_2 contrast that relies on the exchange properties of a proton on a CEST CA with bulk water. In a CEST experiment, saturation of a specific MR frequency corresponding to an exchangeable proton on the CA results in decreased bulk water signal due

TABLE 2
Mn(II)-Porphyrin Complexes for T_1-Weighted MR Imaging

Structure	Notes	Refs.
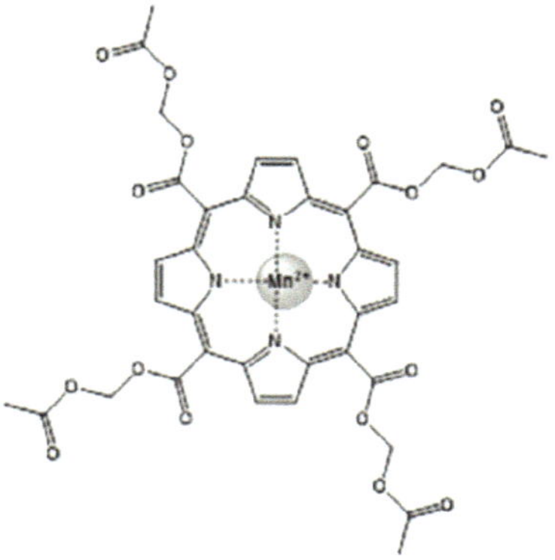	***Generation 1*** MnAMP Excellent cellular permeability and signal production Challenging synthetic scalability	[52, 53]
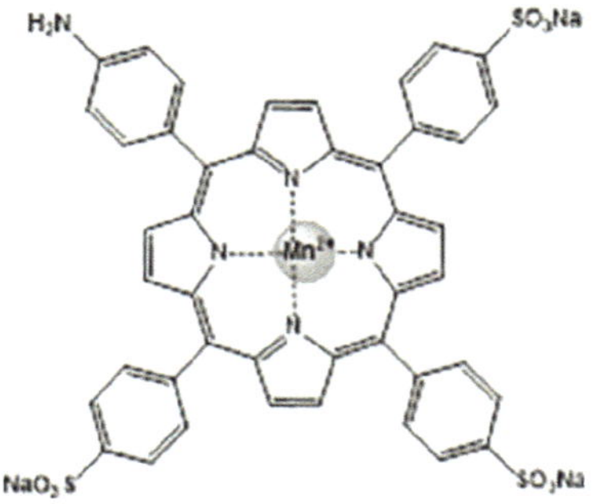	***Generation 2*** $MnPNH_2$ Facile 2-step synthesis Diminished cellular internalization and signal production	[54]
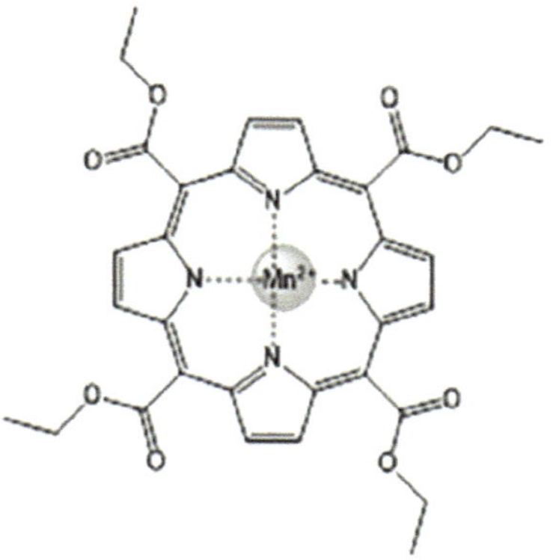	***Generation 3*** MnEtP Facile 2-step synthesis Improved cellular internalization and signal production	[55]

to exchange of the saturated nuclei with water protons [56]. Paramagnetic CEST (ParaCEST) agents contain a paramagnetic ion that shifts the MR frequency of the exchangeable proton away from biological frequencies to promote selective saturation of just the one signal [57, 58]. As an added benefit, different paramagnetic ions shift proton signals to different frequencies, allowing simultaneous detection of multiple probes [59].

Nicholls et al. utilized this strategy to study the co-injection of endothelial cells with NSCs to improve regenerative properties following a stroke. Little is understood about how the interplay between the two cell types results in more complete regeneration than NSCs alone [60]. To explore this, Nicholls et al. labeled endothelial cells with a europium ParaCEST CA (Eu-HPDO3A) and labeled the NSC population with an ytterbium ParaCEST CA (Yb-HPDO3A) through electroporation (Figure 1).

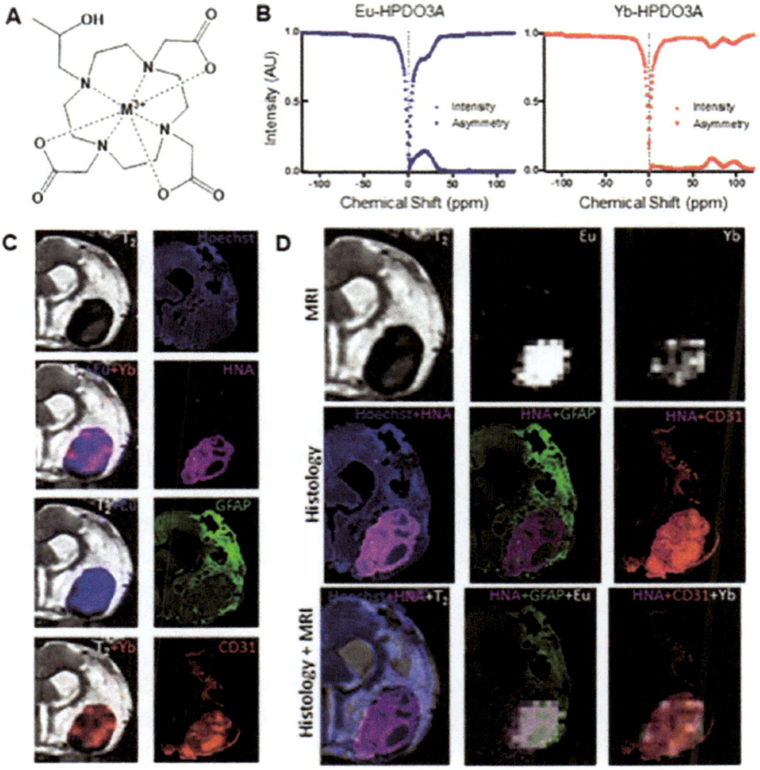

FIGURE 1 ParaCEST tracking endothelial cell and NSC populations in a rat model of stroke. (a) Chemical structure of M(III)-HPDO3A. Eu(III) chelates were used to label transplanted endothelial cells, and Yb(III) chelates were used to label NSCs. (b) z-spectra of agent solutions showing signal intensity compared to the reference at −120 ppm for both Eu-HPDO3A (Eu) and Yb-HPDO3A (Yb). Asymmetry highlights the regions of chemical shift that can be exploited to achieve specific imaging of each agent. Eu (18 ppm) and Yb (69 ppm) regions are non-overlapping, hence affording selective imaging. (c) There is a good correspondence between the distribution of transplanted cells as visualized by paraCEST and the histological marker for human cells (Human Nuclei Antigen, HNA). The Eu (18 ppm) image indicated a fairly homogenous distribution of NSCs, which was paralleled by its histological validation using GFAP as a marker within the transplant area. Yb (97 ppm) imaging also truthfully reflects the macroscopic distribution of transplanted ECs, as detected by CD31. (d) To further validate the accuracy of the MR images, partially transparent histological overlays of transplanted cells (HNA) with NSCs (GFAP), and ECs (CD31) specific markers were co-registered based on landmark identification to the relevant MRI images. This further highlights the regional specificity of the MR images as well as a correct mapping of relative cell distribution [61].

Because the lanthanide identification creates unique proton saturation frequencies, the two cell populations were visualized simultaneously in the stroke cavity of a rat model [61]. A longitudinal study was not performed, but this study highlights a unique advantage for ParaCEST imaging in stem cell tracking.

An interesting future application for this dual-agent system is the ability to create activatable probes that will shift saturation frequency upon interaction with a molecular marker. A marker could be chosen to report back on stem cell differentiation or another process by which therapeutic effects could be imaged alongside cellular localization. ParaCEST agents would be superior to T_1 and T_2 agents for this application as it avoids the uncertainty of gray-scale images. A change in signal is observed as a shift in the exchange frequency, not a slightly brighter or darker image that cannot be accurately quantified. However, this benefit must be cautiously weighted with some of the inherent disadvantages of ParaCEST. Like T_1 and T_2 CAs, ParaCEST suffers from low sensitivity [62]. Additionally, the specific absorption rate of a ParaCEST agents presents a major challenge. The power required to detect ParaCEST signal over time can have a negative impact on the surrounding tissue viability [63].

2.2 Nanoparticle Contrast Agents

The lack of sensitivity of MRI as an imaging modality presents a challenge for molecular CAs. A concern will always be having enough agents in tissue to detect, especially as cells divide and the intracellular CA concentration is diluted. Nanoparticle strategies have emerged as an alternative because their core can be very densely packed with paramagnetic metal ions with bulk magnetic properties. Here, we discuss advances in nanoparticles that contain metal ions in their cores for MRI tracking of stem cells.

2.2.1 Gd(III) Particle Core

Even though several monomer gadolinium agents have received clinical approval, nanoparticle formulations with Gd(III) ions at the core are far less abundant. In one study, Gd(III) complexes chelated by three 2,4-hexanedione ligands were emulsified in wax and Brij78 surfactant to produce a densely packed Gd(III) core. Cellular internalization and signal contrast were superior to that of Gd-DTPA at identical concentrations *in vitro,* and cellular function of MSCs was not significantly hindered [64]. This study did not perform any *in vivo* tracking studies, and a follow-up study was not found. It is likely that the non-covalent nanoemulsion would not be stable in the intracellular milieu long-term, and the hexanedione ligands would not be thermodynamically stable for preventing free Gd(III) ion escape. This highlights the challenges of translating favorable T_1 molecular imaging properties of Gd(III) into nanoparticle formulations.

2.2.2 Superparamagnetic Iron Oxide Nanoparticles

The dense core structures of iron oxide are the most clinically used MRI nanoparticle structures and function as T_2 CAs. Two superparamagnetic iron oxide

(SPIO) formulations, ferumoxides (Feridex® and Endorem®) and ferucarbotran (Resovist®), have received approval in the US and Europe [65]. Clinical investigations of SPIOs have been largely halted due to financial considerations, but many preclinical studies have proceeded to evaluate SPIOs for stem cell tracking [66]. SPIOs have been extensively reviewed for stem cell tracking elsewhere in terms of particle coating to promote cellular internalization [24, 67–69]. Herein, we highlight a few studies with a focus on the longitudinal aspect of stem cell tracking.

The first few studies performed imaging out to 2–4 weeks post cell transplantation in models of stroke and brain/spinal cord injury. On the low end of this time frame, Reddy et al. coated SPIOs with chitosan and successfully imaged labeled MSCs up to 16 days [70]. Another study labeled SPIOs with poly-L-lysine prior to incubation with MSCs. Cells were transplanted into a model of traumatic brain injury and imaged out to 3 weeks [71]. Jiang et al. incubated NSCs with uncoated SPIOs and imaged transplanted cells out to 4 weeks in a rat ischemic model [72]. In a more comprehensive study, ESCs, MSCs, and CD34+ cells were labeled with Endorem® for tracking in models of spinal cord lesions, stroke, and spinal cord injury, respectively. Signal was observed out to 30 days [73]. One study went further, tracking NSCs labeled with ferumoxides coated with protamine sulfate in a rat model of ischemia for as long as 9 weeks [74]. Obenaus et al. went even further out to 58 weeks, successfully imaging SPIO labeled NSCs in a rat model of ischemic injury with no adverse side effects observed [75].

As the stem cell tracking field transitions from asking the first question of "Can we image the specific cell population?" into the second question of "Can we report on functional activity?", imaging strategies need to be compatible with long-term imaging studies. Although these examples show that cells labeled with SPIOs can be imaged on longer time scales, it becomes apparent that inconsistencies in the literature make drawing comparisons between different labeling strategies difficult. These papers discuss how long they chose to image for, but they do not provide any commentary on the time point at which signal enhancement can no longer be detected.

It has become apparent that Gd(III) and SPIOs are the leading choice CAs for MRI. To compare Gd(III) and SPIOs directly, one study focused on the functional aspect of being able to detect viable from non-viable MSCs with either Gd(III) or SPIOs [76]. Guenoun et al. found that non-viable cells labeled with Gd(III) exhibit significantly less contrast enhancement than viable, proliferative cells. In contrast, non-viable and viable MSCs labeled with SPIOs could not be easily distinguished, as non-viable cells produce the same signal void as viable ones. The paper concludes that this is a valuable method for monitoring if a transplant is going to be rejected. Furthermore, the group noticed that the transplanted cells exhibited rapid tumor-like proliferation at the transplant site through 15 days. This was not expected and not wanted. This poignantly illustrates the general necessity for tracking stem cell therapies, as well as within the context of each individual tracking strategy to report on cellular behavior. It is possible that the presence of the tracking agents caused the over-proliferation.

2.2.3 Cobalt Alloy Nanoparticles

Nanoparticles with cobalt cores were found to have superior influences on water molecules as T_2 CAs at smaller core sizes compared to iron oxide [77]. This sensitivity increase is due to the higher saturation magnetization of cobalt over iron [78]. Meng et al. labeled NSCs with CoPt alloy nanoparticles coated in a variety of shells to protect cells from any harmful toxicity. CoPt particles did not hinder NSC viability, proliferation, or neuronal differentiation. They did not test the particles in a traditional *in vivo* study, but labeled cells were transplanted into a spinal cord slice culture and imaged [79]. Although a follow-up study was not found, initial testing of these particles shows promise for further preclinical investigations. However, another study found that CoZn ferrite ($Co_{0.5}Zn_{0.5}Fe_2O_4$) particles were not suitable for MR imaging compared to their purely iron oxide ferrite counterparts due to decreased proliferation in labeled cells, as well as inferior relaxation rates [80]. Further optimization of cobalt alloy nanoparticles will be instrumental for determining any future utility in stem cell tracking.

3 NUCLEAR MEDICINE AND CT IMAGING

Positron emission tomography (PET), single-photon emission computed tomography (SPECT), and CT are three other main clinical imaging modalities. Unlike MRI, all three techniques require some form of ionizing radiation to produce a signal. For PET, a radiotracer CA undergoes beta decay through positron emission from the nucleus. The positron collides with an electron in the surrounding tissue, causing a detectable annihilation event that releases two photons. Common organic radiotracers contain ^{11}C, ^{13}N, ^{15}O or ^{18}F [81]. Although SPECT similarly relies on the radioactive decay of a radiotracer, the difference is that the wavelength emitted falls in the gamma region of the electromagnetic spectrum [82]. CT, on the other hand, functions as a three-dimensional X-ray. X-rays are emitted from the instrument itself and are detected opposite of where they are emitted, producing an image that shows where the beams were blocked by objects like bone [83]. For both PET and SPECT, a CA is a radiotracer itself, whereas CT CAs are anything that block X-ray beams better than the surrounding tissue. Although the use of ionizing radiation is not ideal, these techniques are far superior to MRI in terms of sensitivity, significantly decreasing the amount of CA required for detection [32]. Since the risk of radiation exposure is considered low in many clinical settings when compared to the ability to reliably diagnose cancer and other diseases, these techniques should not be excluded outright as potential choices for stem cell tracking. However, as discussed throughout this section, careful consideration must be given to the limitations of these techniques in long-term studies.

3.1 Molecular PET/SPECT Agents

^{18}F is the most commonly used radioisotope for PET with a half-life of around 2 hours. Radioisotopes of metal ions have also gained interest in the clinic due to ideal decay lifetimes for typical imaging needs [84, 85]. One example of this is ^{68}Ga with a half-life of around 1 hour. A patient injected with a ^{68}Ga radiotracer could reasonably undergo an imaging scan within a 30 minute–1 hour time frame, and within a few hours the majority of harmful ionization coming from decay will be complete [86]. However, a short lifetime is not ideal for a stem cell tracking study as it would be impossible to collect a longitudinal experiment once the number of molecules emitting has fallen below the limit of detection. A summary of the radiotracers discussed in this section can be found in Table 3.

Early studies evaluated the efficacy of ^{64}Cu-pyruvaldehyde bis(N^4-methylthiosemi carbazone) (^{64}Cu-PTSM) complexes with a slightly longer half-life around 13 hours [87]. These studies revealed that stem cells were only detectable for less than one day due to rapid clearance of the agent coupled with the still relatively short

TABLE 3
Molecular Radiotracers for Nuclear Medicine Imaging

Structure	Notes	Refs.
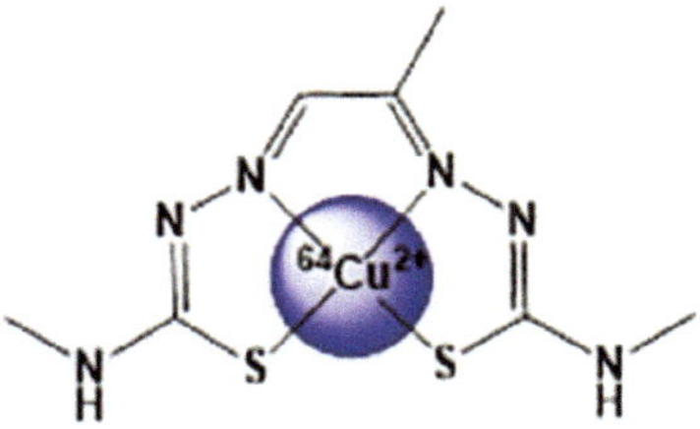	**^{64}Cu-PTSM** PET-imaging agent $T_{1/2} = 13$ hours	[87–89]
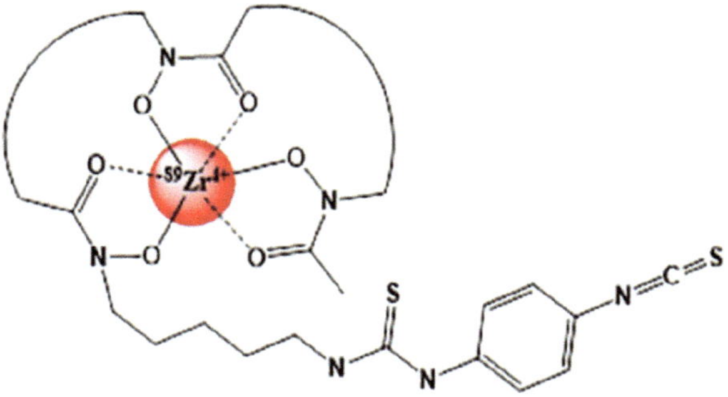	**^{89}Zr-DBN** (simplified structure) PET-imaging agent $T_{1/2} = 3.5$ days NCS functionality allows covalent linkage to cellular membrane	[90]
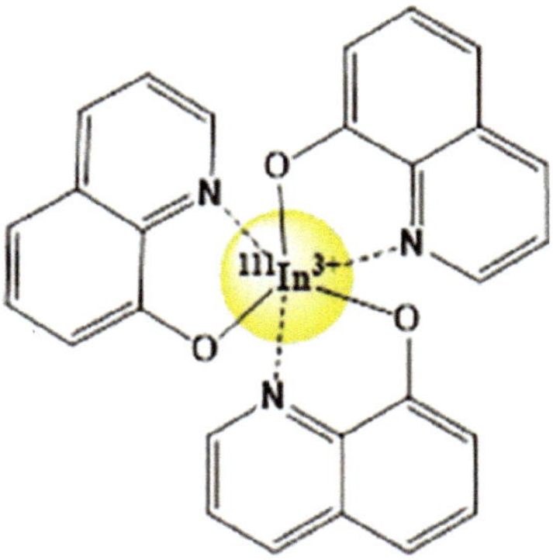	**^{111}In-oxyquinoline** SPECT-imaging agent $T_{1/2} = 2.5$ days Rapid efflux prevents utility in stem cell tracking	[93]

half-life [88, 89] As a solution to this, Bansal et al. evaluated the tracing capability of ^{89}Zr-DBN-labeled MSCs where labeling was achieved by direct conjugation to the cellular membrane surface [90]. ^{89}Zr has a half-life of around 3.5 days, an improvement from the scale of hours, but still not sufficient for weeks to months-long studies [91]. However, a signal was observed out to 7 days post-injection and likely would have been observed a few days longer if the study had continued [90].

Radiotracers that emit gamma radiation often have longer half-lives than positron emitters, making SPECT a potentially better methodology for stem cell tracking [85]. However, the cost/benefit of more harmful gamma radiation must also be considered in both the short and long terms. ^{111}In, for example, has a half-life of around 2.5 days. This half-life is on a similar time scale to ^{89}Zr and has been used rather extensively in clinical imaging of cardiovascular diseases (see review [92].) ^{111}In is often chelated by oxyquinoline for stem cell tracking, but these complexes are rapidly effluxed and require lipophilic delivery vehicles. As a recent solution, Krekorian et al. chelated ^{111}In ions to poly(lactic-co-glycolic acid) nanoparticles and found superior *in vivo* signal generation compared to traditional oxyquinoline formulations for tracking transplanted dendritic cells [93]. Although this strategy was evaluated in a model of lung infection, it could also be applied successfully to other stem cell therapies.

3.2 Gold Nanoparticle-Enhanced CT

While useful in shorter-term imaging studies due to excellent sensitivity, both PET and SPECT imaging techniques ultimately fall short in terms of longitudinal imaging due to the inevitability of radio decay past a limit of detection. Additionally, the ability to monitor specific stem cell populations from nearby tissue is significantly limited by the poor spatial resolution associated with PET and SPECT (1–2 mm vs 25–100 μM in MRI) [32]. CT is a widely used imaging technique due to its cost effectiveness, high spatial resolution (50–200 μM), and ease of use, with an estimated 70 million scans performed annually in the US alone [32, 94]. It is broadly clinically available and a promising candidate for future translation of stem cell-tracking techniques. CAs in CT function by attenuating X-ray beams. Therefore, metal ions with high molecular mass densely packed into nanoparticle cores are the best choices for labeling stem cells for *in vivo* tracking by CT.

Gold nanoparticles (AuNPs) are ideal CT CAs. The retention of AuNPs intracellularly provides an additional benefit of improvement in longitudinal imaging potential over PET and SPECT [94]. As an example, Betzer et al. coated AuNPs in glucose to track MSCs in a rat model of depression. Cells were transplanted into rat brains by direct stereotaxic injection and were detectable using CT imaging as early as 24 hours and as late as 1-month post-injection. Results from a functional study revealed that behavioral signs of depression were decreased in animals that received the AuNP-labeled stem cells [95]. Ongoing work from this group focuses on the optimization of MSC-labeling protocols [6]. This comprehensive study highlights the importance of both longitudinal imaging studies and functional assays to evaluate stem cell therapeutic efficacy. Overall, AuNP-enhanced CT is valuable for stem cell tracking, and as will be discussed in Section 5, is an ideal platform for more advanced multi-modal strategies.

4 OPTICAL IMAGING

In contrast to PET, SPECT, and CT, optical imaging does not rely on ionizing radiation to produce a signal [96]. It relies solely on light. Laser light is shone on tissue that contains a light-absorbing CA. The energy is absorbed into the CA, causing wavelength-dependent electronic excitation. Fluorescent molecules relax back to their ground state using a process that releases a photon in the ultraviolet-visible (UV-vis) spectra. The photon is then detected by the instrument by its unique wavelength [97]. Similar to PET and SPECT, the body does not contain many endogenous fluorophores, making background signals low and positive signals easily detected in an optical scan. Fluorescence is also very sensitive, capable of detecting small amounts of CA [32]. However, depth penetration through biological tissue is a major limiting factor, complicating the clinical utility of optical imaging. To date, the only clinically approved fluorescence technologies are for image-guided surgery [98, 99]. As such, fluorescent CAs are not ideal candidates for stem cell tracking clinically, but do have utility in preclinical models where tissues are thinner. Most fluorescence applications for stem cell therapy have been reserved for multi-modal strategies, which we will discuss in Section 5.

4.1 Two-Photon Molecular Agents

As discussed above, the most limiting factor for optical imaging in biological systems is highly limited depth penetration. Of all the wavelengths in the visible spectrum, lower energy red light penetrates the deepest. However, molecules that absorb red and near-IR light are less common than absorbers of higher energy wavelengths. Two-photon absorption, or non-linear absorption, occurs when two photons of a longer wavelength (low energy) are absorbed simultaneously to induce electronic excitation that should only occur with a shorter wavelength (higher energy) [100]. Another advantage of two-photon CAs is reduced signal interference from the excitation beam. The excitation wavelength is significantly shifted from the emission wavelength and is easily filtered out in image acquisition. This technique promotes a depth penetration greater than 500 μM, as opposed to less than 100 μM for traditional one-photon absorption [101]. To put this into perspective, this is still only as deep as the skin on our eyelids is thick.

Cyclometallated Ir(III) complexes readily undergo two-photon absorption [102, 103]. However, their utility in stem cell tracking has previously been further compromised by toxicity both *in vitro* and *in vivo* [104]. To overcome this, Li et al. formulated Ir(III)-[Bis92-methyl dibenzo[f,h]quinoxaline)(acetylacetone)] ($Ir(MDQ)_2acac$, Figure 2) molecular two-photon absorption CAs into a nanoparticle that would better protect the biological tissues from any Ir(III) induced toxicity. NSCs were labeled with these agents prior to subcutaneous implantation into the back of an athymic mouse model. Although this study did not observe the migration of transplanted cells past the injection point, they successfully imaged out to 21 days with a strong signal still observed [104]. Molecular fluorophore alone is unlikely to translate to a clinical cell tracking study, but it paves the way for promising utility of these complexes in multi-modal agents or studies that do not require great depth penetration, like skin tissue regeneration.

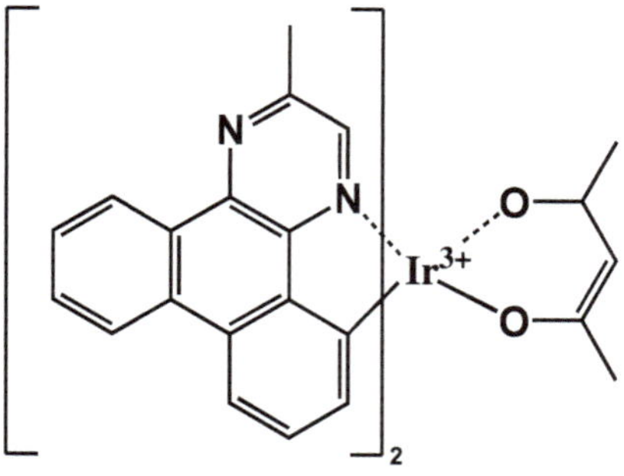

FIGURE 2 Structure of Ir(III)-$(MDQ)_2$acac, a molecular two-photon absorber used in stem cell tracking [104]. Nanoparticle delivery assistance promotes imaging out to 21 days following subcutaneous injection.

4.2 Quantum Dots

Quantum dots (QDs) are fluorescent semiconducting nanocrystals that have a modular structure and tunable band gaps. As such, QDs can be synthesized to absorb/emit anywhere in the UV-vis spectrum [105]. The most commonly studied QDs for cell tracking applications contain a CdSe/ZnS core-shell structure [106–109]. Although some of these studies were imaged successfully out to 8 weeks, structures containing Cd are known to be quite toxic, inhibiting their utility in stem cell tracking beyond depth penetration considerations [110].

As an attempt to retain the longitudinal properties of CdSe QDs but mitigate unwanted toxicity, Chen et al. synthesized Ag_2S QDs and fully characterized their biocompatibility with MSCs by assessing cytotoxicity, apoptosis induction, DNA damage, and differentiation potential. Labeled MSCs were then transplanted into a mouse model of liver failure by tail vein injection, where a signal was observed in the liver for 30 days post-injection [111]. This study highlights perfectly the need for tracking studies, as a signal was also observed strongly in the lungs. This is an undesired location for transplanted MSCs to locate and would require either further optimization of the transplantation protocol to avoid off-target localization, or longer-term imaging studies to confirm that localization in the lungs is not problematic.

5 MULTI-MODAL IMAGING STRATEGIES

The appeal of multi-modal imaging lies in the ability to combine the best aspects of different imaging modalities to overcome the inherent disadvantages of each on its own. As imaging techniques and strategies advance, this also promotes the ability to simultaneously monitor stem cell localization with one modality and function with another. It also presents the opportunity to compound the benefits of both molecular and nanoparticle approaches.

5.1 Molecular MRI/Optical Agents

The most common multi-modal strategies for preclinical stem cell tracking combine the *in vivo* safety profile and depth penetration of MRI with *in vivo* and *ex vivo* validation using fluorescence. As mentioned previously, fluorescence strategies are not well-translated into the clinic, but are nonetheless valuable tools in preclinical investigations. Preliminary studies evaluated the combination of non-covalently tethered Gd-DTPA and fluorescent organic molecules [112, 113]. The main drawback of this strategy is the uncertainty that the MRI CA and the fluorescent probe will remain together in the same tissue if they are not covalently linked. One response to this was to chemically couple both Gd-DTPA and a fluorescent dye to a silica nanoparticle, improving both cellular internalization and co-localization confidence [113].

A similar strategy that has gained much attention for stem cell tracking is the gadolinium-rhodamine dextran (GRID) conjugate. Gd-DTPA and rhodamine are chemically coupled to a dextran backbone, promoting cellular internalization and colocalization of the MRI agent and fluorophore. One of the first studies in 2002 with this conjugate labeled NSCs with GRID prior to injection into a rat model of global ischemia. Transplanted cells were detectable out to 14 days *ex vivo* [114]. In a follow-up in 2004, GRID labeled NSCs were injected directly into the brains of rats 3 months following an induced stroke. The cells were transplanted into the right side of the brain, and migration to the lesion on the left side was tracked for 14 days (Figure 3) [115].

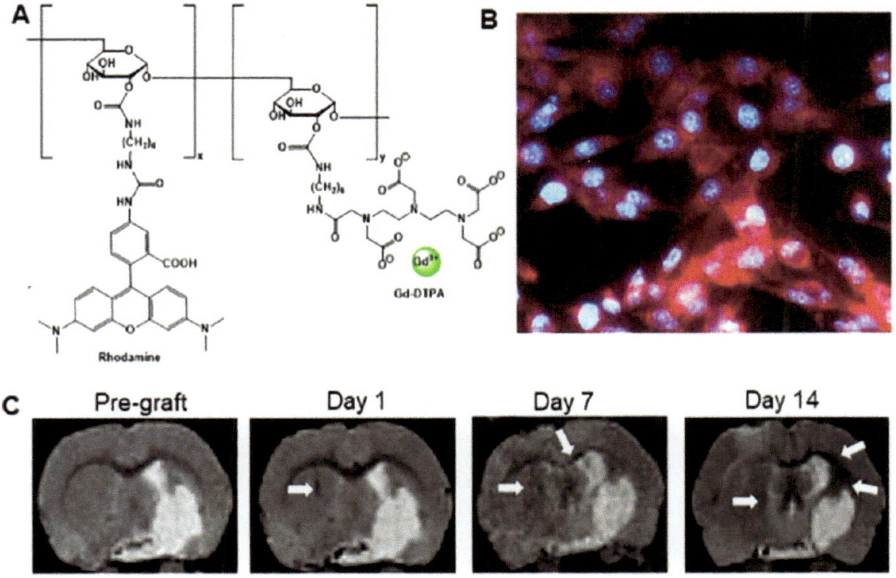

FIGURE 3 Gadolinium rhodamine dextran (GRID) utilized to track NSCs in a rat model of stroke by Modo et al. (a) Structure of GRID. (b) Validation of NSC labeling *in vitro*. (c) *In vivo* tracking of NSC migration from the site of injection to the lesion site over 14 days. NSCs indeed hone to the site of injury as visualized by MRI [115].

While GRID showed initial promise for stem cell tracking, a chronic behavioral study in 2009 revealed that early successful tracking does not necessarily correlate to successful therapeutic efficacy. Animals were monitored for 1-year post-implantation to evaluate changes in brain lesion size over time following a stroke. The researchers found that stem cells labeled with PKH26 red fluorescent dye reduced lesion size by approximately 35%, but lesion sizes of GRID-labeled therapies showed no significant improvement from non-treated controls [116]. They postulate this is due to Gd(III) toxicity associated with thermodynamic instability of Gd-DTPA, which would be heightened by the retention of the imaging agent due to dextran conjugation. These results further confirm the necessity of functional analysis of the actual therapeutic component in the presence of any given imaging agent.

Another molecular combination of MRI and fluorescence for stem cell tracking is Gadophrin-2, a complex that contains two linearly chelated Gd(III) ions and a Cu(II) porphyrin fluorophore (Figure 4). Although not tested in a disease model, peripheral blood cells were successfully labeled with Gadophrin-2 and imaged *in vivo* 24 hours post-injection. The conclusions of this study emphasize the utility of this conjugate to evaluate cellular localization based on transplantation technique

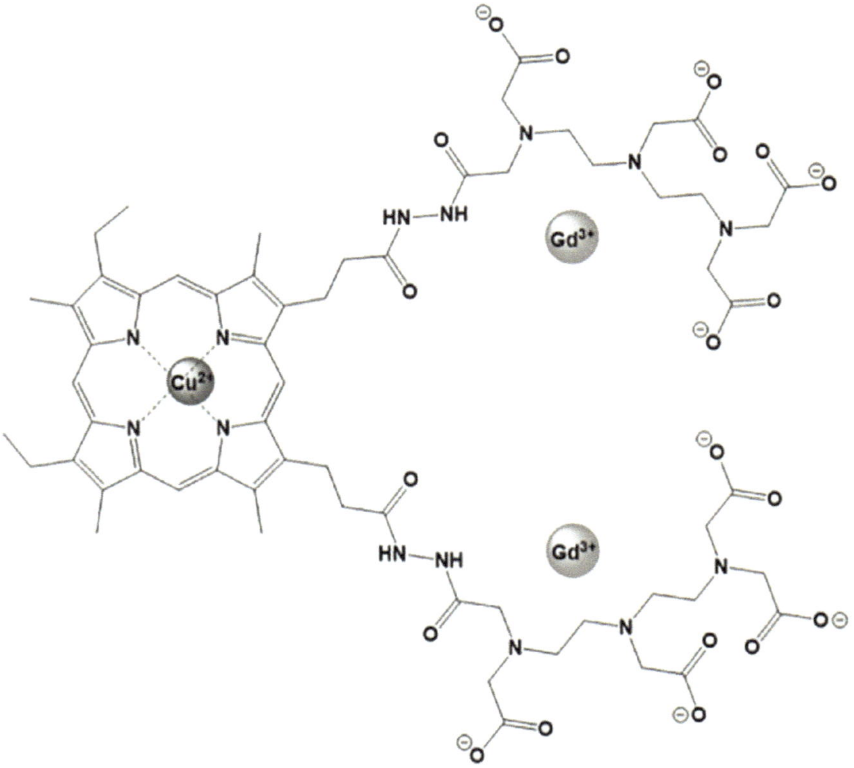

FIGURE 4 Structure of Gadophrin-2. Two units of Gd-DTPA are conjugated to a Cu(II) porphyrin for dual-modal molecular MRI/optical imaging. Analysis reveals that this conjugate has utility for short-term optimization of cellular transplantation methods [117].

and desired localization target [117]. This analysis can be applied to any of the aforementioned Gd-DTPA/fluorescent strategies, as both components present complications for clinical utility. The discussion in this study raises an excellent point that even complications for clinical translation do not completely annihilate preclinical value. Proper care and rigorous evaluation must be taken, though, for any agent seeking to reach clinical use.

5.2 Molecular PET/MRI Agents

While seemingly an unlikely combination, a unique study performed by Lewis et al. in 2015 sought to take advantage of divalent metal transporters (DMTs) for the uptake of ^{52}Mn/Mn for dual-modal PET/MRI imaging. Specifically, NSCs were engineered to overexpress DMT-1 and directly injected into the brains of rats [118]. The premise is that the over-expression of these ion channels would improve the uptake of systemically delivered ^{52}Mn/Mn for PET/MRI imaging that does not require labeling of stem cells prior to injection. In this regard, the Mn agents can be delivered multiple times throughout a longitudinal imaging study and then cleared by the body to avoid issues associated with prolonged retention. An increased signal was observed in engineered cells *in vivo*, though not reliably to the extent that was hypothesized. Additionally, the study does not report any consideration on the impact of engineering NSCs on proliferation and differentiation potential [118]. Follow-ups will be necessary to optimize this technique for potential clinical translation. The initial promise of the overall idea and results do suggest this optimization is worth pursuing.

5.3 Nanoparticle Strategies

As opposed to purely molecular strategies, the possibilities for nanoparticle-based multimodal CAs are almost endless. SPIOs, AuNPs, and even QDs that we have discussed previously are modular platforms with functionalizable surfaces where any number of molecular MRI, PET/SPECT, or optical CAs could be added to build a dual or multimodal agent. Herein, we will focus on a few examples of SPIO and AuNP platforms.

SPIOs have been combined with organic fluorescent dyes to label MSCs for proof-of-concept *in vivo* detection, as well as in stem cell therapies following stroke and myocardial infarction [119–121]. Similarly to the purely molecular combinations of MRI and optical imaging, the depth penetration and excellent resolution of MRI pair well with the sensitivity of fluorescence. That said, negative T_2 contrast will always be inferior to positive T_1 contrast. As we have discussed previously, nanoparticle T_1 agents are more challenging to generate. However, one example of successfully generating a nanoparticle-based T_1 MRI/optical agent came as a response to the long-term therapeutic failure of GRID-labeled stem cells. The Modo group loaded AuNPs with Gd(III)-DOTA macrocyclic chelates chemically coupled to DNA strands. (Figure 5) The DNA imparts stability to the AuNPs to promote cellular internalization and retention, with Gd(III)-DOTA providing T_1 MRI contrast. The DNA strands were also capped with a Cy3 dye for dual-modal MRI/fluorescence imaging. Excellent *in vivo* contrast was observed between labeled and unlabeled cells, and there was no initial evidence for hinderance of cell functionality [122]. Although the CT contrast

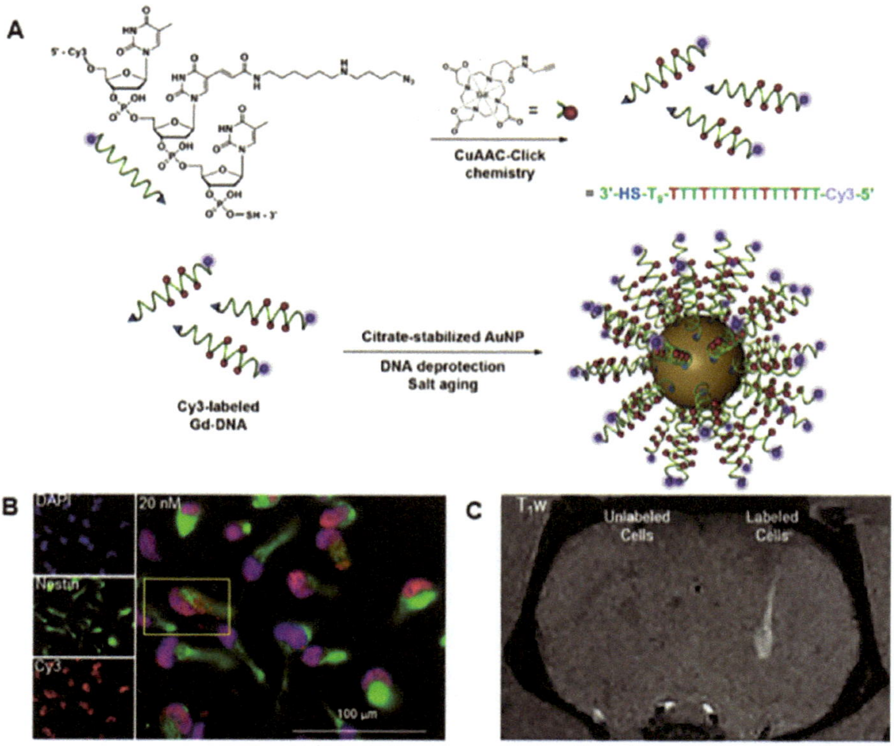

FIGURE 5 A dual-modal MRI/optical nanoparticle strategy for stem cell tracking. (a) Schematic for the formulation of DNA-capped AuNPs for dual-modal MRI/optical imaging. MRI enhancement is gained from Gd(III)-DOTA CAs chemically coupled to the DNA backbone. (b) Validation of internalization *in vitro* using confocal microscopy. Red signal represents endosomal located Cy3/AuNPs. (c) Contrast enhancement is observed in labeled cells over non-labeled cells at the injection site or a rat brain [122].

enhancement of the AuNP was not evaluated and a longitudinal study was not performed, this study amplifies the benefits that multi-modal nanoparticle strategies may exhibit over purely molecular strategies. Mainly for MRI, higher concentrations of Gd(III) agents are internalized, improving the overall signal enhancement and bypassing sensitivity challenges.

CT contrast by AuNPs has shown promise as a single modality of stem cell tracking, encouraging its utility in dual-modal formulations. Combinations of CT and optical imaging have been explored through the functionalization of AuNPs with fluorescent molecules [123]. In one study, the near IR dye indocyanine green was loaded onto AuNPs to label MSCs for transplantation into a model of lung injury, and cells were successfully detected out to 3 weeks [124]. By histological staining, both labeled and unlabeled MSCs effectively reduced the inflammatory response associated with the injury compared to non-treated animals [125]. This study presents a successful example of cell labeling not compromising therapeutic efficacy, and showcases the importance of performing these functional assays.

6 CONCLUDING REMARKS

Stem cell therapy has emerged as a promising therapeutic strategy for regenerative medicine where traditional drug-based therapies have been elusive. However, the future clinical translation of stem cell therapies requires non-invasive tracking methods to evaluate the longer-term localization and function of the implanted cells employed during preclinical and clinical investigations. This chapter has specifically discussed tracking methods involving metal-ion-based CAs for MRI (Section 2), PET, SPECT, CT (Section 3), Optical Imaging (Section 4), and Combinations of These Modalities (Section 5). It is imperative when evaluating each modality/CA pair to consider the potential influence of the agent on stem cell behavior, the ability to perform longitudinal imaging studies, as well as the actual efficacy of the therapy.

After evaluation of the studies presented in this chapter, we conclude that for optimal clinical translation of long-term tracking agents, MRI and CT are superior to PET, SPECT, and optical techniques. This is primarily because PET and SPECT radiotracers are undetectable once sufficient decay has occurred, and fluorescence imaging has very limited clinical utility due to poor tissue penetration. However, we find that optical components of multi-modal imaging strategies have high utility in initial preclinical investigations as they may impart sensitivity specifically to an MRI CA for localization validation. Human cells transplanted inside the human body cannot be tracked post-transplantation with human-specific markers. Pre-labeling the cells can distinguish transplanted cells from existing tissue.

When considering molecular vs. nanoparticle platforms, we find nanoparticle strategies to carry many advantages. The first is the ability to incorporate multiple imaging modalities more modularly. Molecular agents are limited by synthetic coupling needed to ensure proper co-localization at all times. In contrast, the large surface area of nanoparticles can adsorb high amounts of molecular CAs with imaging modality that differs from the nanoparticle itself. This allows for the conjugation of two or more imaging modalities into one system very easily, creating an avenue for more readily incorporating activatable agents to report on therapeutic efficacy. Nanoparticles may be further functionalized with ligands for improved cellular internalization and retention for longitudinal imaging studies.

Although it seems that the "perfect" CA for stem cell tracking is yet to be tested, there are discrepancies across studies in the literature that make comparing previously evaluated strategies difficult. For example, many studies do not record at what time points the signal is no longer detected *in vivo*. Even fewer studies performed analyses of actual stem cell therapeutic function in the presence of metallo-CAs. As the field progresses and tracking strategies are chosen for translation into clinical investigations, these analyses will be imperative.

ACKNOWLEDGMENTS

Meghan W. Dukes would like to acknowledge the National Cancer Institute and the National Institute of Health under the F31 Ruth L. Kirschstein National Research Service Award grant number 1F31CA236175. We further thank these institutes for grant number R01NS115571, along with the National Institute of Neurological

Disorders and Stoke for grant numbers R21NS118399 and R01NS122768. Michel Modo would like to thank T. Kevin Hitchens for the MRI physics support to optimally images MR CAs.

ABBREVIATIONS AND DEFINITIONS

acac	acetylacetone
AMP	acetoxymethoxycarbonylporphyrin
AuNP	gold nanoparticle
CA	contrast agent
CEST	chemical exchange saturation transfer
CT	computed tomography
DBN	desferrioxamine-NCS
DMT	divalent metal transporter
DOTA	1, 4, 7, 10-tetraazacyclododecane
DPDP	dipyridoxaldiphosphate
DTPA	diethylenetriaminepentaacidic acid
ESC	embryonic stem cell
EtP	5, 10, 15, 20-tetrakis (ethoxycarbonyl) porphyrin
FDA	Food and Drug Administration
GRID	gadolinium rhodamine dextran
iPS	Induced pluripotent stem cell
MDQ	Bis(2-methyldibenzo[f,h]quinoxaline
MRI	magnetic resonance imaging
MSC	mesenchymal stem cell
NSC	neuronal stem cell
ParaCEST	paramagnetic chemical exchange saturation transfer
PET	positron emission tomography
PNH2	5-(4-aminophenyl)-10, 15, 20-tris(4-sulfonatophenyl) porphyrin
PTSM	pyruvaldehyde bis(*N*4-methylsemicarbazone
QD	quantum dots
SPECT	single photon emission computed tomography
SPIO	superparamagnetic iron oxide
UV-Vis	Ultraviolet-visible

REFERENCES

1. R. Pal, N. K. Venkataramana, A. Bansal, S. Balaraju, M. Jan, R. Chandra, A. Dixit, A. Rauthan, U. Murgod, S. Totey, *Cytotherapy* **2009**, *11*, 897–911.
2. J. C. Ra, I. S. Shin, S. H. Kim, S. K. Kang, B. C. Kang, H. Y. Lee, Y. J. Kim, J. Y. Jo, E. J. Yoon, H. J. Choi, E. Kwon, *Stem Cells Dev.* **2011**, *20*, 1297–1308.
3. T. O'Brien, F. P. Barry, *Mayo Clin. Proc.* **2009**, *84*, 859–861.
4. A. Abbott, *Nature* **2011**, *471*, 280–280.
5. A. Andrzejewska, S. Dabrowska, B. Lukomska, M. Janowski, *Adv. Sci.* **2021**, *8*, 2002944–2002944.
6. O. Betzer, R. Meir, T. Dreifuss, K. Shamalov, M. Motiei, A. Shwartz, K. Baranes, C. J. Cohen, N. Shraga-Heled, R. Ofir, G. Yadid, R. Popovtzer, *Sci. Rep.* **2015**, *5*, 15400.

7. T. Negoro, H. Okura, M. Maehata, S. Hayashi, S. Yoshida, N. Takada, A. Matsuyama, *NPJ Regen. Med.* **2019**, *4*, 20.
8. Z. Gao, L. Zhang, J. Hu, Y. Sun, *Nanomed.: Nanotechnol. Biol. Med.* **2013**, *9*, 174–184.
9. D. Dapkute, M. Pleckaitis, D. Bulotiene, D. Daunoravicius, R. Rotomskis, V. Karabanovas, *ACS Appl. Mater. Interfaces* **2021**, *13*, 43937–43951.
10. S. E. Freedman, *Semin. Oncol. Nurs.* **1988**, *4*, 3–8.
11. C. Nabhan, J. Mehta, M. S. Tallman, *Bone Marrow Transplant* **2001**, *28*, 219–226.
12. M. Demagalhaes-Silverman, A. D. Donnenberg, S. M. Pincus, E. D. Ball, *Cell Transplant.* **1993**, *2*, 75–98.
13. G. J. Bittle, D. Morales, K. B. Deatrick, N. Parchment, P. Saha, R. Mishra, S. Sharma, N. Pietris, A. Vasilenko, C. Bor, C. Ambastha, M. Gunasekaran, D. Li, S. Kaushal, *Circ. Res.* **2018**, *123*, 288–300.
14. M. B. Britten, N. D. Abolmaali, B. Assmus, R. Lehmann, J. Honold, J. Schmitt, T. J. Vogl, H. Martin, V. Schächinger, S. Dimmeler, A. M. Zeiher, *Circulation* **2003**, *108*, 2212–2218.
15. R. Rikhtegar, M. Pezeshkian, S. Dolati, N. Safaie, A. Afrasiabi Rad, M. Mahdipour, M. Nouri, A. R. Jodati, M. Yousefi, *Biomed. Pharmacother.* **2019**, *109*, 304–313.
16. J.-W. Chang, S.-P. Hung, H.-H. Wu, W.-M. Wu, A.-H. Yang, H.-L. Tsai, L.-Y. Yang, O. K. Lee, *Cell Transplant.* **2011**, *20*, 245–258.
17. W. J. Ennis, A. Sui, A. Bartholomew, *Adv. Wound Care* **2013**, *2*, 369–378.
18. I. Garitaonandia, R. Gonzalez, G. Sherman, A. Semechkin, A. Evans, R. Kern, *Stem Cells Dev.* **2018**, *27*, 951–957.
19. A. Uccelli, A. Laroni, M. S. Freedman, *Lancet Neurol.* **2011**, *10*, 649–656.
20. M. Li, K. Guo, S. Ikehara, *Int. J. Mol. Sci.* **2014**, *15*, 19226–19238.
21. O. Y. Bang, J. S. Lee, P. H. Lee, G. Lee, *Ann. Neurol.* **2005**, *57*, 874–882.
22. M. Barba, G. Di Taranto, W. Lattanzi, *Expert Opin. Biol. Ther.* **2017**, *17*, 677–689.
23. J. V. Frangioni, R. J. Hajjar, *Circulation* **2004**, *110*, 3378–3383.
24. J. W. M. Bulte, H. E. Daldrup-Link, *Radiology* **2018**, *289*, 604–615.
25. X. Zhang, T. Lei, H. Du, *Stem Cell Res. Ther.* **2021**, *12*, 457.
26. D. A. Rosman, R. Duszak, W. Wang, D. R. Hughes, A. B. Rosenkrantz, *Am. J. Roentgenol.* **2017**, *210*, 364–368.
27. A. Berger, *BMJ* **2002**, *324*, 35–35.
28. F. Bloch, *Phys. Rev.* **1946**, *70*, 460–474.
29. J. Wahsner, E. M. Gale, A. Rodríguez-Rodríguez, P. Caravan, *Chem. Rev.* **2019**, *119*, 957–1057.
30. A. Hengerer, J. Grimm, *Biomed. Imaging Interv. J.* **2006**, *2*, e8.
31. M. E. Ladd, P. Bachert, M. Meyerspeer, E. Moser, A. M. Nagel, D. G. Norris, S. Schmitter, O. Speck, S. Straub, M. Zaiss, *Prog. Nucl. Magn. Reson. Spectrosc.* **2018**, *109*, 1–50.
32. F.-M. Lu, Z. Yuan, *Quant. Imaging Med. Surg.* **2015**, *5*, 433–447.
33. V. C. Pierre, M. J. Allen, P. Caravan, *J. Biol. Inorg. Chem.* **2014**, *19*, 127–131.
34. V. M. Runge, *Investig. Radiol.* **2017**, *52*, 317–323.
35. V. Jacques, S. Dumas, W. C. Sun, J. S. Troughton, M. T. Greenfield, P. Caravan, *Investig. Radiol.* **2010**, *45*, 613–624.
36. K. Geng, Z. X. Yang, D. Huang, M. Yi, Y. Jia, G. Yan, X. Cheng, R. Wu, *Mol. Med. Rep.* **2015**, *11*, 954–960.
37. J. Guenoun, G. A. Koning, G. Doeswijk, L. Bosman, P. A. Wielopolski, G. P. Krestin, M. R. Bernsen, *Cell Transplant* **2012**, *21*, 191–205.
38. Y. Liu, Z.-J. He, B. Xu, Q.-Z. Wu, G. Liu, H. Zhu, Q. Zhong, D. Y. Deng, H. Ai, Q. Yue, Y. Wei, S. Jun, G. Zhou, Q.-Y. Gong, *Brain Res.* **2011**, *1391*, 24–35.
39. E. Di Gregorio, E. Gianolio, R. Stefania, G. Barutello, G. Digilio, S. Aime, *Anal. Chem.* **2013**, *85*, 5627–5631.

40. Z. Zhou, Z.-R. Lu, *Wiley Interdiscip. Rev. Nanomed. Nanobiotechnol.* **2013**, *5*, 1–18.
41. J. B. Lansman, *J. Gen. Physiol.* **1990**, *95*, 679–696.
42. B. A. Biagi, J. J. Enyeart, *Am. J. Physiol.* **1990**, *259*, C515–C520.
43. H. S. Thomsen, *Eur. Radiol.* **2006**, *16*, 2619–2621.
44. W.-W. Cai, L.-J. Wang, S.-J. Li, X.-P. Zhang, T.-T. Li, Y.-H. Wang, X. Yang, J. Xie, J.-D. Li, S.-J. Liu, W. Xu, S. He, Z. Cheng, Q.-L. Fan, R.-P. Zhang, *J. Biomed. Mater. Res. Part A* **2017**, *105*, 131–137.
45. Y.-D. Xiao, R. Paudel, J. Liu, C. Ma, Z.-S. Zhang, S.-K. Zhou, *Int. J. Mol. Med.* **2016**, *38*, 1319–1326.
46. M. Zhang, X. Liu, J. Huang, L. Wang, H. Shen, Y. Luo, Z. Li, H. Zhang, Z. Deng, Z. Zhang, *Nanomedicine* **2018**, *14*, 2475–2483.
47. P. Zhang, Y. Zhang, B. Li, H. Zhang, H. Lin, Z. Deng, B. Tan, *J. Pept. Sci.* **2018**, *24*, e3077.
48. F. L. Giesel, M. Stroick, M. Griebe, H. Tröster, C. W. von der Lieth, M. Requardt, M. Rius, M. Essig, H.-U. Kauczor, M. G. Hennerici, M. Fatar, *Investig. Radiol.* **2006**, *41*, 868–873.
49. D. Pan, A. H. Schmieder, S. A. Wickline, G. M. Lanza, *Tetrahedron* **2011**, *67*, 8431–8444.
50. A. Bertin, A. I. Michou-Gallani, J. L. Gallani, D. Felder-Flesch, *Toxicol. In Vitro* **2010**, *24*, 1386–1394.
51. M. Thuen, M. Berry, T. B. Pedersen, P. E. Goa, M. Summerfield, O. Haraldseth, A. Sandvig, C. Brekken, *J. Magn. Reson. Imaging* **2008**, *28*, 855–865.
52. S. Loai, I. Haedicke, Z. Mirzaei, C. A. Simmons, X.-A. Zhang, H. L. Cheng, *J. Magn. Reson. Imaging* **2016**, *44*, 1456–1463.
53. I. E. Haedicke, T. Li, Y. L. K. Zhu, F. Martinez, A. M. Hamilton, D. H. Murrell, J. T. Nofiele, H.-L. M. Cheng, T. J. Scholl, P. J. Foster, X.-A. Zhang, *Chem. Sci.* **2016**, *7*, 4308–4317.
54. A. Venter, D. A. Szulc, S. Loai, T. Ganesh, I. E. Haedicke, H.-L. M. Cheng, *Sci. Rep.* **2018**, *8*, 12129.
55. I. E. Haedicke, S. Loai, H.-L. M. Cheng, *Contrast Media Mol. Imaging* **2019**, *2019*, 3475786.
56. G. Liu, M. M. Ali, B. Yoo, M. A. Griswold, J. A. Tkach, M. D. Pagel, *Magn. Reson. Med.* **2009**, *61*, 399–408.
57. S. Aime, A. Barge, D. Delli Castelli, F. Fedeli, A. Mortillaro, F. U. Nielsen, E. Terreno, *Magn. Reson. Med.* **2002**, *47*, 639–648.
58. M. T. McMahon, A. A. Gilad, J. Zhou, P. Z. Sun, J. W. Bulte, P. C. van Zijl, *Magn. Reson. Med.* **2006**, *55*, 836–847.
59. I. Hancu, W. T. Dixon, M. Woods, E. Vinogradov, A. D. Sherry, R. E. Lenkinski, *Acta Radiol.* **2010**, *51*, 910–923.
60. N. Nakagomi, T. Nakagomi, S. Kubo, A. Nakano-Doi, O. Saino, M. Takata, H. Yoshikawa, D. M. Stern, T. Matsuyama, A. Taguchi, *Stem Cells* **2009**, *27*, 2185–2195.
61. F. J. Nicholls, W. Ling, G. Ferrauto, S. Aime, M. Modo, *Sci. Rep.* **2015**, *5*, 14597–14597.
62. M. Modo, J. Kolosnjaj-Tabi, F. Nicholls, W. Ling, C. Wilhelm, O. Debarge, F. Gazeau, O. Clement, *Contrast Media Mol. Imaging* **2013**, *8*, 439–455.
63. E. Vinogradov, A. D. Sherry, R. E. Lenkinski, *J. Magn. Reson.* **2013**, *229*, 155–172.
64. C.-L. Tseng, I. L. Shih, L. Stobinski, F.-H. Lin, *Biomaterials* **2010**, *31*, 5427–5435.
65. F. Azzabi, M. Rottmar, V. Jovaisaite, M. Rudin, T. Sulser, A. Boss, D. Eberli, *Tissue Eng. – C: Methods* **2015**, *21*, 182–191.
66. Y. X. Wang, *Quant. Imaging Med. Surg.* **2011**, *1*, 35–40.
67. M. Mahmoudi, H. Hosseinkhani, M. Hosseinkhani, S. Boutry, A. Simchi, W. S. Journeay, K. Subramani, S. Laurent, *Chem. Rev.* **2011**, *111*, 253–280.
68. M. P. Nucci, I. S. Filgueiras, J. M. Ferreira, F. A. de Oliveira, L. P. Nucci, J. B. Mamani, G. N. A. Rego, L. F. Gamarra, *World J. Stem Cells* **2020**, *12*, 381–405.

69. M. R. Bernsen, J. Guenoun, S. T. van Tiel, G. P. Krestin, *Br. J. Radiol.* **2015**, *88*, 20150375.
70. A. M. Reddy, B. K. Kwak, H. J. Shim, C. Ahn, H. S. Lee, Y. J. Suh, E. S. Park, *J. Korean Med. Sci.* **2010**, *25*, 211–219.
71. S. K. Mishra, S. Khushu, A. K. Singh, G. Gangenahalli, *Stem Cell Rev. Rep.* **2018**, *14*, 888–900.
72. L. Jiang, R. Li, H. Tang, J. Zhong, H. Sun, W. Tang, H. Wang, J. Zhu, *Cell Transplantation*, *28* (**2018**) 747–755.
73. E. Sykova, P. Jendelova, in *Progress in Brain Research*, Eds J. T. Weber, A. I. R. Maas, Elsevier, Amsterdam, Netherlands, 2007, pp. 367–383.
74. C. Da-Jeong, M. Hyeyoung, L. Yong Hyun, L. Nayeon, J.L. Hong, J. Iksoo, L. Hyunseung, H. Tae-Sun, O. Seung-Hun, S. Dong Ah, U. K. Seung, H. Kwan Soo, S. Jihwan, *IJSC* **2012**, *5*, 79–83.
75. A. Obenaus, N. Dilmac, B. Tone, H. R. Tian, R. Hartman, M. Digicaylioglu, E. Y. Snyder, S. Ashwal, *Ann. Neurol.* **2011**, *69*, 282–291.
76. J. Guenoun, A. Ruggiero, G. Doeswijk, R. C. Janssens, G. A. Koning, G. Kotek, G. P. Krestin, M. R. Bernsen, *Contrast Media Mol. Imaging* **2013**, *8*, 165–174.
77. L. M. Parkes, R. Hodgson, T. Lu Le, D. Tung Le, I. Robinson, D. G. Fernig, N. T. Thanh, *Contrast Media Mol. Imaging* **2008**, *3*, 150–156.
78. M. Edmundson, N. T. K. Thanh, B. Song, *Theranostics* **2013**, *3*, 573–582.
79. X. Meng, H. C. Seton, L. T. Lu, I. A. Prior, N. T. K. Thanh, B. Song, *Nanoscale* **2011**, *3*, 977–984.
80. K. Jiráková, M. Šeneklová, D. Jirák, K. Turnovcová, M. Vosmanská, M. Babič, D. Horák, P. Veverka, P. Jendelová, *Int. J. Nanomed.* **2016**, *11*, 6267–6281.
81. A. Berger, *BMJ* **2003**, *326*, 1449–1449.
82. F. J. Beekman, C. Kamphuis, M. A. King, P. P. van Rijk, M. A. Viergever, *Comput. Med. Imaging Graph.* **2001**, *25*, 135–146.
83. H. Jiang, F. Thomas, *J. Med. Imaging* **2021**, *8*, 1–24.
84. M. A. Said, M. Musarudin, N. F. Zulkaffli, *Ann. Nucl. Med.* **2020**, *34*, 884–891.
85. S. Adak, R. Bhalla, K. K. V. Raj, S. Mandal, R. Pickett, S. K. Luthra, *Radiochim. Acta* **2012**, *100*, 95–107.
86. V. S. Suzanne, J. Marian, H. Vanessa, Radioisotopes — *Applications in Bio-Medical Sciences*, **2011**, Production and Selection of metal PET Radioisotopes for Molecular Imaging, 199–224.
87. C. J. Anderson, R. Ferdani, *Cancer Biother. Radiopharm.* **2009**, *24*, 379–393.
88. N. Adonai, K. N. Nguyen, J. Walsh, M. Iyer, T. Toyokuni, M. E. Phelps, T. McCarthy, D. W. McCarthy, S. S. Gambhir, *Proc. Natl. Acad. Sci.* **2002**, *99*, 3030.
89. K. Chen, Z. Miao, Z. Cheng, *J. Nucl. Med.* **2011**, *52*, 521.
90. A. Bansal, M. K. Pandey, Y. E. Demirhan, J. J. Nesbitt, R. J. Crespo-Diaz, A. Terzic, A. Behfar, T. R. DeGrado, *EJNMMI Res.* **2015**, *5*, 19.
91. G. W. Severin, J. W. Engle, T. E. Barnhart, R. J. Nickles, *J. Med. Chem.* **2011**, *7*, 389–394.
92. C. Kawai, A. Matsumori, T. Nishimura, K. Endo, *Kaku Igaku* **1990**, *27*, 1419–1432.
93. M. Krekorian, G. G. W. Sandker, K. R. G. Cortenbach, O. Tagit, N. K. van Riessen, R. Raavé, M. Srinivas, C. G. Figdor, S. Heskamp, E. H. J. G. Aarntzen, *Bioconjug. Chem.* **2021**, *32*, 1802–1811.
94. R. Meir, R. Popovtzer, *WIREs Nanomed. Nanobiotechnol.* **2018**, *10*, e1480.
95. O. Betzer, A. Shwartz, M. Motiei, G. Kazimirsky, I. Gispan, E. Damti, C. Brodie, G. Yadid, R. Popovtzer, *ACS Nano* **2014**, *8*, 9274–9285.
96. K. Nguyen Phan, D. Trinh, T. Thanh, H. Chuduc, *Int. J. Sci. Res.* **2015**, *4*, 145–148.
97. J. W. Lichtman, J.-A. Conchello, *Nat. Methods* **2005**, *2*, 910–919.
98. S. van Keulen, N. Nishio, S. Fakurnejad, A. Birkeland, B. A. Martin, G. Lu, Q. Zhou, S. U. Chirita, T. Forouzanfar, A. D. Colevas, N. S. van den Berg, E. L. Rosenthal, *J. Nucl. Med.* **2019**, *60*, 758–763.

99. T. Nagaya, Y. A. Nakamura, P. L. Choyke, H. Kobayashi, *Front. Oncol.* **2017**, *7*, 314–314.
100. P. T. C. So, C. Y. Dong, B. R. Masters, K. M. Berland, *Annu. Rev. Biomed. Eng.* **2000**, *2*, 399–429.
101. H. M. Kim, C. Jung, B. R. Kim, S.-Y. Jung, J. H. Hong, Y.-G. Ko, K. J. Lee, B. R. Cho, *Angew. Chem. Int. Ed.* **2007**, *46*, 3460–3463.
102. J. Massue, J. Olesiak-Banska, E. Jeanneau, C. Aronica, K. Matczyszyn, M. Samoc, C. Monnereau, C. Andraud, *Inorg. Chem.* **2013**, *52*, 10705–10707.
103. Y. You, S. Cho, W. Nam, *Inorg. Chem.* **2014**, *53*, 1804–1815.
104. D. Li, X. Yan, Y. Hu, Y. Liu, R. Guo, M. Liao, B. Shao, Q. Tang, X. Guo, R. Chai, Q. Zhang, M. Tang, *ACS Biomater.Sci. Eng.* **2019**, *5*, 1561–1568.
105. S. Pleskova, E. Mikheeva, E. Gornostaeva, in *Cellular and Molecular Toxicology of Nanoparticles*, Eds Q. Saquib, M. Faisal, A. A. Al-Khedhairy, A. A. Alatar, Springer International Publishing, Cham, 2018, pp. 323–334.
106. A. B. Rosen, D. J. Kelly, A. J. Schuldt, J. Lu, I. A. Potapova, S. V. Doronin, K. J. Robichaud, R. B. Robinson, M. R. Rosen, P. R. Brink, G. R. Gaudette, I. S. Cohen, *Stem Cells* **2007**, *25*, 2128–2138.
107. H. Yukawa, M. Watanabe, N. Kaji, Y. Okamoto, M. Tokeshi, Y. Miyamoto, H. Noguchi, Y. Baba, S. Hayashi, *Biomaterials* **2012**, *33*, 2177–2186.
108. H. Yukawa, Y. Baba, *Anal. Chem.* **2017**, *89*, 2671–2681.
109. J.-S. Ni, Y. Li, W. Yue, B. Liu, K. Li, *Theranostics* **2020**, *10*, 1923–1947.
110. K. C. Nguyen, P. Rippstein, A. F. Tayabali, W. G. Willmore, *Toxicol. Sci.* **2015**, *146*, 31–42.
111. G. Chen, F. Tian, Y. Zhang, Y. Zhang, C. Li, Q. Wang, *Adv. Funct. Mater.* **2014**, *24*, 2481–2488.
112. J. Shen, L.-N. Cheng, X.-M. Zhong, X.-H. Duan, R.-M. Guo, G.-B. Hong, *Eur. J. Radiol.* **2010**, *75*, 397–405.
113. J.-K. Hsiao, C.-P. Tsai, T.-H. Chung, Y. Hung, M. Yao, H.-M. Liu, C.-Y. Mou, C.-S. Yang, Y.-C. Chen, D.-M. Huang, *Small* **2008**, *4*, 1445–1452.
114. M. Modo, D. Cash, K. Mellodew, S. C. R. Williams, S. E. Fraser, T. J. Meade, J. Price, H. Hodges, *NeuroImage* **2002**, *17*, 803–811.
115. M. Modo, K. Mellodew, D. Cash, S. E. Fraser, T. J. Meade, J. Price, S. C. R. Williams, *NeuroImage* **2004**, *21*, 311–317.
116. M. Modo, J. S. Beech, T. J. Meade, S. C. R. Williams, J. Price, *NeuroImage* **2009**, *47* (Suppl 2), T133–T142.
117. H. E. Daldrup-Link, M. Rudelius, S. Metz, G. Piontek, B. Pichler, M. Settles, U. Heinzmann, J. Schlegel, R. A. J. Oostendorp, E. J. Rummeny, *Eur. J. Nucl. Med. Mol. Imaging* **2004**, *31*, 1312–1321.
118. C. M. Lewis, S. A. Graves, R. Hernandez, H. F. Valdovinos, T. E. Barnhart, W. Cai, M. E. Meyerand, R. J. Nickles, M. Suzuki, *Theranostics* **2015**, *5*, 227–239.
119. C. K. Sung, K. A. Hong, S. Lin, Y. Lee, J. Cha, J.-K. Lee, C. P. Hong, B. S. Han, S. I. Jung, S. H. Kim, K. S. Yoon, *Korean J. Radiol.* **2009**, *10*, 613–622.
120. S. Lim, H. Y. Yoon, H. J. Jang, S. Song, W. Kim, J. Park, K. E. Lee, S. Jeon, S. Lee, D. K. Lim, B. S. Kim, D. E. Kim, K. Kim, *ACS Nano* **2019**, *13*, 10991–11007.
121. Y. Li, Y. Yao, Z. Sheng, Y. Yang, G. Ma, *Int. J. Nanomed.* **2011**, *6*, 815–823.
122. F. J. Nicholls, M. W. Rotz, H. Ghuman, K. W. MacRenaris, T. J. Meade, M. Modo, *Biomaterials* **2016**, *77*, 291–306.
123. H. Bao, Y. Xia, C. Yu, X. Ning, X. Liu, H. Fu, Z. Chen, J. Huang, Z. Zhang, *Small* **2019**, *15*, e1904314.
124. X. Ning, H. Bao, X. Liu, H. Fu, W. Wang, J. Huang, Z. Zhang, *Nanoscale* **2019**, *11*, 20932–20941.
125. J. Huang, J. Huang, X. Ning, W. Luo, M. Chen, Z. Wang, W. Zhang, Z. Zhang, J. Chao, *J. Mater. Chem. B* **2020**, *8*, 1713–1727.

10 Optical and Electrochemical Metal-Based Sensors in Biological Systems

Patrick S. Barber
Department of Chemistry, University of West Florida, Pensacola, FL 32514, USA

Ana de Bettencourt-Dias
Department of Chemistry, University of Nevada, Reno, Reno, NV 89557, USA

Jorge H.S.K. Monteiro
Department of Chemistry, California Polytechnic State University, Humboldt, Arcata, CA 95521, USA

CONTENTS

DOI: 10.1201/9781003229971-10

Abstract

Bio-sensors provide the opportunity for *in situ* and *in vivo* sensing of a large variety of species, and are thus important in the medical field, for environmental monitoring, and the food and agricultural industries. In this chapter, we provide a brief introduction to and review the recent literature on optical and electrochemical bio-sensors that rely on metals or metal ions as signal generators or signal amplifiers.

KEYWORDS

Bio-sensors; Colorimetric Sensors; Electrochemical Sensors; Luminescent Sensors; Metal Ion Detection; Plasmonic Sensors

1 INTRODUCTION

Medical diagnosis and insights from studying physiological processes have improved diagnosis and therapy. These fields rely heavily on our ability to measure high-throughput drug screening or overall quantities, such as pH or temperature and sense species formed or consumed during physiological processes [1]. Similar measurements facilitate our understanding of biological processes, environmental monitoring, and quality control in the food industry [2, 3]. Devices that enable these measurements are referred to as bio-sensors. They have become an active area of research and new advancements since the successful development in the 1970s of glass fiber optics and ion-selective electrodes [4–7]. The latter are electrodes whose response depends on the activity of a free ion in solution but do not signal the presence of the complexed ion. Thus, these types of electrodes and systems that leverage changes in optical signals, either colorimetric or luminescent, are useful for metal ion detection and analysis. These have been long-standing goals for many areas of chemistry, biology, and environmental science. Sensing of species of interest without sophisticated instrumentation and sample pretreatment enables real-time and on-site monitoring. For example, there is significant interest in *in vivo* detecting and quantitating H_2O_2, a reactive oxygen species involved in signal transduction, cellular communication, enzymatic reactions, and internal body stability [8, 9], *Escherichia coli* and other pathogens in foods [10–12], pesticides in milk and other foods [13, 14], glucose in fermentation processes [15, 16], and blood [17, 18], as well as the presence of the human immunodeficiency virus and, more recently, coronaviruses in human samples [19–22]. A significant number of sensors rely heavily on metals or metal ions as the source or amplifier of the sensing signal. When developing bio-sensors, the critical areas of research typically explored are high specificity and low limits of detection.

Herein we survey recent literature pertaining to bio-sensors that take advantage of the properties of metal compounds to quantitate biologically relevant species or metal ions *in situ* or *in vivo*.

2 OPTICAL BIO-SENSORS FOR VISIBLE ANALYTE DETECTION

2.1 Signal Detection through Light Emission

2.1.1 Transition Metal Complexes and Metal-Organic Frameworks

Transition metal complexes have been studied for their use as bio-sensors and probes due to their high photoluminescence yields, large Stokes shifts, and increased photostability compared to organic fluorophores [23]. The most well-studied transition metal complexes are based on middle-to-late transition metals, such as iridium, ruthenium, platinum, and gold. A key advantage of metal complexes is their long-lived phosphorescence lifetimes that allow for time-gated emission detection, thus removing the short-lived background fluorescence or autofluorescence from biological samples [24]. Transition metal complex-based bio-sensors have been used to sense a variety of analytes, including anions [25] and in cancer theranostics, a combination of therapy and diagnostics [26]. Here, we summarize some recent examples from the literature.

Cyclometalated Ir(III) complexes have emerged as a promising area of bioimaging and bio-sensors over the last decade [27]. You, Cho, and Nam described a design strategy for cyclometalated Ir(III) complexes as phosphorescence sensors for biological metal ions [28]. Through targeted synthetic modification of the ligands (Scheme 1) and ideal phosphorescence properties, careful tuning of the resulting emission properties is possible, as shown in Figure 1. Absorption bands range from 300 to 450 nm, while the emission bands range from 475 to 650 nm, leading to controlled light emission across much of the visible spectrum. A metal-to-ligand charge transfer transition is responsible for the phosphorescence emission of the complexes, with quantum yields ranging from 0.21% to 23% in degassed acetonitrile. The authors demonstrated that Irdfppy (Scheme 1) readily enters HeLa cells localizing within the mitochondria.

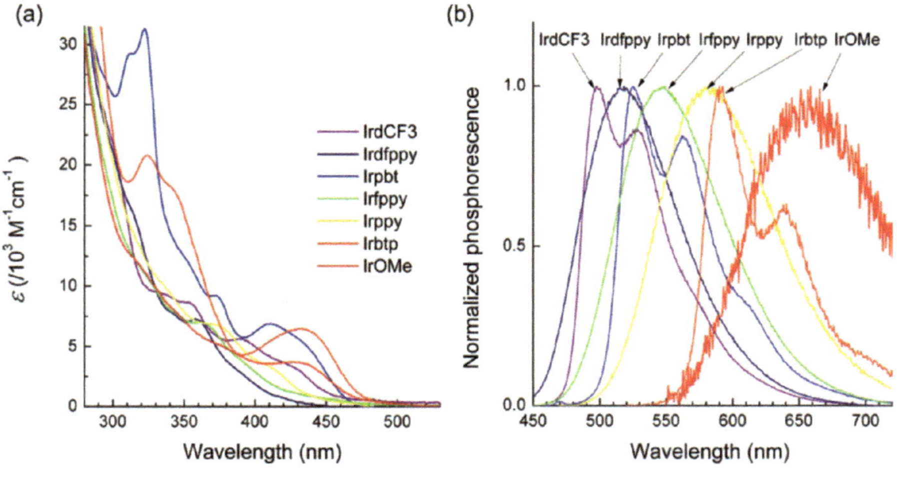

FIGURE 1 (a) UV–vis absorption and (b) normalized phosphorescence spectra of Ir(III) complexes (10 μM, argon-saturated CH_3CN) at room temperature. (Reprinted with permission from ref. 28. Copyright 2013 American Chemical Society.)

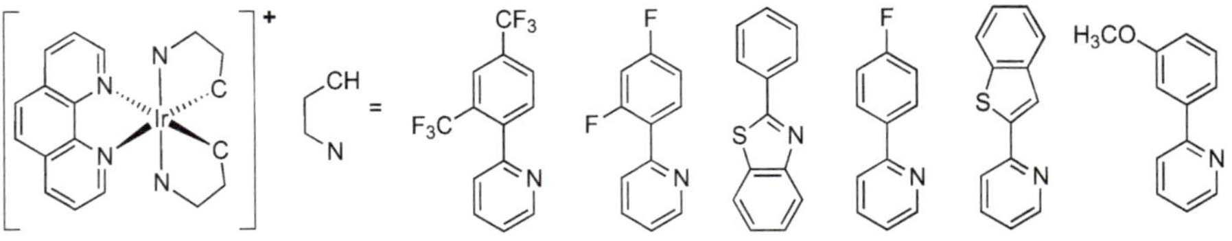

IrdCF$_3$ Irdfppy Irpbt Irfppy Irbtp IrOMe

SCHEME 1 Structures of Ir(III) complexes with selected cyclometalating ligands.

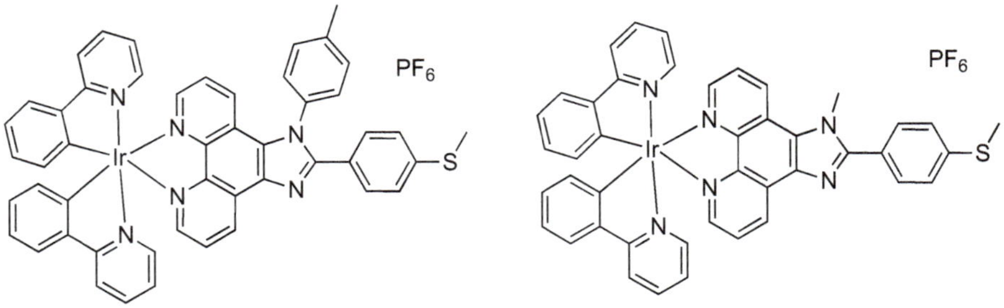

IrL1 IrL2

SCHEME 2 Structures of cyclometalated complexes **IrL1** and **IrL2** containing sulfide functional groups for the detection of IO_4^-.

Furthermore, a 'turn-on' of the phosphorescence was noted upon treating the cells with Zn^{II}, demonstrating that this compound is a potential sensor for labile zinc pools in mitochondria.

Wang and co-workers developed two cyclometalated Ir(III) complexes to detect periodate in living cells [29]. As periodate (IO_4^-) can oxidize sulfide to sulfoxide, the complexes shown in Scheme 2 were designed to contain a sulfide functional group as an oxidative trigger for the modulation of the luminescence of the complexes.

IrL1, the more efficient emitter, showed a good linear response to IO_4^- in the range 0.2–1.7 μM, with a limit of detection (LOD) of 0.077 μM. The selectivity of IO_4^- over other essential biological ions and molecules was good, and time-resolved luminescence spectra enabled detecting IO_4^- in the presence of common organic dyes used in cell imaging. The authors successfully monitored IO_4^- in living HeLa cells with **IrL1** with excellent sensitivity and selectivity with only moderate cell toxicity.

Metal-organic frameworks (MOFs) are a class of porous and crystalline coordination polymers. Initially explored for their tunable porosities and adsorption capacities, their diverse structural flexibility, and post-synthetic modification opportunities have encouraged extensive research in a wide range of applications, including gas storage and heterogeneous catalysis [30]. The discovery of luminescent MOFs has led to substantial interest in their use as chemical sensors. The luminescence in transition metal MOFs usually originates from the organic linkers, while lanthanide ion MOFs exhibit metal-centered luminescence due to the antenna effect (described below). Postsynthetic modification of MOFs allows for additional structural diversity with increased targeted sensing [31]. Utilizing luminescent MOFs as bio-sensors is a thriving area of research, showing detection strategies for nucleic acids and proteins [32], viruses [33], and many others [30, 34–36]. Some recent examples are discussed here.

Song and co-workers designed a Zn-based MOF using 3-(3,5-dicarboxylphenyl)-5-(pyrid-4-yl)-1H-1,2,4-triazole as the linker for the detection of aristolochic acids in biological fluids. The Zn-MOF showed excellent structural stability and bright blue luminescence. Detection studies of aristolochic acids in solution containing the main components of serum and urine showed excellent selectivity with a luminescence quenching mechanism. The LOD was 4nM with a short 2-minute response time. These results led the authors to evaluate the Zn-MOF as a portable, paper-based luminescent sensor. Measurements in human serum and urine showed good accuracy and great sensing with simple operation and rapid response.

Developing materials with long persistent afterglow due to long luminescence lifetimes beyond tens of ms can yield emissions visible to the naked eye. A recent report by Yang and co-workers describes two Zn-MOFs containing terephthalate that exhibit room-temperature afterglow emission as long as 0.47s [37]. Further, the phosphorescence of the Zn-MOFs was temperature sensitive, and the reversible emission intensity could be cycled through high and low temperatures. The authors found that the emission maximum could be changed by introducing guest solvent molecules, particularly pyridine, making it a potential sensor for this molecule.

Wan and co-workers explored physiological temperature sensing by loading a two-photon luminescent dye in a MOF [38]. The researchers prepared indium and 4,4′,4″-benzene-1,3,5-triyl-tribenzoic acid-based MOF, ZJU-28. The two-photon dye, 4-[4-(diphenylamino)styryl]-1-dodecylpyridinium, was encapsulated into the MOF with different loadings (13–30wt%) using cation exchange. The composites exhibited strong emission at 650nm upon excitation at 1,064nm at all temperatures. A linear decrease in the emission intensity was found between 20 and 60°C as a result of increased vibrational quenching. The systems displayed relative sensitivities of 1.51%–3.39%/°C across the physiological temperature range and an LOD of 0.013°C. When evaluated for cytotoxicity in rat pheochromocytoma (PC12) cells, the composites showed lower toxicity than the dye alone, illustrating how loading a dye into a MOF has the potential to reduce toxicity and thus provide a promising platform for other dyes.

Zhao and co-workers [39] reported a nanoMOF system that integrates the pH-sensitive organic dye monocarboxyl-containing fluorescein isothiocyanate, the O_2 indicator platinum *meso*-tetra(4-carboxyphenyl)porphyrin, and a reference dye 1,3,6,8-tetra(4-carboxylphenyl)pyrene into UiO-66, a Zr 1,4-dicarboxybenzene-based MOF. The system could be excited through fluorescence resonance energy transfer at a single wavelength to simultaneously and ratiometrically image intracellular pH and O_2. The pH range for measurement was 4.5–8.2, with the most sensitive behavior at 6.5 and an LOD of 0.0459mg/L. Oxygen levels were detected and showed a good linear response in the range 0–30 mg/L with an LOD of 0.1314 mg/L. The nanoMOF sensor displayed low cytotoxicity in HeLa cells, good biocompatibility, and the ability to simultaneously image and quantify the intracellular pH and O_2 levels.

2.1.2 Lanthanide Ion Complexes and Metal-Organic Frameworks

Due to the unique luminescence properties of lanthanide (Ln^{III}) ions, most sensing using these elements is based on their luminescence [40–49]. This is advantageous due to the high sensitivity, fast response, and capability of emitting in the UV-Vis-NIR range of the electromagnetic spectrum by changing the metal center [50–52]. The emission is based on *f-f* transitions; the 4f are core orbitals, and thus the transitions are line-like, leading to emission with high color purity. Long luminescence lifetimes are observed due to their forbidden nature, enabling time-gated emission spectroscopy that improves the signal-to-noise ratio [53, 54]. However, the direct excitation is inefficient, and the antenna effect is used to sensitize the emission more efficiently through a ligand chromophore [51, 55–57]. Excitation through the ligand leads to a beneficial significant Stokes shift of sensitized emission. In addition, the use of ligand chromophores enables tuning of the chemical and spectroscopic properties of the complexes through judicious ligand functionalization [42, 51, 53, 58–60].

Due to lengthy organic synthesis and moderate stability in aqueous solution, the number of reported bio-sensors based on Ln^{III} luminescent complexes is still low compared to lanthanide-based MOFs (LnMOFs) [23, 25, 61–66]. For example, a Tb^{III} complex containing the ligand ciprofloxacin was used to sense spermine. This biomarker relates to the presence of malignant tumors [67]. The sensing mechanism is based on the decomplexation of Tb^{III} in the presence of spermine resulting in a decrease in the Tb^{III}-centered emission (Figures 2a and b); the sensor showed a linear range for the concentration of spermine from 2 to 180 μM (Figure 2c) and an LOD of 0.17 μM.

LnMOFs have been explored more frequently in biosensing due to their more straightforward synthesis, higher solution stability than Ln^{III} complexes, and the possibility of crystal structure engineering by adjusting synthetic parameters [68–71]. Sensing studies of proteins [72], metals [73], L-kynurenine [74], phosphatase activity [75], *trans,trans*-muconic acid [76], distance between proteins [77], formaldehyde [78], and pheocromocytoma [68] have been reported. One of the challenges in the design of Ln^{III} bio-sensors is the reusability of the sensor. Because most of the Ln^{III} bio-sensors rely on changes in the emission intensity caused by the replacement of coordinated ligands (for example, coordinated water replaced by the analyte), once the analyte binds to the metal, the process is hard to reverse, and the sensor cannot be reused. Yan and Qu proposed a recyclable Tb^{III}, Zn^{II} heterobimetallic MOF (Tb(III)@MOF-SO_3^-) bio-sensor where sensing is due to the collisional quenching of the Tb(III) emission [76]. The Tb(III)@MOF-SO_3^- [76] is stable over a wide range of pH and capable of detecting *trans,trans*-muconic acid, one of the biomarkers for the presence of benzene in living organisms. The sensor's response time was 10 minutes, with an LOD of 0.1 μg/mL and linear response in the range 0–20 μg/mL.

As the sensing of *trans,trans*-muconic acid is based on collisional quenching, the bio-sensor could be reused up to four times after washing with water [76]. Biosensors that rely on monitoring changes in the intensity of one transition require

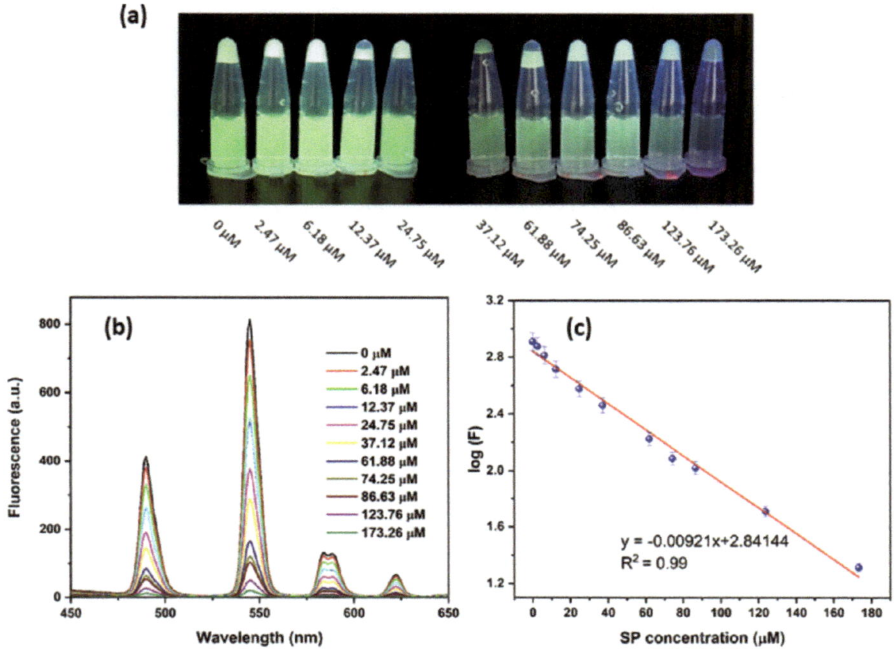

FIGURE 2 (a) Tb(III) luminescence and (b) emission spectra under different spermine concentrations. $[\text{spermine}]=0-173.25\,\mu\text{M}$. (c) Linear relationship between log(I) and [spermine]. The Tb(III) $^5D_4 \rightarrow {}^7F_5$ transition was used. (Reproduced from Ref. [66].)

the measurement of a calibration curve before each sample. As a result, they are prone to errors caused by using different equipment, acquisition parameters, and the concentration of the probe. On the other hand, ratiometric sensors use the ratio between two emission bands in the same spectrum. In this case, the sensing is independent of experimental conditions. Therefore, it does not require determining a calibration curve before each measurement [68, 78]. Li, Lin, and co-workers proposed engineering the structure of an Eu^{III}, Zn^{II} heterobimetallic MOF for ratiometric biosensing for vanillylmandelic acid, a biomarker for pheochromocytoma [26]. The use of nitric acid during the synthesis of the MOF slows down the crystallization process, allowing control of the porosity of the structure, which influences the sensitivity of the sensor (Figure 3a). The sensing of vanillylmandelic acid is based on the ratio of the ligand and Eu^{III} emission intensity bands. It shows a high sensitivity for small changes in vanillylmandelic acid concentration (Figure 3a) that can also be observed by the naked eye (Figure 3b). The sensing mechanism is based on an exciplex formed in the presence of vanillylmandelic acid that decreases the energy transfer rate ligand → Eu^{III}, and consequently, the Eu^{III} emission intensity. The bio-sensor was tested in actual urine samples and showed accurate results as a proof-of-concept.

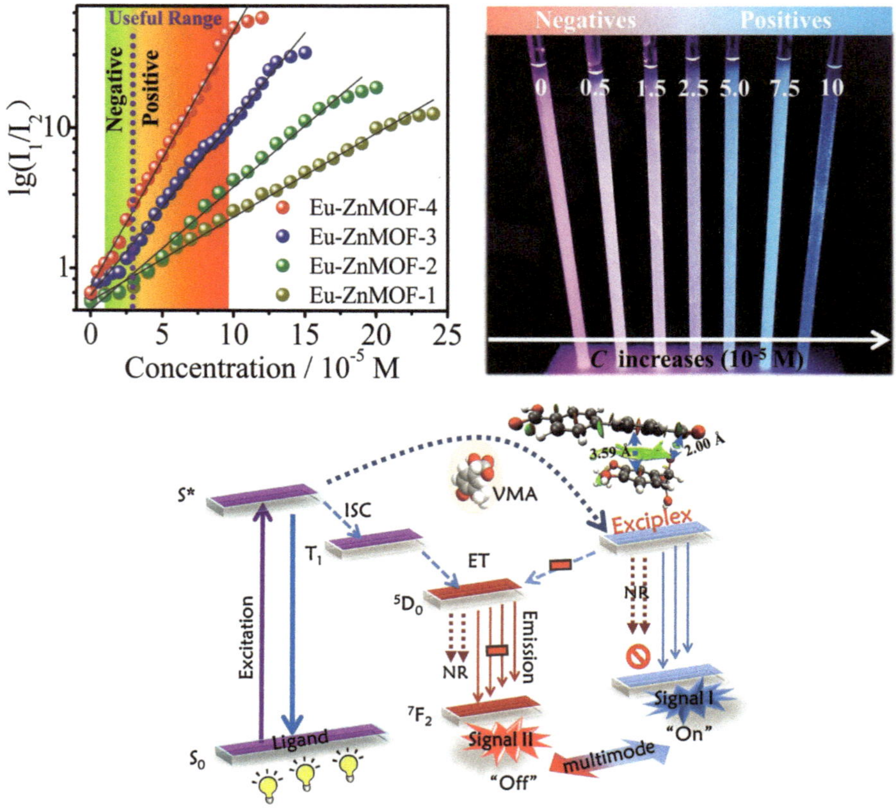

FIGURE 3 (a) Relationship log(I_1/I_2) and concentration of vanillylmandelic acid for the different Eu-ZnMOF-*n*. I_1 and I_2 are the intensities of the ligand- (~433 nm) and Eu(III)- ($^5D_0 \rightarrow {}^7F_2$, ~615 nm) centered transitions, respectively. *n* depends on the amount of nitric acid used; for $n = 1, 2, 3$, and 4 the amounts of nitric acid used were 0.75, 1.0, 1.5, and 2.0 mmol, respectively. (b) Naked-eye luminescence of the Eu-ZnMOF solutions at different concentrations of vanillylmandelic acid. (c) Energy diagram shows the ligand, Eu(III), and exciplex levels. *S* denotes singlet; *T* triplet; *ISC* intersystem crossing; *NR* non-radiative transitions. (Provided by the authors [67].)

2.1.3 Nanomaterials

Nanomaterials represent a broad class of materials that show excellent properties for applications in biosensing and imaging. Unfortunately, one of the most studied materials, nanoparticles (NPs) with sizes between 1 and 100 nm, has suffered due to the concerning side effect of accumulation in the liver and spleen [79]. Fortunately, nanoclusters (NCs), composed of several to tens of atoms and size <2 nm, and quantum clusters or sub-nanoclusters (QCs), composed of only a few atoms and sub-nm in size, are small enough to overcome the toxicity build-up of NPs by being freely excreted from the body. Here we described some recent efforts for luminescent biosensing.

The sensing of Hg^{2+} was explored with deoxyribonucleic acid (DNA)-templated silver nanoclusters (DNA-AgNCs) by Yourston and co-workers [80]. The authors prepared a bio-sensor through a combination of AgNCs and cytosine-rich single-stranded oligonucleotides, resulting in reproducible engineering of the NCs and unique fluorescence properties that extend beyond the visible spectrum. The DNA-templating strategy also allows precise control of the optical properties by selecting the hairpin-loop (HL) size associated with the cytosine-containing DNA oligonucleotides folding process. AgNCs can be prepared with a single or multiple emissive state depending on the selection of the size of the HL. After the synthetic strategy was optimized, the cytosines were replaced with thymines in the HL, which resulted in the targeted AgNCs sensors for Hg^{2+}. The researchers studied a series of four prepared sensors AgNCs@HL-T_MC_N (where M = number of thymines and N = number of cytosines): AgNCs@HL-T_0C_{13}, AgNCs@HL-T_2C_{11}, AgNCs@HL-T_4C_9, and AgNCs@HL-T_6C_7. All sensors showed detection of Hg^{2+} through luminescence quenching of the ~640 nm emission. While the linear detection range was small, the best performance came from AgNCs@HL-T_6C_7 with a range of 0–25 μM Hg^{2+}.

Muhammed and co-workers used luminescent gold QCs for Cu^{2+} and glutathione detection and cellular imaging [81]. The AuQCs were prepared from AuNPs using the reductive capacity of bovine serum albumin (BSA) at pH 12. The BSA-capped AuQCs (Au_{QC}@BSA) showed an emission maximum at 660 nm in the solid state and aqueous solution with a quantum yield of ~4%. The average size of Au_{QC}@BSA was found to be ~1 nm through high-resolution transmission electron microscopy and assigned a cluster size of 38 atoms based on mass spectrometry data. The luminescence of Au_{QC}@BSA was studied in the presence of a variety of metal ions and found to be highly selective for Cu^{2+} through luminescence quenching. Furthermore, adding glutathione to sequester the Cu^{2+} bound to the Au_{QC}@BSA resulted in a complete recovery of the luminescence. Lastly, the presence of AgNPs in solution with Au_{QC}@BSA led to the first reported example of metal-enhanced luminescence for QCs with a ninefold increase in the luminescence intensity.

Cai and co-workers prepared bimetallic copper-cerium NCs for their use as biosensors [82]. They were designed through a glutathione templating strategy, and when dispersed in an isopropanol and water mixture, they resulted in the self-assembly of a nanowire network. The addition of cysteine to the non-emissive network led to a strong red fluorescence 'turn-on.' The nanowire network was highly selective for cysteine over all other tested biologically relevant analytes. A linear detection range was observed between 10 and 32,000 nM with an LOD of 3 nM.

In addition to MOFs and NPs, two-dimensional (2D) transition metal materials are currently being investigated due to their large surface areas and ultra-thin planar structure [83]. Transition metal dichalcogenides (TMDCs) are an advanced family of graphene cognate/nanosheets, represented by the formula MX_2 and a structure that consists of a hexagonal layer of transition metal (M) and two layers of chalcogen (X) [84]. Additionally, transition metal oxides, such as the widely used manganese oxide and a new class of materials called MXenes, have shown unique electrical, optical, and chemical properties. MXenes are 2D-layered early transition metal carbides, nitrides, and carbonitrides with the general formula $M_{n+1}X_nT_x$, where M denotes early transition metals, X denotes C or N, T denotes surface functional groups (–OH, –F, =O) and n is an integer (n = 1–3) [85].

Jannat and co-workers showed the design and preparation of 2D TMDCs based on p-type tin monosulfide (SnS) nanoflakes for chemical sensing [86]. The material was a few layers thick and with submicrometer lateral dimensions. Due to the quantum confinement effect, the SnS nanoflakes had interesting excitation wavelength-dependent photoluminescence properties, which correlates to the nanoflakes' polydispersity. To test the chemical sensing potential, NO_2 gas was introduced to the 2D SnS resulting in a substantial luminescence enhancement. In addition, the physisorption of the NO_2 molecules, a strong electron acceptor, results in the increase of body hole concentrations, therefore increasing the luminescence intensity. The 2D SnS nanosheets were deposited on a low-cost resistive transducing substrate to test practical application. A range of concentrations of NO_2 was measured between 150 and 3,750 ppb at an operating temperature of 60°C and resulted in an LOD of ~17 ppb. Other commonly found gases, such as H_2S, SO_2, and CO, did not enhance the luminescence, thus supporting the high selectivity of the sensor and yielding a promising high-performance environmental sensor.

2.2 Colorimetric Signals Enable Naked Eye Detection

Colorimetric sensors operate on the principle of an induced visual color change through the interaction of the sensor and the analyte due to electronic property changes of the system. Researchers have used this technique to provide qualitative and quantitative detection of various analytes, including inorganic [87] and organic species [88], and environmentally [89] and biologically [88] important species. Advantages of colorimetric sensors include short response time and naked-eye detection, low cost, and simplicity in use e.g., minimal or no instrumentation required. This type of sensor has been prepared using many platforms, such as organic dyes [90], inorganic complexes [25], nanomaterials [91], and polymers [92]. One particularly relevant platform for colorimetric detection is microfluidics–a low-cost, low-reagent/sample, disposable, and rapid detection technology, also known as "lab-on-a-chip" [93, 94]. Here, we describe recent efforts in the preparation of colorimetric bio-sensors utilizing metal-based platforms or for sensing of metal ions.

Nucleic acid-based enzymes have been studied for their metal-dependent activities for several decades [95]. Researchers utilizing deoxyribozymes, also known as catalytic DNA, DNA enzymes, or DNAzymes, can use a combinatorial approach that incorporates the targeted analyte of interest to screen quickly large DNAzyme pools for metal affinity and selectivity. Additional screenings at reduced concentrations of the analyte can identify DNAzymes with higher metal affinity. By combining potentially competing analytes, one can identify a DNAzyme with higher analyte selectivity. Large amounts of the selected and sequenced DNAzyme can be produced relatively quickly and cheaply, given the current status of DNA synthesis. Immobilizing DNAzymes to a substrate provides opportunities for increased stability and analyte sensitivity, ease of purification of the sensor, and the ability to prepare reusable sensors [95]. Some recent examples from the literature use these techniques for highly sensitive and selective detection of metal ions.

Li and co-workers reported the development of a DNAzyme-based colorimetric bio-sensor assay to detect Cd^{2+} and Hg^{2+} [97]. The authors prepared two magnetic beads (MB) modified with metal-specific DNA sequences, one for Cd^{2+} (Cd-MB) and one for Hg^{2+} (Hg-MB). The Cd-MBs were modified with a Cd^{2+}-specific DNAzyme, and the Hg-MBs were modified with split G-quadruplex fragments containing a poly-thymine region. The Cd^{2+}-dependent DNAzyme assay in the presence of

Cd^{2+} results in releasing a G-quadruplex fragment that catalyzes the oxidation of 2,2'-azino-bis(3-ethylbenzothiazoline-6-sulfonic acid) (ABTS), leading to an intense colorimetric signal. Conversely, the presence of Hg^{2+} ions within the Hg^{2+}-specific assay causes inhibition of G-quadruplex formation, thus decreasing the catalytic oxidation of ABTS and reducing the colorimetric signal. Both assays show a good linear relationship between absorbance response and concentration of metal ions. The linear dynamic range was 1–100 and 0–400 nM, and the LODs were 1.9 and 19.5 nM for Cd^{2+} and Hg^{2+}, respectively. The authors found that assaying mixed metal samples for Cd^{2+} resulted in false positives due to Hg^{2+} binding to the available thiol groups. To overcome this challenge during the Cd^{2+} assay, pretreatment of the sample with Hg-MB and removal of the accumulated Hg^{2+} eliminated false positives, yielding a combined bio-sensor assay with excellent selectivity and high sensitivity for both metal ions.

With the understanding that thymine-rich single-stranded DNA (ssDNA) possesses selective binding to Hg^{2+}, Chen and co-workers developed a colorimetric assay based on combining bare gold nanoparticles (AuNPs) and these ssDNA moieties [98]. The assay combines a naked-eye colorimetric detection strategy with light-scattering optical properties of colloidal NPs to cause both a visual color change and the Tyndall effect. The instability of bare NPs shows a weakened Tyndall effect over time as the NPs tend to aggregate. In the absence of Hg^{2+}, the developed assay begins as a solution of two types of short thiolated ssDNAs and their long complementary ssDNA. With the addition of bare AuNPs, the ssDNAs weakly stabilize the bare AuNPs through self-assembly on the NP's surface due to gold-thiol (Au-S) interactions and/or van der Waals forces displaying a weak Tyndall effect. However, in the presence of Hg^{2+}, the solution of ssDNAs forms a stable, rigid double-stranded DNA structure because of the thymine–mercury(II)–thymine interactions. The addition of the bare AuNPs into this solution results in a significantly enhanced Tyndall signal appropriate for qualitative and quantitative detection of Hg^{2+} ions. As a result of the Tyndall effect, the assay can be made portable using a standard red laser pointer and a smartphone. The authors compared the assay results for the detection of Hg^{2+} ions between collecting data on a spectrophotometer and collecting data using the combination of a laser pointer and a smartphone. Quantitation of several Hg^{2+} solutions showed a linear range of 625–5,000 nM with an LOD of ~87 nM for the spectrometer and a range of 156–2,500 nM with an LOD of ~25 nM with the laser pointer/smartphone system. It is important to note that the images collected with the smartphone were analyzed with software to quantitate the data. The assay was used for a variety of samples and showed excellent selectivity when compared to other expected metal ions in real-world water samples.

The use of NPs in sensors, particularly those made of Au and Ag, is well-known and documented [99, 100]. Their optical and electronic properties are well suited for colorimetric detection and the ease of chemical modification makes them attractive for a wide range of biological, chemical, and ionic analytes. In one recent example, researchers prepared a AuNP-based bio-sensor for the detection of severe acquired respiratory syndrome coronavirus-2 (SARS-CoV-2) spike antigen [101]. The surface of the AuNPs was modified with a SARS-CoV-2 spike antibody, resulting in a highly specific bio-sensor for the antigen over other antigens, such as influenza A (i.e., H1N1), Middle Eastern respiratory syndrome coronavirus, and *Streptococcus pneumoniae* in high concentrations. The colorimetric behavior of the sensor showed naked-eye detection in the presence of the SARS-CoV-2 antigen with a linear correlation between the absorption maximum and the concentration of the antigen up to 1,000 ng/mL and an LOD of 48 ng/mL. Improved quantitation was explored through electrochemical detection (see below). Overall, this simple, cost-effective bio-sensor displayed quick detection times (~10 minutes) and excellent selectivity.

Metal NPs have recently been found to display intrinsic enzyme-mimicking activity and have therefore been extensively explored to replace natural enzymes that suffer from high costs and low operational stability. Analyte sensing using enzymes is well-known, and one of the most common techniques is enzyme-linked immunosorbent assay. The strategy entails using the presence of an analyte to modulate the catalytic activity of the enzyme and thus alter the chemical reaction responsible for the visual color change. Metal NPs that display enzyme-like behavior, or nanozymes, are currently being studied for a variety of sensing applications including ions, molecules, nucleic acids, proteins, and cancer cells [101]. Maity and co-workers prepared and studied a copper(I) oxide nanozyme for biosensing and phenolic oxidation [103]. The Cu_2O nanozyme was prepared through a simple microwave-assisted synthesis in a polyol medium, which yielded a glycol-capped Cu_2O nanostructure. Along with the study of the catalytic behavior, the authors investigated the colorimetric detection of the biomolecules epinephrine, dopamine, and acetylcholinesterase enzyme. The linear range of the sensor was explored between 0 and 0.2 mM, and LODs were determined for epinephrine, dopamine, and acetylcholinesterase enzyme at 10, 6.5 μM, and 2.5 pM, respectively.

As mentioned above, natural enzymes possess some disadvantages when being applied to applications requiring high operational stability, such as sensing of biologically important molecules like glucose. Enzymes typically degrade or denature at high temperatures, extreme pH, or over long operation times. Wang and co-workers [104] developed a nanocage-based zeolite imidazolium framework (NC-ZIF) to overcome the fragile nature and leakage of enzymes observed in commercial glucose monitoring systems. The researchers prepared the nanocage-based framework by incorporating glucose oxidase and hemin into ZIF-67 to use as a template for epitaxial growth of ZIF-8, generating a core-shell composite. The core ZIF-67 was selectively etched to transform the interior into a nanocage, resulting in a dual confinement strategy combining the outer shell properties of the ZIF-8 and the interior cage confinement of the two enzymes. The prepared NC-ZIF showed excellent stability under a variety of denaturing conditions. When evaluated for glucose detection, the selectivity toward glucose was high in the presence of common interfering molecules, such as fructose, dopamine, and lactose. The authors found a good linear relationship between the colorimetric signal and the concentration of glucose between 1 and 20 mM and an LOD of 10 μM.

2D materials have excellent properties and potential for colorimetric sensors. Materials, such as graphene, TMDCs/oxides, MXenes, and 2D metal nanoplates, have all been explored for their colorimetric sensing of a variety of substances, including DNA, proteins, metal ions, and more, and have been reviewed by Zhu and co-workers [105]. A more recent example reports the development of a sensing array for proteins using ultrathin ruthenium nanosheets (Ru NSs) with engineered peroxidase-like activity [106]. Tang and co-workers synthesized Ru NSs through a salt-assisted strategy and found a relationship between the crystallinity of the produced nanosheets and their affinity toward H_2O_2. By adjusting the identity of the salt, in this case KBr and KNO_3, during the synthesis, the crystallinity could be optimized to control the peroxidase-like activity of the material. A sensing array was built out of several Ru NSs, each specifically prepared for different catalytic activities and thus different binding affinities of substrates. The authors tested this sensing array with selected model molecules containing thiol groups, and then with selected proteins to yield excellent discrimination within the testing samples based on the measured catalytic inhibition effect of the catalyst. Colorimetric detection is based on the oxidation of 3,3',5,5'-tetramethylbenzidine (TMB), which is initially colorless but becomes blue as it is oxidized. Quantitative detection is possible through standard absorption measurements or through image capture by a smartphone.

2.3 Sensing Using Shifts in the Plasmon Bands

Due to the absence of a bandgap between valence and conducting bands, the high electron mobility in Au and Ag results in a collective free-electron oscillation at quantized frequencies on the surface of these metals. It produces an oscillating electric field called a plasmon, as shown in Figure 4. The intensity of the electric field generated is strongly dependent on the shape of the NPs [107, 108] and has been extensively reviewed [109–111]. Au and Ag are the most explored metals for plasmonic applications [112–114] with Pt, Ti, and other metals also explored as alternatives [115, 116]. Plasmonic sensors are based on shifts in the plasmon band (surface plasmon resonance (SPR), localized SPR (LSPR), or coupled LSPR) or enhancement of external signals (surface-enhanced fluorescence, surface-enhanced Raman scattering, or surface-enhanced infrared absorption) [112]. The advantages of plasmonic sensors are high sensitivity, fast processing time, simple operation, real-time data, label-free detection, simple sample preparation, high reusability, and non-expensive instrumentation, allowing the development of inexpensive point-of-care (POC) systems [117].

When used as a bio-sensor, the sensor must be specific to avoid false positives or false negatives. Different strategies have been used to enhance the specificity of plasmonic-based sensors, such as immunosensors, DNA recognition, antigens, cell-based biosensing, or molecular imprinted polymer-based sensors [117].

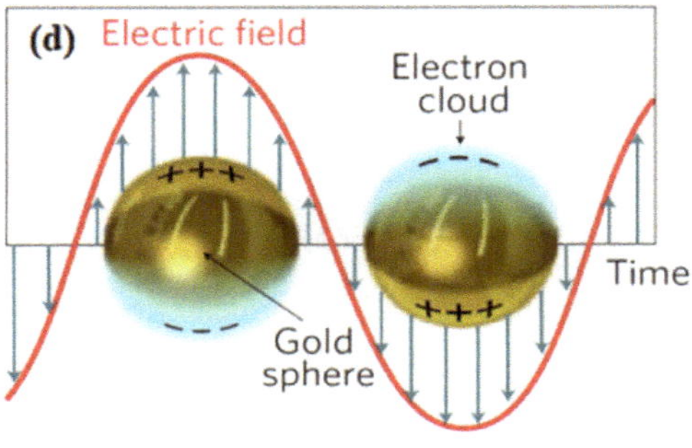

FIGURE 4 Electric field produced by the free-electron oscillation in a metal NP. (Reprinted from Ref. [117], Copyright 2021, with permission from Elsevier.)

Plasmon-based sensors have been used in the detection of small molecules, such as glucose and cholesterol [119], nitrites [120], microRNA (miRNA) biomarkers [121–123], bacteria [124, 125], and small animals [126]. Multi-analyte detection is a desirable characteristic because it allows the determination of the concentration of important biomolecules that might correlate with specific diseases. The challenge in this setup is to avoid interference between the analytes. Zhang and co-workers covered an optical fiber core with a Au film. Half of the film was decorated with 48 nm-diameter Au NPs, resulting in two distinct SPR bands [119]. The AuNPs were then decorated with β-cyclodextrin, resulting in selective sensing of cholesterol, while the Au film was decorated with *p*-mercaptophenylboronic acid, resulting in selective sensing of glucose. The sensor detected glucose and cholesterol in the range 0–1.0 mM and 0–50 nM with LODs of 0.78 μM and 0.012 nM, respectively [119]. Common methods used for diagnosing cancer rely on detecting a biomarker that is overexpressed in the presence of the disease. This method sometimes results in false positives or negatives due to interference caused by the lack of specificity. For example, in the diagnosis of prostate cancer using prostate-specific antigen as a biomarker, 2/3 of the positive tests are false positives, and 1/5 of the tests are false negatives. Specific sequences of miRNAs, usually 19–22 nucleotides in length, are commonly dysregulated in cancer patients. Thus, detecting and quantifying miRNAs has been a very successful method for high accuracy diagnosis [127–131]. Slaughter and co-workers developed a sensor for miR-21 based on Au triangular nanoprisms (AuTNPS) covalently bonded to a tapered optical fiber [122]. To ensure specificity to miRNA-21 and avoid non-specific adsorption of endogenous molecules, the AuTNPs were decorated with ssDNA and thiol-terminated polyethyleneglycol (PEG-SH), respectively. The sensor was capable of distinguishing between patients with and without a high risk of prostate cancer, and did not require pretreatment of the sample. With a short response time of 4 hours, and an LOD in the aM range, this sensor is a strong candidate for future applications as POC devices (Figure 5).

Because the plasmon band is sensitive to the shape of the NPs, the effects of the NP morphology on the sensitivity of the sensors have been investigated [120, 123]. Sim and co-workers used ssDNA for connecting Au nanospheres [132], and then Au nanorods were grown using the ssDNA as a template. The use of ssDNA allows molecular control of the distance and thus molecular control of the plasmon band. The Au NP was used to sense miRNA sequences that are produced when a patient has the potential to develop Alzheimer's disease. The sensor showed an LOD 10^3-fold lower than a similar sensor [133], and was capable of distinguishing between healthy patients and those with Alzheimer's disease.

3 ELECTROCHEMICAL BIO-SENSORS

Electrochemical detection techniques offer low detection limits, high reproducibility, sensitivity, selectivity, low cost, and simple operation, which includes portable devices for field and POC testing [134–138]. In addition, electrochemical techniques provide fast response times and, importantly, enable measurements in highly turbid samples [7] that frequently pose challenges in optical bio-sensors. The techniques used in the electrochemical sensors can be cataloged according to the signal used

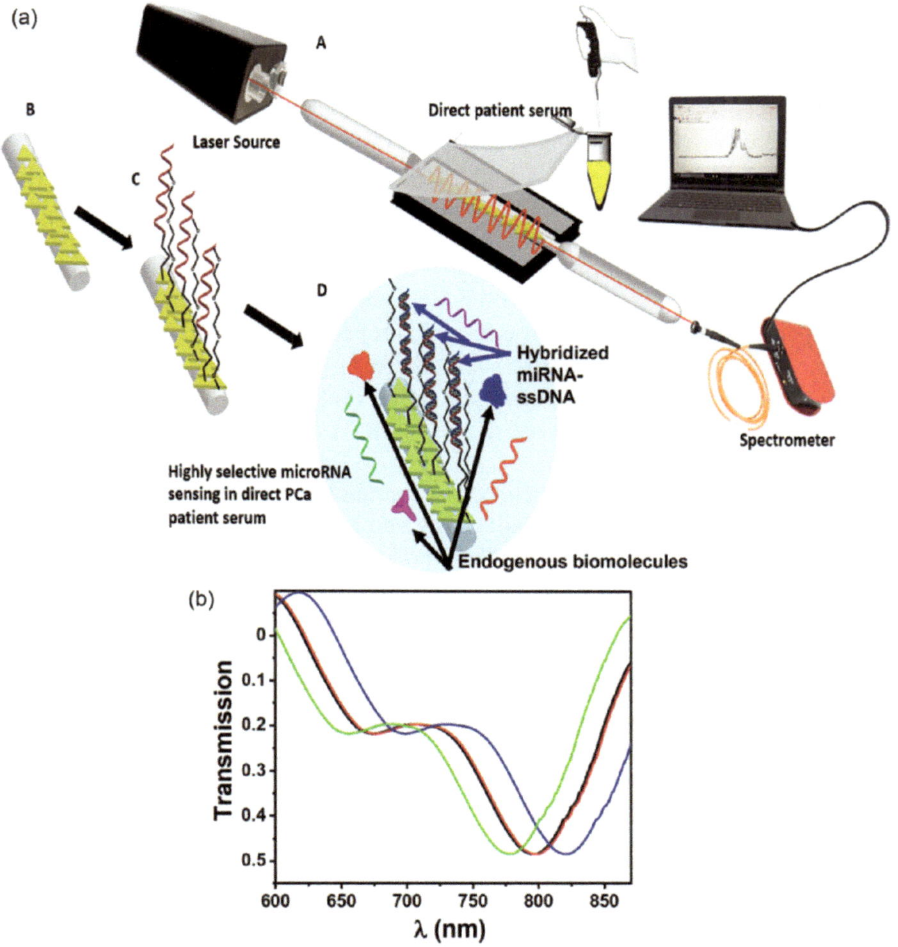

FIGURE 5 (a) A. Schematic representation of the bio-sensor. B. AuTNPs covalently bonded to the tapered optical fiber. C. AuTNPs decorated with ssDNA and PEG-SH, D. Hybridization with miR-21 highlighting the selectivity. (b) Plasmon band transmission spectra of the AuTNPs bio-sensor (peak at ~777 nm, green line), AuTNP decorated with ssDNA and PEG-SH (peak at ~799 nm, black curve), after incubation with a mixture of miRNAs ([miRNA] = 100 nM, peak at ~799 nm, red curve), and after miRNA-21 hybridization (peak at ~825 nm, blue curve). (Reproduced from Ref. [121].)

for detection as impedance [139], amperometry [1], voltammetry, as well as electrochemiluminescence [140] and photoelectrochemistry bio-sensors [134, 141]. The electrochemical properties of low-cost electrodes, such as glassy carbon electrodes (GCEs) or paper-based electrodes (PEs), are most frequently enhanced through functionalization with metals or metal-containing materials, due to their inherent redox properties. The systems considered have nanoparticles, nanoclusters, MOFs as well as liquid metals (LMs), among others, as the active electrochemical species.

Noble metals display excellent catalytic activity and high conductivity, and are thus of interest in catalytic signal amplification [142]. When used as nanoclusters, entities smaller than nanoparticles, there is a higher per atom efficiency due to the larger number of active sites, leading to systems with higher performance and reduced cost [137, 138]. Several developments have also been achieved with metal-functionalized graphene, graphene oxide (GO), and other nanocarbon materials. These systems are based on atomic layers of carbon atoms or nano-sized carbon atom constructs that display unique charge-transfer capabilities [143]. Their characteristics can be enhanced through surface immobilization of redox-active species, such as metal- and metal-oxide nanoparticles, which prevent agglomeration of the graphene-based layers and can dramatically alter their electronic properties [144, 145].

In the following, we summarize selected recent developments in electrochemical bio-sensors with varying metal-based active systems for biologically important species, such as dopamine, viruses such as SARS-CoV-2, miRNAs, and hydrogen peroxide.

LMs have found a unique niche in the preparation of bio-sensors. They have the inherent advantage of excellent conductivity combined with mechanical flexibility, which enables their deployment in flexible electrochemical bio-sensors. The LM eutectic gallium-indium was used to reduce GO to reduced GO (rGO). A resulting core-shell particulate assembly of LM-rGO (namely LM droplets covered in a shell of rGO flakes, where the LM enabled the preservation of the rGO during the electrochemical process), was applied onto GCEs or PEs in standard electrochemical cells or electrochemical paper-based analytical devices. These systems displayed increased charge-transfer ability and were used to sense dopamine in human serum samples. Baharfar and co-workers [146] found that dopamine could be detected selectively with LODs as low as 30 and 100 nM for GCE- and paper electrode-based assemblies, respectively. A similar strategy of a composite with gallium as the LM was adopted by Lim and co-workers [147]. To take advantage of the electrochemical properties of the LM and stabilize it against oxidation, the authors assembled a nanocomposite multi-layer device with a polymer-encapsulated LM wire in direct contact with a poly(diallyldimethylammonium chloride)-wrapped carbon nanotube (CNT), in turn decorated with AuNPs. The latter further reduced the overall impedance of the system, leading to a device with long-term stability under physiological conditions and a linear response to dopamine in the range 0.0875–1 μM and an LOD of 23 nM. Yang and co-workers [148] developed an electrochemiluminescent bio-sensor that used AuNPs deposited onto an indium-doped tin oxide electrode to improve the stability and sensitivity of the electrochemical signal. This bio-sensor showed high sensitivity and specificity for the detection of alkaline phosphatase (ALP) with an LOD as low as 3.7 aM. This enabled the authors to detect ALP within a single Hep G2 cell, a significant step toward single biomolecule detection.

As outlined above, MOFs have found extensive application in biosensing, due to their inherent properties of stability, porosity, flexibility, and opportunities for post-synthetic modification. Decorating a GCE with a multipedal nanomachine capped with Fe-MIL-88-NH_2-based MOFs, which were electrochemically modified to form capping Prussian Blue particles that are electroactive due to the presence of the Fe^{2+}/Fe^{3+} ions, led to a bio-sensor for miRNA21 with an increased electrochemical response. Bao and co-workers [149] reported a linear dependence of the differential pulse voltammogram current with the logarithm of the concentration of miRNA21 between 10 aM and 10 pM. The signal was selective and did not suffer from interference from miRNA-141, miRNA-155, miRNA-205, and miRNA let-7A. In human serum, full recovery of miRNA-21 from MCF-7 and HeLa cancer cells was demonstrated. A Cu-benzenehexathiol-based electrically conducting MOF was used as well by Chen and co-workers [150] as an electrochemical sensor for biologically relevant H_2O_2. The oriented films could be grown at the aqueous/organic interface with a thickness of ~17 nm and transferred onto a solid support. The film displayed a metallic-like conductivity of ~1414 S/cm, owing to a significant overlap of the Fermi level with the conduction and valence bands. An on-chip micro-bio-sensor with Cr/Au electrodes showed a linear relationship between the current density and the concentration of H_2O_2 in the range between 0 and 400 μM. It was not affected by the introduction of interfering species, such as Na_2S, glutathione, or L-cysteine, among others. Cui and co-workers also took advantage of a Cu-based MOF-immobilized hemin to quantify H_2O_2 [151]. The Cu-based MOF efficiently prevented the dimerization and oxidative self-degradation of hemin. When deposited on GCEs, the hemin-Cu-MOF modified electrode displayed improved interfacial electron transfer and catalytic activity promoted by the Cu-MOF active sites, with a wide range of linear response toward H_2O_2 between 0.01 and 5.0 mM and a low LOD of 4.14 μM. The sensing of H_2O_2 was not impacted by the presence of interfering substances, such as glycine, ascorbic acid, and uric acid. H_2O_2 recoveries in human serum samples of normal and cancer patients were in the ranges 88.00%–110.00% and 98.47%–107.70%, respectively. Jia and co-workers [152] took advantage of the ability of Zr-based MOFs with phosphate groups to immobilize aptamers. They developed a Zr-based poly-MOF, namely poly-UiO-66, which was doped with AgNPs to yield polyUiO-66@AgNPs that displayed enhanced electrochemical activity and biocompatibility conveyed by the metallic NPs. The surface functional groups of this hybrid composite successfully enabled the adsorption of either an H1N1-targeted antibody or of an aptamer of SARS-CoV-2 (Figure 6). Upon deposition of the antibody- or aptamer-adsorbed composites onto an electrode, there was an LOD for the H1N1 virus of 49.4 and of 18.2 fg/mL for SARS-CoV-2. Antigen-antibody interactions are used as well to detect exosomes, which, in analogy to the SARS-CoV-2 virus, display specific proteins highly expressed on the vesicle surface.

Liu and co-workers [153] developed an exosome-sensitive sensor based on the UiO-66 Zr-MOF that easily recognizes phosphate groups in the exosome strands. The MOF was used to coat paper, which was deposited onto soft plastic and connected to a screen-printed electrode. In the presence of exosomes and designed hairpin probes, a hybridization chain reaction was triggered, which lead to oxidation of TMB and a reduction of the electrochemical signal. The resulting system displayed an exosome LOD of 5×10^3 particles/mL and recoveries around 100% of exosomes from MCF-7 in fetal bovine serum.

As mentioned above, metals, metal clusters, and metal-oxides can positively influence the chemical and conductivity properties of graphene- and other carbon-based nanomaterials. Wu and co-workers [154] used Pt clusters to sense H_2O_2. The clusters adsorbed on the surface of hollow carbon spheres presented different ratios of

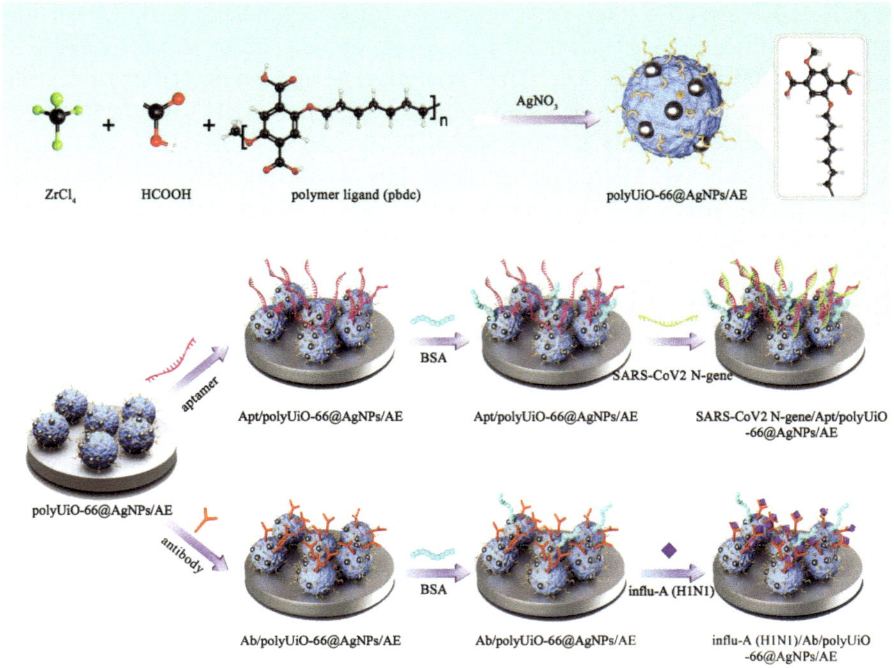

FIGURE 6 Schematic assembly of polyUiO-66@AgNP-based materials from $ZrCl_4$, formic acid, and polymer in the presence of $AgNO_3$, with subsequent deposition onto an electrode surface, which, upon treatment with bovine serum albumin (BSA) and H1N1-antibodies or COVID-19 aptamers, form bio-sensors that detect influenza A (H1N1) and SARS-CoV-2 (COVID-19) viruses. (Courtesy of the authors [152].)

Pt(0)/Pt(II) sites, and were drop-cast on the surface of GCEs. With these systems, the authors were able to determine that one cell of human non-small lung cancer A549 generates 6.86×10^{11} molecules of H_2O_2. Screen-printed carbon electrodes were modified with composites of CNTs and WO_3. The CNTs wrapped around the 55-nm-diameter WO_3 particles, leading to an electrode system with a low charge-transfer resistance. Modification of this surface with SARS-CoV-2 particles embedded in a conducting polymer, followed by the subsequent removal of the virus to generate a virus-imprinted matrix, led to an impedimetric assay for SARS-CoV-2. For this assay, Hussein and co-workers [155] reported an LOD of 57 pg/mL and a limit of quantification of 175 pg/mL. The assay was selective for SARS-CoV-2 without significant response from other respiratory viruses, such as influenzas A and B. AuNPs surface-functionalized with a SARS-CoV-2 spike antibody were used by Karakus and co-workers [101] as a virus bio-sensor. This sensor functions as a colorimetric detector. As the NPs aggregate in the presence of the spike antigen due to antibody-antigen interaction with a color red-shift, it allows for visual detection with a limit of 48 ng/mL as summarized above. When simply deposited onto a screen-printed gold electrode, the system showed a linear voltammetric response selective to the spike antigen in the range 1 pg/mL–10 ng/mL. A recovery of ~100% was achieved in a spiked saliva sample.

4 CONCLUSION

In this chapter, we have reviewed recent work on biosensors which take advantage of the light-emitting or colors of metals or metal ions, or their unique redox properties, to function as the signaling agents or as signal amplifiers in the presence of the analytes of interest. The research discussed shows a high degree of sophistication in sensor design, frequently a collaboration between different fields of science, enabling detection of biologically relevant species in several instances down to atomolar concentrations, and even single cell or single molecule recognition. These examples show a significant level of scientific and engineering maturity; yet, they also show an abundance of opportunities for the continued development of increasingly sensitive and accurate sensors for deployment in medicine, environmental analysis, agriculture and the food industry.

ACKNOWLEDGMENT

Support of AdBD's research endeavors by the National Science Foundation (CHE-1800392) is gratefully acknowledged.

ABBREVIATIONS

A549	human non-small lung cancer cells
ABTS	2,2′-azino-bis(3-ethylbenzothiazoline-6-sulfonic acid)
ALP	alkaline phosphatase
AuNPs	gold nanoparticles
AuTNP	gold triangular nanoprism
BSA	bovine serum albumin
CNT	carbon nanotube
DNA	deoxyribonucleic acid
DNA-AgNCs	DNA-templated silver nanoclusters
DNAzymes	deoxyribozymes
Fe-MIL-88-NH$_2$	iron-based metal-organic framework
GCE	glassy carbon electrode
GO	graphene oxide
H1N1	human influenza A virus
HeLa	human cervical cancer cells
HL	hairpin-loop
LM	liquid metal
LOD	limit of detection
LSPR	localized surface plasmon resonance
MB	magnetic beads
MCF-7	human breast cancer cells
miRNA	micro ribonucleic acid
MOF	metal-organic framework
MXenes	two-dimensional transition metal carbides, nitrides, or carbonitrides

NC	nanocluster
NC-ZIF	nanocage-based zeolite imidazolium framework
NP	nanoparticle
PEG-SH	thiol-terminated polyethyleneglycol
POC	point-of-care
polyUiO-66	zirconium-based polymer-embedded UiO-66 metal-organic framework
QC	quantum cluster
rGO	reduced graphene oxide
Ru NSs	ruthenium nanosheets
SARS-CoV-2	severe acute respiratory syndrome coronavirus 2
SPR	surface plasmon resonance
ssDNA	single-stranded deoxyribonucleic acid
TMB	3,3′, 5,5′-tetramethylbenzidine
TMDC	transition metal dichalcogenides
UiO-66	zirconium-terephthalate-based metal-organic framework
ZIF	zeolite imidazolium framework
ZJU-28	indium and 4,4′,4″-benzene-1,3,5-triyl-tribenzoic acid-framework

REFERENCES

1. S. J. Sadeghi, In *Encyclopedia of Biophysics*, Ed G. C. K. Roberts, Springer, Berlin, Heidelberg, 2013, 61–67.
2. B. R. Eggins, *Chemical Sensors and Biosensors*, Wiley, Chichester, 2002.
3. C. Dincer, R. Bruch, E. Costa-Rama, M. T. Fernández-Abedul, A. Merkoçi, A. Manz, G. A. Urban, F. Güder, *Adv. Mater.* **2019**, *31*, 1806739.
4. A. P. F. Turner, I. Karube, G. S. Wilson, *Biosensors – Fundamentals and Applications*, Oxford University Press, Oxford, 1987.
5. F.-G. Banica, *Chemical Sensors and Biosensors: Fundamentals and Applications.* Wiley: 2012.
6. P. Mehrotra, *J. Oral Biol. & Craniofac. Res.* **2016**, *6*, 153–156.
7. K. Cammann, *Fresenius Z. anal. Chem.* **1977**, *287*, 1–9.
8. L. Giannoudi, E. V. Piletska, S. A. Piletsky, In *Biotechnological Applications of Photosynthetic Proteins: Biochips, Biosensors and Biodevices*, Springer, Boston, MA, 2006.
9. J. Meier, E. M. Hofferber, J. A. Stapleton, N. M. Iversion, *Chemosensors* **2019**, *7*, 64.
10. P. Arora, A. Sindhu, N. Dilbaghi, A. Chaudhury, *Biosens. Bioelectron.* **2011**, *28*, 1–12.
11. A. Huang, Z. Qiu, M. Jin, Z. Shen, Z. Chen, X. Wang, J.-W. Li, *Int. J. Food Microbiol.* **2014**, *185*, 27–32.
12. D. Brandão, S. Liébana, M. I. Pividori, *New Biotech.* **2015**, *32*, 511–520.
13. V. Dasriya, J. Ritu, S. Ranveer, V. Dhundale, N. Kumar, H. V. Raghu, *Sci. Rep.* **2021**, *11*, 18855.
14. R. Zamora-Sequeira, R. Starbird-Pérez, O. Rojas-Carillo, S. Vargas-Villalobos, *Molecules* **2019**, *24*, 2659.
15. K. Pontius, D. Semenova, Y. E. Silina, K. V. Gernaey, H. Junicke, *Front. Bioeng. Biotechnol.* **2020**, *8*, 436.
16. I. E. Tothill, J. D. Newman, S. F. White, A. P. F. Turner, *Enzyme Microb. Technol.* **1997**, *20*, 590–596.

17. L. Johnston, G. Wang, K. Hu, C. Qian, G. Liu, *Front. Bioeng. Biotechnol.* **2021**, *9*, 733810.
18. P. Martinkova, M. Pohanka, *Anal. Lett.* **2015**, *48*, 2509–2532.
19. L. Farzin, M. Shamsipur, L. Samandari, S. Sheibani, *Talanta* **2020**, *206*, 120201.
20. E. R. Gray, V. Turbé, V. E. Lawson, R. H. Page, Z. C. Cook, R. B. Ferns, E. Nastouli, D. Pillay, H. Yatsuda, D. Athey, R. A. McKendry, *NPF Dig. Med.* **2018**, *1*, 35.
21. M. Asif, M. Ajmal, G. Ashraf, N. Muhammad, A. Aziz, T. Iftikhar, J. Wang, H. Liu, *Curr. Opin. Electrochem.* **2020**, *23*, 174–184.
22. S. Behera, G. Rana, S. Satapathy, M. Mohanty, S. Pradhan, M. K. Panda, R. Ningthoujam, B. N. Hazarika, Y. D. Singh, *Sensors Intern.* **2020**, *1*, 100054.
23. Y. Ning, G.-Q. Jin, M.-X. Wang, S. Gao, J.-L. Zhang, *Curr. Opin. Chem. Biol.* **2022**, *66*, 102097.
24. K. Y. Zhang, Q. Yu, H. Wei, S. Liu, Q. Zhao, W. Huang, *Chem. Rev.* **2018**, *118*, 1770–1839.
25. A. B. Aletti, D. M. Gillen, T. Gunnlaugsson, *Coord. Chem. Rev.* **2018**, *354*, 98–120.
26. C.-N. Ko, G. Li, C.-H. Leung, D.-L. Ma, *Coord. Chem. Rev.* **2019**, *381*, 79–103.
27. C. Caporale, M. Massi, *Coord. Chem. Rev.* **2018**, *363*, 71–91.
28. Y. You, S. Cho, W. Nam, *Inorg. Chem.* **2014**, *53*(4), 1804–1815.
29. W. Wang, L. Lu, K.-J. Wu, J. Liu, C.-H. Leung, C.-Y. Wong, D.-L. Ma, *Sens. Actuat. Chem.* **2019**, *288*, 392–398.
30. M. Pamei, A. Puzari, *Nano-Struct. Nano-Objects* **2019**, *19*, 100364.
31. M. Kaur, S. Kumar, M. Yusuf, J. Lee, R. J. C. Brown, K.-H. Kim, A. K. Malik, *Coord. Chem. Rev.* **2021**, *449*, 214214.
32. Q. Zhang, C.-F. Wang, Y.-K. Lv, *Analyst* **2018**, *143*(18), 4221–4229.
33. Y. Wang, Y. Hu, Q. He, J. Yan, H. Xiong, N. Wen, S. Cai, D. Peng, Y. Liu, Z. Liu, *Biosens. Bioelectron.* **2020**, *169*, 112604.
34. B. Mohan, S. Kumar, H. Xi, S. Ma, Z. Tao, T. Xing, H. You, Y. Zhang, P. Ren, *Biosens. Bioelectron.* **2022**, *197*, 113738.
35. D. Kukkar, K. Vellingiri, K.-H. Kim, A. Deep, *Sens. Actuat. B Chem.* **2018**, *273*, 1346–1370.
36. L. Shi, N. Li, D. Wang, M. Fan, S. Zhang, Z. Gong, *TrAC Trends Anal. Chem.* **2021**, *134*, 116131.
37. X. Yang, D. Yan, *Chem. Sci.* **2016**, *7*(7), 4519–4526.
38. Zhao, L.; Yang, J.; Gong, M.; Zhang, Y.; Gu, J., Single wavelength excited multi-channel nanoMOF sensor for simultaneous and ratiometric imaging of intracellular pH and O_2. *J. Mater. Chem. C* **2020**, *8*(11), 3904–3913.
39. Y. Wan, T. Xia, Y. Cui, Y. Yang, G. Qian, *ChemPlusChem* **2017**, *82*(11), 1320–1325.
40. J.-C. G. Bünzli, *Chem. Rev.* **2010**, *110*(5), 2729–55.
41. R. Arppe-Tabbara, M. R. Carro-Temboury, C. Hempel, T. Vosch, T. J. Sorensen, *Chem. Eur. J.* **2018**, *24*(46), 11885–11889.
42. J. Monteiro, D. Machado, L. M. de Hollanda, M. Lancellotti, F. A. Sigoli, A. de Bettencourt-Dias, *Chem. Commun.* **2017**, *53*(86), 11818–11821.
43. Y. Ning, S. Chen, H. Chen, J.-X. Wang, S. He, Y.-W. Liu, Z. Cheng, J.-L. Zhang, *Inorg. Chem. Front.* **2019**, *6*(8), 1962–1967.
44. Y. Ning, S. Cheng, J.-X. Wang, Y.-W. Liu, W. Feng, F. Li, J.-L. Zhang, *Chem. Sci.* **2019**, *10*(15), 4227–4235.
45. Y. Ning, J. Tang, Y.-W. Liu, J. Jing, Y. Sun, J.-L. Zhang, *Chem. Sci.* **2018**, *9*(15), 3742–3753.
46. H. Li, C. Xie, R. Lan, S. Zha, C.-F. Chan, W.-Y. Wong, K.-L. Ho, B. D. Chan, Y. Luo, J.-X. Zhang, G.-L. Law, W. C. S. Tai, J.-C. G. Bünzli, K.-L. Wong, *J. Med. Chem.* **2017**, *60*(21), 8923–8932.
47. D. Li, S. He, Y. Wu, J. Liu, Q. Liu, B. Chang, Q. Zhang, Z. Xiang, Y. Yuan, C. Jian, A. Yu, Z. Cheng, *Adv. Sci.* **2019**, *6*(23), 1902042–1902042.

48. L. Yang, K. Zhang, S. Bi, J.-J. Zhu, *ACS Appl. Mater. Interfaces* **2019**, *11*(42), 38459–38466.
49. T. Liang, Z. Li, P. Wang, F. Zhao, J. Liu, Z. Liu, *J. Am. Chem. Soc.* **2018**, *140*(44), 14696–14703.
50. S. V. Eliseeva, J.-C. G. Bünzli, *Chem. Soc. Rev.* **2010**, *39*(1), 189–227.
51. J.-C. G. Bünzli, S. V. Eliseeva, In *Lanthanide Luminescence: Photophysical, Analytical and Biological Aspects*, Eds P. Hänninen, H. Härmä, Springer, Berlin, 2011, pp. 1–46.
52. A. de Bettencourt-Dias, In *Luminescence of Lanthanide Ions in Coordination Compounds and Nanomaterials*, John Wiley & Sons Ltd, Chichester, 2014, pp. 1–48.
53. J.-C. G. Bünzli, *Coord. Chem. Rev.* **2015**, *293–294*, 19–47.
54. J.-C. G. Bünzli, C. Piguet, *Chem. Soc. Rev.* **2005**, *34*(12), 1048–1077.
55. A. de Bettencourt-Dias, P. S. Barber, S. Viswanathan, D. T. de Lill, A. Rollett, G. Ling, S. Altun, *Inorg. Chem.* **2010**, *49*(19), 8848–61.
56. J. H. S. K. Monteiro, A. de Bettencourt-Dias, F. A. Sigoli, *Inorg. Chem.* **2017**, *56*(2), 709–712.
57. J. H. S. K. Monteiro, N. R. Fetto, M. J. Tucker, A. de Bettencourt-Dias, *Inorg. Chem.* **2020**, *59*(5), 3193–3199.
58. G. L. Law, R. Pal, L. O. Palsson, D. Parker, K. L. Wong, *Chem. Commun.* **2009**, (47), 7321–3.
59. E. Pershagen, J. Nordholm, K. E. Borbas, *J. Am. Chem. Soc.* **2012**, *134*(24), 9832–5.
60. C. Szijjarto, E. Pershagen, N. O. Ilchenko, K. E. Borbas, *Chemistry* **2013**, *19*(9), 3099–109.
61. E. Brunet, O. Juanes, J. C. Rodriguez-Ubis, *Curr. Chem. Biol.* **2007**, *1*(1), 11–39.
62. S. M. Kanan, A. Malkawi, *Comm. Inorg. Chem.* **2021**, *41*(1), 1–66.
63. O. Kotova, S. Comby, T. Gunnlaugsson, *Chem. Commun.* **2011**, *47*(24), 6810–6812.
64. C. Wang, Z. Li, H. Ju, *Anal. Chem.* **2021**, *93*(44), 14878–14884.
65. Y. Wang, S. Lin, J. Luo, R. Huang, H. Cai, W. Yan, H. Yang, *Nanomaterials* **2018**, *8*(10), 796/1–796/14.
66. L. Wang, Y. Chen, *Sens. Actuat. Chem.* **2022**, *350*, 130842.
67. N. Ngoc Nghia, B. The Huy, P. Thanh Phong, J. S. Han, D. H. Kwon, Y.-I. Lee, *PLoS One* **2021**, *16*(5), e0251306.
68. J.-N. Hao, D. Niu, J. Gu, S. Lin, Y. Li, J. Shi, *Adv. Mater.* **2020**, *32*(23), 2000791.
69. Y. Wang, B. Liu, X. Shen, H. Arandiyan, T. Zhao, Y. Li, M. Garbrecht, Z. Su, L. Han, A. Tricoli, C. Zhao, *Adv. Energy Mater.* **2021**, *11*(16), 2003759.
70. X. Yin, A. Alsuwaidi, X. Zhang, *Micropor. Mesopor. Mater.* **2022**, *330*, 111633.
71. Y. Kim, S. Huh, *Cryst. Eng. Comm* **2016**, *18*(20), 3524–3550.
72. E. Al-Enezi, A. Vakurov, A. Eades, M. Ding, G. Jose, S. Saha, P. Millner, *Sensors* **2021**, *21*(3), 831.
73. B. Zhang, X. Dong, Q. Zhou, S. Lu, X. Zhang, Y. Liao, Y. Yang, H. Wang, *Carbohydr. Polym.* **2021**, *263*, 117986.
74. T. Tang, M. Liu, Z. Chen, X. Wang, C. Lai, L. Ding, C. Zeng, *J. Rare Earths* **2021**, *40*(3), 415–420.
75. H.-H. Zeng, F. Liu, Z.-Q. Peng, K. Yu, L.-Q. Rong, Y. Wang, P. Wu, R.-P. Liang, J.-D. Qiu, *ACS Appl. Nano Mater.* **2020**, *3*(3), 2336–2345.
76. X.-L. Qu, B. Yan, *Inorg. Chem.* **2018**, *57*(13), 7815–7824.
77. C. Léger, A. Yahia-Ammar, K. Susumu, I. L. Medintz, A. Urvoas, M. Valerio-Lepiniec, P. Minard, N. Hildebrandt, *ACS Nano* **2020**, *14*(5), 5956–5967.
78. Y. Wang, G. Zhang, F. Zhang, T. Chu, Y. Yang, *Sensors Actuat. Chem.* **2017**, *251*, 667–673.
79. X.-R. Song, N. Goswami, H.-H. Yang, J. Xie, *Analyst* **2016**, *141*(11), 3126–3140.
80. L. Yourston, P. Dhoqina, N. Marshall, R. Mahmud, E. Kuether, A. V. Krasnoslobodtsev, *Processes* **2021**, *9*(10), 1699.

81. M. A. Habeeb Muhammed, P. K. Verma, S. K. Pal, A. Retnakumari, M. Koyakutty, S. Nair, T. Pradeep, *Chem. Euro. J.* **2010**, *16*(33), 10103–10112, S10103/1–S10103/8.
82. Y. Cai, J. Wang, A. Liu, *Sensor Actuat. Chem.* **2022**, *356*, 131356.
83. N. Chauhan, K. Saxena, M. Tikadar, U. Jain, *Nanotechnol. Precis. Eng.* **2021**, *4*(4), 045003.
84. D. Presutti, T. Agarwal, A. Zarepour, N. Celikkin, S. Hooshmand, C. Nayak, M. Ghomi, A. Zarrabi, M. Costantini, B. Behera, T. K. Maiti, *Materials* **2022**, *15*(1), 337.
85. A. Khunger, N. Kaur, Y. K. Mishra, G. Ram Chaudhary, A. Kaushik, *Mater. Lett.* **2021**, *304*, 130656.
86. A. Jannat, F. Haque, K. Xu, C. Zhou, B. Y. Zhang, N. Syed, M. Mohiuddin, K. A. Messalea, X. Li, S. L. Gras, X. Wen, Z. Fei, E. Haque, S. Walia, T. Daeneke, A. Zavabeti, J. Z. Ou, *ACS Appl. Mater. Interfaces* **2019**, *11*(45), 42462–42468.
87. P. A. Gale, C. Caltagirone, *Coord. Chem. Rev.* **2018**, *354*, 2–27.
88. V. S. A. Piriya, P. Joseph, S. C. G. K. Daniel, S. Lakshmanan, T. Kinoshita, S. Muthusamy, *Mater. Sci. Eng. Mater. Biol. Appl.* **2017**, *78*, 1231–1245.
89. G. Alberti, C. Zanoni, L. R. Magnaghi, R. Biesuz, *Int. J. Environ. Res. Public Health* **2020**, *17*(22), 8331–8353.
90. N. A. Rakow, K. S. Suslick, *Nature* **2000**, *406*, 710–713.
91. A. Ghasemi, N. Rabiee, S. Ahmadi, S. Hashemzadeh, F. Lolasi, M. Bozorgomid, A. Kalbasi, B. Nasseri, A. Shiralizadeh Dezfuli, A. R. Aref, M. Karimi, M. R. Hamblin, *Analyst* **2018**, *143*(14), 3249–3283.
92. M. Xu, B. R. Bunes, L. Zang, *ACS Appl. Mater. Interfaces* **2011**, *3*(3), 642–7.
93. E. J. Maxwell, A. D. Mazzeo, G. M. Whitesides, *MRS Bull.* **2013**, *38*(4), 309–314.
94. G. M. Whitesides, *Nature* **2006**, *442*, 368–373.
95. J. Liu, Z. Cao, Y. Lu, *Chem. Rev.* **2009**, *109*, 1948–1998.
96. S. Khan, B. Burciu, C. D. M. Filipe, Y. Li, K. Dellinger, T. F. Didar, *ACS Nano* **2021**, *15*(9), 13943–13969.
97. D. Li, S. Ling, X. Cheng, Z. Yang, B. Lv, *Anal. Bioanal. Chem.* **2021**, *413*(28), 7081–7091.
98. X. Chen, Y. Sun, X. Mo, Q. Gao, Y. Deng, M. Hu, J. Zou, J. Nie, Y. Zhang, *RSC Adv.* **2021**, *11*(58), 36859–36865.
99. L. Polavarapu, J. Pérez-Juste, Q.-H. Xu, L. M. Liz-Marzán, *J. Mater. Chem. C* **2014**, *2*, 7460–7476.
100. C. C. Chang, C. P. Chen, T. H. Wu, C. H. Yang, C. W. Lin, C. Y. Chen, *Nanomaterials* **2019**, *9*(6), 861.
101. E. Karakus, E. Erdemir, N. Demirbilek, L. Liv, *Anal. Chim. Acta* **2021**, *1182*, 338939.
102. Y. Huang, J. Ren, X. Qu, *Chem. Rev.* **2019**, *119*(6), 4357–4412.
103. T. Maity, S. Jain, M. Solra, S. Barman, S. Rana, *ACS Sustain. Chem. Eng.* **2022**, *10*(4).
104. Q. Wang, M. Chen, C. Xiong, X. Zhu, C. Chen, F. Zhou, Y. Dong, Y. Wang, J. Xu, Y. Li, J. Liu, H. Zhang, B. Ye, H. Zhou, Y. Wu, *Biosens. Bioelectron.* **2022**, *196*, 113695.
105. D. Zhu, B. Liu, G. Wei, *Biosensors* **2021**, *11*(8), 259.
106. Y. Tang, Y. Wu, W. Xu, L. Jiao, Y. Chen, M. Sha, H.-R. Ye, W. Gu, C. Zhu, *Anal. Chem.* **2021**, *94*(2), 1022–1028.
107. S. Cui, C. Tian, J. Mao, W. Wu, Y. Fu, *Opt. Commun.* **2022**, *506*, 127548.
108. M. Beiderman, A. Ashkenazy, E. Segal, M. Motiei, A. Salomon, T. Sadan, D. Fixler, R. Popovtzer, *ACS Omega* **2021**, *6*(43), 29071–29077.
109. K. J. Major, C. De, S. O. Obare, *Plasmonics* **2009**, *4*(1), 61–78.
110. L. M. Liz-Marzán, *Langmuir* **2006**, *22*(1), 32–41.
111. T. K. Sau, A. L. Rogach, *Adv. Mater.* **2010**, *22*(16), 1781–1804.
112. J. N. Anker, W. P. Hall, O. Lyandres, N. C. Shah, J. Zhao, R. P. Van Duyne, *Nat. Mater.* **2008**, *7*(6), 442–453.

113. H. Liu, P. Zhao, W. Xiu, L. Zhang, P. Zhu, S. Ge, J. Yu, *Sensors Actuat. Chem.* **2022**, *355*, 131264.
114. K. Shi, N. Na, J. Ouyang, *Analyst* **2022**, *147*(4), 604–613.
115. Z.-H. Li, Y. Chen, Y. Sun, X.-Z. Zhang, *ACS Nano* **2021**, *15*(3), 5189–5200.
116. F. Zhou, F. Qin, Z. Yi, W. Yao, Z. Liu, X. Wu, P. Wu, *Phys. Chem. Chem. Phys.* **2021**, *23*(31), 17041–17048.
117. A. M. Shrivastav, U. Cvelbar, I. Abdulhalim, *Commun. Biol.* **2021**, *4*(1), 70.
118. L. Cui, J. Wang, M. Sun, *Rev. Phys.* **2021**, *6*, 100054.
119. W.-L. Zheng, Y.-n. Zhang, L.-k. Li, X.-G. Li, Y. Zhao, *Biosens. Bioelectron.* **2022**, *198*, 113798.
120. Z. Yang, Z. Lin, J. Yang, J. Wang, J. Yue, B. Liu, L. Jiang, *Appl. Surface Sci.* **2022**, *579*, 152130.
121. Q. Zhang, H. H. Yan, C. Ru, F. Zhu, H. Y. Zou, P. F. Gao, C. Z. Huang, J. Wang, *Biosensors Bioelectron.* **2022**, *201*, 113942.
122. T. Liyanage, B. Alharbi, L. Quan, A. Esquela-Kerscher, G. Slaughter, *ACS Omega* **2022**, *7*(2), 2411–2418.
123. S. Song, J. U. Lee, M. J. Jeon, S. Kim, S. J. Sim, *Biosens. Bioelectron.* **2022**, *199*, 113864.
124. W. Zhao, D. Zhang, T. Zhou, J. Huang, Y. Wang, B. Li, L. Chen, J. Yang, Y. Liu, *Sensors Actuat. Chem.* **2022**, *350*, 130879.
125. D. Han, X. Li, X. Bian, J. Wang, L. Kong, S. Ding, Y. Yan, *Sensors Actuat. Chem.* **2022**, *355*, 131120.
126. V. Lucarelli, D. Colbert, S. Li, M. Cumming, W. Linklater, J. Mitchell, J. Travas-Sejdic, A. Kralicek, *Talanta* **2022**, *240*, 123073.
127. J. Lu, G. Getz, E. A. Miska, E. Alvarez-Saavedra, J. Lamb, D. Peck, A. Sweet-Cordero, B. L. Ebert, R. H. Mak, A. A. Ferrando, J. R. Downing, T. Jacks, H. R. Horvitz, T. R. Golub, *Nature* **2005**, *435*, 834–838.
128. E. Yarali, E. Eksin, H. Torul, A. Ganguly, U. Tamer, P. Papakonstantinou, A. Erdem, *Talanta* **2022**, *241*, 123233.
129. Z. Zhang, Y. Wu, N. Lin, S. Yin, Z. Meng, *ACS Appl. Mater. Interf.* **2022**, *14*(6), 7717–7730.
130. S. Kim, S. Park, Y. S. Cho, Y. Kim, J. H. Tae, T. I. No, J. S. Shim, Y. Jeong, S. H. Kang, K. H. Lee, *ACS Sensors* **2021**, *6*(3), 833–841.
131. L. Nicolè, F. Cappello, R. Cappellesso, C. J. VandenBussche, A. Fassina, *Cancer Cytopathol.* **2019**, *127*(8), 493–500.
132. T. Peng, X. Li, K. Li, Z. Nie, W. Tan, *ACS Appl. Mater. Interfaces* **2020**, *12*(13), 14741–14760.
133. M. A. Mori, R. G. Ludwig, R. Garcia-Martin, B. B. Brandão, C. R. Kahn, *Cell Metabol.* **2019**, *30*(4), 656–673.
134. C. Zhu, G. Yang, H. Li, D. Du, Y. Lin, *Anal. Chem.* **2015**, *87*, 230–249.
135. M. S. Khan, S. K. Misra, K. Dighe, Z. Wang, A. S. Schwartz-Duval, D. Sar, D. Pan, *Biosens. Bioelectron.* **2018**, *110*, 132–140.
136. M. S. Khan, K. Dighe, Z. Wang, I. Srivastava, A. S. Schwartz-Duval, S. K. Misra, D. Pan, *Analyst* **2019**, *144*, 1448–1457.
137. L. A. Layqah, S. Eissa, *Microchim. Acta* **2019**, *186*, 224.
138. Y. Sun, M. Luo, X. Meng, J. Xiang, L. Wang, Q. Ren, S. Guo, *Anal. Chem.* **2017**, *89*, 3761–3767.
139. B. Chakraborty, C. Roychaudhuri, *IEEE Sensors J.* **2021**, *21*(16), 17629–17642.
140. E. Martínez-Periñán, C. Gutiérrez-Sanchez, T. García-Mendiola, E. Lorenzo, *Biosensors* **2020**, *10*, 118.
141. F. A. Settle, *Handbook of Instrumental Techniques for Analytical Chemistry*, Taylor & Francis, Upper Saddle River, NJ, 1997.

142. C. Zhu, D. Du, A. Eychmüller, Y. Lin, *Chem. Rev.* **2015**, *115*, 8896–8943.
143. J. M. Goran, E. N. H. Phan, C. A. Favela, K. J. Stevenson, *Anal. Chem.* **2015**, *87*, 5989–5996.
144. S. Kumar, S. D. Bukkitgar, S. Singh, Pratibha, V. Singh, K. R. Reddy, N. P. Shetti, C. V. Reddy, V. Sadhu, S. Naveen, *Chem. Select* **2019**, *4*, 5322–5337.
145. N. Baig, T. A. Saleh, *Microchim. Acta* **2018**, *185*, 283.
146. M. Baharfar, M. Mayyas, M. Rahbar, F.-M. Allioux, J. Tang, Y. Wang, Z. Cao, F. Centurion, R. Jalili, G. Liu, K. Kalantar-Zadeh, *ACS Nano* **2021**, *15*(12), 19661–19671.
147. T. Lim, H. Zhang, *ACS Appl. Nano Mater.* **2021**, *4*(11), 12690–12701.
148. X.-Y. Yang, Y.-Y. Bai, Y.-Y. Huangfu, W.-J. Guo, Y.-J. Yang, D.-W. Pang, Z.-L. Zhang, *Anal. Chem.* **2021**, *93*, 1757–1763.
149. T. Bao, R. Fu, Y. Jiang, W. Wen, X. Zhang, S. Wang, *Anal. Chem.* **2021**, *93*(40), 13475–13484.
150. X. Chen, J. Dong, K. Chi, L. Wang, F. Xiao, S. Wang, Y. Zhao, Y. Liu, *Adv. Funct. Mater.* **2021**, *31*(51), 2102855.
151. H. Cui, S. Cui, S. Zhang, Q. Tian, Y. Liu, P. Zhang, M. Wang, J. Zhang, X. Li, *Analyst* **2021**, *146*(19), 5951–5961.
152. Q. Jia, Y. Lou, F. Rong, S. Zhang, M. Wang, L. He, Z. Zhang, M. Du, *J. Mater. Chem. C* **2021**, *9*(40), 14190–14200.
153. X. Liu, X. Gao, L. Yang, Y. Zhao, F. Li, *Anal. Chem.* **2021**, *93*(34), 11792–11799.
154. N. Wu, L. Jiao, S. Song, X. Wei, X. Cai, J. Huang, M. Sha, W. Gu, W. Song, C. Zhu, *Anal. Chem.* **2021**, *93*(48), 15982–15989.
155. H. A. Hussein, A. Kandeil, M. Gomaa, R. Mohamed El Nashar, I. M. El-Sherbiny, R. Y. A. Hassan, *ACS Sensors* **2021**, *6*(11), 4098–4107.

Index

Note: **Bold** page numbers refer to tables and *italic* page numbers refer to figures.